# FOUNDATIONS OF COLLEGE CHEMISTRY THE ALTERNATE EDITION

*The Brooks/Cole Series in Chemistry*

FOUNDATIONS OF COLLEGE CHEMISTRY, Fourth Edition, by Morris Hein

FOUNDATIONS OF COLLEGE CHEMISTRY IN THE LAB, Fourth Edition, by Morris Hein, Leo R. Best, and Robert L. Miner

Study Guide for FOUNDATIONS OF COLLEGE CHEMISTRY, Fourth Edition, by Peter Scott

COLLEGE CHEMISTRY: AN INTRODUCTION TO INORGANIC, ORGANIC, AND BIOCHEMISTRY, Second Edition, by Morris Hein and Leo R. Best

COLLEGE CHEMISTRY IN THE LABORATORY, by Morris Hein, Leo R. Best, and Robert L. Miner

Study Guide for COLLEGE CHEMISTRY: AN INTRODUCTION TO INORGANIC, ORGANIC, AND BIOCHEMISTRY, Second Edition, by Peter Scott

FOUNDATIONS OF COLLEGE CHEMISTRY: THE ALTERNATE EDITION, by Morris Hein

Study Guide for FOUNDATIONS OF COLLEGE CHEMISTRY: THE ALTERNATE EDITION, by Peter Scott

* Note: FOUNDATIONS OF COLLEGE CHEMISTRY IN THE LAB, Fourth Edition, is compatible with FOUNDATIONS OF COLLEGE CHEMISTRY: THE ALTERNATE EDITION

# FOUNDATIONS OF COLLEGE CHEMISTRY THE ALTERNATE EDITION

Morris Hein
Mount San Antonio College

Brooks/Cole Publishing Company,
Monterey, California
A Division of Wadsworth, Inc.

Printed in the United States of America
10 9 8 7 6 5 4 3 2 1

Library of Congress Cataloging in Publication Data

Hein, Morris.
Foundations of college chemistry.

Includes index.
1. Chemistry. I. Title.
QD33.H45 1980 540 80-259
ISBN 0-8185-0402-1

Acquisition Editor: *James F. Leisy, Jr.*
Manuscript Editor: *Phyllis Niklas*
Production Editor: *Joan Marsh*
Interior and Cover Design: *Jamie Sue Brooks*
Cover Photograph: *© Batista Moon Studio*
Cover Assistance: *Gordon Williams, Monterey Peninsula College*
Illustrations: *Cyndie Jo Clark*
Typesetting: *Syntax*

TO EDNA

# Preface

The original purpose of *Foundations of College Chemistry* was to instruct students in the basic concepts of chemistry so that they would be qualified to enter courses in general college chemistry. This is still the primary purpose of the alternate edition. The text remains designed for one semester or two quarter courses in beginning or preparatory chemistry.

A survey of the schools using *Foundations of College Chemistry* showed that many of them use only the material in the first 17 chapters. The main reason for the alternate edition is to provide a shorter, less expensive text for these courses. For those courses that include an introduction to organic and biochemistry, the fourth edition of *Foundations of College Chemistry* is still available.

*Foundations of College Chemistry: The Alternate Edition* is written for students who have a limited science background. The material is developed on the assumption that the student has not had a previous course in chemistry. An important guiding objective is to provide students with a textbook that they can read, understand, and study by themselves. Accordingly, numerous changes have been made to further improve the clarity and readability of the material therein.

Because they are inexperienced in the use of technical and abstract scientific content, students often feel apprehensive about undertaking the study of chemistry. As an experienced teacher of these students, I understand their feelings and therefore have used several helpful aids to ease students into confidently facing the technical subject matter. Many students also need considerable practice to develop computational skills and quantitative reasoning. Consequently, many sample problems are given, which are carefully solved and explained in a step-by-step fashion. In nearly all cases, illustrated problems are solved by the conversion-factor dimensional analysis method. This method is an effective tool and requires no mathematics beyond arithmetic and elementary algebra.

The following features are intended to serve as learning aids as well as to make the text convenient to use:

- A list of achievement goals is given at the beginning of each chapter to serve as a guide to the student.
- Each new term that is defined is identified by bold faced type and is also printed in color in the margin.
- A second color is used to point out and emphasize noteworthy aspects of the figures, tables, equations, and summaries given in the text.

- Illustrative problems and examples are set up and worked by logical step-by-step procedures. The conversion-factor dimensional analysis approach is used for nearly all numerical problems.
- Questions and problems are given in several groups at the end of each chapter. The questions and problems within each group are arranged in the order of the related material in the chapter.
- Answers to all numerical problems are given in Appendix V. These answers should be used by students to verify their own calculations.
- Nearly all chapters have a self-evaluation question containing a fairly large number of "correct" or "not correct" statements. Answers to these questions are also in Appendix V.
- Material on inorganic nomenclature appears in several chapters, but for convenient reference, the basic rules for naming inorganic compounds are brought together in Chapter 8.
- For quick reference a list of the names and formulas of common inorganic ions is given on the inside front cover.

I wish to acknowledge the contributions of the following for their professional and critical reviews and suggestions: Professors David Adams of North Shore Community College, Eleanor Behrman of the University of Cincinnati, Keith Biever of Bellevue Community College, Walter Brooks of Santa Ana College, Tom Crablet of Cabrillo College, W. L. Felty of Pennsylvania State University, Wilkes-Barre Campus, Andrew Glaid of Duquesne University, Stewart Karp of Long Island University, Thomas L. McCarley of Iowa Western Community College, Daniel Meloon of State University of New York at Buffalo, Beatrice Paige of Antelope Valley College, Jim Ritchey of California State University, Sacramento, Sally Solomon of Drexel University, and C. M. Wilkerson of Western Kentucky University.

I express my gratitude to Robert L. Miner of Mount San Antonio College, who prepared the Solutions Manual for the text, and I also thank Peter C. Scott of Linn-Benton Community College, who prepared the Study Guide.

I also wish to express my gratitude to all my colleagues in the Mount San Antonio College Chemistry Department for their many helpful suggestions. I would be remiss if I did not mention the cooperation and help received from the editorial and production staff of Brooks/Cole Publishing Company.

Last, but certainly no least, I am forever grateful to my wife, Edna, for her continued support and understanding and to whom I fondly dedicate this volume.

*Morris Hein*

# Contents

# FOUNDATIONS OF COLLEGE CHEMISTRY THE ALTERNATE EDITION

# 1 Introduction

## 1.1 The Nature of Chemistry

What is chemistry? A popular dictionary gives this definition: Chemistry is the science of the composition, structure, properties, and reactions of matter, especially of atomic and molecular systems. Another and somewhat simpler dictionary definition is: Chemistry is the science dealing with the composition of substances and the transformations that they undergo. Neither of these definitions is entirely adequate. Chemistry, along with the closely related science of physics, is a very fundamental branch of science. The scope of chemistry, as implied by the definitions just quoted, is extremely broad—it includes the whole universe and everything, animate and inanimate, to be found in the universe. Chemistry is concerned not only with the composition and changes of composition of matter, but equally importantly, it deals with the energy and energy changes associated with matter. Through chemistry we seek to learn and to understand the general principles that govern the behavior of all matter.

The chemist, like other scientists, observes nature at work and attempts to unlock its secrets: What makes a rose red? Why is sugar sweet? Why is water wet? Why is carbon monoxide poisonous? Why do people wither with age? Problems such as these—some of which have been solved, some of which are still to be solved—are part of what we call chemistry.

A chemist may interpret natural phenomena, devise experiments that will reveal the composition and structure of complex substances, study methods for improving natural processes, or, sometimes, synthesize substances unknown in nature. Ultimately, the efforts of successful chemists advance the frontiers of knowledge and at the same time contribute to the well-being of humanity. Chemistry helps us to understand nature; however, one need not be a professional chemist or scientist to enjoy natural phenomena. Nature and its beauty, its simplicity within complexity, is for all to appreciate.

The body of chemical knowledge is so vast that no one can hope to master it all even in a lifetime of study. However, many of the basic concepts can be learned in a relatively short period of time. These basic concepts have become part of the education required for many professionals, including agriculturists, biologists, dental hygienists, dentists, medical technologists, microbiologists, nurses, nutritionists, pharmacists, physicians, and veterinarians to name a few.

## 1.2 History of Chemistry

From the earliest times, people have practiced empirical chemistry. Ancient civilizations were practicing the art of chemistry in such processes as winemaking, glass-making, pottery, dyeing, and elementary metallurgy. The early Egyptians, for example, had considerable knowledge of certain chemical processes. Excavations into ancient tombs dated about 3000 B.C. have uncovered workings of gold, silver, copper, iron, pottery from clay, glass beads, beautiful dyes and paints, as well as bodies of Egyptian kings in unbelievably well-preserved states. Many other cultures made significant developments in chemistry. However, all these developments were empirical; that is, they were achieved by trial and error and did not rest upon any valid theory of matter.

Philosophical ideas relating to the properties of matter (chemistry) did not develop as early as those relating to astronomy and mathematics. The ancient Greek philosophers made great strides in philosophical speculation concerning materialistic ideas about chemistry. They led the way to placing chemistry on a highly intellectual, scientific basis. They first introduced the concepts of elements, atoms, shapes of atoms, chemical combinations, and so on. The Greeks believed that there were four elements—earth, air, fire, and water—and that all matter was derived from these elements. The Greek philosophers had very keen minds and perhaps came very close to establishing chemistry on a sound basis similar to the one that was to develop about 2000 years later. The main shortcoming of the Greek approach to scientific work was a failure to carry out systematic experimentation.

The Greek civilization declined and was succeeded by the Roman civilization. The Romans were outstanding in military, political, and economic affairs. They continued to practice empirical chemical arts such as metallurgy, enameling, glass-making, and pottery-making, but they did very little to advance new and theoretical knowledge. Eventually, the Roman civilization declined and was succeeded in Europe by the Dark Ages. During this period, European civilization and learning were at a very low ebb.

In the Middle East and in North Africa knowledge did not decline during the Dark Ages as it did in Western Europe. During this period Arabic cultures made contributions that were of great value to the later development of modern chemistry. In particular, the Arabic number system, including the use of zero, gained acceptance; the branch of mathematics known as *algebra* was developed; and alchemy, a sort of pseudochemistry, was practiced extensively.

One of the more interesting periods in the history of chemistry was that of the alchemists (500–1600 A.D.). People have long had a lust for gold, and in those days gold was considered the ultimate, most perfect metal formed in nature. The principal goals of the alchemists were to find a method of prolonging human life indefinitely and to change the base metals—such as iron, zinc, and copper—into gold. They searched for a universal solvent to transmute base metals into gold and for the "philosopher's stone" to rid the body of all diseases and to renew life. In the course of their labors, they learned a great deal of chemistry. Unfortunately, much of their work was done secretly because of the mysticism that shrouded their activity, and very few records remain.

Although the alchemists were not guided by sound theoretical reasoning and were clearly not in the intellectual class of the Greek philosophers, they did something that the philosophers had not considered worthwhile. They subjected various materials to prescribed treatments under what might be loosely described as laboratory methods. These manipulations, carried out in alchemical laboratories, not only uncovered many facts of nature but paved the way for the systematic experimentation that is characteristic of modern science.

Alchemy began to decline in the 16th century when Paracelsus (1493–1541), a Swiss physician and outspoken revolutionary leader in chemistry, strongly advocated that the objectives of chemistry be directed toward the needs of medicine and the curing of human ailments. He openly condemned the mercenary efforts of alchemists to convert cheaper metals to gold.

But the real beginning of modern science can be traced to astronomy during the Renaissance. Nicolaus Copernicus (1473–1543), a Polish astronomer, succeeded in upsetting the generally accepted belief in a geocentric universe. Although not all the Greek philosophers had believed that the sun and the stars revolved about the earth, the geocentric concept had come to be accepted without question. The heliocentric (sun-centered) universe concept of Copernicus was based on direct astronomical observation and represented a radical departure from the concepts handed down from Greek and Roman times. The ideas of Copernicus and the invention of the telescope stimulated additional work in astronomy. This work, especially that of Galileo Galilei (1564–1642) and Johannes Kepler (1571–1630), led directly to a rational explanation of the general laws of motion by Sir Isaac Newton (1642–1727) from about 1665 to 1685.

Modern chemistry was slower to develop than astronomy and physics; it began in the 17th and 18th centuries when Joseph Priestley (1733–1804), who discovered oxygen in 1774, and Robert Boyle (1627–1691) began to record and publish the results of their experiments and to discuss their theories openly. Boyle, who has been called the founder of modern chemistry, was one of the first to practice chemistry as a true science. He believed in the experimental method. In his most important book, *The Skyptical Chemist*, he clearly distinguished between an element and a compound or mixture. Boyle is best known today for the gas law that bears his name. A French chemist, Antoine Lavoisier (1743–1794), placed the science on a firm foundation with experiments in which he used a chemical balance to make quantitative measurements of the weights of substances involved in chemical reactions.

The use of the chemical balance by Lavoisier and others later in the 18th century was almost as revolutionary in chemistry as the use of the telescope had been in astronomy. Thereafter, chemistry was a highly quantitative experimental science. Lavoisier also contributed greatly to the organization of chemical data, to chemical nomenclature, and to the establishment of the Law of Conservation of Mass in chemical changes. During the period from 1803 to 1810, John Dalton (1766–1844), an English schoolteacher, advanced his atomic theory. This theory (see Section 5.2) placed the atomistic concept of matter on a valid rational basis. It remains today as a tremendously important general concept of modern science.

Since the time of Dalton, knowledge of chemistry has advanced in great strides, with the most rapid advancement occurring at the end of the 19th century and during the 20th century. Especially outstanding achievements have been made in determining the structure of the atom, understanding the biochemical fundamentals of life, developing chemical technology, and the mass production of chemicals and related products.

## 1.3 The Branches of Chemistry

Chemistry may be broadly classified into two main branches: *organic* chemistry and *inorganic* chemistry. Organic chemistry is concerned with compounds containing the element carbon. The term *organic* was originally derived from the chemistry of living organisms—plants and animals. Inorganic chemistry deals with all the other elements as well as with some carbon compounds. Substances classified as inorganic are derived mainly from mineral sources rather than from animal or vegetable sources.

Other subdivisions of chemistry, such as analytical chemistry, physical chemistry, biochemistry, electrochemistry, geochemistry, and radiochemistry, may be considered specialized fields of, or auxiliary fields to, the two main branches.

Chemical engineering is the branch of engineering that deals with the development, design, and operation of chemical processes. A chemical engineer generally begins with a chemist's laboratory-scale process and develops it into an industrial-scale operation.

## 1.4 Relationship of Chemistry to Other Sciences and Industry

Besides being a science in its own right, chemistry is the servant of other sciences and industry. Chemical principles contribute to the study of physics, biology, agriculture, engineering, medicine, space research, oceanography, and many other sciences. Chemistry and physics are overlapping sciences, since both are based on the properties and behavior of matter. Biological processes are chemical in nature. The metabolism of food to provide energy to living organisms is a chemical process. Knowledge of molecular structure of proteins, hormones, enzymes, and the nucleic acids is assisting biologists in their investigations of the composition, development, and reproduction of living cells.

Chemistry is playing an important role in alleviating the growing shortage of food in the world. Agricultural production has been increased with the use of chemical fertilizers, pesticides, and improved varieties of seeds. Chemical refrigerants make possible the frozen food industry that preserves large amounts of food that might otherwise spoil. Chemistry is also producing synthetic nutrients, but much remains to be done as the world population multiplies with respect to the land available for cultivation. Expanding energy needs have

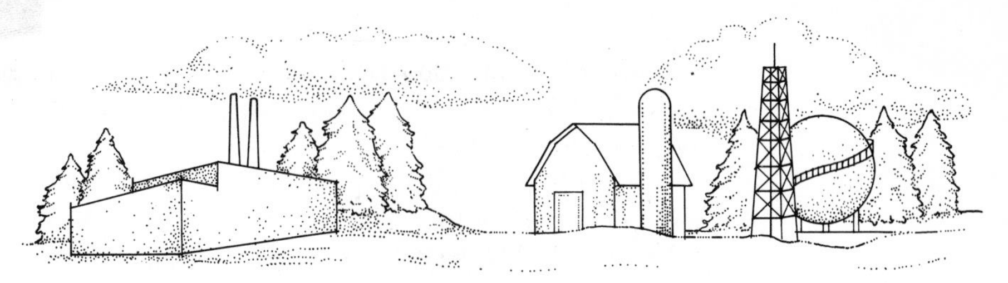

**ABUNDANT RAW MATERIALS**
from mine, forest, sea, air, farm, oil, brine, and gas wells

**THE CHEMICAL INDUSTRY**
in more than 13,500 plants in the United States
converts these raw materials into more than 10,000
**CHEMICALS**
such as acids and alkalies, salts, organic compounds, solvents, compressed gases, pigments, and dyes, which are used

**BY THE CHEMICAL INDUSTRY ITSELF**
**To Produce**

Cosmetics
Detergents and soap
Drugs and medicines
Dyes and inks
Explosives
Fertilizers
Paints
Pesticides
Plastic materials
Sanitizing chemicals
Synthetic fibers
Synthetic rubber
And many others

**BY OTHER INDUSTRIES**
**In the Production of**

*Durable Goods*

Aircraft and equipment
Building materials
Electrical equipment
Hardware
Machinery
Metal products
Motor vehicles and equipment
Other products of
metal, glass, paper, and wood

*Nondurable Goods*

Beverages
Food products
Leather and leather products
Packaging
Paper and paper products
Petroleum and coal products
Rubber products
Textiles

**THE ULTIMATE MARKET**
**(Fundamental human needs)**

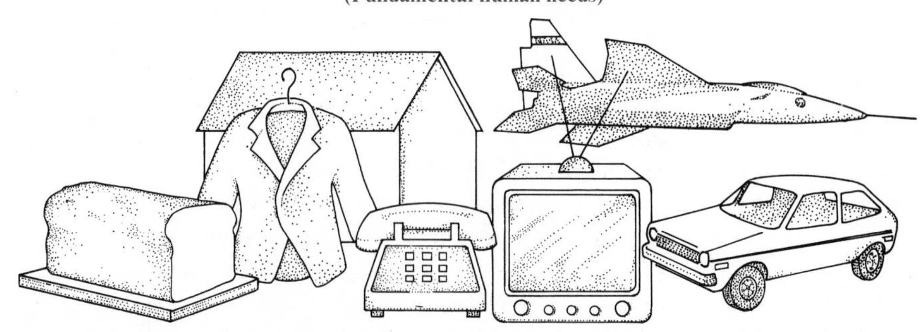

Health, food, clothing, shelter, transportation, communication, defense, and other needs

*Figure 1.1.* Broad scope of the chemical industry today. (Courtesy Manufacturing Chemists Association.)

brought about difficult environmental problems in the form of air and water pollution. Chemists as well as other scientists are working diligently to alleviate these problems.

Advances in medicine and chemotherapy, through the development of new drugs, have contributed to prolonged life and the relief of human suffering. More than 90% of the drugs and pharmaceuticals being used in the United States today have been developed commercially within the past 40 years. The entire plastics and polymer industry, unknown 50 years ago, has revolutionized the packaging and textile industries and is producing more durable and useful construction materials. Energy derived from chemical processes is used universally for heating, lighting, and transportation. There is virtually no industry that is not dependent on chemicals—for example, the petroleum, steel, rubber, pharmaceutical, electronic, transportation, cosmetic, garment, aircraft, and television industries (the list could go on and on). Figure 1.1 illustrates the conversion of natural resources by the chemical industry into useful products for commerce, industry, and human needs.

## 1.5 Scientific Method

Chemistry as a science or field of knowledge is concerned with ideas and concepts relating to the behavior of matter. Although these concepts are abstract, their application has had an extraordinarily concrete impact on human culture. This impact is due to modern technology, which may be said to have begun about 200 years ago and has continued to grow at an accelerating rate ever since.

There is a very important difference between the science of chemistry and technology. The science represents a basic body of knowledge; technology represents the physical application of this knowledge to the real world in which we live.

Why has the science of chemistry and its associated technology flourished so abundantly in the last two centuries? Is it because we are growing more intelligent? No, there is absolutely no reason to believe that the general level of human intelligence is any higher today than it was 1000 years ago in the Dark Ages. The use of the scientific method is usually credited with being the most important single factor in the amazing development of chemistry and technology. Although complete agreement is lacking on exactly what is meant by "using the scientific method," the general approach is as follows:

1. Collect facts or data that are relevant to the problem or question at hand. This is usually done by planned experimentation.
2. Analyze the data to find trends (regularities) that are pertinent to the problem. Formulate a hypothesis that will account for the data that have been accumulated and that can be tested by further experimentation.
3. Plan and do additional experiments to test the hypothesis. Such experiments extend beyond the range that is covered in Step 1.
4. Modify the hypothesis as necessary so that it is compatible with all the pertinent experimental data.

Confusion sometimes arises regarding the exact meanings of the words *hypothesis, theory*, and *law*. A well-established hypothesis is often called a theory. Hypotheses and theories explain natural phenomena, whereas scientific laws are simple statements of natural phenomena to which no exceptions are known.

While the four steps listed above are a broad outline of the general procedure that is followed in much scientific work, they do not provide a recipe for doing chemistry or any other science. But chemistry is an experimental science; and much of its progress has been due to application of the scientific method through systematic research. Occasionally, a great discovery is made by accident, but the majority of scientific achievements are accomplished by well-planned experiments.

Many theories and laws are studied in chemistry. They make the study of chemistry or any science easier, because they summarize a particular aspect of the science. Although the student will see that theories advanced by great thinkers have been subject to change, this change does not mean that their contributions are of lesser significance than the discoveries of today. Change is the natural evolution of scientific knowledge.

## 1.6 How to Study Chemistry

How do you as a student approach a subject such as chemistry, with its unfamiliar terminology, symbols, formulas, theories, and laws? All the normally accepted habits of good study are applicable to the study of chemistry. Budget your study time and spend it wisely. In particular, you can spend your study time more profitably in regular, relatively short periods rather than in one prolonged cram session.

Chemistry has its own language, and learning this language is of prime importance to the successful study of chemistry. Chemistry is a subject of many facts. At first you will simply have to memorize some of them. However, you will also learn these facts by referring to them frequently in your studies and by repetitive use. For example, you must learn the symbols of 30 or 40 common elements in order to be able to write chemical formulas and equations. As with the alphabet, repetitive use of these symbols will soon make them part of your vocabulary.

Careful reading of assigned material cannot be overemphasized. You should read each chapter at least twice. The first time, read the chapter rapidly, noting especially topic headings, diagrams, and other outstanding features. Then read more thoroughly and deliberately for better understanding. It may be profitable to underline and abstract material during the second reading. Isolated reading may be sufficient for learning some subjects, but it is not sufficient for learning chemistry. During the lectures, become an active mental participant and try to think along with your instructor—do not just occupy a seat. Lecture and laboratory sessions will be much more meaningful if you have already read the assigned material.

Your studies must include a good deal of written chemistry. Chemical symbolism, equations, problem solving, and so on, require much written practice for proficiency. One does not become an accomplished pianist by merely reading or listening to music—it takes practice. One does not become a good baseball player by reading the rules and watching baseball games—it takes practice. So it is with chemistry. One does not become proficient in chemistry by only reading about it—it takes practice.

In solving a numerical problem, you should read the problem carefully to determine what is being asked. Then develop a plan for solving the problem. It is a good idea to start by writing down something—a formula, a diagram, an equation, the data given in the problem. This will give you something to work with, to think about, to modify—and finally to expand into an answer. When you have arrived at an answer, consider it carefully to make sure that it is a reasonable one. The solutions to problems should be recorded in a neat, orderly, stepwise fashion. Fewer errors and time saved are the rewards of a neat and orderly approach to problem solving. If you need to read and study still further for complete understanding, do it!

# 2 Standards for Measurement

*After studying Chapter 2 you should be able to:*

1. Understand the terms listed in Question A at the end of the chapter.
2. Differentiate clearly between mass and weight.
3. Know the basic metric units of mass, length, and volume.
4. Give the numerical equivalents of the metric prefixes deci, centi, milli, micro, deka, hecto, kilo, and mega.
5. Express any number in exponential notation form.
6. Express the results of arithmetic operations to the proper number of significant figures.
7. Set up and solve problems by the dimensional analysis, or factor-label, method.
8. Convert any measurement of mass, length, or volume in American units to metric units and vice versa.
9. Make conversions between Fahrenheit, Celsius, and Kelvin temperatures.
10. Differentiate clearly between temperature and heat.
11. Make calculations using the equation

$$\text{calories} = (\text{Grams of substance}) \times (\text{Heat capacity of substance}) \times (\Delta t)$$

12. Calculate density, mass, or volume of an object (or substance) from appropriate data.
13. Calculate the specific gravity when given the density of a substance and vice versa.
14. Recognize the common laboratory measuring instruments illustrated in this chapter.

## 2.1 Mass and Weight

Chemistry is an experimental science. The results of experiments are usually determined by making measurements. In elementary experiments, the quantities that are commonly measured are mass (weight), length, volume, and time. Measurements of electrical and optical quantities may also be needed in more sophisticated experimental work.

mass

Although mass and weight are often used interchangeably, the two words have quite different meanings. The **mass** of a body is defined as the amount of matter in that body. The amount of mass in an object is a fixed and unvarying quantity that is independent of the object's location.

weight

The **weight** of a body is the measure of the earth's gravitational attraction for that body. Unlike its mass, the weight of an object varies in relation to (1) its position on or its distance from the earth and (2) whether the rate of motion of the object is changing with respect to the motion of the earth. Consider an astronaut of mass 70.0 kilograms (154 pounds) who is being shot into a space orbit. At the instant before blastoff the weight of the astronaut is also 70.0 kilograms. As the distance from the earth increases and the rocket turns into an orbiting course, the gravitational pull on the astronaut's body decreases until a state of "weightlessness" is attained. However, the mass of the astronaut's body has remained constant at 70.0 kilograms during the entire process of lift-off and going into orbit.

The mass of an object may be measured on a chemical balance by comparing it with other known masses. Two objects of equal mass will also have equal weights if they are measured in the same place. Thus, under these conditions the terms *mass* and *weight* are used interchangeably. Although the chemical balance is used to determine mass, it is said to *weigh* objects, and we often speak of the *weight* of an object when we really mean its mass.

## 2.2 Measurement, Significant Figures, and Calculations

To understand certain phases of chemistry it is necessary to set up and solve problems. Problem solving requires an understanding of the elementary mathematical operations used to manipulate numbers. Numerical values or data are obtained from measurements made in an experiment. A chemist may use these data to calculate the extent of the physical and chemical changes occurring in the substances that are being studied. By appropriate calculations, the results of an experiment may be compared with those of other experiments and summarized in ways that are meaningful.

When expressing a quantity of something, one must state both the numerical value and the units in which the quantity is expressed. In the statement "1 kilometre contains 1000 metres," kilometre and metre are the units in which this length is expressed.

There is some degree of uncertainty in every experimental measurement—due to inherent limitations of the measuring instrument and in the skill of the experimenter. The value recorded for a measurement should give some indication of its reliability (precision). To express maximum precision, this value should contain all the digits that are known plus one digit that is estimated. This last estimated digit introduces some degree of uncertainty. Because of this uncertainty, every number that expresses a measurement can have only a limited number of digits. These digits, used to express a measured quantity, are known as **significant figures** or **significant digits**.

significant figures

A detailed discussion of significant figures in calculations, exponents, powers of 10, rounding off numbers, large and small numbers, the dimensional analysis method of calculation, and other mathematical operations is given in the Mathematical Review in Appendix I. You are urged to review Appendix I and to study carefully any portions that are not familiar to you. This study may be done at various times during the course as the need for additional knowledge of certain mathematical operations arises.

## 2.3 The Metric System

metric system or SI

The **metric system**, or **International System (SI)**, is a decimal system of units for measurements of mass, length, time, and other physical constants. It is built around a set of basic units and uses factors of 10 to express larger or smaller quantities of these units. To express larger and smaller quantities, prefixes are added to the names of the units. These prefixes represent multiples of 10, making the metric system a total decimal system of measurements. Table 2.1 shows the names, symbols, and numerical values of the prefixes. These are also shown in Appendix III. Some of the more commonly used prefixes are

kilo One thousand (1000) times the unit expressed
deci One-tenth (0.1) of the unit expressed
centi One-hundredth (0.01) of the unit expressed
milli One-thousandth (0.001) of the unit expressed
micro One-millionth (0.000001) of the unit expressed

Examples are

1 kilometre = 1000 metres
1 kilogram = 1000 grams
1 microsecond = 0.000001 second

*Table 2.1.* Prefixes used in the metric system and their numerical values.

| Prefix | Symbol | Numerical value |
|---|---|---|
| tera | T | 1,000,000,000,000 or $10^{12}$ |
| giga | G | 1,000,000,000 or $10^{9}$ |
| mega | M | 1,000,000 or $10^{6}$ |
| kilo | k | 1,000 or $10^{3}$ |
| hecto | h | 100 or $10^{2}$ |
| deka | da | 10 or $10^{1}$ |
| deci | d | 0.1 or $10^{-1}$ |
| centi | c | 0.01 or $10^{-2}$ |
| milli | m | 0.001 or $10^{-3}$ |
| micro | $\mu$ | 0.000001 or $10^{-6}$ |
| nano | n | 0.000000001 or $10^{-9}$ |
| pico | p | 0.000000000001 or $10^{-12}$ |
| femto | f | 0.000000000000001 or $10^{-15}$ |
| atto | a | 0.000000000000000001 or $10^{-18}$ |

The seven base units of measurement in the International System are given below. Other units are derived from these base units.

| Quantity | Name of Unit | Symbol |
|---|---|---|
| Length | Metre | m |
| Mass | Kilogram | kg |
| Time | Second | s |
| Electric current | Ampere | A |
| Temperature | Kelvin | K |
| Luminous intensity | Candela | cd |
| Amount of substance | Mole | mol |

The metric system, or International System, is currently used by most of the countries in the world, not only for scientific and technical work, but also in commerce and industry. The United States is currently in the process of changing to the metric system of weights and measurements.

## 2.4 Measurement of Length

Standards for the measurement of length have an interesting historical development. The Old Testament mentions such units as the cubit (the distance from a man's elbow to the tip of his outstretched hand). In ancient Scotland the inch was once defined as a distance equal to the width of a man's thumb.

metre

Reference standards of measurements have undergone continuous improvements in precision. The standard unit of length in the metric system is the **metre**. When the metric system was first introduced in the 1790s, the metre was defined as one ten-millionth of the distance from the equator to the North Pole measured along the meridian passing through Dunkirk, France. In 1889, the metre was redefined as the distance between two engraved lines on a platinum–iridium alloy bar maintained at 0° Celsius. This international metre bar is stored in a vault at Sèvres near Paris. Duplicate metre bars have been made and used as standards by many nations.

By the 1950s, length could be measured with such precision that a new standard was needed. Accordingly, in 1960, by international agreement the metre was again redefined, this time as 1,650,763.73 wavelengths of a particular spectral emission line of krypton-86. This is the reference standard for length presently in use.

A metre is 39.37 inches, a little longer than 1 yard. One metre contains 10 decimetres, 100 centimetres, or 1000 millimetres (see Figure 2.1). A kilometre contains 1000 metres. Table 2.2 shows the relationships of these units.

The angstrom unit ($10^{-8}$ cm) is used extensively in expressing the wavelength of light and in atomic dimensions. Other important relationships are

$$1 \text{ m} = 100 \text{ cm} = 1000 \text{ mm} = 10^{6}\ \mu = 10^{10}\ \text{Å}$$
$$1 \text{ cm} = 10 \text{ mm} = 0.01 \text{ m}$$
$$1 \text{ in.} = 2.54 \text{ cm}$$

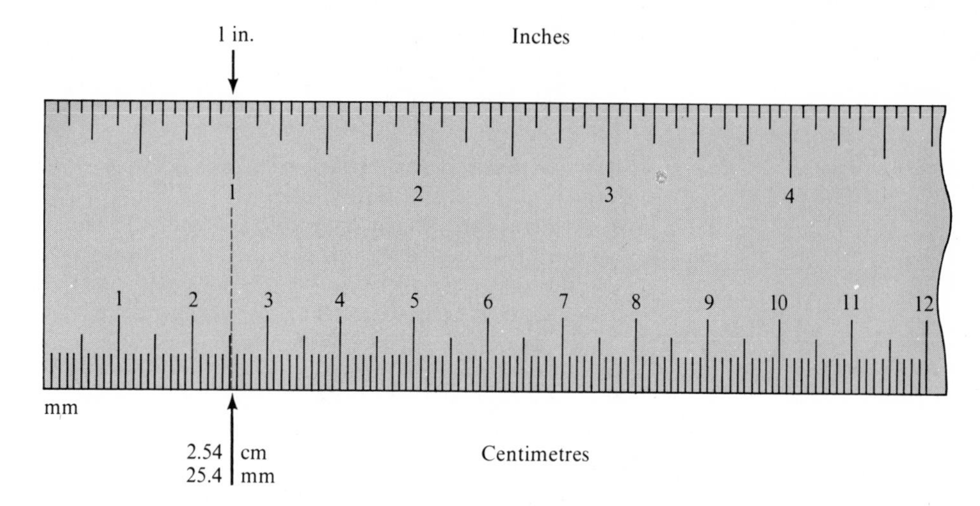


*Figure 2.1.* Comparison of the metric and American systems of length measurement: 2.54 cm = 1 in.

*Table 2.2.* Metric units of length.

| Unit | Abbreviation | Metre equivalent | Exponential equivalent |
|---|---|---|---|
| Kilometre | km | 1000 m | $10^3$ m |
| Metre | m | 1 m | $10^0$ m |
| Decimetre | dm | 0.1 m | $10^{-1}$ m |
| Centimetre | cm | 0.01 m | $10^{-2}$ m |
| Millimetre | mm | 0.001 m | $10^{-3}$ m |
| Micrometre | μm | 0.000001 m | $10^{-6}$ m |
| Micron | μ | 0.000001 m | $10^{-6}$ m |
| Nanometre | nm | 0.000000001 m | $10^{-9}$ m |
| Angstrom | Å | 0.0000000001 m | $10^{-10}$ m |

## 2.5 Problem Solving

One of the most consistently troublesome areas in chemistry involves the solving of mathematical problems. Since many chemical principles are illustrated by mathematical concepts, it is necessary to learn to solve problems dealing with these concepts.

There are usually several methods by which a problem can be solved. But in all methods it is best, especially for beginners, to use a systematic, orderly approach. The dimensional analysis, or factor-label, method (Appendix I, Section 16) is stressed in this book because:

1. It provides a systematic, straightforward way to set up problems.
2. It gives a clear understanding of the principles involved.
3. It helps in learning to organize and evaluate data.

4. It helps to identify errors because unwanted units are not eliminated if the setup of the problem is incorrect.

The basic steps for solving problems are:

1. Read the problem very carefully to determine what is to be solved for. Write down what you are solving for.
2. Tabulate the data given in the problem. Even in tabulating data, it is important to label all factors and measurements with the proper units.
3. Determine which principles are involved and which unit relationships are needed to solve the problem. It may be necessary to refer to certain tables to obtain other data needed. Use sample problems in the text to help set up and solve the problem.
4. Proceed with the necessary mathematical operations. Make certain that the answer contains the proper number of significant figures.
5. Check the answer to see if it is reasonable.
6. Do your work in a neat and organized form.

*Label all factors with the proper units.*

Just a few more words about problem solving. Don't allow any formal method of problem solving to limit your use of common sense and intuition. If the solution to a problem is obvious and seems simpler to you by another method, by all means use it. But in the long run you should be able to solve many otherwise difficult problems by using the dimensional analysis method. Here are some examples of problem solving by the dimensional analysis method.

Suppose you want to change 2.5 metres to an equivalent number of millimetres. Write

$$\text{m} \longrightarrow \text{mm}$$

To accomplish this you need to find a conversion unit that will change the given units (metres) to the desired units (millimetres). Write

$$1 \text{ m} = 1000 \text{ mm}$$

From this statement these two factors can be derived:

$$\frac{1 \text{ m}}{1000 \text{ m}} = 1 \quad \text{and} \quad \frac{1000 \text{ mm}}{1 \text{ m}} = 1$$

These factors are read "1 metre per 1000 millimetres" and "1000 millimetres per 1 metre." They can also be written as 1 m/1000 mm and 1000 mm/1 m. Since either factor is equal to the number 1, any number of metres can be converted to the equivalent number of millimetres by multiplying by 1000 mm/m; and, any number of millimetres can be converted to the equivalent number of metres by multiplying by 1 m/1000 mm. (Any number multiplied by 1 has the same value as the number.) To convert metres to millimetres we choose the factor 1000 mm/1 m because in the multiplication, metres cancel, leaving the answer in the desired units, millimetres.

$$2.5 \not{m} \times \frac{1000 \text{ mm}}{1 \not{m}} = 2500 \text{ mm} \qquad (2.5 \times 10^3 \text{ mm, two significant figures})$$

Note that in making this calculation, units are treated the same as numbers—metres in the numerator are canceled by metres in the denominator.

$$\not{m} \times \frac{\text{mm}}{\not{m}} = \text{mm}$$

Now suppose you need to change 215 centimetres to metres. Write

$$\text{cm} \longrightarrow \text{m}$$

Then write

$$100 \text{ cm} = 1 \text{ m}$$

The two possible conversion factors are

$$\frac{100 \text{ cm}}{\text{m}} = 1 \quad \text{and} \quad \frac{1 \text{ m}}{100 \text{ cm}} = 1$$

Next, set up the calculation choosing the conversion factor that will give the answer in the desired units, metres.

$$215 \not{cm} \times \frac{1 \text{ m}}{100 \not{cm}} = \frac{215 \text{ m}}{100} = 2.15 \text{ m}$$

The dimensional analysis, or factor-label, method used in the preceding work shows how unit conversion factors are derived and used in calculations. After you become more proficient with the terms, you can save steps by writing the factors directly in the calculation. Here are some examples of the conversion from American to metric units.

*Problem 2.1*

How many centimetres are in 2.00 ft?
The stepwise conversion of units from feet to centimetres may be done in this manner: Convert feet to inches; then convert inches to centimetres.

$$\text{ft} \rightarrow \text{in.} \rightarrow \text{cm}$$

The needed conversion factors are

$$\frac{12 \text{ in.}}{1 \text{ ft}} \quad \text{and} \quad \frac{2.54 \text{ cm}}{1 \text{ in.}}$$

$$2.00 \not{ft} \times \frac{12 \text{ in.}}{\not{ft}} = 24.0 \text{ in.}$$

$$24.0 \not{in.} \times \frac{2.54 \text{ cm}}{\not{in.}} = 61.0 \text{ cm} \quad \text{(Answer)}$$

Since 1 ft and 12 in. are considered to be exact numbers, the number of significant figures allowed in the answer is three, based on the number 2.00.

*Problem 2.2*

How many metres are there in a 100 yd football field? The stepwise conversion of units from yards to metres may be done in this manner, using the proper conversion factors.

$$\text{yd} \rightarrow \text{ft} \rightarrow \text{in.} \rightarrow \text{cm} \rightarrow \text{m}$$

$$100\ \cancel{\text{yd}} \times \frac{3\ \text{ft}}{\cancel{\text{yd}}} = 300\ \text{ft} \qquad (3\ \text{ft/yd})$$

$$300\ \cancel{\text{ft}} \times \frac{12\ \text{in.}}{\cancel{\text{ft}}} = 3600\ \text{in.} \qquad (12\ \text{in./ft})$$

$$3600\ \cancel{\text{in.}} \times \frac{2.54\ \text{cm}}{\cancel{\text{in.}}} = 9144\ \text{cm} \qquad (2.54\ \text{cm/in.})$$

$$9144\ \cancel{\text{cm}} \times \frac{1\ \text{m}}{100\ \cancel{\text{cm}}} = 91.4\ \text{m} \qquad (1\ \text{m}/100\ \text{cm}) \qquad \text{(three significant figures)}$$

Problems 2.1 and 2.2 may be solved using a running linear expression, writing down each conversion factor in succession. Very often this saves one or two calculation steps, and numerical values may be reduced to simpler terms leading to simpler calculations. The single linear expressions for Problems 2.1 and 2.2 are

$$2.00\ \cancel{\text{ft}} \times \frac{12\ \cancel{\text{in.}}}{\cancel{\text{ft}}} \times \frac{2.54\ \text{cm}}{\cancel{\text{in.}}} = 61.0\ \text{cm}$$

$$100\ \cancel{\text{yd}} \times \frac{3\ \cancel{\text{ft}}}{\cancel{\text{yd}}} \times \frac{12\ \cancel{\text{in.}}}{\cancel{\text{ft}}} \times \frac{2.54\ \cancel{\text{cm}}}{\cancel{\text{in.}}} \times \frac{1\ \text{m}}{100\ \cancel{\text{cm}}} = 91.4\ \text{m}$$

Using the units alone (Problem 2.2), we see that the stepwise cancellation proceeds in succession until the unit desired is reached.

$$\cancel{\text{yd}} \times \frac{\cancel{\text{ft}}}{\cancel{\text{yd}}} \times \frac{\cancel{\text{in.}}}{\cancel{\text{ft}}} \times \frac{\cancel{\text{cm}}}{\cancel{\text{in.}}} \times \frac{\text{m}}{\cancel{\text{cm}}} = \text{metres}$$

## 2.6 Measurement of Mass

kilogram

The standard unit of mass in the metric system is the **kilogram**. This amount of mass is defined by international agreement as being exactly equal to the mass of a platinum–iridium weight (*Kilogramme de Archive*) kept in a vault at Sèvres. A kilogram contains 1000 grams. Comparing this unit of mass to 1 pound (16 ounces), we find that a kilogram is equal to 2.2 pounds. A pound is equal to 454 grams (0.454 kilogram). The same prefixes used in length measurement are used to indicate larger and smaller gram units (see Table 2.3).

It is convenient to remember that

$$1\ \text{g} = 1000\ \text{mg}$$
$$1\ \text{kg} = 1000\ \text{g}$$
$$1\ \text{kg} = 2.2\ \text{lb}$$
$$1\ \text{lb} = 454\ \text{g}$$

*Table 2.3.* Metric units of mass.

| Unit | Abbreviation | Gram equivalent | Exponential equivalent |
|---|---|---|---|
| Kilogram | kg | 1000 g | $10^3$ g |
| Gram | g | 1 g | $10^0$ g |
| Decigram | dg | 0.1 g | $10^{-1}$ g |
| Centigram | cg | 0.01 g | $10^{-2}$ g |
| Milligram | mg | 0.001 g | $10^{-3}$ g |
| Microgram | μg | 0.000001 g | $10^{-6}$ g |

To change grams to milligrams, multiply grams by the conversion factor 1000 mg/g. The setup for converting 25 g is

$$25\ \not{g} \times \frac{1000\ \text{mg}}{1\ \text{g}} = 25{,}000\ \text{mg} \qquad (2.5 \times 10^4\ \text{mg}) \quad \text{(Answer)}$$

To change milligrams to grams, multiply milligrams by the conversion factor 1 g/1000 mg. For example, to convert 150 mg to grams:

$$150\ \not{mg} \times \frac{1\ \text{g}}{1000\ \not{mg}} = 0.150\ \text{g} \quad \text{(Answer)}$$

Examples of converting weights from English to metric units are shown below.

*Problem 2.3*

A 1.50 lb package of sodium bicarbonate costs 80 cents. How many grams of this substance are in this package?
We are solving for the number of grams equivalent to 1.50 lb. Since 1 lb = 454 g, the factor to convert pounds to grams is 454 g/lb.

$$1.50\ \not{lb} \times \frac{454\ \text{g}}{\not{lb}} = 681\ \text{g} \quad \text{(Answer)}$$

*Note*: The cost of the sodium bicarbonate has no bearing on the question asked in this problem.

*Problem 2.4*

Suppose four ostrich feathers weigh 1.00 lb. Assuming that each feather is equal in weight, how many milligrams does a single feather weigh? The unit conversion in this problem is from 1 lb/4 feathers to milligrams per feather. Since the unit feathers occurs in the denominator of both the starting unit and the desired unit, the unit conversions needed are

$$\text{lb} \rightarrow \text{g} \rightarrow \text{mg}$$

$$\frac{1.00\ \not{lb}}{4\ \text{feathers}} \times \frac{454\ \not{g}}{\not{lb}} \times \frac{1000\ \text{mg}}{\not{g}} = \frac{113{,}500\ \text{mg}}{\text{feather}} \qquad (1.14 \times 10^5\ \text{mg/feather})$$

## 2.7 Measurement of Volume

litre

The metric system unit of volume is the **litre**. The litre is a little larger than a U.S. quart; 1.000 litre equals 1.057 quarts. The most commonly used fractional unit of a litre is the millilitre: 1 litre = 1000 ml, and 946 ml = 1.00 qt. This small unit of volume is also commonly referred to as a cubic centimetre, abbreviated $cm^3$ (or sometimes cc). This is because a litre corresponds to the volume enclosed in a cube measuring exactly 10 cm on an edge. The volume of this cube is determined by multiplying length times width times height—that is, 10 cm × 10 cm × 10 cm = 1000 $cm^3$ (see Figure 2.2). The relationship between a millilitre and a cubic centimetre is that 1 ml equals 1 $cm^3$ exactly (see Table 2.4).

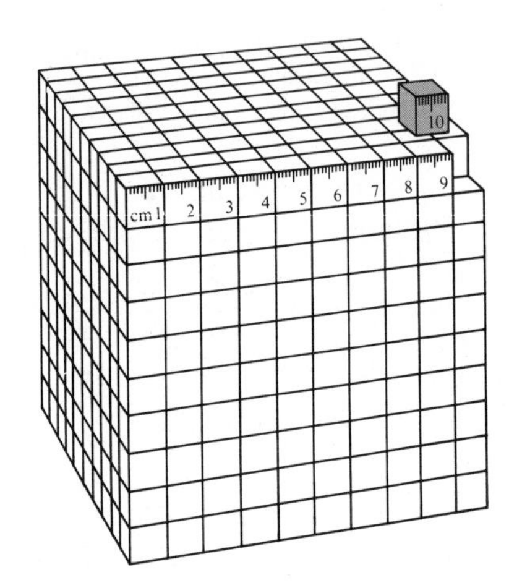


*Figure 2.2.* The large cube, 10.0 cm on a side, has a volume of 1000 $cm^3$, or 1.0 litre. The small cube on top is 1.0 $cm^3$. The large block contains 1000 of these small cubes.

*Table 2.4.* Metric units of volume.

| Unit | Abbreviation | Litre equivalent |
|---|---|---|
| Litre | 1 | 1.0 litre |
| Millilitre | ml | 0.001 litre |
| Cubic centimetre | $cm^3$ or cc | 0.001 litre |

Examples of volume conversions are shown below.

*Problem 2.5*

How many millilitres are contained in 3.5 litres?

The conversion factor to change litres to millilitres is 1000 ml/litre.

$$3.5\ \cancel{\text{litres}} \times \frac{1000\ \text{ml}}{\cancel{\text{litre}}} = 3500\ \text{ml} \qquad (3.5 \times 10^3\ \text{ml})$$

Litres may be changed to millilitres by moving the decimal point three places to the right and changing the units to millilitres.

1.500 litres = 1500 ml

*Problem 2.6*

How many cubic centimetres are in a cube that is 11.1 in. on a side?

First change inches to centimetres.

$$11.1\ \cancel{\text{in.}} \times \frac{2.54\ \text{cm}}{\cancel{\text{in.}}} = 28.2\ \text{cm on a side}$$

Then change to cubic volume (Length × Width × Height).

$$28.2\ \text{cm} \times 28.2\ \text{cm} \times 28.2\ \text{cm} = 22{,}426\ \text{cm}^3 \qquad (2.24 \times 10^4\ \text{cm}^3)$$

---

Table 2.5 summarizes the units and conversion factors that are used most often.

*Table 2.5.* Most often used units and their equivalents.

|  |  |
|---|---|
| 1 m | = 1000 mm |
| 1 m | = 39.37 in. |
| 1 cm | = 10 mm |
| 2.54 cm | = 1 in. |
| 1 g | = 1000 mg |
| 1 kg | = 1000 g |
| 454 g | = 1 lb |
| 1 litre | = 1000 ml |
| 1 ml | = 1 $\text{cm}^3$ or cc |
| 946 ml | = 1 qt |

## 2.8 Temperature Scales

The *temperature* of a system measures how hot or cold that system is and can be expressed by several different temperature scales. Three commonly used temperature scales are the Celsius (centigrade) scale, the Kelvin (absolute) scale, and the Fahrenheit scale. A unit of temperature on each of these scales is called a *degree*, although the size of the degree varies. The symbol for the degree is ° and it is placed as a superscript after the number and before the temperature scales for Celsius and Fahrenheit. Thus, 100°C means 100 *degrees Celsius.* The degree sign is not used with Kelvin temperatures.

Degrees Celsius (centigrade) = °C
Degrees Kelvin (absolute) = K
Degrees Fahrenheit = °F

The Celsius scale is based on dividing the interval between the freezing and boiling temperatures of water into 100 equal parts, or degrees. The freezing point of water is assigned a temperature of 0°C and the boiling point of water a temperature of 100°C. The Kelvin temperature scale is also known as the absolute temperature scale because 0 K is the lowest possible temperature theoretically attainable. The Kelvin zero is 273.16 degrees below the Celsius zero. Kelvin degrees are equal in size to Celsius degrees. The freezing point of water on the Kelvin scale is 273.16 K (usually rounded to 273 K). On the Fahrenheit scale there are 180 degrees between the freezing and boiling temperatures of water. On this scale, the freezing point of water is 32°F and the boiling point is 212°F.

$$0°\text{C} \cong 273\ \text{K} \cong 32°\text{F}$$

The three scales are compared in Figure 2.3. Although absolute zero is the lower limit of temperature on these scales, there is no known upper limit to temperature. (Temperatures of several million degrees are known to exist in the sun and in other stars.)

By examining Figure 2.3 we can see that there are 100 Celsius degrees and 100 Kelvin degrees between the freezing and boiling points of water. But there are 180 Fahrenheit degrees between these two temperatures. Hence, the size of a degree on the Celsius scale is the same as the size of a degree on the Kelvin scale, but the Celsius degree corresponds to 1.8 degrees on the Fahrenheit scale. From these data, mathematical formulas have been derived to convert a temperature on one scale to the corresponding temperature on another scale. These formulas are

$$K = {}^{\circ}C + 273 \tag{1}$$

$$^{\circ}F = (1.8 \times {}^{\circ}C) + 32 \tag{2}$$

$$^{\circ}C = \frac{(^{\circ}F - 32)}{1.8} \tag{3}$$

Interpretation: Formula (1) states that the addition of 273 to the degrees Celsius converts the temperature to degrees Kelvin. Formula (2) states that to obtain the Fahrenheit temperature corresponding to a given Celsius temperature we multiply the degrees Celsius by 1.8 and then add 32. Formula (3) states that to

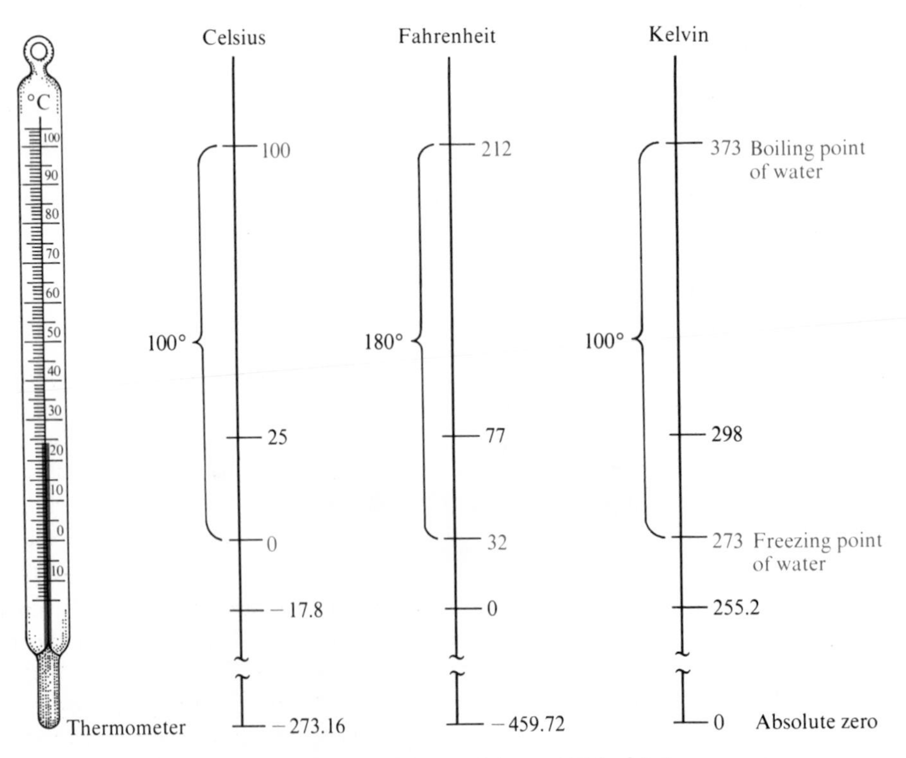


*Figure 2.3.* Comparison of Celsius, Kelvin, and Fahrenheit temperature scales.

obtain the corresponding Celsius temperature we subtract 32 from the degrees Fahrenheit and then divide this figure by 1.8.

Examples of temperature conversions follow.

*Problem 2.7*

The temperature at which salt (sodium chloride) melts is 800°C. What is this temperature on the Kelvin and Fahrenheit scales?

We need to calculate K from °C, so we use formula (1) above. We also need to calculate °F from °C; for this we use formula (2).

$$K = °C + 273$$

$$K = 800°C + 273 = 1073\ K$$

$$°F = (1.8 \times °C) + 32$$

$$°F = (1.8 \times 800°C) + 32$$

$$°F = 1440 + 32 = 1472°F$$

$$800°C = 1073\ K = 1472°F$$

*Problem 2.8*

The temperature for December 1 was 110°F, a new record. Calculate this temperature in °C.

Formula (3) applies here.

$$°C = \frac{(°F - 32)}{1.8}$$

$$°C = \frac{(110 - 32)}{1.8} = \frac{78}{1.8} = 43°C$$

*Problem 2.9*

What temperature on the Fahrenheit scale corresponds to −8.0°C? (Be alert to the presence of the minus sign in this problem.)

$$°F = (1.8 \times °C) + 32$$

$$°F = [1.8 \times (-8.0)] + 32 = -14.4 + 32$$

$$°F = 17.6$$

Temperatures used in this book are in degrees Celsius (°C) unless specified otherwise.

## 2.9 Heat and Temperature

heat

temperature

**Heat** is a form of energy associated with the motion of small particles of matter. Heat is associated with a quantity of energy within a system or a quantity of energy supplied to a system. **Temperature** is a measure of the intensity of heat, or how hot a system is, regardless of its size. Heat always flows from a region of high temperature to one of lower temperature.

calorie

kilocalorie

The unit of heat commonly used in chemical systems is the calorie. A **calorie** (cal) is the quantity of heat required to change the temperature of 1 g water 1°C, usually measured from 14.5 to 15.5°C. The **kilocalorie** (kcal), also known as the nutritional or large Calorie (spelled with a capital C and abbreviated Cal), is equal to 1000 small calories. Temperature is measured in degrees and heat quantity is measured in calories.

The difference in the meanings of the terms *heat* and *temperature* can be seen by this example: Visualize two beakers, A and B. Beaker A contains 100 g water at 20°C, and beaker B contains 200 g water also at 20°C. The beakers are now heated until the temperature of the water in each reaches 30°C. The *temperature* of the water in the beakers was raised by exactly the same amount, 10°C. Yet twice as much *heat* (2000 cal) was required by the water in beaker B than was required by the water in beaker A (1000 cal).

heat capacity

The **heat capacity**, also known as *specific heat*, of any substance is the quantity of heat (in calories) required to change the temperature of 1 g of that substance by 1°C. It follows from the definition of the calorie that the heat capacity of water is 1 cal/g°C. The heat capacity of water is high compared to most substances. Aluminum and copper, for example, have heat capacities of 0.215 and 0.0921 cal/g°C, respectively (see Table 2.6). The relation of mass, heat capacity, temperature change ($\Delta t$), and quantity of heat lost or gained by a system is expressed by this general equation:

$$\begin{pmatrix}\text{Grams of}\\ \text{substance}\end{pmatrix} \times \begin{pmatrix}\text{Heat capacity}\\ \text{of substance}\end{pmatrix} \times \Delta t = \text{calories} \qquad (1)$$

Thus, the amount of heat needed to warm 200 g water 10°C can be calculated as follows:

$$200\text{ g} \times \frac{1.00\text{ cal}}{\cancel{\text{g}}\,^{\circ}\cancel{\text{C}}} \times 10\,^{\circ}\cancel{\text{C}} = 2000\text{ cal}$$

*Table 2.6.* Heat capacity of selected substances.

| Substance | Heat capacity cal/g°C |
|---|---|
| Water | 1.00 |
| Ethyl alcohol | 0.511 |
| Ice | 0.492 |
| Aluminum | 0.215 |
| Iron | 0.113 |
| Copper | 0.0921 |
| Gold | 0.0312 |
| Lead | 0.0305 |

The application of equation (1) to the solution of a more complicated example is illustrated by Problem 2.10.

*Problem 2.10*

The heat capacity of aluminum (Al) is 0.215 cal/g°C. How many grams of water ($H_2O$) can be raised from 24.0°C to 33.0°C when a 10.0 g ingot of Al at 143°C is dropped into the water?

The heat lost or gained by a system is given by equation (1). For our problem, we have

$\Delta t$ = Change in temperature (°C) of Al

Heat lost by Al = Heat gained by $H_2O$

The Al ingot is dropped into water and is allowed to cool to 33.0°C. Therefore, its temperature change ($\Delta t$) is 110°C (143 − 33.0).

$$\text{Heat lost by Al} = 10.0 \text{ g Al} \times 0.215 \text{ cal/g°C} \times 110\text{°C} = 236 \text{ cal}$$

$$\text{Heat gained by } H_2O = 236 \text{ cal}$$

Solving equation (1) for grams of water, we obtain

$$\text{g } H_2O = \frac{\text{cal}}{\text{Heat capacity} \times \Delta t}$$

$$\Delta t = 9.0\text{°C} \quad \text{(change in water temperature from 24.0°C to 33.0°C)}$$

$$\text{g } H_2O = \frac{236 \text{ cal}}{1 \text{ cal/g°C} \times 9.0\text{°C}}$$

$$\text{g } H_2O = 26 \text{ g}$$

## 2.10 Tools for Measurement

Common measuring instruments used in chemical laboratories are illustrated in Figures 2.4 and 2.5. A balance is used to measure mass. Balances are obtainable that will weigh objects to the nearest microgram. The choice of the balance depends on the accuracy required and the amount of material being

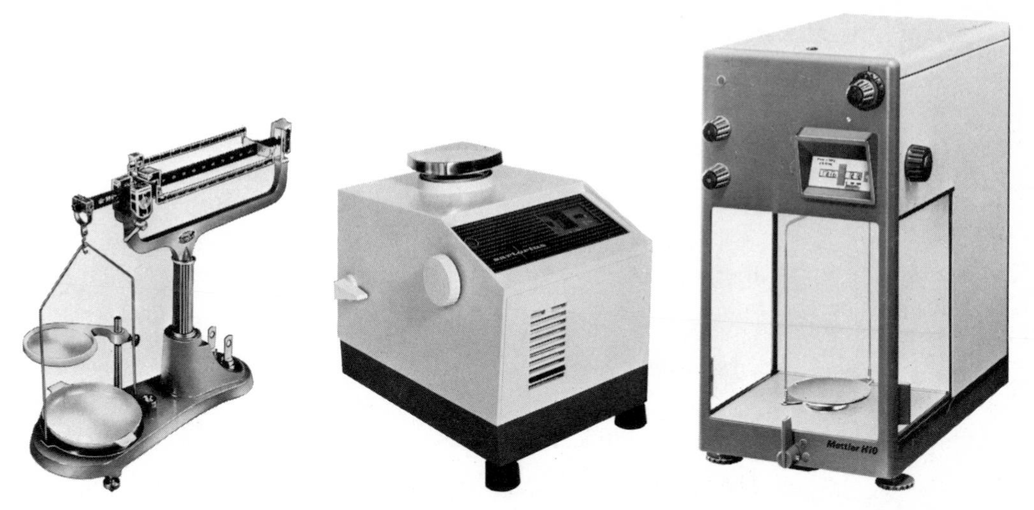


*Figure 2.4.* At left is a triple-beam balance. This balance has three calibrated horizontal beams, each fitted with a specific movable weight. The weight of the object placed on the pan is determined by moving the weights along the beams until the swinging beam is in balance, as shown by the indicator on the right. (Courtesy Ohaus Scale Company.) In the center is a single-pan, top-loading, rapid-weighing balance with direct read-out to the nearest milligram (for example, 125.456 g). (Courtesy Brinkmann Instruments Inc.) At right is a single-pan analytical balance for high-precision weighing. The precision of this balance is 0.1 mg (0.0001 g). (Courtesy Mettler Instrument Corporation.)

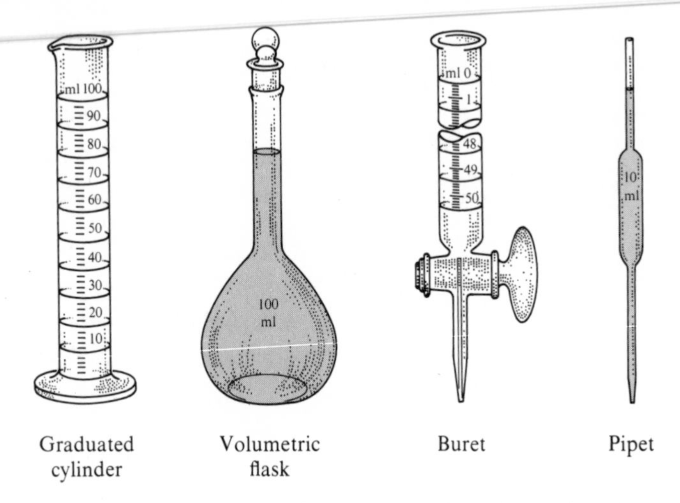


*Figure 2.5.* Calibrated glassware for measuring the volume of liquids.

weighed. Three standard balances are shown in Figure 2.4: a triple-beam balance with precision up to 0.01 g; a single-pan, top-loading balance with a precision of 0.001 g (1 mg); a single-pan analytical balance with a precision up to 0.0001 g. Automatic-recording balances are also available.

The most common instruments for measuring liquids are the graduated cylinder, volumetric flask, buret, and pipet, which are shown in Figure 2.5. These calibrated pieces are usually made of glass and are available in various sizes.

The common laboratory tool for measuring temperature is a thermometer (see Figure 2.3).

## 2.11 Density

density

**Density** ($d$) is the ratio of the mass of a substance to the volume occupied by that mass; it is the mass per unit of volume and is given by the equation

$$d = \frac{\text{Mass}}{\text{Volume}}$$

Density is a physical characteristic of a substance and may be used as an aid to its identification. When the density of a solid or a liquid is given, the mass is usually expressed in grams and the volume in millilitres or cubic centimetres.

$$d = \frac{\text{Mass}}{\text{Volume}} = \frac{\text{g}}{\text{ml}} \quad \text{or} \quad d = \frac{\text{g}}{\text{cm}^3}$$

Since the volume of a substance (especially liquids and gases) varies with temperature, it is important to state the temperature along with the density. For example, the volume of 1.0000 g water at 4°C is 1.0000 ml, while at 20°C it is 1.0018 ml, and at 80°C it is 1.0290 ml. Density, therefore, also varies with temperature.

The density of water at 4°C is 1.0000 g/ml but at 80°C the density of water is 0.9718 g/ml.

$$d^{4°C} = \frac{1.0000 \text{ g}}{1.0000 \text{ ml}} = 1.0000 \text{ g/ml}$$

$$d^{80°C} = \frac{1.0000 \text{ g}}{1.0290 \text{ ml}} = 0.9718 \text{ g/ml}$$

Densities for liquids and solids are usually represented in terms of grams per millilitre or grams per cubic centimetre. The density of gases, however, is normally expressed in terms of grams per litre (g/litre). Unless otherwise stated, gas densities are given for 0°C and 1 atmosphere pressure (discussed further in Chapter 12). Table 2.7 lists the densities of a number of common materials.

*Table 2.7.* Densities of some selected materials. For comparing densities, the density of water is the reference for solids and liquids; air is the reference for gases.

| Liquids and solids | | Gases | |
|---|---|---|---|
| Substance | Density (g/ml at 20°C) | Substance | Density (g/litre at 0°C) |
| Wood (Douglas fir) | 0.512 | Hydrogen | 0.090 |
| Ethyl alcohol | 0.79 | Helium | 0.178 |
| Cottonseed oil | 0.926 | Methane | 0.714 |
| **Water (4°C)** | **1.0000** | Ammonia | 0.771 |
| Sugar | 1.59 | Neon | 0.90 |
| Carbon tetrachloride | 1.595 | Carbon monoxide | 1.25 |
| Magnesium | 1.74 | Nitrogen | 1.251 |
| Sulfuric acid | 1.84 | **Air** | **1.293** |
| Sulfur | 2.07 | Oxygen | 1.429 |
| Salt | 2.16 | Hydrogen chloride | 1.63 |
| Aluminum | 2.70 | Argon | 1.78 |
| Silver | 10.5 | Carbon dioxide | 1.963 |
| Lead | 11.34 | Chlorine | 3.17 |
| Mercury | 13.55 | | |
| Gold | 19.3 | | |

Suppose that water, carbon tetrachloride, and cottonseed oil are successively poured into a graduated cylinder. The result is a layered three-liquid system (Figure 2.6). Can we predict the order of the liquid layers? Yes, by looking up the liquid densities in Table 2.7. Carbon tetrachloride has the greatest density (1.595 g/ml) and cottonseed oil has the lowest density (0.926 g/ml). Carbon tetrachloride will, therefore, form the bottom layer and cottonseed oil the top layer. Water, with a density between the other two liquids, will, of course, form the middle layer. This information can also be determined readily by experiment. Add a few millilitres of carbon tetrachloride to a beaker of water. The carbon tetrachloride, being more dense than the water, will sink. Cottonseed oil, being less dense than water, will float when added to water.

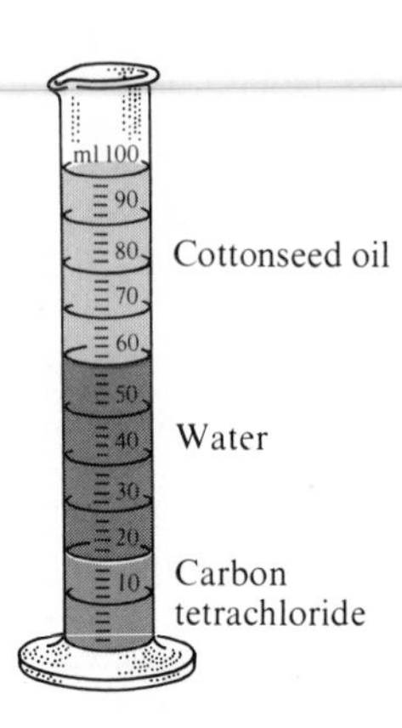


*Figure 2.6.* Relative density of liquids. When three immiscible (not capable of mixing) liquids are poured together, the liquid with the highest density will be the bottom layer. In the case of cottonseed oil, water, and carbon tetrachloride, cottonseed oil is the top layer.

Direct comparisons of density in this manner can be made only with liquids that are *immiscible* (do not dissolve in one another).

The density of air at 0°C is approximately 1.293 g/litre. Gases with densities less than this value are said to be "lighter than air." A helium-filled balloon will rise rapidly in air because the density of helium is only 0.178 g/litre.

When an insoluble solid object is dropped into water, the object will sink or float—depending on its density. If the object is less dense than water, it will float, displacing a *mass* of water equal to the mass of the object. If the object is more dense than water, it will sink, displacing a *volume* of water equal to the volume of the object. This information can be utilized to determine the volume (and density) of irregularly shaped objects.

Sample calculations of density problems follow.

*Problem 2.11*

What is the density of a mineral if 427 g of the mineral occupy a volume of 35.0 ml? We need to solve for density, so we start by writing the formula for calculating density.

$$d = \frac{\text{Mass}}{\text{Volume}}$$

Then we substitute the data given in the problem into the equation and solve.

Mass = 427 g    Volume = 35.0 ml

$$d = \frac{\text{Mass}}{\text{Volume}} = \frac{427\text{ g}}{35.0\text{ ml}} = 12.2\text{ g/ml} \quad \text{(Answer)}$$

---

*Problem 2.12*

The density of gold is 19.3 g/ml. What is the mass of 25.0 ml of gold?

(a) First write the formula for calculating density.

$$d = \frac{\text{Mass}}{\text{Volume}}$$

(b) We need to solve for mass; therefore, we solve the density equation to obtain mass on one side by itself.

$$\text{Mass} = \text{Volume} \times d$$

(c) Substitute data given in the problem and calculate.

$$\text{Mass} = 25.0\ \cancel{\text{ml}} \times \frac{19.3\ \text{g}}{\cancel{\text{ml}}} = 482\ \text{g}$$

---

*Problem 2.13*

The water level in a graduated cylinder stands at 20.0 ml before and at 26.2 ml after a 16.74 g metal bolt is submerged in the water. (a) What is the volume of the bolt? (b) What is the density of the bolt?

(a)
$$\begin{array}{rl} 26.2\ \text{ml} = & \text{Volume of water plus bolt} \\ -20.0\ \text{ml} = & \text{Volume of water} \\ \hline 6.2\ \text{ml} = & \text{Volume of bolt} \quad \text{(Answer)} \end{array}$$

(b)
$$d = \frac{\text{Mass of bolt}}{\text{Volume of bolt}} = \frac{16.74\ \text{g}}{6.2\ \text{ml}} = 2.7\ \text{g/ml} \quad \text{(Answer)}$$

---

## 2.12 Specific Gravity

specific gravity

The **specific gravity** (sp gr) of a substance is a ratio of the density of that substance to the density of another substance. Water is usually used as the reference standard for solids and liquids.

$$\text{sp gr} = \frac{\text{Density of a liquid or solid}}{\text{Density of water}} \quad \text{or} \quad \frac{\text{Density of a gas}}{\text{Density of air}}$$

Specific gravity has no units but is a number that compares the density of a liquid or a solid with that of water, or the density of a gas with that of air.

*Problem 2.14*

What is the specific gravity of mercury with respect to water at 4°C? (Density of water at 4°C is 1.000 g/ml.)

$$\text{sp gr} = \frac{\text{Density of mercury}}{\text{Density of water}} = \frac{13.55\ \text{g/ml}}{1.000\ \text{g/ml}}$$

$$\text{sp gr of mercury} = 13.55$$

The value for the specific gravity of mercury (13.55) tells us that, per unit volume, mercury is 13.55 times as heavy as water. Do you think that you could readily lift a litre (approximately 1 quart) of mercury?

---

hydrometer

A **hydrometer** consists of a weighted bulb at the end of a sealed, calibrated tube. This instrument is used to measure the specific gravity of a liquid (see Figure 2.7). When a hydrometer is floated in a liquid, the specific gravity is indicated on the scale at the surface of the liquid.

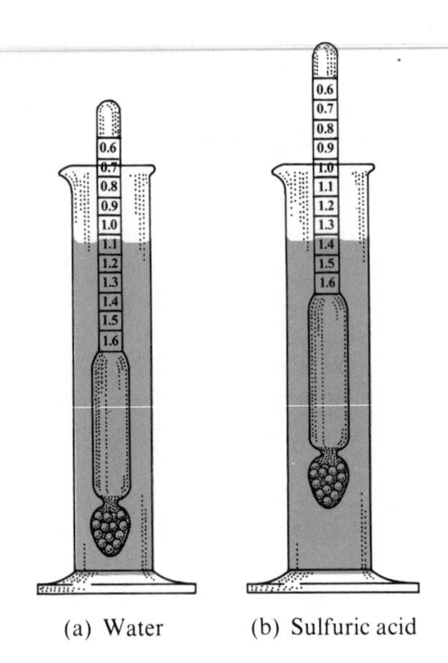


(a) Water (b) Sulfuric acid

*Figure 2.7.* Specific gravity determination using hydrometers. The hydrometer in (a) is floating in water, showing a specific gravity of 1.0. The hydrometer in (b) is floating in dilute sulfuric acid (battery acid), showing a specific gravity of 1.3.

## Questions

A. *Review the meanings of the new terms introduced in this chapter. The terms listed in Section A of each set of Questions are new terms defined in the chapter. They appear in boldface type and occur in the chapter in the order listed in Question A.*

1. Mass
2. Weight
3. Significant figures
4. Metric system, or SI
5. Metre
6. Kilogram
7. Litre
8. Heat
9. Temperature
10. A calorie
11. Kilocalorie
12. Heat capacity
13. Density
14. Specific gravity
15. Hydrometer

B. *Answers to the following questions will be found in tables and figures.*

1. Use Table 2.2 to determine how many centimetres make 1 km.
2. What is the temperature difference in Celsius degrees between 0°C and 0°F?
3. Why do you suppose the top ends of the pipet and the volumetric flask are narrower than the bulk of these volumetric instruments? (See Figure 2.5.)
4. Refer to Table 2.7 and describe the arrangement you would see when these three immiscible materials are placed in a 100 ml graduated cylinder: 135.5 g mercury, 25 ml cottonseed oil, and a cube of sulfur measuring 2.0 cm on an edge. Mercury and cottonseed oil are liquids.
5. Arrange the following materials in order of increasing density: lead, sulfur, gold, wood, and water.
6. Would argon be a satisfactory lifting gas for a balloon? Explain.

C. *Review questions.*

1. Why is the metric system of weights and measurements more desirable than the American system?

2. What are the abbreviated symbols for the following?
   (a) Gram (b) Milligram (c) Kilogram (d) Megagram (e) Decimetre (f) Centimetre (g) Cubic centimetre (h) Nanometre (i) Millilitre (j) Kilolitre
3. In a number, when is zero significant and when is it not significant? See Appendix I.
4. What are the three rules for rounding off a number? See Appendix I.
5. Suppose you had a litre of water in a flask, a quart bottle, and a balance with a capacity of 200 g. Describe how you would go about determining the volume of the water required to fill the quart bottle. Assume the density of water is 1.00 g/ml.
6. Distinguish between heat and temperature.
7. Will aluminum or iron become hotter when 100 cal of energy are added to 10 g samples of each of these metals?
8. Ice floats in water and sinks in ethyl alcohol. What information does this give you about the density of ice?
9. Which of the following statements are correct?
   (a) The prefix *milli* indicates one-millionth of the unit expressed.
   (b) The quantity 10 cm is equal to 100 mm.
   (c) The number 383.263 reduced to four significant figures becomes 383.3.
   (d) The number of significant figures in the number 29,004 is three.
   (e) The sum of 24.928 g + 2.126 g should contain five significant figures.
   (f) The product of 14.63 cm × 2.50 cm should contain three significant figures.
   (g) One microsecond is $10^{-6}$ second.
   (h) One thousand metres is a shorter distance than 1000 yards.
   (i) The number 0.002894 expressed in scientific notation is $2.894 \times 10^{-3}$.
   (j) $2.0 \times 10^4 \times 6.0 \times 10^6 = 1.2 \times 10^{11}$
   (k) $5.0 \times 10^5 \times 5.0 \times 10^{-3} = 2.5 \times 10^9$
   (l) One degree on the Celsius scale is equal to one degree on the Kelvin scale and to 1.8 degrees on the Fahrenheit scale.
   (m) The direction of heat flow is from cold to hot.
   (n) A calorie is a unit of temperature.
   (o) Temperature is a form of energy.
   (p) The density of water at 4°C is 1.00 g/ml.
   (q) A graduated cylinder would be a more accurate instrument for measuring 10.0 ml water than would a pipet.
   (r) A hydrometer is an instrument for measuring the specific gravity of liquids.

*D. Review problems.*

1. How many significant figures are in each of the following numbers? (See Mathematical Review in Appendix I.)
   (a) 0.007 (b) 22.2 (c) 0.1002 (d) 300.0 (e) 345,409 (f) 82.060 (g) 0.0283 (h) 0.0720
2. Round off the following numbers to four significant figures. (See Mathematical Review in Appendix I.)
   (a) 3.00051 (b) 9.3775 (c) 41.127 (d) 25.5555 (e) 2144.4 (f) 82.365 (g) 19.995
3. Express each of the following numbers in exponential notation (as a power of 10). (See Mathematical Review in Appendix I.)
   (a) 847 (b) 0.000586 (c) 22,400 (d) 0.088 (e) 0.0000611 (f) 4286 (g) 0.0650
4. Solve the following mathematical problems. (See Mathematical Review in Appendix I.)

(a) $23.89 + 13.0 + 1.3 =$
(b) $33.04 + 9.009 + 106.8 =$
(c) $15.3 \times 6.82 =$
(d) $2.90 \times 29.0 \times 290 =$
(e) $\frac{5}{6} \times \frac{2}{3} =$
(f) Change to decimal fractions: $\frac{3}{7}, \frac{11}{15}, \frac{6}{9}, \frac{98}{125}$
(g) $\frac{1}{3} + \frac{5}{9} =$
(h) $2.5(3.0X + 12) =$
(i) $\frac{(°F - 32)}{1.8}$, where $°F = 200$
(j) $124.36 \div 6.40$
(k) $\frac{0.2386}{0.2550}$

5. Show calculations for the following conversions:
(a) 12.0 cm to m
(b) 142 m to km
(c) 2.5 cm to Å
(d) 42.4 cm to mm
(e) 12.0 in. to cm
(f) 5.00 miles to km
(g) 2.10 m to cm
(h) 3.0 km to m
(i) 10 Å to cm
(j) 400 mm to cm
(k) 22.0 cm to in.
(l) 70.0 km to miles
6. An automobile travelling at 55 miles per hour is moving at what speed in kilometres per hour?
7. The speed of light in a vacuum is $3.00 \times 10^{10}$ cm per second. Calculate the speed of light in miles per second.
8. The sun is approximately 92 million miles from the earth. How many seconds will it take light to travel from the sun to the earth if the velocity of light is $3.00 \times 10^{10}$ cm per second?
9. Oil spreads in a thin layer on water and is commonly called an "oil slick." How much area in square metres ($m^2$) will 100 $cm^3$ of oil cover if it forms a layer 5 Å in thickness?
10. Show calculations for the following conversions:
(a) 1.200 g to mg
(b) 454 g to kg
(c) 1000 mg to kg
(d) 2.55 lb to g
(e) 50 mg to g
(f) 2.2 kg to g
(g) 0.350 kg to mg
(h) 25.6 g to lb
11. How many kilograms does a 170 lb man weigh?
12. The usual aspirin tablet contains 5.0 grains of aspirin. How many grams of aspirin are in one tablet (1 grain = 1/7000 lb)?
13. The price of gold varies greatly and has been almost as high as $400 per ounce. What is the value of 250 g of gold at $325 per ounce?
14. The largest nugget of gold on record was found in 1872 in New South Wales, Australia, and weighed 93.3 kg. What was the volume of this nugget in cubic centimeters? In litres?
15. At 35 cents per litre, how much will it cost to fill a 15 gal tank with gasoline?
16. A French automobile manufacturer claims that its sedan uses only 6.0 litres of

gasoline per 100 km. How many miles per gallon of gasoline could be expected from the car?

17. An adult ruby-throated hummingbird has an average weight of 3.2 g, whereas an adult California condor may attain a weight of 21 lb. How many times heavier than the hummingbird is the condor?
18. The average weight of the heart of a human baby is about 1 oz. What is this weight in milligrams?
19. More sulfuric acid is manufactured in the United States than any other chemical; the annual production is $6.4 \times 10^{10}$ lb. What is the average daily production of sulfuric acid in tons? In kilograms?
20. Control of automobile exhaust emission of oxides of nitrogen began with the 1971 car models. The Clean Air Act of 1970 states that these emissions must be reduced from 6.2 g per mile to 0.62 g per mile. What will be the average daily emission (in grams) of oxides of nitrogen in a city having 6 million automobiles, each driving an average of 10,000 miles per year, when this goal is reached?
21. Show calculations for the following conversions:
    (a) 145 ml to litres
    (b) 5.00 in.$^3$ to cubic centimetres
    (c) 150 gal to litres
    (d) 2.50 litres to millilitres
    (e) 6.00 litres to millilitres
    (f) 58.0 cm$^3$ to cubic inches
    (g) 2.50 litres to gallons
    (h) 22.4 litres to millilitres
22. Calculate the volume, in litres, of a box 1.8 m long, 16 cm wide, and 55 mm deep.
23. At a price of $1.15 per gallon, what will it cost to fill a 50 litre tank with gasoline?
24. An aquarium has the following dimensions: 30 in. long by 10 in. wide by 20 in. high. How much water will it take to fill the aquarium three-fourths full? Express your answer in both litres and gallons.
25. Show calculations for the following conversions:
    (a) 140°F to °C
    (b) 0°F to °C
    (c) 0°F to K
    (d) −12°F to °C
    (e) 25°C to °F
    (f) −12°C to °F
    (g) 273°C to K
    (h) 0°C to K
26. Normal body temperature for humans is 37.0°C. What is this temperature on the Fahrenheit scale?
27. Which is colder, −90°C or −135°F?
28. (a) At what temperature are the Fahrenheit and Celsius scales exactly equal?
    (b) At what temperature are they numerically equal but opposite in sign?
29. How many calories of heat are required to raise the temperature of 100 g water from 20°C to 50°C?
30. How many calories are required to raise the temperature of 100 g iron from 20°C to 50°C?
31. A 20.0 g piece of copper at 203°C is dropped into 80.0 g water at 25.0°C. The water temperature rises to 29.0°C. Calculate the heat capacity (specific heat) of copper. Assume all the heat lost by the copper is transferred to the water.
32. Assuming no heat losses by the system, what will be the final temperature when 50 g water at 10°C is mixed with 10 g water at 50°C?
33. Calculate the density of a liquid if 16.60 ml of the liquid weighs 17.25 g.
34. A 25.0 ml sample of bromine weighs 78.0 g. What is the density of bromine?
35. When a 15.6 g piece of chromium metal was placed into a graduated cylinder containing 25.0 ml water, the water level rose to 27.2 ml. Calculate the density of the chromium.
36. Concentrated hydrochloric acid has a density of 1.19 g/ml. Calculate the weight, in grams, of 1.00 litre of this acid.

37. Thirty-five millilitres of ethyl alcohol (density 0.79 g/ml) is added to a graduated cylinder that weighs 44.28 g. What will be the weight of the cylinder plus the alcohol?
38. What weight of mercury (density 13.6 g/ml) will occupy a volume of 50.0 ml?
39. You are given three cubes, A, B, and C; one is magnesium, one is aluminum, and the other is silver. All three cubes weigh the same, but cube A has a volume of 29.0 ml, cube B a volume of 18.8 ml, and cube C a volume of 4.81 ml. Identify cubes A, B, and C.
40. A cube of aluminum weighs 500 g. What will be the weight of a cube of silver of the same dimensions?
41. Twenty-five millilitres of water at 90°C weighs 24.12 g. Calculate the density of water at this temperature.
42. Calculate (a) the density and (b) the specific gravity of a solid that weighs 160 g and has a volume of 50.0 ml.
43. Which liquid will occupy the greater volume, 100 g water or 100 g ethyl alcohol? Explain!
44. A solution made by adding 143 g sulfuric acid to 500 ml water had a volume of 554 ml.
    (a) What value will a hydrometer read when placed in this solution?
    (b) What volume of concentrated sulfuric acid ($d = 1.84$ g/ml) was added?

# 3 Properties of Matter

*After studying Chapter 3 you should be able to:*

1. Understand the new terms listed in Question A at the end of the chapter.
2. Identify the three physical states of matter and list the physical properties that characterize each state.
3. Distinguish between the physical and chemical properties of matter.
4. Classify changes undergone by matter as being either physical or chemical changes.
5. Distinguish between substances and mixtures.
6. Distinguish between kinetic and potential energy.
7. State the Law of Conservation of Mass.
8. State the Law of Conservation of Energy.
9. Explain why the laws dealing with the conservation of mass and of energy may be combined into a single more accurate general statement.
10. Calculate the percent composition of compounds from the weights of the elements involved in a chemical reaction or vice versa.

## 3.1 Matter Defined

matter

The entire universe consists of matter and energy. Every day we come into contact with countless kinds of matter. Air, food, water, rocks, soil, glass, this book—all are different types of matter. Broadly defined, **matter** is *anything* that has mass and occupies space.

Matter may be quite invisible. If an apparently empty test tube is submerged mouth downward in a beaker of water, the water rises only slightly into the tube. The water cannot rise further because the tube is filled with invisible matter—air (see Figure 3.1).

To the eye matter appears to be continuous and unbroken. However, it is actually discontinuous and is composed of discrete, tiny particles called *atoms*.

## 3.2 Physical States of Matter

solid

Matter exists in three physical states: solid, liquid, and gas. A **solid** is characterized by having a definite shape and volume, with particles that cohere rigidly to one another. The shape of a solid may be independent of its container.

*Figure 3.1.* An apparently empty test tube is submerged, mouth downward, in water. Only a small volume of water rises into the tube, which is actually filled with air. This experiment proves that air, which is matter, occupies space.

For example, a crystal of sulfur has the same shape and volume whether it is placed in a beaker or simply laid on a glass plate.

Most commonly occurring solids, such as salt, sugar, quartz, and metals, are *crystalline*. Crystalline materials exist in regular, recurring geometric patterns. Solids such as plastics, glass, and gels, because they do not have any particular regular internal geometric form, are called **amorphous** solids. ("Amorphous" means without shape or form.) Figure 3.2 illustrates three crystalline solids—salt, quartz, and gypsum.

amorphous

A **liquid** is characterized by having a definite volume, but not a definite shape, with particles that cohere firmly but not rigidly. Although held together by strong attractive forces, the particles are able to move freely but remain in close contact with each other. Particle mobility gives the liquid fluidity and causes it to take the shape of the container in which it is stored. Figure 3.3 shows the same amount of liquid in differently shaped containers.

liquid

A **gas** is characterized by having no fixed shape, with particles that are moving independently of each other. The particles in the gaseous state have gained enough energy to overcome the attractive forces holding them together as liquids or solids. A gas presses continuously and in all directions upon the walls of any container. Because of this quality, a gas completely fills a container. The particles of a gas are relatively far apart, compared to those of solids and liquids. The actual volume of the gas particles is usually very small in comparison to the volume of space occupied by the gas. A gas therefore may be compressed into a very small volume or expanded practically indefinitely. Liquids cannot be compressed to any great extent, and solids are even less compressible than liquids.

gas

When a bottle of ammonia solution is opened in one corner of the laboratory, you can soon smell its familiar odor in all parts of the room. The ammonia gas escaping from the solution demonstrates that gaseous particles move freely and rapidly, and tend to permeate the entire area into which they are released.

Although matter is discontinuous, attractive forces exist that hold the particles together and give matter its appearance of continuity. These attractive

*Figure 3.2.* These three naturally occurring substances are examples of regular geometric formations that are characteristic of crystalline solids: (a) salt; (b) quartz; (c) gypsum.

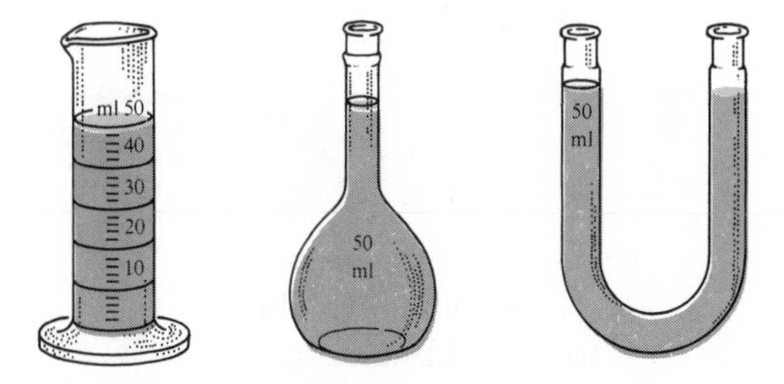


*Figure 3.3.* Liquids have the property of fluidity and assume the shape of their container, as illustrated in each of the three different calibrated containers.

forces are strongest in solids, giving them rigidity; they are weaker in liquids, but still strong enough to hold liquids to definite volumes. In gases the attractive forces are so weak that the particles of a gas are practically independent of each other. Table 3.1 lists a number of common materials that exist as solids, liquids, and gases. Table 3.2 summarizes comparative properties of solids, liquids, and gases.

*Table 3.1.* Common materials in the solid, liquid, and gaseous states of matter.

| Solids | Liquids | Gases |
|---|---|---|
| Aluminum | Alcohol | Acetylene |
| Copper | Blood | Air |
| Gold | Gasoline | Butane |
| Polyethylene | Honey | Carbon dioxide |
| Salt | Mercury | Chlorine |
| Sand | Oil | Helium |
| Steel | Vinegar | Methane |
| Sulfur | Water | Oxygen |

*Table 3.2.* Physical properties of solids, liquids, and gases.

| State | Shape | Volume | Particles | Compressibility |
|---|---|---|---|---|
| Solid | Definite | Definite | Rigidly cohering; tightly packed | Very slight |
| Liquid | Indefinite | Definite | Mobile; cohering | Slight |
| Gas | Indefinite | Indefinite | Independent of each other and relatively far apart | High |

## 3.3 Substances and Mixtures

homogeneous

heterogeneous

phase

The term *matter* refers to the total concept of material things. There are thousands of distinct and different kinds of matter. Upon closely examining different samples of matter, we can observe them to be either homogeneous or heterogeneous. By homogeneous we mean uniform in appearance when observed by the unaided eye or through a microscope. Matter that has identical properties throughout is **homogeneous**. Matter consisting of two or more physically distinct phases is **heterogeneous.** A **phase** is a homogeneous part of a system separated from other parts by physical boundaries. A system of ice and water is heterogeneous, containing both solid and liquid phases, although each physical state of water is uniform in composition and is homogeneous. Whenever we have a system in which definite boundaries exist between the components, no matter whether they are in the solid, liquid, or gaseous states, the system has more than one phase and is heterogeneous. Thus, when we first put a spoonful of sugar into water, there exist a solid and a liquid phase, and the system is heterogeneous. After the sugar has been stirred and dissolved, the system has only one phase and is homogeneous.

substance

A **substance** is a particular kind of matter that is homogeneous and has a definite, fixed composition. Substances, sometimes known as pure substances, occur in two forms: elements and compounds. Several examples of elements and compounds are copper, gold, oxygen, salt, sugar, and water. Elements and compounds are discussed in more detail in Chapter 4.

mixture

Matter that contains two or more substances mixed together is known as a **mixture.** Mixtures are variable in composition and may be either homogeneous or heterogeneous. When sugar is dissolved in water, a sugar solution is formed. All parts of this solution are sweet and contain both substances, sugar and water, uniformly mixed. Solutions are homogeneous mixtures. Air is a homogeneous mixture (solution) of several gases. If we examine ordinary concrete, granite, iron ore, or other naturally occurring mineral deposits, we observe them to be heterogeneous mixtures of several different substances. Of course, it is very easy to prepare a heterogeneous mixture simply by physically mixing two or more substances, such as sugar and salt. We will consider mixtures again in Chapter 4. Figure 3.4 illustrates the relationship between homogeneous and heterogeneous matter.

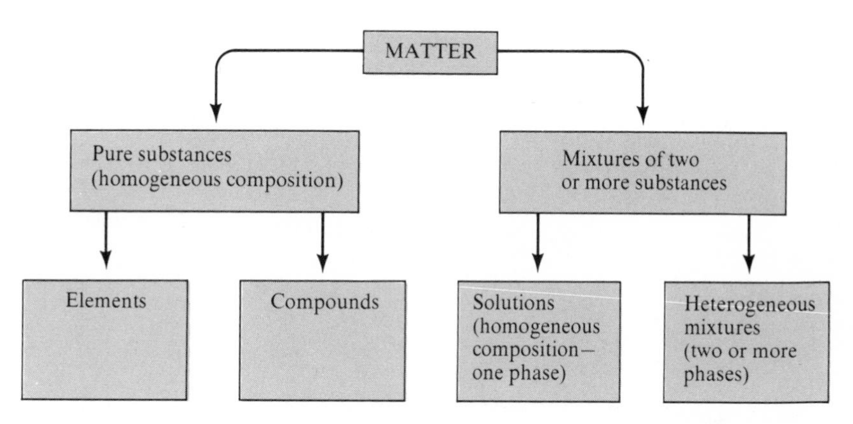


*Figure 3.4.* Forms of matter: A pure substance is always homogeneous. A mixture always contains two or more substances and may be either homogeneous (a solution) or heterogeneous.

## 3.4 Properties of Substances

properties

physical properties

chemical properties

How do we recognize substances? Each substance has a set of **properties** that is characteristic of that substance and gives it a unique identity. Properties are the personality traits of substances and are classified as either physical or chemical. **Physical properties** are the inherent characteristics of a substance that may be determined without altering the composition of that substance; they are associated with its physical existence. Common physical properties are color, taste, odor, state of matter (solid, liquid, or gas), density, melting point, and boiling point. **Chemical properties** describe the ability of a substance to form new substances, either by reaction with other substances or by decomposition.

We can select a few of the physical and chemical properties of chlorine as an example. Physically, chlorine is a gas about 2.4 times heavier than air. It is yellowish-green in color and has a disagreeable odor. Chemically, chlorine will not burn but will support the combustion of certain other substances. It can be used as a bleaching agent, as a disinfectant for water, and in many chlorinated substances such as refrigerants and insecticides. When chlorine combines with the metal sodium, it forms a salt, sodium chloride. These properties, among others, help to characterize and identify chlorine.

Substances, then, are recognized and differentiated by their properties. Table 3.3 lists four substances and tabulates several of their common physical properties. Information about common physical properties, such as given in Table 3.3, is readily available in handbooks of chemistry and physics. Scientists don't pretend to know all the answers or to remember voluminous amounts of data, but it is important for them to know where to look for data in the literature. Handbooks are one of the most widely used resources for scientific data.

*Table 3.3.* Physical properties of chlorine, water, sugar, and acetic acid.

| Substance | Color | Odor | Taste | Physical state | Boiling point (°C) | Freezing point (°C) |
|---|---|---|---|---|---|---|
| Chlorine | Yellowish-green | Sharp, suffocating | Sharp, sour | Gas | −34.6 | −101.6 |
| Water | Colorless | Odorless | Tasteless | Liquid | 100.0 | 0.0 |
| Sugar | White | Odorless | Sweet | Solid | Decomposes 170–186 | — |
| Acetic acid | Colorless | Like vinegar | Sour | Liquid | 118.0 | 16.7 |

## 3.5 Physical Changes

physical change

Matter can undergo two types of changes, physical and chemical. **Physical changes** are mainly changes in physical properties (such as size, shape, density) or state of matter without an accompanying change in composition. The changing of ice into water and water into steam are physical changes from one state of matter into another. No new substances are formed in these physical changes (see Figure 3.5).

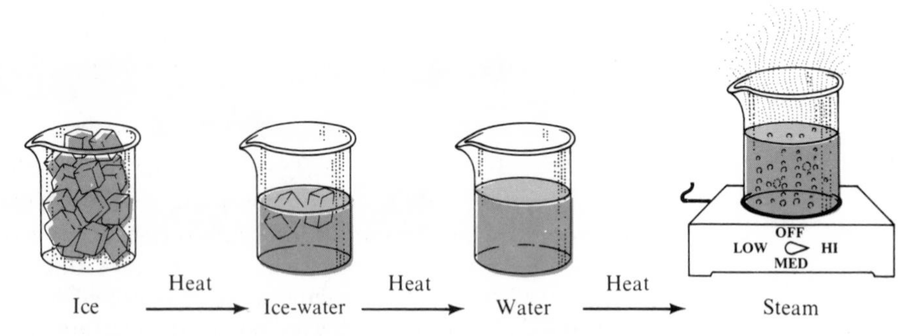


*Figure 3.5.* Physical changes in the appearance and state of water.

If we heat a platinum wire in a burner flame, the wire will become red hot. It returns to its original silvery, metallic form after cooling. The platinum undergoes a physical change in appearance while in the flame, but its composition remains the same.

## 3.6 Chemical Changes

chemical change

In a **chemical change**, new substances are formed that have different properties and composition from the original material. The new substances need not in any way resemble the initial material.

If a clean copper wire is heated in a burner flame, a change in the appearance of the wire is readily noted after it cools. The copper no longer has its characteristic color, but now appears black. The black material is copper(II) oxide, a new substance formed when copper is combined chemically with oxygen in the air during the heating process. The wire before heating was essentially 100% copper, whereas the black copper(II) oxide contains only 79.9% copper, the rest being oxygen (see Figure 3.6). When both platinum and copper are heated under the conditions described, platinum, which does not readily combine with oxygen, changes only physically, but copper changes chemically as well as physically.

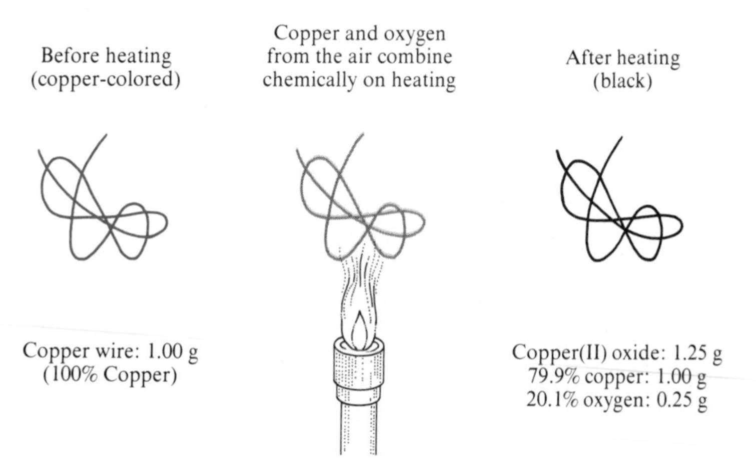


*Figure 3.6.* Chemical change: formation of copper(II) oxide from copper and oxygen.

Mercuric oxide is an orange-red powder which, when subjected to high temperature (500–600°C), decomposes into a colorless gas (oxygen) and a silvery, liquid metal (mercury). The composition of both of these products, as well as their physical appearances, is noticeably different from that of the starting compound. When mercuric oxide is heated in a test tube (see Figure 3.7), small globules of mercury are observed collecting on the cooler part of the tube. Evidence of the oxygen formed is observed when a glowing wood splint,

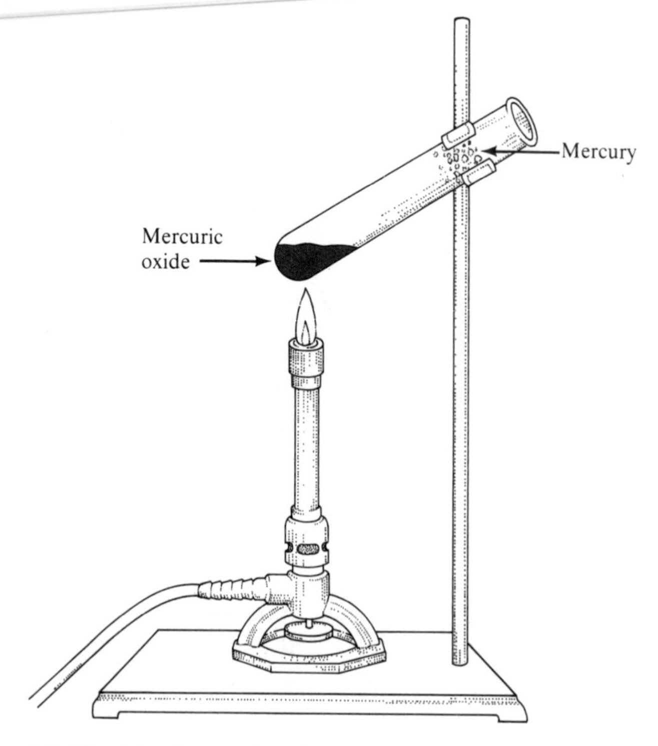


*Figure 3.7.* Heating of mercuric oxide causes it to decompose into mercury and oxygen. Observation of the mercury and oxygen with properties different from mercuric oxide is evidence that a chemical change has occurred.

lowered into the tube, bursts into flame. Oxygen supports and intensifies the combustion of the wood. From these observations, we can conclude that a chemical change has taken place.

Chemists have devised *chemical equations* as a shorthand method of expressing chemical changes. The two examples of chemical changes presented above may be represented by the following word equations:

$$\text{Copper} + \text{Oxygen} \xrightarrow{\Delta} \text{Copper(II) oxide} \quad \text{(Cupric oxide)} \qquad (1)$$

$$\text{Mercuric oxide} \xrightarrow{\Delta} \text{Mercury} + \text{Oxygen} \qquad (2)$$

Equation (1) states: copper plus oxygen when heated produce copper(II) oxide. Equation (2) states: mercuric oxide when heated produces mercury plus oxygen. The arrow means "produces"; it points to the products. The delta sign (Δ) represents heat. The starting substances (copper, oxygen, and mercuric oxide) are called the *reactants*; and the substances produced (copper(II) oxide, mercury, and oxygen) are called the *products*. In later chapters, equations are presented in a still more abbreviated form, with symbols used for each substance.

Physical change inevitably accompanies a chemical change. Table 3.4 lists common physical and chemical changes. In the examples given in the

*Table 3.4.* Examples of processes involving physical or chemical changes.

| Process taking place | Type of change | Accompanying physical changes |
|---|---|---|
| Rusting of iron | Chemical | Shiny, bright metal changes to reddish-brown rust |
| Boiling of water | Physical | Liquid changes to vapor |
| Burning of sulfur in air | Chemical | Yellow solid sulfur changes to gaseous, choking sulfur dioxide |
| Melting of lead | Physical | Solid changes to liquid |
| Combustion of gasoline | Chemical | Liquid gasoline burns to gaseous carbon monoxide, carbon dioxide, and water |
| Cutting of a diamond | Physical | Small diamonds are made from a larger diamond |
| Sawing of wood | Physical | Smaller pieces of wood plus sawdust are made from a larger piece of wood |
| Burning of wood | Chemical | Wood burns to ashes and gaseous carbon dioxide and water |
| Heating of glass | Physical | Solid becomes pliable during heating and the glass may change its shape |

table, you will note that wherever a chemical change occurs, a physical change occurs also. However, wherever a physical change is listed, only a physical change occurs.

## 3.7 Conservation of Mass

Law of Conservation of Mass

The **Law of Conservation of Mass** states that there is no detectable change in the total mass of the substances involved in a chemical change. This law, tested by extensive laboratory experimentation, is the basis for the quantitative weight relationships among reactants and products.

The decomposition of mercuric oxide into mercury and oxygen illustrates this law. One hundred grams of mercuric oxide decomposes into 92.6 g of mercury and 7.4 g of oxygen.

$$\underset{100.0\text{ g}}{\text{Mercuric oxide}} \longrightarrow \underset{92.6\text{ g}}{\text{Magnesium}} + \underset{7.4\text{ g}}{\text{Oxygen}}$$

| 100 g Reactant | | 100 g Products |
|---|---|---|

Sealed within the ordinary photographic flashbulb are fine wires of magnesium (a metal) and oxygen (a gas). When these reactants are energized, they combine chemically, producing magnesium oxide, together with a blinding white light and considerable heat. The chemical change may be represented by this equation:

Magnesium + Oxygen ⟶ Magnesium oxide + Heat + Light

When weighed before and after the chemical change, as illustrated in Figure 3.8, the bulb shows no increase or decrease in weight.

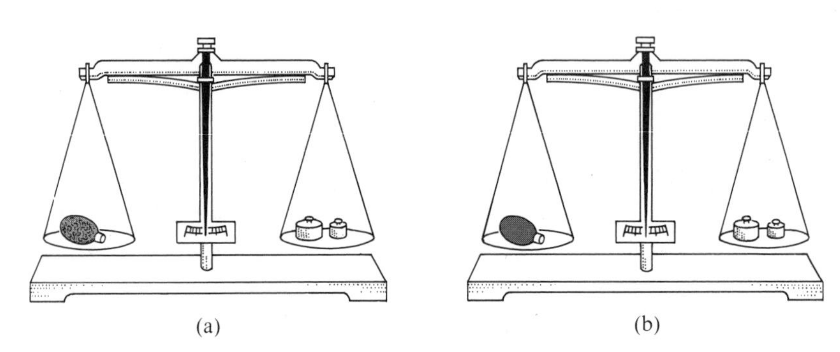


*Figure 3.8.* The flashbulb, containing magnesium and oxygen, weighs the same before (a) and after (b) the bulb is flashed. When the bulb is flashed, a chemical change occurs. The original substances are changed into the white powder, magnesium oxide.

⊙ *Mass of reactants = Mass of products*

## 3.8 Energy

From the prehistoric discovery that fire could be used to warm shelters and cook food to the modern-day discovery that nuclear reactors can be used to produce vast amounts of controlled energy, man's progress has been directed by the ability to harness, produce, and utilize energy. **Energy** is the capacity of matter to do work. Energy exists in several forms; some of the more common forms are mechanical, chemical, electrical, heat, nuclear, and radiant or light energy. Matter can have both potential and kinetic energy.

energy

**Potential energy** is stored energy, or energy an object possesses due to its relative position. For example, a ball located 20 feet above the ground has more potential energy than when located 10 feet above the ground, and will bounce higher when allowed to fall. Water backed up behind a dam represents potential energy that can be converted into useful work in the form of electrical energy. Gasoline represents a source of stored chemical potential energy that can be released during combustion.

potential energy

**Kinetic energy** is the energy that matter possesses due to its motion. When the water behind the dam is released and allowed to flow, its potential energy is changed into kinetic energy, which may be used to drive generators and produce electricity. All moving bodies possess kinetic energy. The pressure exerted by a

kinetic energy

confined gas is due to the kinetic energy of rapidly moving gas particles. We all know the results when two moving vehicles collide—their kinetic energy is expended in the crash that occurs.

Energy may be converted from one form to another form. Some kinds of energy can be converted to other forms easily and efficiently. For example, mechanical energy can be converted to electrical energy with an electric generator at better than 90% efficiency. On the other hand, solar energy has thus far been directly converted to electrical energy at an efficiency of about 15%.

## 3.9 Energy in Chemical Changes

In all chemical changes, matter either absorbs or releases energy. Chemical changes can be used to produce different forms of energy. Electrical energy to start automobiles is produced by chemical changes in the lead storage battery. Light energy for photographic purposes occurs as a flash during the chemical change in the magnesium flashbulb. Heat and light energies are released from the combustion of fuels. All the energy needed for our life processes—breathing, muscle contraction, blood circulation, and so on—is produced by chemical changes occurring within the cells of the body.

Conversely, energy is used to cause chemical changes. For example, a chemical change occurs in the electroplating of metals when electrical energy is passed through a salt solution in which the metal is submerged. A chemical change also occurs when radiant energy from the sun is utilized by plants in the process of photosynthesis. And, as we saw, a chemical change occurs when heat causes mercuric oxide to decompose into mercury and oxygen. Chemical changes are often used primarily to produce energy rather than to produce new substances. The heat or thrust generated during the combustion of fuels is more important than the new substances formed.

## 3.10 Conservation of Energy

An energy transformation occurs whenever there is a chemical change. If energy is absorbed during the change, the products will have more chemical or potential energy than the reactants. Conversely, if energy is given off in a chemical change, the products will have less chemical or potential energy than the reactants. Water, for example, can be decomposed in an electrolytic cell. Electrical energy is absorbed in the decomposition, and the products—hydrogen and oxygen—have a greater chemical or potential energy level than that of water. This potential energy is released in the form of heat and light when the hydrogen and oxygen are burned to form water again (see Figure 3.9). Thus, energy can be changed from one form to another or from one substance to another and therefore is not lost.

The energy changes occurring in many systems have been thoroughly studied by many investigators. No system has been found to acquire energy

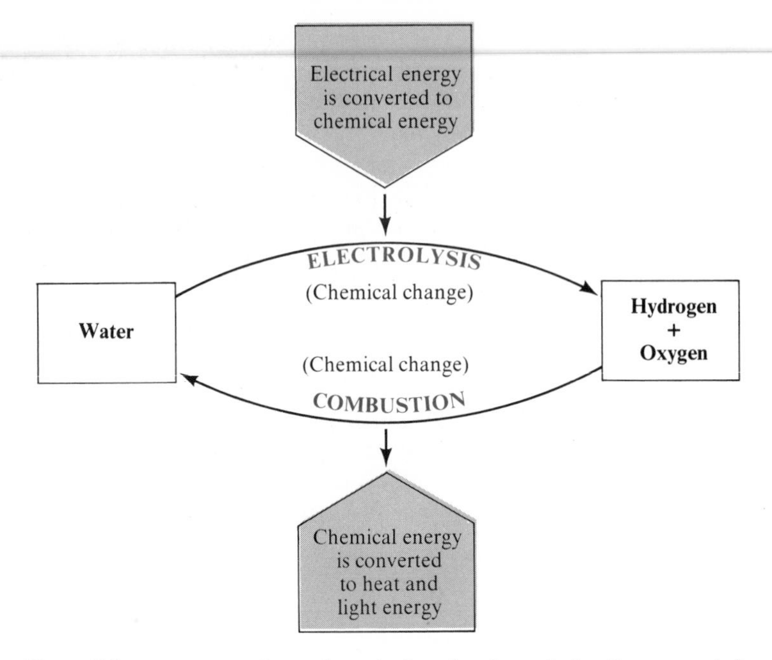


*Figure 3.9.* Energy transformations during the electrolysis of water and the combustion of hydrogen and oxygen. Electrical energy is converted to chemical energy in the electrolysis, and chemical energy is converted to heat and light energy in the combustion.

except at the expense of energy possessed by another system. This is stated in other words as the **Law of Conservation of Energy**: Energy can be neither created nor destroyed, though it may be transformed from one form to another.

Law of Conservation of Energy

## 3.11 Interchangeability of Matter and Energy

Sections 3.7–3.10 dealt with matter and energy. The two are clearly related; any attempt to deal with one inevitably involves the other. The nature of this relationship eluded the most able scientists until the beginning of the 20th century. Then, in 1905, Albert Einstein (Figure 3.10) presented one of the most original scientific concepts ever devised.

Einstein stated that the quantity of energy ($E$) equivalent to the mass ($m$) could be calculated by the equation $E = mc^2$, where $m$ is in grams and $c$ is the velocity of light ($3.0 \times 10^{10}$ cm/s). According to Einstein's equation, whenever energy is absorbed or released by a substance, there must be a loss or gain of mass. Although the energy changes in chemical reactions are measurable and may appear to be large, the amounts are relatively small. The accompanying difference in mass between reactants and products in chemical changes is so small that it cannot be detected by available measuring instruments. According to Einstein's equation, $2.2 \times 10^7$ cal ($9 \times 10^{14}$ ergs) of energy are equivalent to 0.000001 g (1 microgram) of mass. In a more practical sense, when $2.8 \times 10^3$ g

*Figure 3.10.* Albert Einstein (1879–1955), world-renowned physicist and author of the theory of relativity and the interrelationship between matter and energy: $E = mc^2$. (Courtesy of The Bettmann Archive.)

of carbon are burned to carbon dioxide, $2.2 \times 10^7$ cal of energy are released. Of this very large amount of carbon only about one millionth of a gram, which is $3.6 \times 10^{-8}$% of the starting mass, is converted to energy. Therefore, in actual practice we may treat the reactants and products of chemical changes as having constant mass. However, because mass and energy are interchangeable, the two laws dealing with the conservation of matter may be combined into a single and generally more accurate statement:

*The total amount of mass and energy remains constant during chemical change.*

## Questions

A. *Review the meanings of the new terms introduced in this chapter.*

1. Matter
2. Solid
3. Amorphous
4. Liquid
5. Gas
6. Homogeneous
7. Heterogeneous
8. Phase
9. Substance
10. Mixture
11. Properties
12. Physical properties
13. Chemical properties
14. Physical change
15. Chemical change
16. Law of Conservation of Mass
17. Energy
18. Potential energy
19. Kinetic energy
20. Law of Conservation of Energy

B. *Answers to the following questions will be found in tables and figures.*

1. Name three liquids listed in Table 3.1 that are mixtures.
2. Which of the gases listed in Table 3.1 is not a pure substance?
3. What physical properties do solids and liquids have in common? (See Table 3.2.)
4. In what physical state will acetic acid exist at 150°C? (See Table 3.3.)

5. In what physical state will water exist at 293 K? (See Table 3.3.)
6. From Figure 3.1 what evidence can you find that gases occupy space?
7. What effect does the absorption of heat energy have on mercuric oxide? (See Figure 3.7.)
8. What physical changes occur to the matter in the flashbulb of Figure 3.8 when the bulb is flashed?

C. *Review questions.*

1. List three substances in each of the three physical states of matter.
2. Explain why a gas can be compressed and why a liquid cannot be compressed appreciably.
3. In terms of the properties of the ultimate particles explain why a liquid can be poured but a solid cannot be poured.
4. When the stopper is removed from a partly filled bottle containing solid and liquid acetic acid at 16.7°C, a strong vinegar-like odor is noticeable immediately. How many acetic acid phases must be present in the bottle? Explain.
5. Is the system enclosed in the bottle of Question 4 homogeneous or heterogeneous? Explain.
6. Distinguish between physical and chemical properties of matter.
7. Is a system containing only water necessarily homogeneous? Explain.
8. Is a system containing only one substance necessarily homogeneous? Explain.
9. Distinguish between physical and chemical changes.
10. Classify the following as primarily physical or chemical changes:
    (a) Boiling water
    (b) Freezing ice cream
    (c) Boiling an egg
    (d) Homogenizing milk
    (e) Souring milk
11. Reread Section 3.4 and list those properties given for chlorine that are physical and those that are chemical.
12. Cite the evidence demonstrating that the heating of mercuric oxide brings about a chemical change.
13. Distinguish between potential and kinetic energy.
14. Is chemical energy potential or kinetic?
15. In an ordinary chemical change, why can we consider that mass is neither lost nor gained (for practical purposes)?
16. When the flashbulb of Figure 3.8 is flashed, energy is given off to the surroundings. Explain why the apparent mass of the bulb was the same after flashing as it was before, although according to Einstein, energy is equivalent to mass.
17. Which of the following statements are correct? (Try to answer this question without referring to the text.)
    (a) Liquids are the most compact state of matter.
    (b) Matter in the solid state is discontinuous—that is, it is made up of discrete particles.
    (c) Seawater, although homogeneous, is considered to be a mixture.
    (d) Any system that consists of two or more phases is heterogeneous.
    (e) A solution, although it contains dissolved material, is considered to be homogeneous.
    (f) Boiling water represents a chemical change because no change in composition occurs.
    (g) All of the following represent chemical change: baking a cake, frying an egg, leaves changing color, iron changing to rust.
    (h) All of the following represent physical change: breaking a stick, melting wax, folding a napkin, burning hydrogen to form water.

(i) A stretched rubber band possesses kinetic energy.
(j) An automobile rolling down a hill possesses both kinetic and potential energy.

D. *Review problems.*

1. Calculate the boiling point of chlorine in degrees Fahrenheit (see Table 3.3).
2. What weight of mercury can be obtained from 75.0 g of mercuric oxide?
3. Given the chemical reaction

   Magnesium + Oxygen → Magnesium oxide

   When 9.50 g of magnesium is heated in air, 15.75 g of magnesium oxide is produced.
   (a) What weight of oxygen has combined with the magnesium?
   (b) What percentage of the magnesium oxide is magnesium?
4. If a nickel weighs about 5.0 g:
   (a) How many calories would be released by the complete conversion of a nickel to energy?
   (b) If $3.03 \times 10^5$ cal are needed to heat a gallon of water from room temperature (20°C) to boiling (100°C), how many gallons of water could be heated to the boiling point by the energy from part (a)?

# 4 Elements and Compounds

*After studying Chapter 4 you should be able to:*

1. Understand the terms listed in Question A at the end of the chapter.
2. List in order of abundance the five most abundant elements in the earth's crust, seawater, and atmosphere.
3. List in order of abundance the six most abundant elements in the human body.
4. Classify common materials as elements, compounds, or mixtures.
5. Write the symbols when given the names or write the names when given the symbols of the common elements listed in Table 4.3.
6. State the Law of Definite Composition.
7. Understand how symbols, including subscripts and parentheses, are used to write chemical formulas.
8. Differentiate among atoms, molecules, and ions.
9. List the characteristics of metals and nonmetals.
10. Name binary compounds from their formulas.
11. Balance simple chemical equations when the formulas are given.
12. List the elements that occur as diatomic molecules.

## 4.1 Elements

element

All the words in the English dictionary are formed from an alphabet consisting of only 26 letters. All known substances on earth—and most probably in the universe, too—are formed from a sort of "chemical alphabet" consisting of 106 known elements. An **element** is a fundamental or elementary substance that cannot be broken down, by chemical means, to simpler substances. Elements are the basic building blocks of all substances. The elements are numbered in order of increasing complexity beginning with hydrogen, number 1. Of the first 92 elements, 88 are known to occur in nature. The other four—technetium (43), promethium (61), astatine (85), and francium (87)—either do not occur in nature or have only transitory existences resulting from radioactive decay. With the exception of number 94, elements 93–106 are not known to occur naturally, but have been synthesized—usually in very small quantities—in laboratories. The discovery of trace amounts of element 94 (plutonium) in nature has been reported recently. Element 106 was reported to have been synthesized in 1974. No elements other than those on the earth have been detected on other bodies in the universe.

Most substances can be decomposed into two or more other simpler substances. We have seen that mercuric oxide can be decomposed into mercury and oxygen and that water can be decomposed into hydrogen and oxygen. Sugar may be decomposed into carbon, hydrogen, and oxygen. Table salt is easily decomposed into sodium and chlorine. An element, however, cannot be decomposed into simpler substances by ordinary chemical changes.

atom

If we could take a small piece of an element, say copper, and divide it and subdivide it into smaller and smaller particles, we finally would come to a single unit of copper that we could no longer divide and still have copper. This ultimate particle, the smallest particle of an element that can exist, is called an **atom**. An atom is also the smallest unit of an element that can enter into a chemical reaction. Chapter 5 describes the smaller subatomic particles that make up atoms, but these particles no longer have the properties of elements.

## 4.2 Distribution of Elements

Elements are distributed very unequally in nature. Table 4.1 shows that ten of the elements make up about 99% of the weight of the earth's crust, seawater, and the atmosphere (see Figure 4.1). Oxygen, the most abundant of these, constitutes about 50% of this mass. This list does not include the mantle

*Table 4.1.* Distribution of the elements in the earth's crust, seawater, and atmosphere.

| Element | Weight percent | Element | Weight percent |
|---|---|---|---|
| Oxygen | 49.20 | Chlorine | 0.19 |
| Silicon | 25.67 | Phosphorus | 0.11 |
| Aluminum | 7.50 | Manganese | 0.09 |
| Iron | 4.71 | Carbon | 0.08 |
| Calcium | 3.39 | Sulfur | 0.06 |
| Sodium | 2.63 | Barium | 0.04 |
| Potassium | 2.40 | Nitrogen | 0.03 |
| Magnesium | 1.93 | Fluorine | 0.03 |
| Hydrogen | 0.87 | | |
| Titanium | 0.58 | All others | 0.47 |

and the core of the earth, which are believed to be composed of metallic iron and nickel. The average distribution of the elements in the human body is shown in Table 4.2. Note again the high percentage of oxygen.

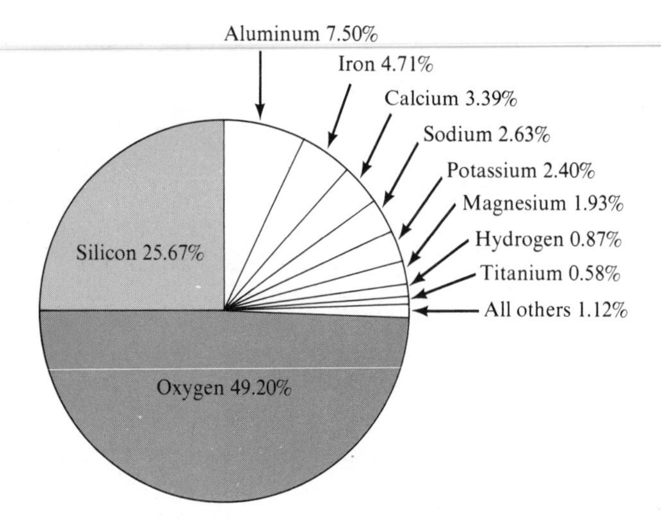


*Figure 4.1.* Weight percent of the elements in the earth's crust, seawater, and atmosphere.

*Table 4.2.* Average elemental composition of the human body.

| Element | Weight percent |
|---|---|
| Oxygen | 65.0 |
| Carbon | 18.0 |
| Hydrogen | 10.0 |
| Nitrogen | 3.0 |
| Calcium | 2.0 |
| Phosphorus | 1.0 |
| Traces of several other elements | 1.0 |

## 4.3 Names of the Elements

The names of the elements came to us from various sources. Many are derived from early Greek, Latin, or German words that generally described some property of the element. For example, iodine is taken from the Greek word *iodes*, meaning violet-like. Iodine, indeed, is violet in the vapor state. The name of the metal bismuth had its origin from the German words *weisse masse*, which means white mass. Miners called it *wismat*; it was later changed to *bismat*, and finally to bismuth. Some elements are named for the location of their discovery—for example, germanium, discovered in 1886 by Winkler, a German chemist. Others are named in commemoration of famous scientists, such as einsteinium and curium, for Albert Einstein and Marie Curie, respectively.

## 4.4 Symbols of the Elements

symbol

We all recognize Mr., N.Y., and St. as abbreviations for "mister," "New York," and "street." In like manner, chemists have assigned specific abbreviations to each element; these are called **symbols** of the elements. Fourteen of the elements have a single letter as their symbol, and all the others have two letters. The symbol stands for the element itself, for one atom of the element, and, as we shall see later, for a particular quantity of the element.

Rules governing symbols of elements are as follows:

1. Symbols are composed of one or two letters.
2. If one letter is used, it is capitalized.
3. If two letters are used, the first is capitalized and the second is a lowercase letter.

*Examples:* Sulfur S
Barium Ba

The symbols and names of all the elements are given in the table on the inside back cover of this book. Table 4.3 lists the more commonly used symbols. If we examine this table carefully, we note that most of the symbols start with the same letter as the name of the element that is represented. A number of symbols, however, appear to have no connection with the names of the elements they represent (see Table 4.4). These symbols have been carried over from earlier names (usually in Latin) of the elements and are so firmly implanted in the literature that their use is continued today.

Special care must be used in writing symbols. Begin each with a capital letter and use a lowercase second letter if needed. For example, consider Co, the symbol for the element cobalt. If through error CO (capital C and capital O) is written, the two elements carbon and oxygen (the *formula* for carbon monoxide) are represented instead of the single element cobalt. Another example of

*Table 4.3.* Symbols of the most common elements.

| Element | Symbol | Element | Symbol | Element | Symbol |
|---|---|---|---|---|---|
| Aluminum | Al | Fluorine | F | Phosphorus | P |
| Antimony | Sb | Gold | Au | Platinum | Pt |
| Argon | Ar | Helium | He | Potassium | K |
| Arsenic | As | Hydrogen | H | Radium | Ra |
| Barium | Ba | Iodine | I | Silicon | Si |
| Bismuth | Bi | Iron | Fe | Silver | Ag |
| Boron | B | Lead | Pb | Sodium | Na |
| Bromine | Br | Lithium | Li | Strontium | Sr |
| Cadmium | Cd | Magnesium | Mg | Sulfur | S |
| Calcium | Ca | Manganese | Mn | Tin | Sn |
| Carbon | C | Mercury | Hg | Titanium | Ti |
| Chlorine | Cl | Neon | Ne | Tungsten | W |
| Chromium | Cr | Nickel | Ni | Uranium | U |
| Cobalt | Co | Nitrogen | N | Zinc | Zn |
| Copper | Cu | Oxygen | O | | |

*Table 4.4.* Symbols of the elements derived from early names. These symbols are in use today, even though they do not appear to correspond to the current name of the element.

| Present name | Symbol | Former name |
|---|---|---|
| Antimony | Sb | Stibium |
| Copper | Cu | Cuprum |
| Gold | Au | Aurum |
| Iron | Fe | Ferrum |
| Lead | Pb | Plumbum |
| Mercury | Hg | Hydrargyrum |
| Potassium | K | Kalium |
| Silver | Ag | Argentum |
| Sodium | Na | Natrium |
| Tin | Sn | Stannum |
| Tungsten | W | Wolfram |

the need for care in writing symbols is the symbol Ca for calcium versus Co for cobalt.

A knowledge of symbols is essential for writing chemical formulas and equations. You should begin to learn symbols immediately since they will be used extensively in the remainder of this book and in any future chemistry courses you may take. One way to learn the symbols is to practice a few minutes a day by making side-by-side lists of names and symbols and then covering each list alternately and writing the corresponding name or symbol. Initially, it is a good plan to learn the symbols of the most common elements shown in Table 4.3.

The experiments of alchemists paved the way for the development of chemistry. Alchemists surrounded their work in mysticism partly by devising a system of symbols known only to practitioners of alchemy (see Figure 4.2). In the early 1800s the Swedish chemist J. J. Berzelius (1779–1848) made a great contribution to chemistry by devising the present system of symbols using letters of the alphabet.

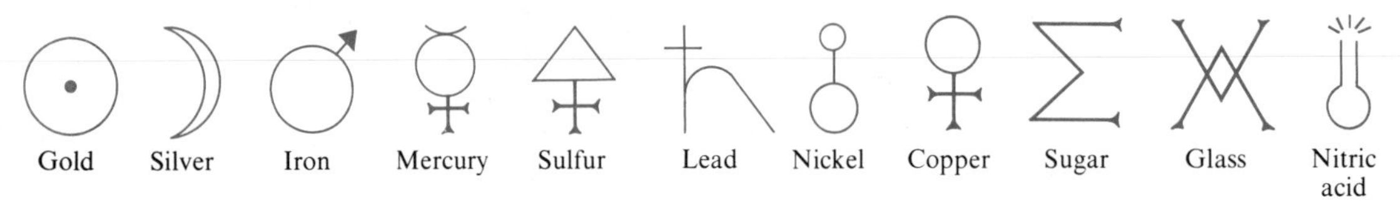


*Figure 4.2.* Some typical alchemists' symbols.

## 4.5 Compounds

compound

**Compounds**, unlike elements, can be decomposed chemically into simpler substances—that is, into other compounds and/or elements.

The atoms of the elements that form a compound are combined in whole-number ratios, never as fractional atoms. Compounds exist either as molecules

molecule

or as ions. A **molecule** is the smallest uncharged individual unit of a compound formed by the union of two or more atoms. If we subdivide a drop of water into smaller and smaller particles, we ultimately obtain a single molecule of water consisting of two hydrogen atoms bonded to one oxygen atom. This molecule cannot be further subdivided without destroying the water and forming the elements hydrogen and oxygen. Thus, a molecule of water is the smallest unit of the substance water.

ion

An **ion** is a positive or negative electrically charged atom or group of atoms. The ions in a compound are held together in a crystalline structure by the attractive forces that exist between their positive and negative charges. Compounds consisting of ions do not exist as molecules. The formula of such a compound usually represents the simplest ratio of the charged atoms or ions that exist in the substance. Sodium chloride (table salt) is such a substance. It consists of sodium ions (positively charged sodium atoms) and chloride ions (negatively charged chlorine atoms). The two types of compounds, *molecular* and *ionic*, are illustrated in Figure 4.3.

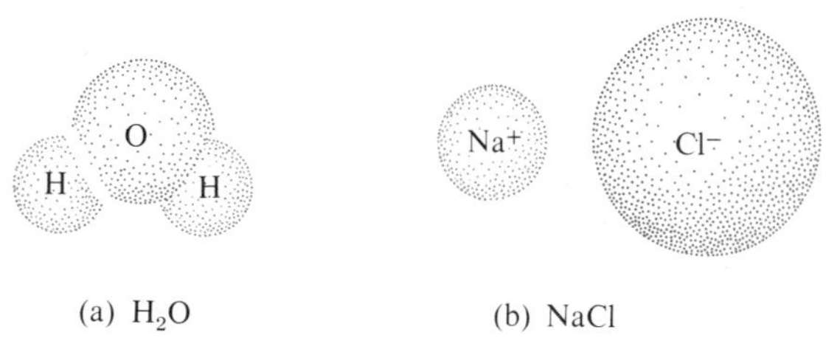


*Figure 4.3.* Representation of molecular and ionic (nonmolecular) compounds. (a) Two hydrogen atoms combined with an oxygen atom to form a molecule of water. (b) A positively charged sodium ion and a negatively charged chloride ion form the compound sodium chloride.

The compound carbon monoxide (CO) is composed of carbon and oxygen in the ratio of one atom of carbon to one atom of oxygen. Hydrogen chloride (HCl) contains a ratio of one atom of hydrogen to one atom of chlorine. Compounds may contain more than one atom of the same element. Methane (natural gas, $CH_4$) is composed of a ratio of one atom of carbon to four atoms of hydrogen; ordinary table sugar (sucrose, $C_{12}H_{22}O_{11}$) contains a ratio of 12 atoms of carbon to 22 atoms of hydrogen to 11 atoms of oxygen. These atoms are held together in the compound by *chemical bonds*.

| Substance | Each molecule composed of | Formula |
|---|---|---|
| Carbon monoxide | 1 carbon atom + 1 oxygen atom | $CO$ |
| Hydrogen chloride | 1 hydrogen atom + 1 chlorine atom | $HCl$ |
| Methane | 1 carbon atom + 4 hydrogen atoms | $CH_4$ |
| Sugar (sucrose) | 12 carbon atoms + 22 hydrogen atoms + 11 oxygen atoms | $C_{12}H_{22}O_{11}$ |
| Water | 2 hydrogen atoms + 1 oxygen atom | $H_2O$ |

There are about 4 million known compounds, with no end in sight as to the number that can and will be prepared in the future. Each compound is unique and has characteristic physical and chemical properties. Let us consider in some detail two compounds—water and mercuric oxide. Water is a colorless, odorless, tasteless liquid that can be changed to a solid (ice) at 0°C and to a gas (steam) at 100°C. It is composed of two atoms of hydrogen and one atom of oxygen per molecule, which represents 11.2% hydrogen and 88.8% oxygen by weight. Water reacts chemically with sodium to produce hydrogen gas and sodium hydroxide, with lime to produce calcium hydroxide, and with sulfur trioxide to produce sulfuric acid. No other compound has all these exact physical and chemical properties; they are characteristic of water alone.

Mercuric oxide is a dense, orange-red powder composed of a ratio of one atom of mercury to one atom of oxygen. Its composition by weight is 92.6% mercury and 7.4% oxygen. When it is heated to temperatures greater than 360°C, a colorless gas, oxygen, and a silvery liquid metal, mercury, are produced. These are specific physical and chemical properties belonging to mercuric oxide and to no other substance. Thus, a compound may be identified and distinguished from all other compounds by its characteristic properties.

## 4.6 Law of Definite Composition of Compounds

Many experiments extending over a long period of time have established the fact that a specific compound always contains the same elements in a fixed proportion by weight. For example, water contains 11.2% hydrogen and 88.8% oxygen by weight. Water will always contain hydrogen and oxygen in this fixed weight ratio. The fact that water contains hydrogen and oxygen in this particular ratio does not mean that hydrogen and oxygen cannot combine in some other ratio. However, the resulting compound will not be water. In fact, hydrogen peroxide is made up of two atoms of hydrogen and two atoms of oxygen per molecule and contains 5.9% hydrogen and 94.1% oxygen by weight; its properties are markedly different from those of water.

| | Water | Hydrogen peroxide |
|---|---|---|
| Percent H | 11.2 | 5.9 |
| Percent O | 88.8 | 94.1 |
| Atomic composition | 2 H + 1 O | 2 H + 2 O |

Law of Definite Composition

The **Law of Definite Composition** states: A compound always contains two or more elements combined in a definite proportion by weight. The reliability of this law, which in essence states that the composition of a substance will always be the same no matter what its origin or how it is formed, is the cornerstone of chemical science.

## 4.7 Chemical Formulas

chemical formulas

In a manner similar to the use of symbols for elements, chemists use **chemical formulas** as abbreviations for compounds. The chemical formula represents two or more elements that are in chemical combination. Sodium chloride contains one atom of sodium per atom of chlorine; its formula is NaCl. The formula for water is $H_2O$; it shows that a molecule of water contains two atoms of hydrogen and one atom of oxygen.

The formula of a compound tells us which elements it is composed of and how many atoms of each element are present in a formula unit. For example, a molecule of sulfuric acid is composed of two atoms of hydrogen, one atom of sulfur, and four atoms of oxygen. We can express this compound as HHSOOOO, but the usual formula for writing sulfuric acid is $H_2SO_4$. The formula may be expressed verbally as "H-two-S-O-four." Characteristics of chemical formulas are summarized below.

1. The formula of a compound contains the symbols of all the elements in the compound.
2. When the formula contains one atom of an element, the symbol of that element represents that one atom. The number one (1) is not used as a subscript to indicate one atom of an element.
3. When the formula contains more than one atom of an element, the number of atoms is indicated by a subscript written to the right of the symbol of that atom. For example, the two (2) in $H_2O$ indicates two atoms of H in the formula.
4. When the formula contains more than one of a group of atoms that occurs as a unit, parentheses are placed around the group and the number of units of the group are indicated by a subscript placed to the right of the parentheses. Consider

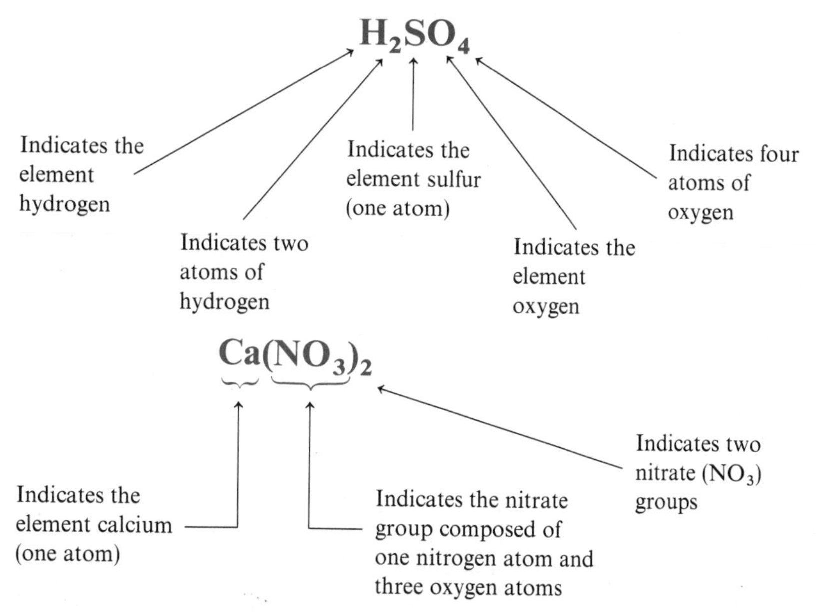


*Figure 4.4.* Explanation of the formulas $H_2SO_4$ and $Ca(NO_3)_2$.

the nitrate group, $NO_3^-$. In the formula for sodium nitrate, $NaNO_3$, there is only one nitrate group; therefore, no parentheses are needed. In calcium nitrate, $Ca(NO_3)_2$, there are two nitrate groups, and parentheses and the subscript 2 are used to indicate this. There are a total of nine atoms in $Ca(NO_3)_2$—one Ca, two N, and six O atoms.

5. Formulas written as $H_2O$, $H_2SO_4$, $Ca(NO_3)_2$, $C_{12}H_{22}O_{11}$ show only the number and kind of each atom contained in the compound; they do not show the arrangement of the atoms in the compound or how they are chemically bonded to one another.

Figure 4.4 illustrates how symbols and numbers are used in chemical formulas. There is more extensive use of formulas in later chapters.

## 4.8 Mixtures

Single substances—elements or compounds—seldom occur naturally in the pure state. Air is a mixture of gases; seawater is a mixture containing a variety of dissolved minerals; ordinary soil is a complex mixture of minerals and various organic materials.

How is a mixture distinguished from a pure substance? A mixture (see Section 3.3) always contains two or more substances that can be present in varying concentrations. Let us consider an example of a homogeneous mixture and an example of a heterogeneous mixture. Homogeneous mixtures (solutions) containing either 5% or 10% salt in water can be prepared by simply mixing the correct amounts of salt and water. These mixtures can be separated by boiling away the water, leaving the salt as a residue. The composition of a heterogeneous mixture of sulfur crystals and iron filings can be varied by merely blending in either more sulfur or more iron filings. This mixture can be separated physically, either by using a magnet to attract the iron or by adding carbon disulfide to dissolve the sulfur.

Iron(II) sulfide (FeS) contains 63.5% Fe and 36.5% S by weight. If we mix iron and sulfur in this proportion, do we have iron(II) sulfide? No, it is still a

*Table 4.5.* Comparison of mixtures and compounds.

| | Mixture | Compound |
|---|---|---|
| Composition | May be composed of elements, compounds, or both in variable composition | Composed of two or more elements in a definite, fixed proportion by weight |
| Separation of components | Separation may be made by simple physical or mechanical means | Elements can be separated by chemical changes only |
| Identification of components | Components do not lose their identity | A compound does not resemble the elements from which it is formed |

mixture; the iron is still attracted by a magnet. But if this mixture is heated strongly, a chemical change (reaction) occurs, forming iron(II) sulfide. This is a substance with properties that are different from either those of iron or of sulfur—FeS is neither attracted by a magnet nor dissolved by carbon disulfide. Thus, the properties of the reactants are lost and a compound (a pure substance) is formed. Key differences between mixtures and compounds are summarized in Table 4.5.

## 4.9 *Metals, Nonmetals, and Metalloids*

metal

nonmetal

metalloid

Three primary classifications of the elements are **metals**, **nonmetals**, and **metalloids**. Most of the elements are metals. We are familiar with metals because of their widespread use in tools, materials of construction, automobiles, and so on. But nonmetals are equally useful in our everyday life as major components of such items as clothing, food, fuel, glass, plastics, and wood.

The metallic elements are solids at room temperature (mercury is an exception). They have a high luster, are good conductors of heat and electricity, can be rolled or hammered into sheets (they are *malleable*), and can be drawn into wires (they are *ductile*). In addition, most metals have a high melting point and a high density. Metals familiar to most of us are aluminum, chromium, copper, gold, iron, lead, magnesium, mercury, nickel, platinum, silver, tin, and zinc. Other less familiar but still important metals are calcium, cobalt, potassium, sodium, uranium, and titanium.

Metals have little tendency to combine with each other to form compounds. But many metals readily combine with nonmetals such as chlorine, oxygen, and sulfur to form mainly ionic compounds such as metallic chlorides, oxides, and sulfides, respectively. The more active metals are found in nature combined with other elements as minerals. A few of the less active ones—such as copper, gold, and silver—are sometimes found in a native or free state as well.

Nonmetals, unlike metals, are not lustrous, have relatively low melting points and densities, and are generally poor conductors of heat and electricity. Carbon, phosphorus, sulfur, selenium, and iodine are solids; bromine is a liquid; the rest of the nonmetals are gases. Common nonmetals found uncombined in nature are carbon (graphite and diamond), nitrogen, oxygen, sulfur, and the noble gases (helium, neon, argon, krypton, xenon, and radon).

Nonmetals combine with one another to form molecular compounds such as carbon dioxide ($CO_2$), methane ($CH_4$), butane ($C_4H_{10}$), and sulfur dioxide ($SO_2$). Fluorine, the most reactive nonmetal, combines readily with almost all the other elements.

Several elements (boron, silicon, germanium, arsenic, antimony, tellurium, and polonium) are classified as *metalloids* and have properties that are intermediate between those of metals and those of nonmetals. The intermediate position for these elements is shown in Table 4.6, which classifies all the elements as metals, nonmetals, or metalloids. Certain of the metalloids are the raw materials for the semiconductor devices that make our modern electronics industry possible.

*Table 4.6.* Classification of the elements into metals, metalloids, and nonmetals.

Key

Metals

Metalloids

Nonmetals

| 1 H | | | | | | | | | | | | | | | | 1 H | 2 He |
|---|---|---|---|---|---|---|---|---|---|---|---|---|---|---|---|---|---|
| 3 Li | 4 Be | | | | | | | | | | | 5 B | 6 C | 7 N | 8 O | 9 F | 10 Ne |
| 11 Na | 12 Mg | | | | | | | | | | | 13 Al | 14 Si | 15 P | 16 S | 17 Cl | 18 Ar |
| 19 K | 20 Ca | 21 Sc | 22 Ti | 23 V | 24 Cr | 25 Mn | 26 Fe | 27 Co | 28 Ni | 29 Cu | 30 Zn | 31 Ga | 32 Ge | 33 As | 34 Se | 35 Br | 36 Kr |
| 37 Rb | 38 Sr | 39 Y | 40 Zr | 41 Nb | 42 Mo | 43 Tc | 44 Ru | 45 Rh | 46 Pd | 47 Ag | 48 Cd | 49 In | 50 Sn | 51 Sb | 52 Te | 53 I | 54 Xe |
| 55 Cs | 56 Ba | 57 La | 72 Hf | 73 Ta | 74 W | 75 Re | 76 Os | 77 Ir | 78 Pt | 79 Au | 80 Hg | 81 Tl | 82 Pb | 83 Bi | 84 Po | 85 At | 86 Rn |
| 87 Fr | 88 Ra | 89 Ac | 104 Ku | 105 Ha | 106 – | | | | | | | | | | | | |
| | | | | | | | | | | | | | | | | | |
| | | | | 58 Ce | 59 Pr | 60 Nd | 61 Pm | 62 Sm | 63 Eu | 64 Gd | 65 Tb | 66 Dy | 67 Ho | 68 Er | 69 Tm | 70 Yb | 71 Lu |
| | | | | 90 Th | 91 Pa | 92 U | 93 Np | 94 Pu | 95 Am | 96 Cm | 97 Bk | 98 Cf | 99 Es | 100 Fm | 101 Md | 102 No | 103 Lr |

Seven of the elements (all nonmetals) occur as *diatomic molecules* (consisting of two atoms). These seven elements, together with their formulas and brief descriptions, are listed below.

| Element | Molecular formula | Normal state |
|---|---|---|
| Hydrogen | $H_2$ | Colorless gas |
| Nitrogen | $N_2$ | Colorless gas |
| Oxygen | $O_2$ | Colorless gas |
| Fluorine | $F_2$ | Pale yellow gas |
| Chlorine | $Cl_2$ | Yellow-green gas |
| Bromine | $Br_2$ | Reddish-brown liquid |
| Iodine | $I_2$ | Greyish-black solid |

Whether found free in nature or prepared in the laboratory, the molecules of each of these elements contain two atoms. Their formulas, therefore, are always written to show the molecular composition of the element—that is, $H_2$, $N_2$, $O_2$, $F_2$, $Cl_2$, $Br_2$, and $I_2$.

It is important to understand how symbols are used to designate either an atom or a molecule of an element. Consider the elements hydrogen and oxygen. Hydrogen is found in gases coming from volcanoes and in certain natural gas supplies; it can also be prepared by many different chemical reactions. Regardless of the source, all samples of hydrogen are identical and consist of diatomic molecules. The composition of this molecular hydrogen is expressed by the formula $H_2$. Free oxygen makes up about 21% (by volume) of the air that we breathe. Whether obtained from the air or prepared by chemical reaction, free oxygen also exists as diatomic molecules; its composition is expressed by the formula $O_2$. Now consider hydrogen in the compound water, which has the composition expressed by the formula $H_2O$ (or HOH). Water contains neither free hydrogen ($H_2$) nor free oxygen ($O_2$); the $H_2$ part of the formula $H_2O$ simply tells us that each molecule of water contains two hydrogen atoms combined chemically with one oxygen atom. Symbols and subscripts are used in this way to show the molecular composition of elements and to show the composition of compounds.

## 4.10 Nomenclature and Chemical Equations

### Nomenclature

Knowledge of chemical names and of the writing and balancing of chemical equations is vital to the study of chemistry. This section serves only as an introduction to the naming of compounds and writing equations. More complete details of the systematic methods of naming inorganic compounds are given in Chapter 8, and a more detailed explanation of chemical equations is given in Chapter 10. Refer to these two chapters often, as needed. Neither chapter is intended to be studied only in the sequence given in the text; rather, they are common depositories of information on chemical nomenclature and equations.

We have already used such names as hydrogen chloride (HCl), mercuric oxide (HgO), magnesium oxide (MgO), and carbon dioxide ($CO_2$). Note that all four names end in *ide*. This *ide* ending is characteristic of the names of compounds composed of atoms of two different elements. Compounds composed of two elements are called *binary* compounds. Some compounds contain several atoms of the same element (for example, $CCl_4$, carbon tetrachloride), but as long as there are only two different kinds of atoms, the compound is considered to be binary.

When naming a compound consisting of a metal and a nonmetal, the name of the metal is given first, followed by the name of the nonmetal, which is modified to end in *ide*. There are, of course, exceptions to this rule, and the names of some compounds containing more than two elements end in *ide* (for example, $NH_4Cl$, ammonium chloride; NaOH, sodium hydroxide). Refer to Section 8.3 for more details on naming binary compounds. Examples of binary compounds with names ending in *ide* are given below.

| | | | |
|---|---|---|---|
| NaCl | Sodium chloride | $H_2S$ | Hydrogen sulfide |
| $CO_2$ | Carbon dioxide | $AlBr_3$ | Aluminum bromide |
| NaI | Sodium iodide | $K_2S$ | Potassium sulfide |
| $CaF_2$ | Calcium fluoride | $Mg_3N_2$ | Magnesium nitride |

### Chemical Equations

Chemical changes or reactions result in the formation of substances with compositions that are different from the starting substances. A chemical equation is a shorthand expression for a chemical reaction. Substances in the reaction are represented by their symbols or formulas in the equation. The equation indicates both the reactants (starting substances) and the products, and often shows the conditions necessary to facilitate the chemical change. The reactants are written on the left side and the products on the right side of the equation. An arrow (→) pointing to the products separates the reactants from the products. A plus sign (+) is used to separate one reactant (or product) from another.

*Reactants ⟶ Products*

We will see in Chapter 5 that every atom has a specific mass. In an equation, the symbols or formulas that represent a substance also represent a specific mass of that substance. Since no detectable change in mass results from a chemical change, the mass of the products must equal the mass of the reactants. In representing a chemical change by an equation, this conservation of mass is attained by balancing the equation. After establishing the correct formulas for the reactants and products, an equation is balanced by placing integral numbers (as needed) in front of the formulas of the substances in the equation. We use these numbers to obtain an equation with the same number of atoms of each kinds of element on each side of the equation.

Consider, again, the reaction of metallic copper heated in air. The chemical change may be represented by the following equations:

$$\text{Copper} + \text{Oxygen} \xrightarrow{\Delta} \text{Copper(II) oxide} \qquad (1)$$

$$Cu + O_2 \xrightarrow{\Delta} CuO \qquad \text{(Unbalanced)} \qquad (2)$$

Copper and oxygen are the reactants, and copper(II) oxide is the product. Equation (2) as written is not balanced because there are two oxygen atoms on the left side and only one on the right side. We place a 2 in front of Cu and a 2 in front of CuO to obtain the balanced equation (3):

$$2Cu + O_2 \xrightarrow{\Delta} 2CuO \qquad \text{(Balanced)} \qquad (3)$$

This balanced equation contains 2 Cu atoms and 2 O atoms on both sides of the equation.

A very important factor to remember when balancing equations is that a correct formula of a substance may not be changed for the convenience of balancing the equation. In the unbalanced equation (2) above, we cannot change the formula of CuO to $CuO_2$ to balance the equation, even though by so doing we balance the number of atoms of each element on each side of the equation. The formula $CuO_2$ is not the correct formula for the product. It is also important to be aware that a number in front of a formula multiplies every atom in that formula by that number. Thus,

2 CuO means 2 Cu atoms and 2 O atoms
3 $H_2O$ means 6 H atoms and 3 O atoms
4 $H_2SO_4$ means 8 H atoms, 4 S atoms, and 16 O atoms

## Questions

A. *Review the meanings of the new terms introduced in this chapter.*

1. Element
2. Atom
3. Symbol
4. Compound
5. Molecule
6. Ion
7. Law of Definite Composition
8. Chemical formula
9. Metal
10. Nonmetal
11. Metalloid

B. *Answers to the following questions will be found in tables and figures.*

1. List, in decreasing order of abundance, the six most abundant elements in the human body.
2. Are there more atoms of silicon or hydrogen in the earth's crust, seawater, and atmosphere? Use Table 4.1 and the fact that the mass of a silicon atom is about 28 times that of a hydrogen atom.
3. Why is the symbol for lead Pb instead of Le?
4. Make a list of the names of the elements in Table 4.3. Now see how many of the symbols you know by writing the correct symbol after each name.
5. How many metals are there? Nonmetals? Metalloids? (See Table 4.6.)

C. *Review questions.*

1. What does the symbol of an element stand for?
2. Write down what you believe to be the symbols for the elements argon, lithium, manganese, nickel, nitrogen, platinum, plutonium, and uranium. Now look up the correct symbols and rewrite them, comparing the two sets.
3. Interpret the difference in meanings for each of these pairs:
   (a) Pb and PB (b) Co and CO
4. Distinguish between an element and a compound.
5. Explain why the Law of Definite Composition does not pertain to mixtures.
6. Does the Law of Definite Composition pertain to an element? Discuss.
7. Distinguish between a chemical formula and a symbol.
8. Given the following list of compounds and their formulas, what elements are present in each compound?

| | |
|---|---|
| (a) Potassium iodide | $KI$ |
| (b) Sodium carbonate | $Na_2CO_3$ |
| (c) Aluminum oxide | $Al_2O_3$ |
| (d) Calcium bromide | $CaBr_2$ |
| (e) Carbon tetrachloride | $CCl_4$ |
| (f) Magnesium bromide | $MgBr_2$ |
| (g) Nitric acid | $HNO_3$ |
| (h) Barium sulfate | $BaSO_4$ |
| (i) Aluminum phosphate | $AlPO_4$ |
| (j) Acetic acid | $HC_2H_3O_2$ |

9. Write the formula for each of the following compounds, the composition of which is given after each name:

| | |
|---|---|
| (a) Zinc oxide | 1 atom Zn, 1 atom O |
| (b) Potassium chlorate | 1 atom K, 1 atom Cl, 3 atoms O |
| (c) Sodium hydroxide | 1 atom Na, 1 atom O, 1 atom H |
| (d) Aluminum bromide | 1 atom Al, 3 atoms Br |
| (e) Calcium fluoride | 1 atom Ca, 2 atoms F |
| (f) Lead(II) chromate | 1 atom Pb, 1 atom Cr, 4 atoms O |
| (g) Ethyl alcohol | 2 atoms C, 6 atoms H, 1 atom O |
| (h) Benzene | 6 atoms C, 6 atoms H |

10. Explain the meaning of each symbol and number in the following formulas:
    (a) $H_2O$ (b) $CCl_4$ (c) $Cd(NO_3)_2$ (d) $Cu_2S$ (e) $C_6H_{12}O_6$ (glucose)
11. How many atoms are represented in each of these formulas?
    (a) AgI (d) $K_2Cr_2O_7$ (g) $CCl_2F_2$ (freon)
    (b) $H_3PO_4$ (e) $Cl_2$ (h) $C_6H_8N_2O_2S$ (sulfanilamide)
    (c) $LiNO_3$ (f) $Mg(NO_3)_2$
12. How many atoms of oxygen are contained in one molecule of oxygen?
13. How many atoms of hydrogen and oxygen are contained in one molecule of water? In one molecule of hydrogen peroxide?
14. Write the names and formulas of the elements that exist as diatomic molecules.
15. How many atoms of oxygen are represented in each expression?
    (a) 4 $H_2O$ (b) 3 $CuSO_4$ (c) $H_2O_2$ (d) 3 $Fe(OH)_3$ (e) $Al(ClO_3)_3$
16. Are all mixtures heterogeneous? Explain.
17. Classify each of the following materials as an element, compound, or mixture:

    | | | | |
    |---|---|---|---|
    | Air | Coal | Oil | Magnesium |
    | Brass | Water | Silver | Wine |
    | Cement | Milk | Sugar | Sodium chloride |

18. A white solid, on heating, formed a colorless gas and a yellow solid. Assuming there was no reaction with the air, is the original solid an element or a compound? Explain.
19. Tabulate the properties that characterize metals and nonmetals.
20. Which of the following are diatomic molecules?
    (a) $H_2$ (c) HCl (e) NO (g) $MgCl_2$
    (b) $SO_2$ (d) $H_2O$ (f) $NO_2$
21. Name the following binary compounds. Refer to Chapter 8 if necessary.
    (a) $CaCl_2$ (c) AgCl (e) $RaBr_2$ (g) $H_2S$ (i) $ZnCl_2$ (k) $CaC_2$
    (b) HBr (d) $Al_2S_3$ (f) CdO (h) BN (j) $BaF_2$ (l) KI
22. Which of the compounds listed in Questions 8 and 9 are binary compounds? What is common to the names of these binary compounds?
23. An atom of silver is represented by the symbol Ag; a hydrogen molecule by the formula $H_2$; a water molecule by $H_2O$. Write expressions to represent:
    (a) Five silver atoms
    (b) Four hydrogen molecules
    (c) Three water molecules
24. Balance these equations (all formulas are correct as written):
    (a) $H_2 + Cl_2 \rightarrow HCl$
    (b) $Zn + CuSO_4 \rightarrow Cu + ZnSO_4$
    (c) $HCl + NaOH \rightarrow NaCl + H_2O$
    (d) $Ca + O_2 \rightarrow CaO$
    (e) $Fe + HCl \rightarrow FeCl_2 + H_2$
    (f) $P + I_2 \rightarrow PI_3$
    (g) $MgO + HCl \rightarrow MgCl_2 + H_2O$
    (h) $HNO_3 + Ba(OH)_2 \rightarrow Ba(NO_3)_2 + H_2O$
    (i) $BiCl_3 + H_2S \rightarrow Bi_2S_3 + HCl$
    (j) $Mg_3N_2 + H_2O \rightarrow Mg(OH)_2 + NH_3$
25. Balance the following equations, each of which represents a method of preparing oxygen gas:
    (a) $H_2O_2 \rightarrow H_2O + O_2$
    (b) $KClO_3 \xrightarrow{\Delta} KCl + O_2$
    (c) $KNO_3 \xrightarrow{\Delta} KNO_2 + O_2$

(d) $Na_2O_2 + H_2O \rightarrow NaOH + O_2$

(e) $H_2O \xrightarrow[H_2SO_4]{\text{Electrical energy}} H_2 + O_2$

26. Balance the following equations, each of which represents a method of preparing hydrogen gas:
(a) $Zn + HCl \rightarrow ZnCl_2 + H_2$
(b) $Al + H_2SO_4 \rightarrow Al_2(SO_4)_3 + H_2$
(c) $Na + H_2O \rightarrow NaOH + H_2$
(d) $C + H_2O$ (steam) $\rightarrow CO + H_2$
(e) $Fe + H_2O$ (steam) $\rightarrow Fe_3O_4 + H_2$

D. *Review problems.*

1. Common table salt, NaCl, contains 39.3% sodium and 60.7% chlorine. What weight of sodium is present in 25.0 g of salt?
2. Yellow brass is a homogeneous mixture of 67% copper and 33% zinc. If 50 g of copper is added to 100 g of yellow brass, what will be the new composition?
3. Calcium oxide, CaO, contains 71.5% calcium. What size sample of CaO would contain 12.0 g of calcium?
4. What would be the density of a solution made by mixing 2.50 ml of carbon tetrachloride ($CCl_4$, $d = 1.595$ g/ml) and 3.50 ml of carbon tetrabromide ($CBr_4$, $d = 3.420$ g/ml)? Assume that the volume of the mixed liquids is the sum of the two volumes used.
5. When 4.00 g of calcium and 4.00 g of sulfur were mixed and reacted to give the compound calcium sulfide (CaS), 0.80 g of sulfur remained unreacted.
(a) What percentage of the compound is sulfur?
(b) An atom of which element, Ca or S, has the greater mass? Explain.
(c) How many grams of sulfur will combine with 20.0 g of calcium?
6. Pure gold is too soft a metal for many uses, so it is alloyed to give it more mechanical strength. One particular alloy is made by mixing 60 g of gold, 8.0 g of silver, and 12 g of copper. What carat gold is this alloy if pure gold is considered to be 24 carat?
7. Methane, the chief component of natural gas, has the formula $CH_4$. Each atom of carbon weighs 12 times as much as an atom of hydrogen. Calculate the weight percent of hydrogen in methane.

# 5 Atomic Theory and Structure

*After studying Chapter 5 you should be able to:*

1. Understand the terms listed in Question A at the end of the chapter.
2. State the major provisions of Dalton's atomic theory.
3. Give the names, symbols, charges, and relative masses of the three principal subatomic particles.
4. Describe the atom as conceived by Ernest Rutherford after his alpha scattering experiments.
5. Describe the atom as conceived by Niels Bohr.
6. Discuss the contributions to atomic theory made by Dalton, Thomson, Rutherford, Bohr, Moseley, Chadwick, and Schrödinger.
7. State the Pauli exclusion principle.
8. Determine the maximum number of electrons that can exist in a given main energy level.
9. Determine the atomic number, atomic mass, or number of neutrons of any isotope when given any two of these three terms.
10. Draw the diagram of any isotope of the first 38 elements, showing the composition of the nucleus and the numbers of electrons in the main energy levels.
11. Give the electron structure ($1s^2 2s^2 2p^6$, etc.) for any of the first 38 elements.
12. Explain what is represented by the electron-dot (Lewis-dot) structure of an element.
13. Draw the electron-dot (Lewis-dot) diagrams for any of the first 20 elements.
14. Name and distinguish among the three isotopes of hydrogen.
15. Convert grams or atoms of an element to moles of that element, and vice versa.
16. Understand the relationships among a mole, Avogadro's number, and a gram-atomic weight of an element.

## 5.1 Early Thoughts

The structure of matter has long intrigued and engaged the minds of people. The seed of modern atomic theory was sown during the time of the ancient Greek philosophers. About 440 B.C. Empedocles stated that all matter was composed of four "elements"—earth, air, water, and fire. Democritus (about 470–370 B.C.), one of the early atomistic philosophers, thought that all forms of matter were finitely divisible into invisible particles, which he called atoms. He held that atoms were in constant motion and that they combined with one another in various ways. This purely speculative hypothesis was not based on scientific observations. Shortly thereafter, Aristotle (384–322 B.C.) opposed the theory of Democritus and endorsed and advanced the Empedoclean theory. So strong was the influence of Aristotle that his theory dominated the thinking of scientists and philosophers until the beginning of the 17th century. The term *atom* is derived from the Greek word *atomos*, meaning "indivisible."

## 5.2 Dalton's Atomic Theory

More than 2000 years after Democritus, the English schoolmaster John Dalton (1766–1844) revived the concept of atoms and proposed an atomic theory based on facts and experimental evidence. This theory, described in a series of papers published during the period 1803–1810, rested on the idea of a different kind of atom for each element. The essence of **Dalton's atomic theory** may be summed up as follows:

Dalton's atomic theory

1. Elements are composed of minute, indivisible particles called atoms.
2. Atoms of the same element are alike in mass and size.
3. Atoms of different elements have different masses and sizes.
4. Chemical compounds are formed by the union of two or more atoms of different elements.
5. When atoms combine to form compounds, they do so in simple numerical ratios, such as one to one, two to one, two to three, and so on.
6. Atoms of two elements may combine in different ratios to form more than one compound.

Dalton's atomic theory stands as a landmark in the development of chemistry. The major premises of his theory are still valid today. However, some of the statements must be modified or qualified because investigations since Dalton's time have shown that (1) atoms are composed of subatomic particles; (2) all the atoms of a specific element do not have the same mass; and (3) atoms, under special circumstances, can be decomposed.

## 5.3 Subatomic Parts of the Atom

The concept of the atom—a particle so small that it cannot be seen even with the most powerful microscope—and the subsequent determination of its structure stand among the very greatest creative intellectual human achievements.

When we refer to any visible quantity of an element, we are considering a vast number of identical atoms of that element. But when we refer to an atom of an element, we isolate a single atom from the multitude in order to present that element in its simplest form. Figure 5.1 illustrates the hypothetical isolation of a single copper atom from its crystal lattice.

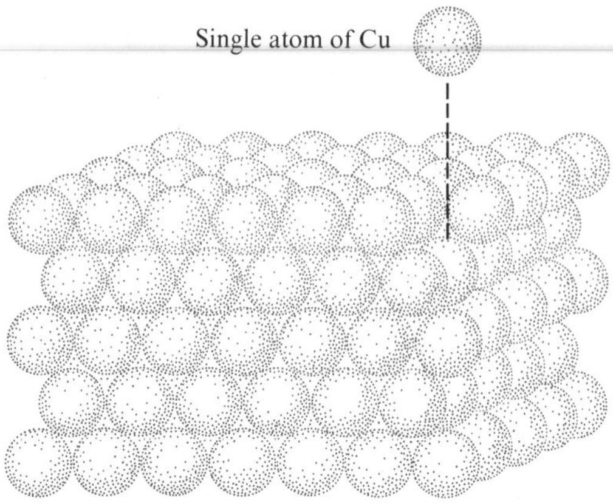

Single atom of Cu

Crystalline structure of copper atoms

*Figure 5.1.* A single atom of copper compared with copper as it occurs in its regular crystalline lattice structure. Billions of atoms are present in even the smallest strand of copper wire.

Let us examine this tiny particle we call the atom. The diameter of a single atom ranges from 1 to 5 angstroms (1 Å = $1 \times 10^{-8}$ cm). Hydrogen, the smallest atom, has a diameter of about 1 Å. To arrive at some idea of how small an atom is, consider this dot (•), which has a diameter of about 1 mm, or $1 \times 10^{7}$ Å. It would take 10 million hydrogen atoms to form a line stretching across this dot. To carry this size illustration a bit further, 10 million of the 1 millimetre dots laid edge to edge would extend for 10,000 metres, or more than 6 miles! As inconceivably small as atoms are, they contain even more minute particles, the **subatomic particles**, such as electrons, protons, and neutrons.

subatomic particles

electron

The experimental discovery of the **electron** ($e^-$) was made in 1897 by J. J. Thomson (1856–1940). The **electron** is a particle with a negative electrical charge and has a mass of $9.107 \times 10^{-28}$ gram. This mass is 1/1837 the mass of a hydrogen atom and corresponds to 0.0005486 atomic mass unit (amu). One atomic mass unit has a mass of $1.660 \times 10^{-24}$ gram. Although the actual electrical charge of an electron is known, its value is too cumbersome for practical use. The electron, therefore, has been assigned a relative electrical charge of $-1$. The size of an electron has not been determined exactly, but its diameter is believed to be less than $10^{-12}$ cm.

proton

Protons were first observed by E. Goldstein (1850–1930) in 1886. However, it was J. J. Thomson who discovered the nature of the proton. He showed that the proton is a particle and he calculated its mass to be about 1837 times that of an electron. The **proton** (p) is a particle with a relative mass of 1 amu and an actual mass of $1.672 \times 10^{-24}$ g. Its charge ($+1$) is equal in magnitude but of opposite sign to the charge on the electron. The mass of a proton is only very slightly less than that of a hydrogen atom.

neutron

The third major subatomic particle was discovered in 1932 by James Chadwick (1891–1974). This particle, the **neutron** (n) bears neither a positive nor

*Table 5.1.* Electrical charge and relative mass of electrons, protons, and neutrons.

| Particle | Symbol | Electrical charge | Relative mass (amu) | Actual mass (g) |
|---|---|---|---|---|
| Electron | $e^-$ | $-1$ | $\frac{1}{1837}$ | $9.107 \times 10^{-28}$ |
| Proton | $p$ | $+1$ | 1 | $1.672 \times 10^{-24}$ |
| Neutron | $n$ | 0 | 1 | $1.675 \times 10^{-24}$ |

a negative charge and has a relative mass of about 1 amu. Its actual mass ($1.675 \times 10^{-24}$ g) is only very slightly greater than that of a proton. The properties of these three subatomic particles are summarized in Table 5.1.

Nearly all the ordinary chemical properties of matter can be explained in terms of atoms consisting of electrons, protons, and neutrons. The discussion of atomic structure that follows is based on the assumption that atoms contain only these principal subatomic particles. Many other subatomic particles such as mesons, positrons, neutrinos, and antiprotons have been discovered. At this time it is not clear whether all these particles are originally present in the atom or whether they are produced by reactions occurring within the nucleus. The field of atomic physics is fascinating and has attracted many young scientists in recent years. This interest has resulted in a great deal of research that is producing a long list of additional subatomic particles. Descriptions of the properties of many of these particles are to be found in recent textbooks on atomic physics and in various articles appearing in *Scientific American* over the past several years.

## 5.4 The Nuclear Atom

The discovery that positively charged particles were present in atoms came soon after the discovery of radioactivity by Henri Becquerel in 1896.

Ernest Rutherford (Figure 5.2) had, by 1907, established that the positively charged alpha particles emitted by certain radioactive elements were ions of the element helium. Rutherford used these alpha particles to establish the nuclear nature of atoms. In some experiments performed in 1911, he directed a stream of positively charged helium ions (alpha particles) at a very thin sheet of gold foil (about 1000 atoms thick). He observed that most of the alpha particles passed through the foil with little or no deflection; but a few of the particles were deflected at large angles, and occasionally one even bounced back from the foil (see Figure 5.3). It was known that like charges repel each other and that an electron with a mass of 1/1837 amu could not have an appreciable effect upon the path of a far more massive (4 amu) alpha particle. Rutherford therefore reasoned that each gold atom must contain a positively charged mass occupying a relatively tiny volume and that when an alpha particle approached close enough to this positive mass, it was deflected. Rutherford spoke of this positively

*Figure 5.2.* Ernest Rutherford (1871–1937), British physicist, who identified two of the three principal rays emanating from radioactive substances. His experiments with alpha particles led to the first laboratory transmutation of an element and to his formulation of the nuclear atom. Rutherford was awarded the Nobel prize in 1908 for his work on transmutation. (Courtesy McGill University.)

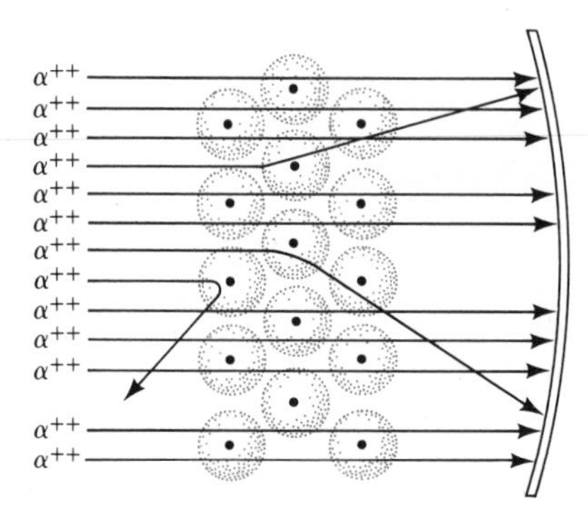


*Figure 5.3.* Diagram representing Rutherford's experiment on alpha particle scattering. Positive alpha particles, emanating from a radioactive source, were directed at a thin metal foil. Diagram illustrates the repulsion of the positive alpha particles by the positive nucleus of the metal atom.

charged mass as the *nucleus* of the atom. Since alpha particles are relatively high in mass, the extent of the deflections—remember some actually bounced back—indicated to Rutherford that the nucleus is relatively very heavy and dense. Since most of the alpha particles passed through the thousand or so gold atoms without any apparent deflection, he further concluded that the bulk of an atom consists of empty space.

When we speak of the mass of an atom, we are, for practical purposes, referring primarily to the mass of the nucleus. This is because the nucleus contains all the protons and neutrons, and these represent more than 99.9% of the total mass of any atom (see Table 5.1). By way of illustration, the largest number of electrons known to exist in an atom is 106. The mass of even 106 electrons represents only about 1/17 of the mass of a single proton or neutron. The mass of an atom, therefore, is primarily determined by the mass of its protons and neutrons.

## 5.5 General Arrangement of Subatomic Particles

The alpha particle scattering experiments of Rutherford established that the atom contains a dense, positively charged nucleus. The later work of Chadwick demonstrated that the atom contains neutrons, which are particles with mass but no charge. Light negatively charged electrons are also present and offset the positive charges in the nucleus. Based on this experimental evidence, a general description of the atom and location of its subatomic particles was devised. Each atom consists of a **nucleus** surrounded by electrons. The nucleus contains protons and neutrons, but electrons are not found in the nucleus. In a neutral atom, the positive charge of the nucleus (due to protons) is exactly offset by the negative electrons. Since the charge on an electron is equal to but of opposite sign to the charge on a proton, a neutral atom must contain exactly the same number of electrons as protons. However, this generalized picture of atomic structure provides no information on the arrangement of electrons within the atom.

nucleus

## 5.6 The Bohr Atom

At high temperatures or when subjected to high voltages, elements in the gaseous state give off colored light. Neon signs illustrate this property of matter very well. When passed through the prism or grating of a spectroscope, the light emitted by a gas appears as a set of bright colored lines (band spectra). These colored lines indicate that the light is being emitted only at certain wavelengths or frequencies that correspond to specific colors. Each element possesses a unique set of these spectral lines that is different from the sets of all the other elements.

In 1912–1913, while studying the line spectra of hydrogen, Niels Bohr (1885–1962), a Danish physicist, made a significant contribution to the rapidly growing knowledge of atomic structure. His research led him to believe that electrons in an atom exist in specific regions at various distances from the nucleus. He also visualized the electrons as rotating in orbits around the nucleus like planets rotating around the sun.

Bohr's first paper in this field dealt with the hydrogen atom, which he described as a single electron rotating in an orbit about a relatively heavy nucleus (see Figure 5.5). He applied the concept of energy quanta, proposed in 1900 by the German physicist Max Planck (1858–1947), to the observed line spectra of hydrogen. Planck stated that energy is never emitted in a continuous stream, but only in small discrete packets called quanta (Latin, *quantus*, how much). Bohr theorized that there are several possible orbits for electrons at different distances from the nucleus. But an electron had to be in one specific orbit or another; it could not exist between orbits. Bohr also stated that when a hydrogen atom absorbed one or more quanta of energy, its electron would "jump" to another orbit at a greater distance from the nucleus.

Bohr was able to account for spectral lines this way. Each orbit corresponds to a different energy level, the one closest to the nucleus representing the lowest or ground-state energy level. Orbits at increasing distances from the nucleus represent the second, third, fourth energy levels. When an electron "falls" from a high-energy orbit to one of lower energy, a quantum of energy in the form of light is emitted, thus giving rise to a spectral line at a specific frequency. A number of orbits exist, and when electrons "fall" different distances, correspondingly different quanta of energy are emitted, producing the several lines visible in the hydrogen spectrum.

Bohr contributed greatly to the advancement of our knowledge of atomic structure by (1) suggesting quantized energy levels for electrons and (2) showing that spectral lines result from the radiation of small increments of energy (Planck's quanta) when electrons shift from one energy level to another.

Much of Bohr's work related to the simplest atom, hydrogen. Difficulties arose when his energy calculations were applied to atoms containing many electrons. Bohr's concept of the atom has been replaced by the quantum mechanics theory. One of the chief differences between these two theories is that in the quantum mechanics theory electrons are not considered to be revolving around the nucleus in orbits, but to occupy *orbitals*—somewhat cloudlike regions surrounding the nucleus and corresponding to energy levels. The concept of electrons being in specific energy levels is still retained in the modern theory.

## 5.7 The Quantum Mechanical Atom

The discussion in this section and Sections 5.8–5.17 describes the ordered system by which the electrons are distributed within the atoms of all the elements. The classifications given in the periodic table (Chapter 6), as well as the chemical properties of the elements, are dependent on their electronic arrangements.

The most important feature of the Bohr atom is the concept of definite energy levels for electrons in atoms. In 1924, the French physicist Louis de Broglie (1892–     ) suggested that moving electrons had the properties of waves as well as mass. In 1926, the Austrian physicist Erwin Schrödinger (1887–1961) introduced a new method of calculation—quantum mechanics, or wave mechanics. Schrödinger's equation, which is a complex mathematical expression, describes an electron as simultaneously having properties of a wave and a particle. Thus, the electron was given dual characteristics—some of its properties are best described in terms of waves (like light), and other properties are described in terms of a particle having mass.

The solution of the Schrödinger equation is complex and we will not concern ourselves with it here. However, the solution leads to the introduction of three quantum numbers that define the probabilities of location and spatial properties of electrons in atoms. An extension of this theory shows that a fourth quantum number (the spin quantum number) is also necessary to define an electron completely.

quantum number

The four **quantum numbers—n, 1, m, and s**—specify the energy and probable location of each electron in an atom:

1. Electrons exist in energy levels at different distances from the nucleus. The principal quantum number **n** indicates the energy levels of the electrons relative to their distance from the nucleus. The number **n** may have any positive integral value up to infinity (**n** = 1, 2, 3, 4, . . . , ∞), but only values of 1–7 have be established for atoms of known elements in their ground state (lowest energy state). Energy level **n** = 1 is closest to the nucleus and is the lowest principal energy level. (See Figure 5.7.)
2. Electrons exist in orbitals that have specific shapes. The principal energy levels (except the first) contain closely grouped sublevels. These sublevels consist of orbitals of specific shape. In quantum mechanics the term **orbital** (not orbit) is used and refers to the region around the nucleus in which we may expect to find a particular electron. The orbital quantum number **l** (ell) specifies the shape of the electron cloud about the nucleus. The four common sublevels (orbitals) normally encountered are designated by the lowercase italic letters *s*, *p*, *d*, and *f*.
3. Electron orbitals have specific orientation in space. The magnetic quantum number **m** designates this orientation. This quantum number accounts for the number of *s*, *p*, *d*, and *f* orbitals that can be present in the principal energy levels. There can be at most one *s* orbital, three *p* orbitals, five *d* orbitals, and seven *f* orbitals in any given principal energy level.
4. An electron spins about its own axis in either a clockwise or counterclockwise direction. The spin quantum number **s** relates to the direction of spin of an electron. Because there are only two possible directions of spin, each orbital, no matter what its designation, can contain a maximum of two electrons. When two electrons occupy the same orbital, they must have opposite spins. When an orbital contains two electrons, the electrons are said to be *paired*.

orbital

Thus, the quantum numbers tell us relatively how far the electrons are located from the nucleus, the shapes of the electron orbitals, the orientation of the orbitals in space, and whether the electrons are paired within a given orbital.

The basic rules and limitations regarding the state of electrons in atoms are the following:

1. In the ground state (lowest energy state) of an atom, electrons tend to occupy orbitals of the lowest possible energy. Thus, an electron will occupy an *s* orbital in the **n** = 1 level before it occupies an *s* orbital in the **n** = 2 level.

2. Each orbital may contain a maximum of two electrons (with opposite spins).
3. No two electrons in an atom can have the same four quantum numbers. This is a statement of the *Pauli exclusion principle*.

The use of quantum numbers in atomic structures is illustrated later in the chapter.

## 5.8 Energy Levels of Electrons

energy levels of electrons

electron shells

All the electrons in an atom are not located the same distance from the nucleus. As pointed out by both the Bohr theory and quantum mechanics, the probability of finding the electrons is greatest at certain specified distances—called **energy levels**—from the nucleus. Energy levels are also referred to as **electron shells** and may contain only a limited number of electrons. Energy levels are numbered, starting with **n** = 1 as the shell nearest the nucleus and going to **n** = 7. They are also identified by the letters *K, L, M, N, O, P, Q*, with *K* equivalent to the first energy level, *L* to the second level, and so on, as follows:

| Energy level | **n** | Letter designation |
|---|---|---|
| First | 1 | *K* |
| Second | 2 | *L* |
| Third | 3 | *M* |
| Fourth | 4 | *N* |
| Fifth | 5 | *O* |
| Sixth | 6 | *P* |
| Seventh | 7 | *Q* |

Each succeeding energy level is located farther away from the nucleus.

The maximum number of electrons that can occupy a specific energy level can be calculated from the formula $2\mathbf{n}^2$, where **n** is the number of the principal energy level. For example, shell *K*, or energy level 1, can have a maximum of two electrons ($2 \times 1^2 = 2$); shell *L*, or energy level 2, can have a maximum

*Table 5.2.* Maximum number of electrons that can occupy each principal energy level.

| Energy level, **n** | Letter designation | Maximum number of electrons in each energy level, $2\mathbf{n}^2$ |
|---|---|---|
| 1 | *K* | $2 \times 1^2 = 2$ |
| 2 | *L* | $2 \times 2^2 = 8$ |
| 3 | *M* | $2 \times 3^2 = 18$ |
| 4 | *N* | $2 \times 4^2 = 32$ |
| 5 | *O* | $2 \times 5^2 = 50$[a] |

[a] The theoretical value of 50 electrons in energy level 5 has never been attained in any element known to date.

of eight electrons ($2 \times 2^2 = 8$), and so on. Table 5.2 shows the maximum number of electrons that can exist in each of the first five energy levels.

## 5.9 Energy Sublevels

The principal energy levels contain sublevels designated by the letters *s*, *p*, *d*, *f*. The *s* sublevel consists of one orbital; the *p* sublevel consists of three orbitals; the *d* sublevel consists of five orbitals; and the *f* sublevel consists of seven orbitals. Since no more than two electrons can exist in an orbital, the maximum numbers of electrons that can exist in the sublevels are 2 in the *s* orbital, 6 in the three *p* orbitals, 10 in the five *d* orbitals, and 14 in the seven *f* orbitals.

| Type of sublevel | Number of orbitals possible | Number of electrons possible |
|---|---|---|
| *s* | 1 | 2 |
| *p* | 3 | 6 |
| *d* | 5 | 10 |
| *f* | 7 | 14 |

The order of energy of the sublevels within a specific principal energy level is the following: *s* electrons are lower in energy than *p* electrons, which are lower than *d* electrons, which are lower than *f* electrons. This may be expressed in the following manner:

Sublevel energy: $s < p < d < f$

The order of energy of the principal energy levels, **n**, is

Principal energy levels: $1 < 2 < 3 < 4 < 5 < 6 < 7$

We can determine what type of sublevels occur in any energy level from the maximum number of electrons that can exist in that energy level. To make this determination, we need to know the maximum number of electrons possible in an energy level and to use two of the rules set forth in Section 5.7: (1) no more than two electrons can occupy one orbital, and (2) an electron will occupy the lowest possible sublevel. The maximum number of electrons in the first energy level is two; both of these are *s* electrons. They are designated as $1s^2$, indicating two *s* electrons in the first energy level. The *s* orbital in the second energy level is written as 2*s*, in the third energy level as 3*s*, and so on. The second energy level, with a maximum of eight electrons, will contain only *s* and *p* electrons—namely, a maximum of two *s* and six *p* electrons. They are designated as $2s^2 2p^6$.

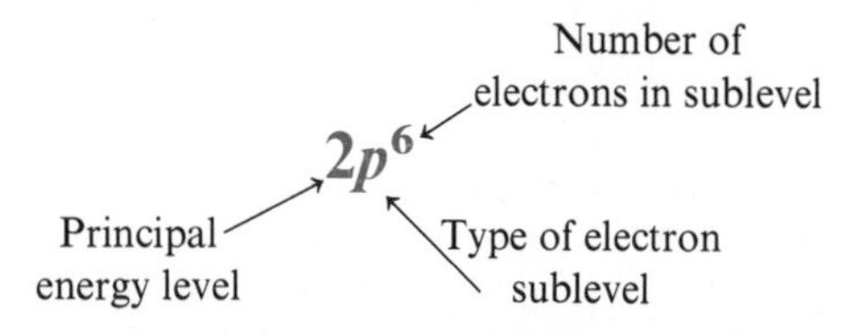

If each orbital contains two electrons, the second energy level can have four orbitals: one *s* orbital and three individual *p* orbitals. These three *p* orbitals are energetically equivalent to each other and are labeled $2p_x$, $2p_y$, and $2p_z$ to indicate their orientation in space (see Figure 5.4). The symbols $3s^2$, $3p^6$, and $3d^{10}$ illustrate the sublevel breakdown of electrons in the third energy level. From this line of reasoning, we can see that if there are sufficient electrons, *f* electrons first appear in the fourth energy level. Table 5.3 shows the type of sublevel electrons and maximum number of orbitals and electrons in each energy level.

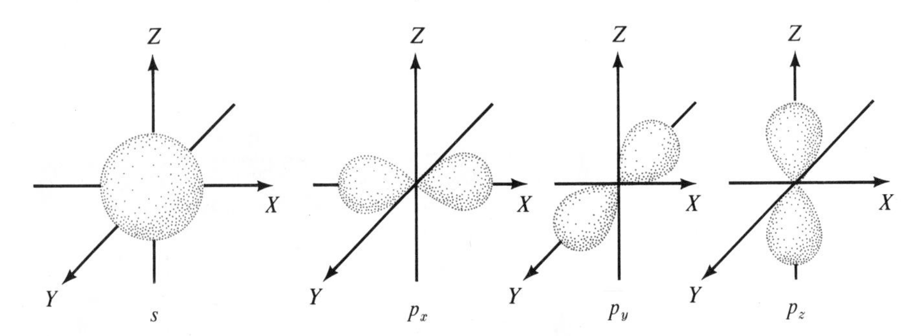


*Figure 5.4.* Perspective representation of the *s*, $p_x$, $p_y$, and $p_z$ atomic orbitals.

*Table 5.3.* Sublevel electrons in each principal energy level and the maximum number of orbitals and electrons in each energy level.

| Energy level | Sublevel electrons | Maximum number of orbitals | Total number of electrons |
|---|---|---|---|
| 1 (*K*) | *s* | 1 | 2 |
| 2 (*L*) | *s*, *p* | 4 | 8 |
| 3 (*M*) | *s*, *p*, *d* | 9 | 18 |
| 4 (*N*) | *s*, *p*, *d*, *f* | 16 | 32 |
| 5[a] (*O*) | *s*, *p*, *d*, *f* | Incomplete | — |
| 6[a] (*P*) | *s*, *p*, *d* | Incomplete | — |
| 7[a] (*Q*) | *s* | Incomplete | — |

[a] Insufficient electrons to complete the shell.

Since the *spdf* atomic orbitals have definite distribution in space, they are represented by particular spatial shapes. At this time we will consider only the *s* and *p* orbitals. The *s* orbitals are spherically symmetrical about the nucleus, as illustrated in Figure 5.4. A 2*s* orbital is a larger sphere than a 1*s* orbital. The *p* orbitals ($p_x, p_y, p_z$) are dumbbell-shaped and are oriented at right angles to each other along the *x*, *y*, and *z* axes in space (see Figure 5.4). An electron has equal probability of being located in either lobe of the *p* orbital. In illustrations such as Figure 5.4, the boundaries of the orbitals enclose the region of the greatest probability (about 90% chance) of finding an electron.

In the ground state of a hydrogen atom, this falls within a sphere having a radius of 0.53 Å.

## 5.10 Atomic Numbers of the Elements

atomic number

The **atomic number** of an element is the number of protons in the nucleus of an atom of that element. The atomic number is a fundamental characteristic and identifies an atom as being a particular element. The presently known elements are numbered consecutively from 1 to 106 to coincide with the number of protons in their nuclei. Thus, hydrogen, element number 1, has one proton; calcium, element number 20, has 20 protons; and uranium, element number 92, has 92 protons in the nucleus. The atomic number tells us not only the amount of positive charge in the nucleus, but also the number of electrons in the neutral atom.

## 5.11 The Simplest Atom—Hydrogen

The hydrogen atom, consisting of a nucleus containing one proton and an electron orbital containing one electron, is the simplest known atom. (Some hydrogen atoms are known to contain one or two neutrons in their nucleus—see Section 5.12 on isotopes.) The electron in hydrogen occupies an *s* orbital in the first energy level. This electron does not move in any definite path but rather in a random motion within its orbital, forming an electron cloud about the nucleus. The diameter of the nucleus is believed to be about $10^{-13}$ cm, and the diameter of the electron orbital to be about $10^{-8}$ cm. Hence, the diameter of the electron orbital of a hydrogen atom is about 100,000 times greater than the diameter of the nucleus.

What we have, then, is a positive nucleus surrounded by an electron cloud formed by an electron in an *s* orbital. The net electrical charge on the hydrogen atom is zero; it is called a *neutral atom*. Figure 5.5 shows two methods of representing a hydrogen atom.

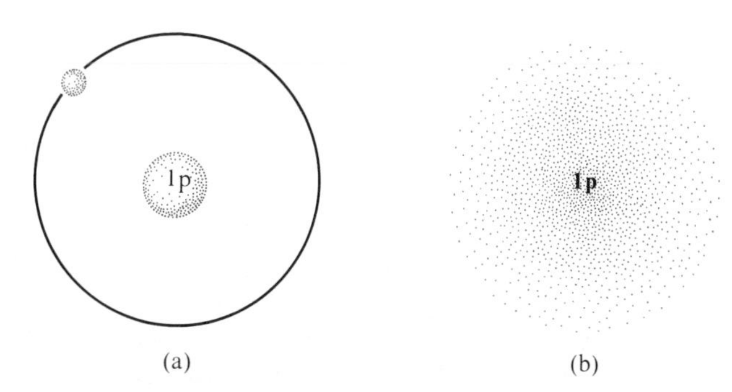


*Figure 5.5.* The hydrogen atom. (a) Illustration of the Bohr description, indicating a discrete electron moving around its nucleus of one proton. (b) Illustration of the quantum mechanical concept, showing the electron orbital as a cloud surrounding the nucleus.

## 5.12 Isotopes of the Elements

Shortly after Rutherford's conception of the nuclear atom, experiments were performed to determine the masses of individual atoms. These experiments showed that the masses of nearly all atoms were greater than could be accounted for by simply adding up the masses of all the protons and electrons that were known to be present. This fact led to the concept of the neutron, a particle with no charge but with a mass about the same as that of a proton. Since this particle has no charge, it was very difficult to detect, and the existence of the neutron was not proven experimentally until 1932. All atomic nuclei except that of the simplest hydrogen atom are now believed to contain neutrons.

All atoms of a given element have the same number of protons, but experimental evidence has shown that, in most cases, all atoms of a given element do not have identical masses. This is because atoms of the same element may have different numbers of neutrons in their nuclei.

isotopes

Atoms of an element having the same atomic number but different atomic masses are called **isotopes** of that element. Atoms of the isotopes of an element, therefore, have the same number of protons and electrons but different numbers of neutrons.

Three isotopes of hydrogen (atomic number 1) are known. Each has one proton in the nucleus and one electron in the first energy level. The first isotope (protium) without a neutron has a mass of 1; the second isotope (deuterium) with one neutron in the nucleus has a mass of 2; the third isotope (tritium) with two neutrons has a mass of 3 (see Figure 5.6).

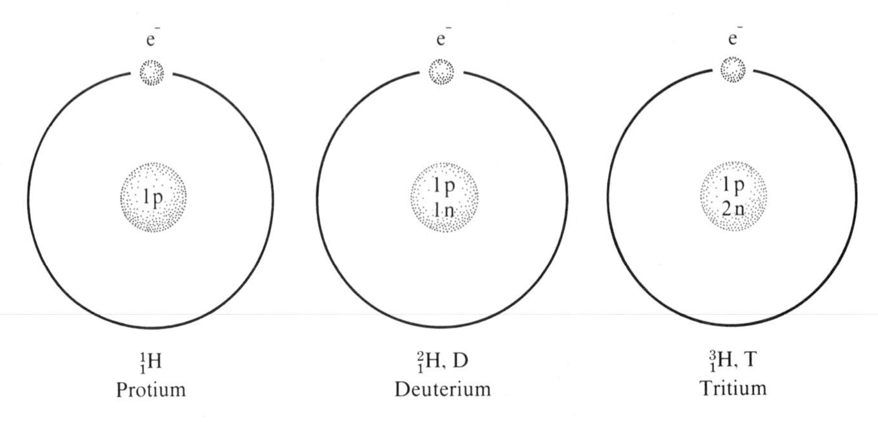


*Figure 5.6.* Diagram of the isotopes of hydrogen. The number of protons (p) and neutrons (n) are shown within the nucleus.

The three isotopes of hydrogen may be represented by the symbols $^1_1H$, $^2_1H$, $^3_1H$, indicating an atomic number of 1 and mass numbers of 1, 2, and 3, respectively. This method of representing atoms is called *isotopic notation.* The subscript number is the atomic number; the superscript number is the *mass number* (total number of protons and neutrons). The hydrogen isotopes may also be referred to as hydrogen-1, hydrogen-2, and hydrogen-3

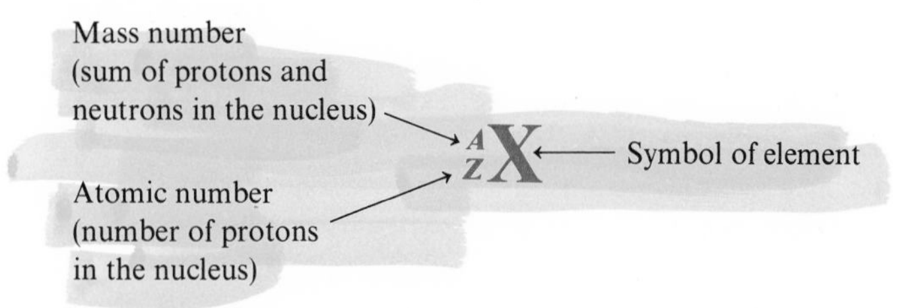


Two or more isotopes are known for all elements. However, not all isotopes are stable; some are radioactive and are continually decomposing to form other elements. For example, of the seven known isotopes of carbon, only two, carbon-12 and carbon-13, are stable. Of the seven known isotopes of oxygen, only three, $^{16}_{8}O$, $^{17}_{8}O$, and $^{18}_{8}O$, are stable. Of the fifteen known isotopes of arsenic, $^{75}_{33}As$ is the only one that is stable.

## 5.13 Atomic Structure of the First Twenty Elements

Starting with hydrogen, and progressing in order of increasing atomic number to helium, lithium, beryllium, and so on, the atoms of each successive element contain one more proton and one more electron than do the atoms of the preceding element. This sequence continues, without exception, throughout the entire list of known elements and is one of the most impressive examples of order in nature.

The number of neutrons also increases as we progress through the elements. But this number, unlike the number of protons and electrons, does not increase in a perfectly uniform manner as we go from elements of low atomic numbers to those of higher atomic numbers. Furthermore, atoms of the same element may contain different numbers of neutrons (see Section 5.12). For example, the predominant isotope of hydrogen contains no neutrons, but two other hydrogen isotopes, containing one and two neutrons, respectively, are known. The predominant isotope of helium (element number 2) has two neutrons, but helium isotopes containing one and four neutrons are known.

The ground-state electronic structures of the first 20 elements fall into a regular pattern. The one hydrogen electron is in the first energy level, and both helium electrons are in the first energy level. The electron structures for hydrogen and helium are written $1s^1$ and $1s^2$, respectively. The maximum number of electrons in the first energy level is two ($2 \times 1^2 = 2$). In lithium (atomic number 3), the third electron is in the $2s$ sublevel of the second energy level. Lithium has the electron structure $1s^2 2s^1$.

In succession, the atoms of beryllium (4), boron (5), carbon (6), nitrogen (7), oxygen (8), fluorine (9), and neon (10) have one more proton and one more electron than the preceding element. Both the first and second energy levels are filled to capacity by the ten electrons of neon, which has two electrons in the first and eight electrons in the second energy level.

| | |
|---|---|
| H | $1s^1$ |
| He | $1s^2$ |
| Li | $1s^2 2s^1$ |
| Be | $1s^2 2s^2$ |
| B | $1s^2 2s^2 2p^1$ |
| C | $1s^2 2s^2 2p^2$ |
| N | $1s^2 2s^2 2p^3$ |
| O | $1s^2 2s^2 2p^4$ |
| F | $1s^2 2s^2 2p^5$ |
| Ne | $1s^2 2s^2 2p^6$ |

Element 11, sodium (Na), has two electrons in the first energy level and eight electrons in the second energy level, with the remaining electron occu-

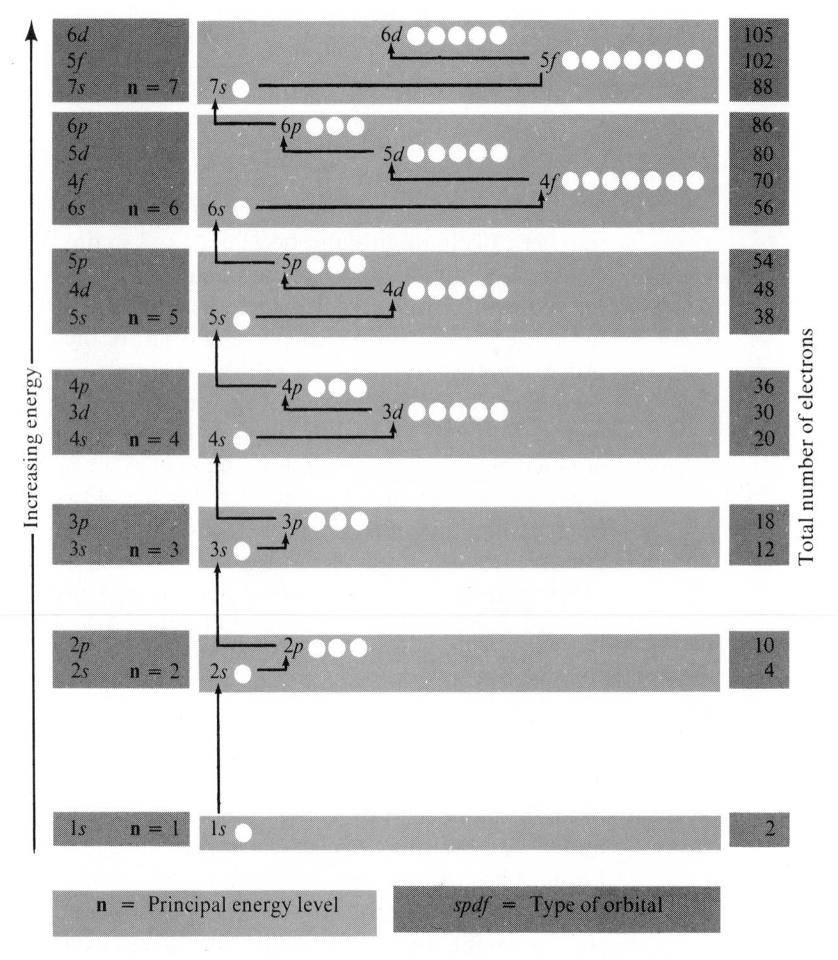


*Figure 5.7.* Order of filling electron orbitals. Each circle represents an orbital, which can contain two electrons. (Some exceptions to this order are known.)

pying the 3*s* orbital in the third energy level. The electron structure of sodium is $1s^2 2s^2 2p^6 3s^1$. Magnesium (12), aluminum (13), silicon (14), phosphorus (15), sulfur (16), chlorine (17), and argon (18) follow in order, each adding one electron to the third energy level up to argon, which has eight electrons in the *M* shell.

The placement of the last electron in potassium and calcium, elements numbers 19 and 20, departs somewhat from the expected order. One might expect that if the third energy level can contain a maximum of 18 electrons (see Table 5.2), electrons would continue to fill this shell until the maximum capacity was reached. However, this is not the case. The 4*s* sublevel is at a lower energy state than the 3*d* sublevel (see Figure 5.7). Hence, in elements 19 and 20, the last electron is found in the 4*s* level. The electron structure for potassium is $1s^2 2s^2 2p^6 3s^2 3p^6 4s^1$. Calcium has an electron structure similar to potassium, except that it has two 4*s* electrons. This break in sequence does not invalidate the formula $2\mathbf{n}^2$, which merely prescribes the maximum number of electrons that each shell may contain, but does not state the order in which the shells are filled. Table 5.4 shows the electron arrangement of the first 20 elements.

*Table 5.4.* Electron arrangement of the first 20 elements.

| Element | Symbol | Number of protons (atomic number) | Total number of electrons | Arrangement of electrons $\mathbf{n} = 1\ 2\ 3\ 4$ |
|---|---|---|---|---|
| Hydrogen | H | 1 | 1 | 1 |
| Helium | He | 2 | 2 | 2 |
| Lithium | Li | 3 | 3 | 2 1 |
| Beryllium | Be | 4 | 4 | 2 2 |
| Boron | B | 5 | 5 | 2 3 |
| Carbon | C | 6 | 6 | 2 4 |
| Nitrogen | N | 7 | 7 | 2 5 |
| Oxygen | O | 8 | 8 | 2 6 |
| Fluorine | F | 9 | 9 | 2 7 |
| Neon | Ne | 10 | 10 | 2 8 |
| Sodium | Na | 11 | 11 | 2 8 1 |
| Magnesium | Mg | 12 | 12 | 2 8 2 |
| Aluminum | Al | 13 | 13 | 2 8 3 |
| Silicon | Si | 14 | 14 | 2 8 4 |
| Phosphorus | P | 15 | 15 | 2 8 5 |
| Sulfur | S | 16 | 16 | 2 8 6 |
| Chlorine | Cl | 17 | 17 | 2 8 7 |
| Argon | Ar | 18 | 18 | 2 8 8 |
| Potassium | K | 19 | 19 | 2 8 8 1 |
| Calcium | Ca | 20 | 20 | 2 8 8 2 |

The relative energies of the electron orbitals are shown in Figure 5.7. The order given can be used to determine the electron distribution in the atoms of the elements, although some exceptions to the pattern are known. Suppose we wish to determine the electron structure of a chlorine atom, which has 17 electrons. Following the order in Figure 5.7, we begin by placing two electrons in the 1*s* orbital, then two electrons in the 2*s* orbital, and six electrons in the 2*p* orbitals. We now have used ten electrons. Finally, we place the next two electrons in the 3*s* orbital and the remaining five electrons in the 3*p* orbitals, which uses all 17 electrons, giving the electron structure for a chlorine atom as $1s^2 2s^2 2p^6 3s^2 3p^5$. The sum of the superscripts equals 17, the number of electrons in the atom. This procedure is summarized below.

Order of orbitals to be filled: $1s2s2p3s3p$
Distribution of the 17 electrons in a chlorine atom: $1s^2 2s^2 2p^6 3s^2 3p^5$

*Problem 5.1*

What is the electron distribution in a phosphorus atom?
First determine the number of electrons contained in a phosphorus atom. The atomic number of phosphorus is 15; therefore, each atom contains 15 protons and 15 electrons. Now tabulate the number of electrons in each principal and subenergy level until all 15 electrons are assigned.

| Sublevel | Number of $e^-$ | Total $e^-$ |
|---|---|---|
| 1*s* orbital | 2 $e^-$ | 2 |
| 2*s* orbital | 2 $e^-$ | 4 |
| 2*p* orbital | 6 $e^-$ | 10 |
| 3*s* orbital | 2 $e^-$ | 12 |
| 3*p* orbital | 3 $e^-$ | 15 |

Therefore, the electron distribution in phosphorus is $1s^2 2s^2 2p^6 3s^2 3p^3$.

## 5.14 Electron Structure of the Elements Beyond Calcium

The elements following calcium have a less regular pattern of adding electrons. The lowest energy level available for the twenty-first electron is the 3*d* level. Thus, scandium (21) has the following electron arrangement: first energy level, two electrons; second energy level, eight electrons; third energy level, nine electrons; fourth energy level, two electrons. The last electron is located in the 3*d* level. The structure for scandium is $1s^2 2s^2 2p^6 3s^2 3p^6 3d^1 4s^2$. The elements following scandium, titanium (22) through copper (29), continue to add *d* electrons until the third energy level has its maximum of 18. Two exceptions in the orderly electron addition are chromium (24) and copper (29), the structures of which are given in Table 5.5. The third energy level of the electrons is first completed in the element copper. Table 5.5 shows the order of filling of the electron orbitals and the electron configuration of all the known elements.

*Table 5.5.* Electron structure of the elements. (For simplicity of expression, symbols of the chemically stable noble gases are used as a portion of the electron structure for the elements beyond neon. For example, the electron structure of a sodium (Na) atom consists of ten electrons, as in neon [Ne], plus a $3s^1$ electron.) Detailed electron structures for the noble gases are given in Table 5.6.

| Element | Atomic number | Electron structure | Element | Atomic number | Electron structure |
|---|---|---|---|---|---|
| H | 1 | $1s^1$ | Ru | 44 | [Kr] $4d^7 5s^1$ |
| He | 2 | $1s^2$ | Rh | 45 | [Kr] $4d^8 5s^1$ |
| Li | 3 | $1s^2 2s^1$ | Pd | 46 | [Kr] $4d^{10}$ |
| Be | 4 | $1s^2 2s^2$ | Ag | 47 | [Kr] $4d^{10} 5s^1$ |
| B | 5 | $1s^2 2s^2 2p^1$ | Cd | 48 | [Kr] $4d^{10} 5s^2$ |
| C | 6 | $1s^2 2s^2 2p^2$ | In | 49 | [Kr] $4d^{10} 5s^2 5p^1$ |
| N | 7 | $1s^2 2s^2 2p^3$ | Sn | 50 | [Kr] $4d^{10} 5s^2 5p^2$ |
| O | 8 | $1s^2 2s^2 2p^4$ | Sb | 51 | [Kr] $4d^{10} 5s^2 5p^3$ |
| F | 9 | $1s^2 2s^2 2p^5$ | Te | 52 | [Kr] $4d^{10} 5s^2 5p^4$ |
| Ne | 10 | $1s^2 2s^2 2p^6$ | I | 53 | [Kr] $4d^{10} 5s^2 5p^5$ |
| Na | 11 | [Ne] $3s^1$ | Xe | 54 | [Kr] $4d^{10} 5s^2 5p^6$ |
| Mg | 12 | [Ne] $3s^2$ | Cs | 55 | [Xe] $6s^1$ |
| Al | 13 | [Ne] $3s^2 3p^1$ | Ba | 56 | [Xe] $6s^2$ |
| Si | 14 | [Ne] $3s^2 3p^2$ | La | 57 | [Xe] $5d^1 6s^2$ |
| P | 15 | [Ne] $3s^2 3p^3$ | Ce | 58 | [Xe] $4f^1 5d^1 6s^2$ |
| S | 16 | [Ne] $3s^2 3p^4$ | Pr | 59 | [Xe] $4f^3 6s^2$ |
| Cl | 17 | [Ne] $3s^2 3p^5$ | Nd | 60 | [Xe] $4f^4 6s^2$ |
| Ar | 18 | [Ne] $3s^2 3p^6$ | Pm | 61 | [Xe] $4f^5 6s^2$ |
| K | 19 | [Ar] $4s^1$ | Sm | 62 | [Xe] $4f^6 6s^2$ |
| Ca | 20 | [Ar] $4s^2$ | Eu | 63 | [Xe] $4f^7 6s^2$ |
| Sc | 21 | [Ar] $3d^1 4s^2$ | Gd | 64 | [Xe] $4f^7 5d^1 6s^2$ |
| Ti | 22 | [Ar] $3d^2 4s^2$ | Tb | 65 | [Xe] $4f^9 6s^2$ |
| V | 23 | [Ar] $3d^3 4s^2$ | Dy | 66 | [Xe] $4f^{10} 6s^2$ |
| Cr | 24 | [Ar] $3d^5 4s^1$ | Ho | 67 | [Xe] $4f^{11} 6s^2$ |
| Mn | 25 | [Ar] $3d^5 4s^2$ | Er | 68 | [Xe] $4f^{12} 6s^2$ |
| Fe | 26 | [Ar] $3d^6 4s^2$ | Tm | 69 | [Xe] $4f^{13} 6s^2$ |
| Co | 27 | [Ar] $3d^7 4s^2$ | Yb | 70 | [Xe] $4f^{14} 6s^2$ |
| Ni | 28 | [Ar] $3d^8 4s^2$ | Lu | 71 | [Xe] $4f^{14} 5d^1 6s^2$ |
| Cu | 29 | [Ar] $3d^{10} 4s^1$ | Hf | 72 | [Xe] $4f^{14} 5d^2 6s^2$ |
| Zn | 30 | [Ar] $3d^{10} 4s^2$ | Ta | 73 | [Xe] $4f^{14} 5d^3 6s^2$ |
| Ga | 31 | [Ar] $3d^{10} 4s^2 4p^1$ | W | 74 | [Xe] $4f^{14} 5d^4 6s^2$ |
| Ge | 32 | [Ar] $3d^{10} 4s^2 4p^2$ | Re | 75 | [Xe] $4f^{14} 5d^5 6s^2$ |
| As | 33 | [Ar] $3d^{10} 4s^2 4p^3$ | Os | 76 | [Xe] $4f^{14} 5d^6 6s^2$ |
| Se | 34 | [Ar] $3d^{10} 4s^2 4p^4$ | Ir | 77 | [Xe] $4f^{14} 5d^7 6s^2$ |
| Br | 35 | [Ar] $3d^{10} 4s^2 4p^5$ | Pt | 78 | [Xe] $4f^{14} 5d^9 6s^1$ |
| Kr | 36 | [Ar] $3d^{10} 4s^2 4p^6$ | Au | 79 | [Xe] $4f^{14} 5d^{10} 6s^1$ |
| Rb | 37 | [Kr] $5s^1$ | Hg | 80 | [Xe] $4f^{14} 5d^{10} 6s^2$ |
| Sr | 38 | [Kr] $5s^2$ | Tl | 81 | [Xe] $4f^{14} 5d^{10} 6s^2 6p^1$ |
| Y | 39 | [Kr] $4d^1 5s^2$ | Pb | 82 | [Xe] $4f^{14} 5d^{10} 6s^2 6p^2$ |
| Zr | 40 | [Kr] $4d^2 5s^2$ | Bi | 83 | [Xe] $4f^{14} 5d^{10} 6s^2 6p^3$ |
| Nb | 41 | [Kr] $4d^4 5s^1$ | Po | 84 | [Xe] $4f^{14} 5d^{10} 6s^2 6p^4$ |
| Mo | 42 | [Kr] $4d^5 5s^1$ | At | 85 | [Xe] $4f^{14} 5d^{10} 6s^2 6p^5$ |
| Tc | 43 | [Kr] $4d^5 5s^2$ | Rn | 86 | [Xe] $4f^{14} 5d^{10} 6s^2 6p^6$ |

*Table 5.5 (continued)*

| Element | Atomic number | Electron structure | Element | Atomic number | Electron structure |
|---|---|---|---|---|---|
| Fr | 87 | [Rn] $7s^1$ | Bk | 97 | [Rn] $5f^97s^2$ |
| Ra | 88 | [Rn] $7s^2$ | Cf | 98 | [Rn] $5f^{10}7s^2$ |
| Ac | 89 | [Rn] $6d^17s^2$ | Es | 99 | [Rn] $5f^{11}7s^2$ |
| Th | 90 | [Rn] $6d^27s^2$ | Fm | 100 | [Rn] $5f^{12}7s^2$ |
| Pa | 91 | [Rn] $5f^26d^17s^2$ | Md | 101 | [Rn] $5f^{13}7s^2$ |
| U | 92 | [Rn] $5f^36d^17s^2$ | No | 102 | [Rn] $5f^{14}7s^2$ |
| Np | 93 | [Rn] $5f^46d^17s^2$ | Lr | 103 | [Rn] $5f^{14}6d^17s^2$ |
| Pu | 94 | [Rn] $5f^67s^2$ | Ku | 104 | [Rn] $5f^{14}6d^27s^2$ |
| Am | 95 | [Rn] $5f^77s^2$ | Ha | 105 | [Rn] $5f^{14}6d^37s^2$ |
| Cm | 96 | [Rn] $5f^76d^17s^2$ | — | 106 | [Rn] $5f^{14}6d^47s^2$ |

## 5.15 Diagramming Atomic Structures

Several methods can be used to diagram atomic structures of atoms, depending on what we are trying to illustrate. When we want to show both the nuclear makeup and the total electron structure of each energy level (without orbital detail), we can use a diagram such as shown in Figure 5.8.

A method of diagramming subenergy levels is shown in Figure 5.9. Each orbital is represented by a circle ◯. When the orbital contains one electron, an arrow (↑) is placed in the circle. A second arrow, pointing downward (↓), indicates the second electron in that orbital.

The diagram for hydrogen is (↑). Helium, with two electrons, is drawn as (↑↓); both electrons are 1*s* electrons. The diagram for lithium shows three electrons in two energy levels, $1s^22s^1$. All four electrons of beryllium are *s* electrons, $1s^22s^2$. Boron has the first *p* electron, which is located in the $2p_x$ orbital. Since it is energetically more difficult for the next *p* electron to pair up with the electron in the $p_x$ orbital than to occupy a second *p* orbital, the second *p* electron in carbon is located in the $2p_y$ orbital. The third *p* electron in nitrogen is still unpaired and is found in the $2p_z$ orbital. The next three electrons pair with each of the 2*p* electrons up to the element neon. Also shown in Figure 5.9 are the equivalent linear expressions for these orbital electron structures.

The electrons in successive elements are found in sublevels of increasing energy. The general sequence of increasing energy of subshells is 1*s*2*s*2*p*3*s*3*p* 4*s*3*d*4*p*5*s*4*d*5*p*6*s*4*f*5*d*6*p*7*s*5*f*6*d*7*p*. Figure 5.10 is a useful mnemonic device for

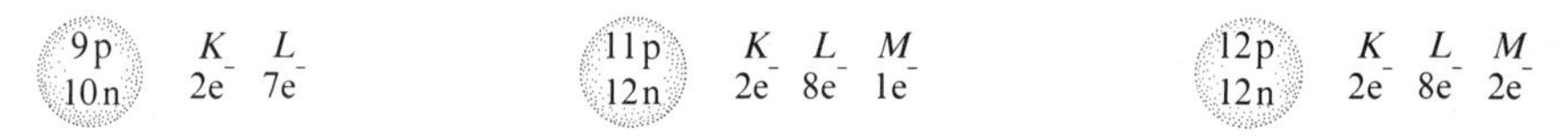


*Figure 5.8.* Atomic structure diagrams of fluorine, sodium, and magnesium atoms. The number of protons and neutrons are shown in the nucleus; outside the nucleus are shown the number of electrons in each principal energy level.

| Element | Orbital electron structure | | | | | | Linear expression of electron structure |
|---|---|---|---|---|---|---|---|
| | $1s$ | $2s$ | $2p_x$ | $2p_y$ | $2p_z$ | $3s$ | |
| H | ↑ | | | | | | $1s^1$ |
| He | ↑↓ | | | | | | $1s^2$ |
| Li | ↑↓ | ↑ | ○ | ○ | ○ | | $1s^2 2s^1$ |
| Be | ↑↓ | ↑↓ | ○ | ○ | ○ | | $1s^2 2s^2$ |
| B | ↑↓ | ↑↓ | ↑ | ○ | ○ | | $1s^2 2s^2 2p_x^1$ |
| C | ↑↓ | ↑↓ | ↑ | ↑ | ○ | | $1s^2 2s^2 2p_x^1 2p_y^1$ |
| N | ↑↓ | ↑↓ | ↑ | ↑ | ↑ | | $1s^2 2s^2 2p_x^1 2p_y^1 2p_z^1$ |
| O | ↑↓ | ↑↓ | ↑↓ | ↑ | ↑ | | $1s^2 2s^2 2p_x^2 2p_y^1 2p_z^1$ |
| F | ↑↓ | ↑↓ | ↑↓ | ↑↓ | ↑ | | $1s^2 2s^2 2p_x^2 2p_y^2 2p_z^1$ |
| Ne | ↑↓ | ↑↓ | ↑↓ | ↑↓ | ↑↓ | | $1s^2 2s^2 2p_x^2 2p_y^2 2p_z^2$ |
| Na | ↑↓ | ↑↓ | ↑↓ | ↑↓ | ↑↓ | ↑ | $1s^2 2s^2 2p_x^2 2p_y^2 2p_z^2 3s^1$ |

*Figure 5.9.* Subenergy-level electron structure of hydrogen through sodium atoms. Each electron is indicated by an arrow placed in the circle, which represents the orbitals.

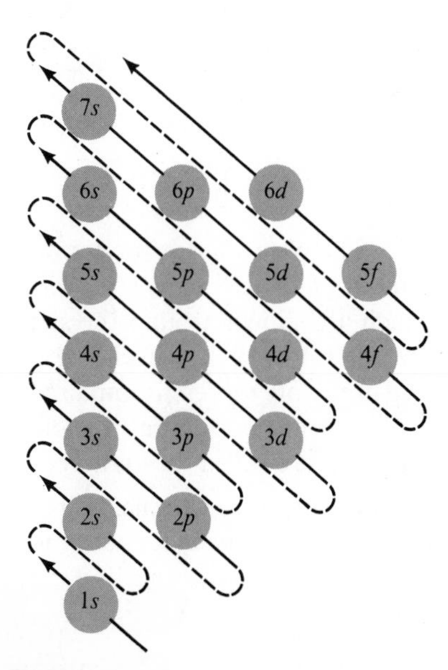


*Figure 5.10.* Approximate order for placing electrons in subenergy levels. Follow the arrows as indicated. Fill each successive subenergy level with the proper number of electrons, starting with $1s$ ($s = 2, p = 6, d = 10, f = 14$ electrons), until all the electrons of an atom have been assigned. (There are a few exceptions to this order—for example, chromium and copper among the first 30 elements.)

writing electron structures. Minor variations from the electron structure predicted by the foregoing general sequence or mnemonic device of Figure 5.10 are found in a number of atoms. Table 5.5 shows the accepted ground-state electron structure for all the elements.

*Problem 5.2*

Diagram the electron structure of a zinc atom and a rubidium atom. Use the $1s^2 2s^2 2p^6$, etc., method.
The atomic number of zinc is 30; therefore, it has 30 protons and 30 electrons in a neutral atom. Using Figure 5.7 or Figure 5.10, tabulate the 30 electrons as follows:

| Orbital | Number of $e^-$ | Total $e^-$ |
|---|---|---|
| $1s$ | $2\ e^-$ | 2 |
| $2s$ | $2\ e^-$ | 4 |
| $2p$ | $6\ e^-$ | 10 |
| $3s$ | $2\ e^-$ | 12 |
| $3p$ | $6\ e^-$ | 18 |
| $4s$ | $2\ e^-$ | 20 |
| $3d$ | $10\ e^-$ | 30 |

The electron distribution in Zn is $1s^2 2s^2 2p^6 3s^2 3p^6 3d^{10} 4s^2$. Check by adding the superscripts, which should equal 30.

The atomic number of rubidium is 37; therefore, it has 37 protons and 37 electrons in a neutral atom. With a little practice, and using either Figure 5.7 or Figure 5.10, the electron structure may be written directly in the linear form. The structure for rubidium is $1s^2 2s^2 2p^6 3s^2 3p^6 3d^{10} 4s^2 4p^6 5s^1$. Check by adding the superscripts, which should equal 37.

## 5.16 Electron-Dot Representation of Atoms

The electron-dot (or Lewis-dot) method of representing atoms uses the symbol of the element, together with dots as electrons. The number of dots, which are shown around the symbol, are equal to the number of electrons in the outermost energy level of the atom. Paired dots represent paired electrons; unpaired dots represent unpaired electrons.

kernel of an atom

The nucleus and all the electrons other than those in the outermost energy level are called the **kernel** of the atom and are represented by the symbol of the element. For example, $:\dot{\mathbf{B}}$ represents a boron atom and tells us that boron has three electrons in its outermost energy level. The kernel of the boron atom includes the nucleus and its $1s^2$ electrons.

$$:\dot{\mathbf{B}}$$

Dots represent electrons in the outer energy level — Symbol represents the kernel of the atom

$:\ddot{\underset{\cdot\cdot}{\mathrm{I}}}\cdot$ indicates an iodine atom, which has seven electrons in its outermost principal energy level. The electron-dot system is used a great deal, not only because of its simplicity of expression, but also because much of the chemistry of the atom is directly associated with the electrons in the outermost energy level. This association is especially true for the first 20 elements and the remaining Group A elements of the periodic table (see Chapter 6). Figure 5.11 shows electron-dot diagrams for the elements hydrogen through calcium.

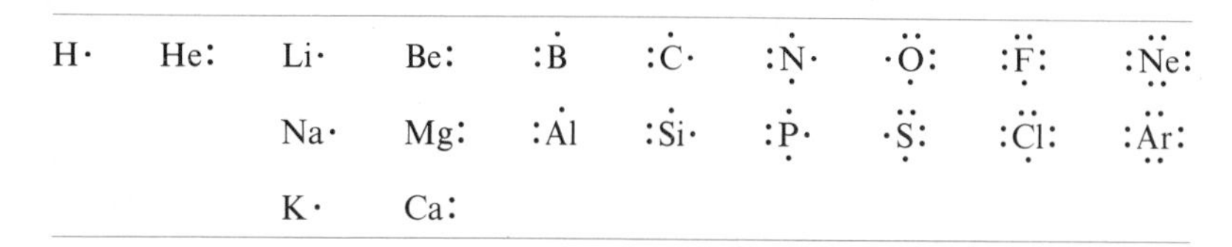

*Figure 5.11.* Electron-dot diagrams of the first 20 elements. Dots represent electrons in the outermost energy level only. Symbol represents the kernel of the atom.

*Problem 5.3*

Write the electron-dot structure for a phosphorus atom.

First establish the electron structure for a phosphorus atom. It is $1s^2 2s^2 2p^6 3s^2 3p^3$. Note that there are five electrons in the outermost principal energy level; they are $3s^2 3p^3$. Write the symbol for phosphorus and place the five electrons as dots around it.

:Ṗ·
 ·

The $3s^2$ electrons are paired and are represented by the paired dots. The $3p^3$ electrons, which are unpaired, are represented by the single dots.

## 5.17 The Noble Gases

noble gases

The family of elements consisting of helium, neon, argon, krypton, xenon, and radon is known as the **noble gases**. The electron structure of these gases has particular interest for the chemist. Until 1961, attempts to prepare compounds of these elements met with failure. Because of their supposed inability to enter into chemical combinations, these elements were formerly called *inert* gases. They have also been called the *rare* gases. Because of their chemical inactivity, the electron structure of the noble gases is considered to be extraordinarily stable.

Each of these elements, except helium, has eight electrons in its outer shell (see Table 5.6). This electron structure is such that the *s* and *p* orbitals in the outer shell are filled with paired electrons, an arrangement that is very stable and chemically unreactive.

Recognition of this structure led to the *rule of eight* principle: The elements, through chemical changes, attempt to attain an electron structure of eight electrons in the outermost energy level identical with that of the chemically stable noble gases. Although the rule of eight principle applies to the chemical behavior of many elements and compounds, it is not universally applicable; some elements do not follow this rule.

All the noble gases are present in the atmosphere. Argon is the most abundant and is found at a concentration of about 1% by volume. The others are present in only trace amounts. Argon was discovered in 1894 by Lord

*Table 5.6.* Arrangement of electrons in the noble gases. Each gas except helium has eight electrons in its outermost energy level.

| | | Electron structure | | | | | |
|---|---|---|---|---|---|---|---|
| Noble gas | Symbol | n = 1 | 2 | 3 | 4 | 5 | 6 |
| Helium | He | $1s^2$ | | | | | |
| Neon | Ne | $1s^2$ | $2s^22p^6$ | | | | |
| Argon | Ar | $1s^2$ | $2s^22p^6$ | $3s^23p^6$ | | | |
| Krypton | Kr | $1s^2$ | $2s^22p^6$ | $3s^23p^63d^{10}$ | $4s^24p^6$ | | |
| Xenon | Xe | $1s^2$ | $2s^22p^6$ | $3s^23p^63d^{10}$ | $4s^24p^64d^{10}$ | $5s^25p^6$ | |
| Radon | Rn | $1s^2$ | $2s^22p^6$ | $3s^23p^63d^{10}$ | $4s^24p^64d^{10}4f^{14}$ | $5s^25p^65d^{10}$ | $6s^26p^6$ |

Raleigh (1842–1919) and Sir William Ramsay (1852–1916). Helium was first observed in the spectrum of the sun during an eclipse in 1868. It was not until 1894 that Ramsay recognized that helium exists in the earth's atmosphere. In 1898 he and his coworker, Morris W. Travers (1872–1961), announced the discovery of neon, krypton, and xenon, having isolated them from liquid air. Friedrich E. Dorn (1848–1916) first identified radon, the heaviest member of the noble gases, as a radioactive gas emanating from the element radium.

Because of its low density and nonflammability, helium has been used for filling balloons and dirigibles. Only hydrogen surpasses helium in lifting power. Helium mixed with oxygen is used by deep-sea divers for breathing. This mixture reduces the danger of acquiring the "bends" (caisson disease), pains and paralysis suffered by divers on returning from the ocean depths to normal atmospheric pressures. Helium is also used in "heliarc welding," where it supplies an inert atmosphere for the welding of active metals such as magnesium. Helium is found in some natural gas wells in the southwestern United States. As a liquid, it is used to study the properties of substances at very low temperatures. The boiling point of liquid helium, 4.2 K, is not far above absolute zero.

We are all familiar with the neon sign, in which a characteristic red color is produced when an electric discharge is passed through a tube filled with neon. This color may be modified by mixing the neon with other gases or by changing the color of the glass. Argon is used primarily in gas-filled electric light bulbs and other types of electronic tubes to provide an inert atmosphere for prolonging tube life. Argon is also used in some welding applications where an inert atmosphere is needed. Krypton and xenon have not been used extensively because of their limited availability. Radon is radioactive; it has been used medicinally in the treatment of cancer.

For many years it was believed that the noble gases could not be made to combine chemically with any other element. Then, in 1962, Neil Bartlett at the University of British Columbia, Vancouver, synthesized the first noble gas compound, xenon hexafluoroplatinate, $XePtF_6$. This outstanding discovery opened a new field in the techniques of preparing noble gas compounds and investigating their chemical bonding and properties. Other compounds of xenon as well as compounds of krypton and radon have been prepared. Some of these are $XeF_2$, $XeF_4$, $XeF_6$, $XeOF_4$, $Xe(OH)_6$, and $KrF_2$.

## 5.18 Atomic Weight

By means of an instrument called a *mass spectrometer*, it is possible to make fairly precise physical determinations of the masses of individual atoms. In the ordinary sense of weighing, single atoms are far too tiny actually to be weighed (the mass of a single hydrogen atom is $1.67 \times 10^{-24}$ g). If we could magnify the mass of such an atom by a factor of 100 billion, it would still require about 300 million of these magnified hydrogen atoms to equal the weight of a single drop of water (0.05 g)!

Because it is not only inconvenient but impractical to express the actual weights of atoms in grams, chemists have devised a useful table of relative atomic weights. The carbon isotope having six protons and six neutrons and designated carbon-12, or $^{12}_{6}C$, was chosen as the standard of reference for atomic weights. (Until 1961, oxygen had been the chemical standard of reference.) The mass of this isotope is assigned a value of exactly 12 atomic mass units (amu). Thus, **1 atomic mass unit** is defined as 1/12 the mass of a carbon-12 atom. In the table of atomic weights all other elements are assigned values proportional to the arbitrary mass assigned to the reference isotope carbon-12. A table of atomic weights is given on the inside back cover of this book. Hydrogen atoms, with a mass of about 1/12 that of a carbon atom, have an average atomic mass of 1.00797 amu on this relative scale. Magnesium atoms, which are about twice as heavy as carbon, have an average mass of 24.312 amu. The average atomic mass of oxygen is 15.9994 amu (usually rounded off to 16.0 for calculations).

atomic mass unit

Since all elements exist as isotopes having different masses, the atomic weight of an element represents the average relative mass of all the naturally occurring isotopes of that element. The atomic weights of the individual isotopes are approximately whole numbers, because the relative masses of the protons and neutrons are approximately 1.0 amu each. Yet we find that the atomic weights given for many of the elements deviate considerably from whole numbers. For example, the atomic weight of rubidium is 85.47 amu, that of copper is 63.54 amu, and that of magnesium is 24.312 amu. The deviation of an atomic weight from a whole number is due mainly to the unequal occurrence of the various isotopes of an element. It is also due partly to the difference between the mass of a free proton or neutron and the mass of these same particles in the nucleus. For example, the two principal isotopes of chlorine are $^{35}_{17}Cl$ and $^{37}_{17}Cl$. It is apparent that chlorine-35 atoms are the more abundant isotope, since the atomic weight of chlorine, 35.453 amu, is closer to 35 than to 37 amu. The actual values of the chlorine isotopes observed by mass spectra determination are shown below.

| Isotope | Isotopic mass (amu) | Abundance (%) | Average atomic mass (amu) |
|---|---|---|---|
| $^{35}_{17}Cl$ | 34.969 | 75.53 | 35.453 |
| $^{37}_{17}Cl$ | 36.966 | 24.47 | |

atomic weight

The **atomic weight** of an element is the average relative mass of the isotopes of that element referred to the atomic mass of carbon-12 (exactly 12.0000 amu).

The relationship of mass number and atomic number is such that if we subtract the atomic number from the mass number, we obtain the number of neutrons in the nucleus of the atom. Table 5.7 shows the application of this method of determining the number of neutrons. For example, the fluorine atom ($^{19}_{9}F$), atomic number 9, having a mass of 19 amu, contains 10 neutrons:

| Mass number | − | Atomic number | = | Number of neutrons |
|---|---|---|---|---|
| 19 | − | 9 | = | 10 |

The atomic weights given in the table on the inside back cover of this book are values accepted by international agreement. There is no need to memorize atomic weights. In most of the calculations needed in this book, their use to the first decimal place will give results of sufficient accuracy.

*Table 5.7.* Calculation of the number of neutrons in an atom by subtracting the atomic number from the mass number.

| | Hydrogen ($^{1}_{1}H$) | Oxygen ($^{16}_{8}O$) | Sulfur ($^{32}_{16}S$) | Fluorine ($^{19}_{9}F$) | Iron ($^{56}_{26}Fe$) |
|---|---|---|---|---|---|
| Mass number | 1 | 16 | 32 | 19 | 56 |
| Atomic number | 1 | 8 | 16 | 9 | 26 |
| Number of neutrons | 0 | 8 | 16 | 10 | 30 |

## 5.19 The Mole

According to the atomic theory set forth by John Dalton, atoms always combine in whole-number ratios. Since individual atoms certainly cannot be weighed, there was a need to establish some weighable unit for comparing the quantities of elements involved in chemical reactions. The working unit devised for this purpose is the **gram-atomic weight** (g-at. wt), which is defined as the number of grams of an element *numerically* equal to its atomic weight. Be very careful to note this distinction: The atomic weight of magnesium, for example, as given in the table, is 24.312. This value indicates the average mass of magnesium atoms relative to the carbon-12 isotope. But the gram-atomic weight of magnesium is 24.312 *grams* of magnesium.

gram-atomic weight

It is known that 1 g-at. wt of magnesium reacts chemically with 1 g-at. wt of sulfur to form magnesium sulfide:

$$Mg + S \longrightarrow MgS$$

Thus, magnesium atoms and sulfur atoms react in a 1:1 atom ratio. It follows, then, that a gram-atomic weight of magnesium must contain the same number of atoms as a gram-atomic weight of sulfur.

Since the actual weight of an atom is very minute, 1 g-at. wt of atoms will contain a very large number of individual atoms. The number of atoms in 1 g-at.

Avogadro's number

wt of any element has been experimentally determined to be $6.02 \times 10^{23}$ (that is, 602,000,000,000,000,000,000,000 atoms). This number is known as **Avogadro's number**. Thus, 1 g-at. wt (24.312 g) of magnesium contains $6.02 \times 10^{23}$ magnesium atoms. A gram-atomic weight of any element—for example, 32.064 g of sulfur, 15.9994 g of oxygen, 1.00797 g of hydrogen, and so on—therefore contains Avogadro's number of atoms. Table 5.8 summarizes these relationships.

*Table 5.8.* Avogadro's number related to the gram-atomic weights of oxygen, hydrogen, sulfur, and magnesium.

| Element | Gram-atomic weight | Avogadro's number (Number of atoms/g-at. wt) |
|---|---|---|
| Oxygen | 15.9994 g | $6.02 \times 10^{23}$ |
| Hydrogen | 1.00797 g | $6.02 \times 10^{23}$ |
| Sulfur | 32.064 g | $6.02 \times 10^{23}$ |
| Magnesium | 24.312 g | $6.02 \times 10^{23}$ |

$$\textit{Avogadro's number} = 6.02 \times 10^{23}$$

It is difficult to imagine how large Avogadro's number really is, but perhaps the following analogy will help express it: If 10,000 people started to count Avogadro's number and each counted at the rate of 100 numbers per minute each minute of the day, it would take them over 1 trillion ($10^{12}$) years to count the total number. So, even the minutest amount of matter contains extremely large numbers of atoms.

mole

Avogadro's number is the basis for an additional very important quantity used to express a particular number of chemical species—such as atoms, molecules, formula units, ions, or electrons. This quantity is the mole. We define a **mole** as an amount of a substance containing the same number of formula units as there are atoms in exactly 12 g of carbon-12. Other definitions are used, but they all relate to a mole being Avogadro's number of formula units of a substance. A formula unit is whatever is indicated by the formula of the substance under consideration—for example, Mg, MgS, $H_2O$, $O_2$, ${}^{75}_{33}As$. The gram-atomic weight of any element contains a mole of atoms of that element.

$$\textit{1 gram-atomic weight} = \textit{1 mole of atoms}$$
$$= \textit{Avogadro's number } (6.02 \times 10^{23}) \textit{ of atoms}$$

Thus, for example, 1 mole of hydrogen atoms represents Avogadro's number of hydrogen atoms and has a mass of 1.008 g (the gram-atomic weight).

The term *mole* is so commonplace in chemical jargon that chemists use it as freely as the words *atom* and *molecule*. The mole is used in conjunction with many different particles, such as atoms, molecules, ions, and electrons, to represent Avogadro's number of these particles. If we can speak of a mole of atoms, we can also speak of a mole of molecules, a mole of electrons, a mole of ions, understanding that in each case we mean $6.02 \times 10^{23}$ formula units of these particles.

## *Avogadro's number of formula units = 1 mole of formula units*

We frequently encounter problems that require interconversions involving the quantities of mass, numbers, and moles of atoms of an element. Conversion factors that may be used for this purpose are

(a) Grams to atoms: $\dfrac{6.02 \times 10^{23} \text{ atoms of the element}}{\text{g-at. wt of the element}}$

(b) Atoms to grams: $\dfrac{\text{g-at. wt of the element}}{6.02 \times 10^{23} \text{ atoms of the element}}$

(c) Grams to moles: (monatomic elements) $\dfrac{1 \text{ mole of the element}}{1 \text{ g-at. wt of the element}}$

(d) Moles to grams: (monatomic elements) $\dfrac{1 \text{ g-at. wt of the element}}{1 \text{ mole of the element}}$

Sample problems follow.

*Problem 5.4*

How many moles of iron does 25.0 g of Fe represent?
The problem requires that we change grams of Fe to moles of Fe. We look up the atomic weight of Fe in the atomic weight table and find it to be 55.8. Then we use the proper conversion factor to obtain moles. The conversion factor is (c) above.

$$\frac{1 \text{ mole Fe}}{55.8 \text{ g Fe}}$$

$$25.0 \text{ g Fe} \times \frac{1 \text{ mole Fe}}{55.8 \text{ g Fe}} = 0.448 \text{ mole Fe} \quad \text{(Answer)}$$

*Problem 5.5*

How many magnesium atoms does 5.00 g of Mg represent?
The problem requires that we change grams of magnesium to atoms of magnesium through the following sequence: Grams Mg → Moles Mg → Atoms Mg. We find the atomic weight of magnesium to be 24.3 and set up the sequence using conversion factors (c) and (a) above.

$$\frac{1 \text{ mole Mg}}{24.3 \text{ g Mg}} \quad \text{and} \quad \frac{6.02 \times 10^{23} \text{ Mg atoms}}{1 \text{ mole Mg}}$$

$$5.00 \text{ g Mg} \times \frac{1 \text{ mole Mg}}{24.3 \text{ g Mg}} \times \frac{6.02 \times 10^{23} \text{ Mg atoms}}{1 \text{ mole Mg}}$$

$$= 1.24 \times 10^{23} \text{ Mg atoms} \quad \text{(Answer)}$$

Thus, there are $1.24 \times 10^{23}$ atoms of Mg in 5.00 g of Mg.

*Problem 5.6*

What is the mass (in grams) of one atom of carbon?
From the table of atomic weights, we see that 1 g-at. wt of carbon is 12.0 g. The factor to

convert atoms to grams therefore is

$$\frac{12.0 \text{ g C}}{6.02 \times 10^{23} \text{ atoms C}}$$

Then

$$1 \text{ atom C} \times \frac{12.0 \text{ g C}}{6.02 \times 10^{23} \text{ atoms C}} = 1.99 \times 10^{-23} \text{ g C} \quad \text{(Answer)}$$

*Problem 5.7*

How much do $3.01 \times 10^{23}$ atoms of sodium weigh?
The information needed to solve this problem is the gram-atomic weight of Na (23.0 g) and the conversion factors

$$\frac{1 \text{ mole}}{6.02 \times 10^{23} \text{ atoms}} \quad \text{and} \quad \frac{1 \text{ g-at. wt}}{1 \text{ mole}}$$

$$3.01 \times 10^{23} \text{ atoms Na} \times \frac{1 \text{ mole Na}}{6.02 \times 10^{23} \text{ atoms Na}} \times \frac{23.0 \text{ g Na}}{1 \text{ mole Na}} = 11.5 \text{ g Na} \quad \text{(Answer)}$$

*Problem 5.8*

How much does Avogadro's number of copper atoms weigh?

Avogadro's number of Cu atoms = 1 mole of Cu atoms

One mole of Cu atoms by definition is the gram-atomic weight of Cu. We therefore look up the atomic weight of Cu in the atomic weight table to find the answer.

Avogadro's number of Cu atoms = 1 mole Cu = 63.5 g Cu (Answer)

Or, by conversion factors,

$$6.02 \times 10^{23} \text{ atoms Cu} \times \frac{1 \text{ mole Cu}}{6.02 \times 10^{23} \text{ atoms Cu}} \times \frac{63.5 \text{ g Cu}}{1 \text{ mole Cu}} = 63.5 \text{ g Cu} \quad \text{(Answer)}$$

## Questions

A. *Review the meanings of the new terms introduced in this chapter.*

1. Dalton's atomic theory
2. Subatomic particles
3. Electron
4. Proton
5. Neutron
6. Nucleus
7. Quantum number
8. Orbital
9. Energy level of electrons
10. Electron shell
11. Atomic number
12. Isotopes
13. Kernel of an atom
14. Noble gas
15. Atomic mass unit
16. Atomic weight
17. Gram-atomic weight
18. Avogadro's number
19. Mole

B. *Answers to the following questions will be found in tables and figures.*

1. What are the atomic numbers of aluminum, platinum, radium, and krypton?
2. How many electron orbitals can be present in the third energy level? What are they?
3. Show the sublevel electron structure ($1s^22s^22p^6$, etc.) for elements of atomic numbers 13, 15, 19, 24, and 36.

4. Using only Table 5.6, write electron structures ($1s^22s^22p^6$, etc.) for elements containing the following number of electrons:
   (a) 21 (b) 33 (c) 52 (d) 55
5. Diagram the atomic structure of the following atoms: F, Ca, Li, K, Br, and Mg (see Figure 5.8).
6. Show electron-dot structures for K, O, Mg, Kr, Br, Co, Si, and Al.
7. Explain the meaning of the following symbols. ${}^{131}_{53}I$, ${}^{235}_{92}U$.

C. *Review questions.*

1. From the point of view of a chemist, what are the essential differences among a proton, a neutron, and an electron?
2. Describe the general arrangement of particles in the atom.
3. What part of the atom contains practically all its mass?
4. What experimental evidence led Rutherford to conclude that:
   (a) The nucleus of the atom contains most of the atomic mass?
   (b) The nucleus of the atom is positively charged?
   (c) The atom consists of mostly empty space?
5. (a) What is an atomic orbital?
   (b) What are the shapes of an *s* orbital and a *p* orbital?
6. What is the Pauli exclusion principle?
7. Under which conditions can a second electron enter an orbital already containing one electron?
8. What is meant when we say that the electron structure of an atom is in its ground state?
9. List the following electron sublevels in order of increasing energy: 2*s*, 2*p*, 4*s*, 1*s*, 3*d*, 3*p*, 4*p*, 3*s*.
10. How many *s* electrons, *p* electrons, and *d* electrons are possible in any electron shell?
11. How many protons are in the nucleus of an atom of each of these elements: H, N, B, Sc, Hg, U, Br, Sn, and Pb?
12. Why is the eleventh electron of the sodium atom located in the third energy level rather than in the second energy level?
13. Why are the last two electrons in calcium located in the fourth energy level rather than in the third energy level?
14. Which atoms have the following structures?
    (a) $1s^22s^22p^63s^2$
    (b) $1s^22s^22p^5$
    (c) $1s^22s^22p^63s^23p^63d^84s^2$
    (d) $1s^22s^22p^63s^23p^63d^54s^2$
    (e) $1s^22s^22p^63s^23p^63d^{10}4s^24p^64d^55s^1$
15. What does the atomic number of an atom represent?
16. Which of the following statements are correct?
    (a) The maximum number of *p* electrons in the first energy level is six.
    (b) A 2*s* electron is in a lower energy state than a 2*p* electron.
    (c) The energy level of a 3*d* electron is higher than a 4*s* electron.
    (d) The electron structure for a carbon atom is $1s^22s^22p_x^2$.
    (e) The $2p_x$, $2p_y$, and $2p_z$ electron orbitals are in the same energy state.
17. In what ways are isotopes alike? In what ways are they different?
18. An atom of an element has a mass number of 108 and has 62 neutrons in its nucleus.
    (a) What is the symbol and name of the element?
    (b) What is its nuclear charge?
19. List the similarities and differences in the three isotopes of hydrogen.

20. What is the nuclear and electron composition of the six naturally occurring isotopes of calcium having mass numbers of 40, 42, 43, 44, 46, and 48?
21. How many electrons are not shown in each electron-dot structure in Si, N, O, Ca, and He atoms?
22. What characterizes the special chemical stability of the noble gases?
23. Which of the following statements about the neutral atoms $^{23}_{11}Na$ and $^{24}_{11}Na$ are true?
    (a) $^{23}_{11}Na$ and $^{24}_{11}Na$ are isotopes.
    (b) $^{24}_{11}Na$ has one more electron than $^{23}_{11}Na$.
    (c) $^{24}_{11}Na$ has one more proton than $^{23}_{11}Na$.
    (d) $^{24}_{11}Na$ has one more neutron than $^{23}_{11}Na$.
    (e) A mole of $^{24}_{11}Na$ contains more atoms than does a mole of $^{23}_{11}Na$.
24. Explain why the atomic weights of elements are not whole numbers.
25. Distinguish between atomic weights expressed in atomic mass units and in grams.
26. What information is needed to calculate the approximate atomic weight of an atom?
27. Which of the isotopes of calcium in Question 20 is the most abundant isotope? Explain your choice.
28. What is the significance of Avogadro's number?
29. Complete the following statements, supplying the proper quantity:
    (a) A mole of oxygen atoms (O) contains ________ atoms.
    (b) A mole of oxygen molecules ($O_2$) contains ________ molecules.
    (c) A mole of oxygen molecules ($O_2$) contains ________ atoms.
    (d) A mole of oxygen atoms (O) weighs ________ grams.
    (e) A mole of oxygen molecules ($O_2$) weighs ________ grams.
30. What contribution did each of the following scientists make to the atomic theory?
    (a) Dalton (c) Rutherford (e) Bohr
    (b) Thomson (d) Chadwick (f) Schrödinger
31. Which of the following statements are correct?
    (a) Hydrogen is the smallest atom.
    (b) A proton is about 1837 times as heavy as an electron.
    (c) The nucleus of an atom contains protons, neutrons, and electrons.
    (d) The Bohr structure of the atom had one electron in each of several orbits surrounding the nucleus.
    (e) There are seven principal electron energy levels.
    (f) The second principal energy level can have four subenergy levels and contain a maximum of eight electrons.
    (g) The *M* energy level can have a maximum of 32 electrons.
    (h) The number of possible *d* electrons in the third energy level is ten.
    (i) The first *f* electron occurs in the fourth principal energy level.
    (j) The 4*s* subenergy level is at a higher energy than the 3*d* subenergy level.
    (k) An element with an atomic number of 29 has 29 protons, 29 neutrons, and 29 electrons.
    (l) An atom of the isotope $^{60}_{26}Fe$ has 34 neutrons in its nucleus.
    (m) $^{2}_{1}H$ is a symbol for the isotope deuterium.
    (n) The electron-dot structure, P·, is used for potassium.
    (o) Atoms of all the noble gases (except helium) have eight electrons in their outermost energy level.
    (p) One gram-atomic weight of any element contains $6.02 \times 10^{23}$ atoms.

(q) The mass of one atom of chlorine is $\frac{35.5 \text{ g}}{6.02 \times 10^{23} \text{ atoms}}$.

(r) A mole of magnesium atoms (24.3 g) contains the same number of atoms as a mole of sodium atoms (23.0 g).

(s) A mole of bromine atoms contains $6.02 \times 10^{23}$ atoms of bromine.

(t) A mole of chlorine molecules ($Cl_2$) contains $6.02 \times 10^{23}$ atoms of chlorine.

*D. Review problems.*

1. Change the following to powers of 10:
   (a) 640,000 (b) 0.000568 (c) $0.01^2$ (d) $5^3$
   (See Mathematical Review in Appendix I.)
2. Express the following as numbers without using powers of 10:
   (a) $4.2 \times 10^6$ (b) $9 \times 10^{-5}$ (c) $10^{-4}$ (d) $35 \times 10^3$
   (See Mathematical Review in Appendix I.)
3. Using the formula $2\mathbf{n}^2$, calculate the maximum number of electrons that can exist in electron shells *K*, *L*, *M*, *N*, *O*, and *P*. Energy levels: $\mathbf{n} = 1, 2, 3, 4, 5$, and 6.
4. How many neutrons are in an atom of $^{40}_{20}Ca$, $^{59}_{28}Ni$, $^{119}_{50}Sn$, $^{207}_{82}Pb$, $^{235}_{92}U$, $^{254}_{102}No$?
5. How many individual atoms are contained in the following?
   (a) 12.0 g Na (c) 160 g C (e) 56.2 g Cd
   (b) 0.75 g P (d) 25.0 g Cu (f) $1.00 \times 10^{-4}$ g H
6. Complete the following table with the appropriate data for each element given:

| Atomic number | Symbol of element | Number of protons | Atomic weight | Gram-atomic weight | Mass (g) of 1 atom |
|---|---|---|---|---|---|
| 9 | | | | | |
| 33 | | | | | |
| 82 | | | | | |

7. Calculate the weight, in grams, of one atom of:
   (a) Sulfur (b) Tin (c) Mercury (d) Helium
8. How many grams are represented by each of the following?
   (a) 1,000,000 atoms of Ar (c) 3.00 g-at. wt of Cl
   (b) 4.50 moles of Al (d) 0.00100 mole of Ag
9. Make the following conversions:
   (a) 2.62 moles of C to grams of C
   (b) 0.150 mole of Cu to grams of Cu
   (c) 22.5 moles of Ag to kilograms of Ag
   (d) 3.0 moles of $Cl_2$ to grams of $Cl_2$
   (e) 125 g of Fe to moles of Fe
   (f) 2.00 g of Sn to moles of Sn
   (g) 28.0 g of $N_2$ to moles of $N_2$
   (h) 25.0 ml Hg ($d = 13.6$ g/ml) to moles of Hg
10. One mole of water contains:
    (a) How many water molecules?
    (b) How many oxygen atoms?
    (c) How many hydrogen atoms?

11. White phosphorus is one of several forms of phosphorus and exists as a waxy solid consisting of $P_4$ molecules. How many atoms are present in 0.50 mole of this phosphorus?
12. How many grams of magnesium contain the same number of atoms as 6.00 g of calcium?
13. One atom of an unknown element is found to have a mass of $2.28 \times 10^{-22}$ g. What is the gram-atomic weight of the element?
14. There are about four billion ($4 \times 10^9$) people on the earth. If 1 mole of dollars were distributed equally among these people, how many dollars would each person receive?
15. A ping-pong ball is about 4 cm in diameter. If a mole of ping-pong balls were laid end to end, what distance, in kilometres, would they cover? How many miles is this?

E. *Review exercises.*

1. Show the atomic structure of the most abundant isotope of cobalt. Of sulfur.
2. Which atom would you expect to have the larger volume, chlorine or bromine? Why?
3. Draw the atomic structure of an atom with an atomic weight of 34 and an atomic number of 16. What element is this?
4. In your own words, describe a chlorine atom.
5. Would you expect the density of the nucleus of an atom to be relatively large or small? Explain.
6. What mass of sodium contains the same number of atoms as 5.00 g of calcium?
7. One atom of an unknown element is found to have a mass of $9.75 \times 10^{-23}$ g. What is the gram-atomic weight of the element?

# 6 The Periodic Arrangement of the Elements

*After studying Chapter 6 you should be able to:*

1. Understand the terms listed under Question A at the end of the chapter.
2. Describe briefly the contributions of Döbereiner, Newlands, Mendeleev, Meyer, and Moseley to the development of the periodic law.
3. State the periodic law in its modern form.
4. Explain why there were blank spaces in Mendeleev's periodic table and how he was able to predict the properties of the elements that belonged in those spaces.
5. Indicate the locations of the metals, nonmetals, metalloids, and noble gases in the periodic table.
6. Indicate in the periodic table the areas where the *s*, *p*, *d*, and *f* orbitals are being filled.
7. Describe how atomic radii vary (a) from left to right in a period and (b) from top to bottom in a group.
8. Describe the changes in outer-level electron structure when (a) moving from left to right in a period and (b) going from top to bottom in a group.
9. List the general characteristics of group properties.
10. Predict formulas of simple compounds formed between Group A elements using the periodic table.
11. Point out how the change in electron structure in going from one transition element to the next differs from that in nontransition elements.

## 6.1 Early Attempts to Classify the Elements

Chemists of the early 19th century had sufficient knowledge of the properties of elements to recognize similarities among groups of elements. J. W. Döbereiner (1780–1849), professor at the University of Jena in Germany, observed the existence of "triads" of similarly behaving elements, in which the middle element had an atomic weight approximating the average of the other two elements. He also noted that many other properties of the central element were approximately the average of the other two elements. Table 6.1 presents comparative data on atomic weight and density for two sets of Döbereiner's triads.

*Table 6.1.* Döbereiner's triads.

| Triads | Atomic weight | Density (g/ml at 4°C) |
|---|---|---|
| Chlorine | 35.5 | 1.56[a] |
| Bromine | 79.9 | 3.12 |
| Iodine | 126.9 | 4.95 |
| Average of chlorine and iodine | 81.2 | 3.26 |
| Calcium | 40.1 | 1.55 |
| Strontium | 87.6 | 2.6 |
| Barium | 137.4 | 3.5 |
| Average of calcium and barium | 88.8 | 2.52 |

[a] Density at −34°C (liquid).

In 1864, J. A. R. Newlands (1837–1898), an English chemist, reported his *Law of Octaves*. In his studies, Newlands observed that when the elements were arranged according to increasing atomic weights, every eighth element had similar properties. (The noble gases were not yet discovered at that time.) But Newlands' theory was ridiculed by his contemporaries in the Royal Chemical Society, and they refused to publish his work. Many years later, however, Newlands was awarded the highest honor of the society for this important contribution to the development of the periodic law.

In 1869, Dmitri Ivanovitch Mendeleev (1834–1907) of Russia and Lothar Meyer (1830–1895) of Germany independently published their periodic arrangements of the elements that were based on increasing atomic weights. Because his arrangement was published slightly earlier and was in a somewhat more useful form than that of Meyer, Mendeleev's name is usually associated with the modern periodic table.

## 6.2 The Periodic Law

Only about 63 elements were known when Mendeleev constructed his table. These elements were arranged so that those with similar chemical properties fitted into columns to form family groups. The arrangement left many gaps between elements. Mendeleev predicted that these spaces would be filled as new elements were discovered. For example, spaces for undiscovered elements were left after calcium, under aluminum, and under silicon. He called these unknown elements eka-boron, eka-aluminum, and eka-silicon. The term *eka* comes from Sanskrit meaning "one" and was used to indicate that the missing element was one place away in the table from the element indicated. Mendeleev even went so far as to predict with high accuracy the physical and chemical properties of these elements yet to be discovered. The three elements, scandium (atomic number 21), gallium (31), and germanium (32) were in fact discovered during Mendeleev's lifetime and were found to have properties agreeing very closely with the predictions that he had made for eka-boron,

*Table 6.2.* Comparison of the properties of eka-silicon predicted by Mendeleev with the properties of germanium.

| Property | Mendeleev's predictions in 1871 for eka-silicon (Es) | Observed properties for germanium (Ge) |
|---|---|---|
| Atomic weight | 72 | 72.6 |
| Color of metal | Dirty gray | Grayish-white |
| Density | 5.5 g/ml | 5.47 g/ml |
| Oxide formula | $EsO_2$ | $GeO_2$ |
| Oxide density | 4.7 g/ml | 4.70 g/ml |
| Chloride formula | $EsCl_4$ | $GeCl_4$ |
| Chloride density | 1.9 g/ml | 1.89 g/ml |
| Boiling temperature of chloride | Under 100°C | 86°C |

eka-aluminum, and eka-silicon. The amazing way in which Mendeleev's predictions were fulfilled is illustrated in Table 6.2, which compares the predicted properties of eka-silicon with those of germanium, discovered by the German chemist C. Winkler in 1886.

Two major additions have been made to the periodic table since Mendeleev's time: (1) a new family of elements, the noble gases, was discovered and added; and (2) elements having atomic numbers greater than 92 have been discovered and fitted into the table.

The original table was based on the premise that the properties of the elements are a periodic function of their atomic weights. However, there were some disturbing discrepancies to this basic premise. For example, the atomic weight for argon is greater than that of potassium. Yet potassium had to be placed after argon since argon is certainly one of the noble gases and potassium behaves like the other alkali metals. These discrepancies were resolved by the work of the British physicist, H. G. J. Moseley (1887–1915) and by the discovery of the existence of isotopes. Moseley noted that the x-ray emission frequencies of the elements increased in a regular, stepwise fashion each time the nuclear charge (atomic number) increased by one unit. This showed that the basis for placing an element in the periodic table should be dependent on atomic number rather than on atomic weight. Atomic weights are the average masses of the naturally occurring mixtures of isotopes of each element. The atomic weight for argon is greater than that of potassium (the next element of higher atomic number) because the average mass of the argon isotopes is greater than the average mass of the potassium isotopes. The current statement of the **periodic law** is:

periodic law

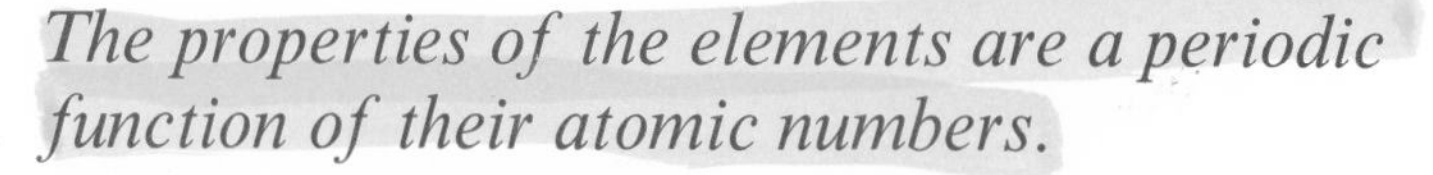
*The properties of the elements are a periodic function of their atomic numbers.*

As the format of the periodic table is studied, it becomes evident that the periodicity of the properties of the elements is due to the recurring similarities of their electron structures.

*Table 6.3.* Periodic table of the elements.

Key: Atomic number — 11; Name — Sodium; Symbol — Na; Electron structure — 2, 8, 1; Atomic weight — 22.9898

[a]Mass number of most stable or best-known isotope

[b]Mass of most commonly long-lived isotope

Atomic weights are based on carbon-12. Atomic weights in parentheses indicate the most stable or best-known isotope. Slight disagreement exists as to the exact electronic configuration of several of the high-atomic-number elements. Names and symbols for elements 104 and 105 are unofficial.

Cell format: atomic number, name, **symbol**, atomic weight (electron structure).

| Period | Group IA | IIA | IIIB | IVB | VB | VIB | VIIB | VIII | VIII | VIII | IB | IIB | IIIA | IVA | VA | VIA | VIIA | Noble gases |
|---|---|---|---|---|---|---|---|---|---|---|---|---|---|---|---|---|---|---|
| 1 | 1 Hydrogen **H** 1.0079 (1) | | | | | | | | | | | | | | | | | 2 Helium **He** 4.00260 (2) |
| 2 | 3 Lithium **Li** 6.941 (2, 1) | 4 Beryllium **Be** 9.01218 (2, 2) | | | | | | | | | | | 5 Boron **B** 10.81 (2, 3) | 6 Carbon **C** 12.011 (2, 4) | 7 Nitrogen **N** 14.0067 (2, 5) | 8 Oxygen **O** 15.9994 (2, 6) | 9 Fluorine **F** 18.9984 (2, 7) | 10 Neon **Ne** 20.179 (2, 8) |
| 3 | 11 Sodium **Na** 22.9898 (2, 8, 1) | 12 Magnesium **Mg** 24.305 (2, 8, 2) | ← Transition elements | | | | | | | | | → | 13 Aluminum **Al** 26.9815 (2, 8, 3) | 14 Silicon **Si** 28.0855 (2, 8, 4) | 15 Phosphorus **P** 30.9738 (2, 8, 5) | 16 Sulfur **S** 32.06 (2, 8, 6) | 17 Chlorine **Cl** 35.453 (2, 8, 7) | 18 Argon **Ar** 39.948 (2, 8, 8) |
| 4 | 19 Potassium **K** 39.0983 (2, 8, 8, 1) | 20 Calcium **Ca** 40.08 (2, 8, 8, 2) | 21 Scandium **Sc** 44.9559 (2, 8, 9, 2) | 22 Titanium **Ti** 47.90 (2, 8, 10, 2) | 23 Vanadium **V** 50.9415 (2, 8, 11, 2) | 24 Chromium **Cr** 51.996 (2, 8, 13, 1) | 25 Manganese **Mn** 54.938 (2, 8, 13, 2) | 26 Iron **Fe** 55.847 (2, 8, 14, 2) | 27 Cobalt **Co** 58.9332 (2, 8, 15, 2) | 28 Nickel **Ni** 58.71 (2, 8, 16, 2) | 29 Copper **Cu** 63.546 (2, 8, 18, 1) | 30 Zinc **Zn** 65.38 (2, 8, 18, 2) | 31 Gallium **Ga** 69.72 (2, 8, 18, 3) | 32 Germanium **Ge** 72.59 (2, 8, 18, 4) | 33 Arsenic **As** 74.922 (2, 8, 18, 5) | 34 Selenium **Se** 78.96 (2, 8, 18, 6) | 35 Bromine **Br** 79.904 (2, 8, 18, 7) | 36 Krypton **Kr** 83.80 (2, 8, 18, 8) |
| 5 | 37 Rubidium **Rb** 85.4678 (2, 8, 18, 8, 1) | 38 Strontium **Sr** 87.62 (2, 8, 18, 8, 2) | 39 Yttrium **Y** 88.9059 (2, 8, 18, 9, 2) | 40 Zirconium **Zr** 91.22 (2, 8, 18, 10, 2) | 41 Niobium **Nb** 92.9064 (2, 8, 18, 12, 1) | 42 Molybdenum **Mo** 95.94 (2, 8, 18, 13, 1) | 43 Technetium **Tc** 98.9062[b] (2, 8, 18, 14, 1) | 44 Ruthenium **Ru** 101.07 (2, 8, 18, 15, 1) | 45 Rhodium **Rh** 102.9055 (2, 8, 18, 16, 1) | 46 Palladium **Pd** 106.4 (2, 8, 18, 18, 0) | 47 Silver **Ag** 107.868 (2, 8, 18, 18, 1) | 48 Cadmium **Cd** 112.41 (2, 8, 18, 18, 2) | 49 Indium **In** 114.82 (2, 8, 18, 18, 3) | 50 Tin **Sn** 118.69 (2, 8, 18, 18, 4) | 51 Antimony **Sb** 121.75 (2, 8, 18, 18, 5) | 52 Telurium **Te** 127.60 (2, 8, 18, 18, 6) | 53 Iodine **I** 126.9045 (2, 8, 18, 18, 7) | 54 Xenon **Xe** 131.30 (2, 8, 18, 18, 8) |
| 6 | 55 Cesium **Cs** 132.905 (2, 8, 18, 18, 8, 1) | 56 Barium **Ba** 137.33 (2, 8, 18, 18, 8, 2) | * 57 Lanthanum **La** 138.9055 (2, 8, 18, 18, 9, 2) | 72 Hafnium **Hf** 178.49 (2, 8, 18, 32, 10, 2) | 73 Tantalum **Ta** 180.9479 (2, 8, 18, 32, 11, 2) | 74 Wolfram (Tungsten) **W** 183.85 (2, 8, 18, 32, 12, 2) | 75 Rhenium **Re** 186.2 (2, 8, 18, 32, 13, 2) | 76 Osmium **Os** 190.2 (2, 8, 18, 32, 14, 2) | 77 Iridium **Ir** 192.2 (2, 8, 18, 32, 17, 0) | 78 Platinum **Pt** 195.09 (2, 8, 18, 32, 17, 1) | 79 Gold **Au** 196.9665 (2, 8, 18, 32, 18, 1) | 80 Mercury **Hg** 200.59 (2, 8, 18, 32, 18, 2) | 81 Thallium **Tl** 204.37 (2, 8, 18, 32, 18, 3) | 82 Lead **Pb** 207.2 (2, 8, 18, 32, 18, 4) | 83 Bismuth **Bi** 208.9084 (2, 8, 18, 32, 18, 5) | 84 Polonium **Po** (210)[a] (2, 8, 18, 32, 18, 6) | 85 Astatine **At** (210)[a] (2, 8, 18, 32, 18, 7) | 86 Radon **Rn** (222)[a] (2, 8, 18, 32, 18, 8) |
| 7 | 87 Francium **Fr** (223)[a] (2, 8, 18, 32, 18, 8, 1) | 88 Radium **Ra** 226.0254[b] (2, 8, 18, 32, 18, 8, 2) | ** 89 Actinium **Ac** (227)[a] (2, 8, 18, 32, 18, 9, 2) | 104 Kurchatovium **Ku** (260)[a] | 105 Hahnium **Ha** (260)[a] | 106 — — | | | | | | | | | | | | |

Inner transition elements

|  | | | | | | | | | | | | | | |
|---|---|---|---|---|---|---|---|---|---|---|---|---|---|---|
| * Lanthanide series 6 | 58 Cerium **Ce** 140.12 (2, 8, 18, 20, 8, 2) | 59 Praseodymium **Pr** 140.9077 (2, 8, 18, 21, 8, 2) | 60 Neodymium **Nd** 144.24 (2, 8, 18, 22, 8, 2) | 61 Promethium **Pm** (145)[a] (2, 8, 18, 23, 8, 2) | 62 Samarium **Sm** 150.4 (2, 8, 18, 24, 8, 2) | 63 Europium **Eu** 151.96 (2, 8, 18, 25, 8, 2) | 64 Gadolinium **Gd** 157.25 (2, 8, 18, 25, 9, 2) | 65 Terbium **Tb** 158.9254 (2, 8, 18, 27, 8, 2) | 66 Dysprosium **Dy** 162.50 (2, 8, 18, 28, 8, 2) | 67 Holmium **Ho** 164.930 (2, 8, 18, 29, 8, 2) | 68 Erbium **Er** 167.26 (2, 8, 18, 30, 8, 2) | 69 Thulium **Tm** 168.9342 (2, 8, 18, 31, 8, 2) | 70 Ytterbium **Yb** 173.04 (2, 8, 18, 32, 8, 2) | 71 Lutetium **Lu** 174.967 (2, 8, 18, 32, 9, 2) |
| ** Actinide series 7 | 90 Thorium **Th** 232.0381[b] (2, 8, 18, 32, 18, 10, 2) | 91 Protactinium **Pa** 231.0359[b] (2, 8, 18, 32, 20, 9, 2) | 92 Uranium **U** 238.029 (2, 8, 18, 32, 21, 9, 2) | 93 Neptunium **Np** 237.048 (2, 8, 18, 32, 22, 9, 2) | 94 Plutonium **Pu** (242)[a] (2, 8, 18, 32, 23, 9, 2) | 95 Americium **Am** (243)[a] (2, 8, 18, 32, 25, 8, 2) | 96 Curium **Cm** (247)[a] (2, 8, 18, 32, 25, 9, 2) | 97 Berkelium **Bk** (249)[a] (2, 8, 18, 32, 26, 9, 2) | 98 Californium **Cf** (251)[a] (2, 8, 18, 32, 27, 9, 2) | 99 Einsteinium **Es** (254)[a] (2, 8, 18, 32, 28, 9, 2) | 100 Fermium **Fm** (253)[a] (2, 8, 18, 32, 29, 9, 2) | 101 Mendelevium **Md** (256)[a] (2, 8, 18, 32, 30, 9, 2) | 102 Nobelium **No** (254)[a] (2, 8, 18, 32, 31, 9, 2) | 103 Lawrencium **Lr** (257)[a] (2, 8, 18, 32, 32, 9, 2) |

## 6.3 Arrangement of the Periodic Table

periodic table

periods of elements

groups or families of elements

The most commonly used **periodic table** is the long form shown in Table 6.3. In this table the elements are arranged horizontally in numerical sequence, according to their atomic numbers; the result is seven horizontal **periods**. Each period, with the exception of the first, starts with an alkali metal and ends with a noble gas. By this arrangement, vertical columns of elements are formed, having identical or similar outer-shell electron structures and thus similar chemical properties. These columns are known as **groups** or **families of elements**.

The heavy zigzag line starting at boron and running diagonally down the table separates the elements into metals and nonmetals. The elements to the right of the line are nonmetallic, and those to the left are metallic. The elements bordering the zigzag line are the metalloids, which show both metallic and nonmetallic properties. With some exceptions, the characteristic electronic arrangement of metals is that their atoms have one, two, or three electrons in the outer energy level, while nonmetals have five, six, or seven electrons in the outer energy level.

With this periodic arrangement, the elements fall into blocks according to the sublevel of electrons that is being filled. The grouping of the elements into *spdf* blocks is shown in Table 6.4. The *s* block comprising Groups IA and IIA has one or two *s* electrons in its outer energy level. The *p* block includes Groups IIIA through VIIA and the noble gases (except helium). In these elements, electrons are filling the *p* sublevel orbitals. The *d* block includes the

*Table 6.4.* Arrangement of the elements into blocks according to the sublevel of electrons being filled in their atomic structure.

| *s* block IA | IIA | IIIB | IVB | VB | VIB | VIIB | VIII | VIII | VIII | IB | IIB | IIIA | IVA | VA | VIA | VIIA | Noble gases |
|---|---|---|---|---|---|---|---|---|---|---|---|---|---|---|---|---|---|
| 1 H | | | | | | | | | | | | | | | | | 2 He |
| 3 Li | 4 Be | | | | | | | | | | | 5 B | 6 C | 7 N | 8 O | 9 F | 10 Ne |
| 11 Na | 12 Mg | | | | | | | | | | | 13 Al | 14 Si | 15 P | 16 S | 17 Cl | 18 Ar |
| 19 K | 20 Ca | 21 Sc | 22 Ti | 23 V | 24 Cr | 25 Mn | 26 Fe | 27 Co | 28 Ni | 29 Cu | 30 Zn | 31 Ga | 32 Ge | 33 As | 34 Se | 35 Br | 36 Kr |
| 37 Rb | 38 Sr | 39 Y | 40 Zr | 41 Nb | 42 Mo | 43 Tc | 44 Ru | 45 Rh | 46 Pd | 47 Ag | 48 Cd | 49 In | 50 Sn | 51 Sb | 52 Te | 53 I | 54 Xe |
| 55 Cs | 56 Ba | 57* La | 72 Hf | 73 Ta | 74 W | 75 Re | 76 Os | 77 Ir | 78 Pt | 79 Au | 80 Hg | 81 Tl | 82 Pb | 83 Bi | 84 Po | 85 At | 86 Rn |
| 87 Fr | 88 Ra | 89** Ac | 104 Ku | 105 Ha | 106 | | | | | | | | | | | | |

*d* block: Transition elements (Groups IIIB through IIB)

*p* block: Groups IIIA through VIIA and Noble gases

*f* block
Inner transition elements

| | | | | | | | | | | | | | |
|---|---|---|---|---|---|---|---|---|---|---|---|---|---|
| *58 Ce | 59 Pr | 60 Nd | 61 Pm | 62 Sm | 63 Eu | 64 Gd | 65 Tb | 66 Dy | 67 Ho | 68 Er | 69 Tm | 70 Yb | 71 Lu |
| **90 Th | 91 Pa | 92 U | 93 Np | 94 Pu | 95 Am | 96 Cm | 97 Bk | 98 Cf | 99 Es | 100 Fm | 101 Md | 102 No | 103 Lr |

*s* block
*p* block
*d* block
*f* block

transition elements of Groups IB through VIIB and Group VIII. The *d* sublevels of electrons are being filled in these elements. The *f* block of elements includes the inner transition series. In the lanthanide series, electrons are filling the 4*f* sublevel. In the actinide series, electrons are filling the 5*f* sublevel.

## 6.4 Periods of Elements

The number of elements in each period is shown in Table 6.5. The first period contains 2 elements, hydrogen and helium, and coincides with the full *K* shell of electrons. Period 2 contains 8 elements, starting with lithium and ending with neon. Period 3 also contains 8 elements, sodium to argon. Periods 4 and 5 each contain 18 elements; period 6 has 32 elements; and period 7, which is incomplete, contains the remaining 20 elements.

*Table 6.5.* The number of elements in each period.

| Period number | Number of elements | Electron orbitals in each period being filled |
|---|---|---|
| 1 | 2 | 1*s* |
| 2 | 8 | 2*s*2*p* |
| 3 | 8 | 3*s*3*p* |
| 4 | 18 | 4*s*3*d*4*p* |
| 5 | 18 | 5*s*4*d*5*p* |
| 6 | 32 | 6*s*4*f*5*d*6*p* |
| 7 | 20 | 7*s*5*f*6*d* |

The first three periods are known as *short periods*; the others, as *long periods*. The number of each period corresponds to the outermost energy level in which electrons are located in the neutral atom. For example, the elements in period 1 contain electrons in the first energy level only; period 2 elements contain electrons in the first and second levels; period 3 elements contain electrons in the first, second, and third levels; and so on. Moving horizontally across periods 2 through 6, we find that the properties of the elements vary from strongly metallic at the beginning to nonmetallic at the end of the period. Starting with the third element of the long periods 4, 5, 6, and 7 (scandium, Sc; yttrium, Y; lanthanum, La; actinium, Ac), the inner shells of *d* and *f* orbital electrons begin to fill in.

In general, the atomic radii of the elements within a period decrease with increasing nuclear charge. Therefore, the size of the atoms becomes progressively smaller from left to right within each period. Because the noble gases do not readily combine with other elements to form compounds, the radii of their atoms are not determined in the same comparative manner as the other elements. However, calculations have shown that the radii of the noble gas atoms are about the same as, or slightly smaller than, the element immediately preceding them. Slight deviations in atomic radii occur in the middle of the long periods of the elements. The elements of period 3 serve to illustrate this principle. The radii are given in angstrom units:

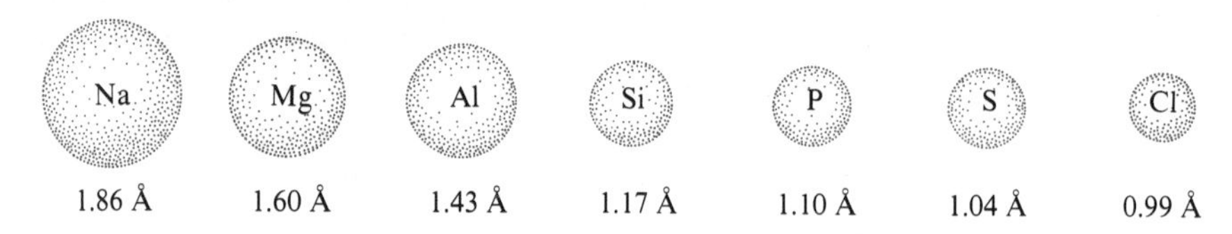


## 6.5 Groups or Families of Elements

The groups, or families, of elements are numbered IA through VIIA, IB through VIIB, VIII, and noble gases.

The elements comprising each family have similar outer energy-level electron structures. In Group A elements, the number of electrons in the outer energy level is identical to the group number. Group IA is known as the *alkali metal* family. Each atom of this family of elements has one *s* electron in its outer energy level. Group IIA atoms, the *alkaline earth metals*, each have two *s* electrons in their outer energy level. All atoms of the *halogen* family, Group VIIA, have an outer energy-level electron structure of $ns^2np^5$. The noble gases (except helium) have an outer energy-level structure of $ns^2np^6$.

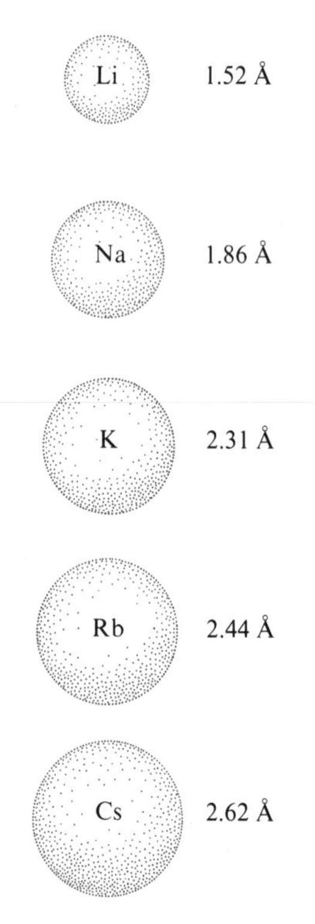

Each alkali metal starts a new period of elements in which the *s* electron occupies a principal energy level one greater than in the previous period. As a result, the size of the atoms of this and other families of elements increases from the top to the bottom of the family. The relative size of the alkali metals is illustrated at the left. The radii are given in angstrom units.

One major distinction between groups lies in the energy level to which the last electron is added. In the elements of Groups IA through VIIA, IB, IIB, and the noble gases, the last electron is added either to an *s* or to a *p* orbital located in the outermost energy level (copper is an exception; see Table 5.5). In the elements of Groups IIIB through VIIB and VIII, the last electron goes to a *d* or to an *f* orbital located in an inner energy level. For example, the last electron in potassium (a fourth-period, Group IA element) is a 4*s* electron located in the fourth energy level (an outer shell); in scandium (a fourth-period Group IIIB element) the last electron added is a 3*d* electron located in the third energy level (an inner shell). Figure 6.1 compares the locations of the last electron added in Group A and B types of elements.

| Group | Noble gas | IA | IIA | IIIB | IVB |
|---|---|---|---|---|---|
| Element | Ar | K | Ca | Sc | Ti |
| Electron structure | 2,8,8 | 2,8,8,1 | 2,8,8,2 | 2,8,9,2 | 2,8,10,2 |
| Energy level | 1,2,3 | 1,2,3,4 | 1,2,3,4 | 1,2,3,4 | 1,2,3,4 |
| Last electron added | ↑ | ↑ | ↑ | ↑ | ↑ |

*Figure 6.1.* Comparison of the placement of the last electron in Group A and Group B elements. The energy level to which the last electron is added is indicated by the arrow—an outer level for Group A elements, an inner level for Group B elements.

The general characteristics of group properties are as follows:

1. The number of electrons in the outer energy level of Groups IA through VIIA, IB, and IIB elements is the same as the group number. The other B groups and Group VIII do not show this characteristic. Each noble gas except helium has eight electrons in its outer energy level.
2. The group number in which an element is located will indicate one of its possible oxidation states, with some exceptions, notably in Group VIII.
3. The groups on the left and in the middle sections of the table tend to be metallic in nature. The groups on the right tend to be nonmetallic.
4. The radii of the elements increase from top to bottom within a particular group (for example, from lithium to francium).
5. Elements at the bottom of a group tend to be more metallic in their properties than those at the top. This tendency is especially noticeable in Groups IVA through VIIA.
6. Elements within an A group have the same number of electrons in their outer shell and show closely related chemical properties.
7. Elements within a B group have some similarity in electron structure and also show some similarities in chemical properties.

## 6.6 Predicting Formulas by Use of the Periodic Table

The periodic table can be used to predict the formulas of simple compounds. As we shall see in Chapter 7, the chemical properties of the elements are dependent on their electrons. Group A elements ordinarily form compounds using only the electrons in their outer energy level. If we examine Group IA, the alkali metals, we see that all of them have one electron in their outer energy level, all follow a noble gas in the table, and all, except lithium, have eight electrons in their next inner shell. These likenesses suggest that there should be a great deal of similarity in the chemistry of these metals, since their chemical properties are vested primarily in their outer-shell electron. And there is similarity—all readily lose their outer electron and attain a noble gas electron structure. In doing this, they form compounds with similar atomic compositions. For example, all the monoxides of Group IA contain two atoms of the alkali metal to one atom of oxygen. Their formulas are $Li_2O$, $Na_2O$, $K_2O$, $Rb_2O$, $Cs_2O$, and $Fr_2O$.

How can we use the table to predict formulas of other compounds? Because of similar electron structures, the elements in a family generally form compounds with the same atomic ratios. This was shown for the oxides of the Group IA metals above. In general, if we know the atomic ratio of a particular compound, say sodium chloride (NaCl), we can predict the atomic ratios and formulas of the other alkali metal chlorides. These formulas are LiCl, KCl, RbCl, CsCl, and FrCl.

In a similar way, if we know that the formula of the oxide of hydrogen is $H_2O$, we predict that the formula of the sulfide will be $H_2S$ because sulfur has the same outer-shell electron structure as oxygen. It must be recognized, however, that these are only predictions; it does not necessarily follow that every element in a group will behave like the others, or even that a predicted compound will actually exist. Knowing the formulas for potassium chlorate, bromate, and iodate to be $KClO_3$, $KBrO_3$, and $KIO_3$, we can correctly predict the corresponding sodium compounds to have the formulas $NaClO_3$, $NaBrO_3$, and $NaIO_3$. Fluorine belongs to the same family of elements (Group VIIA) as chlorine, bromine, and iodine. Therefore, we can predict that the formulas for potassium and sodium fluorates will be $KFO_3$ and $NaFO_3$. These compounds are not known to exist; however, if they did exist, the formulas could very well be correct, for these predictions are based on comparisons with known formulas and/or similar electron structures.

*Problem 6.1*

The formula for magnesium sulfate is $MgSO_4$ and that for potassium sulfate is $K_2SO_4$. Predict the formulas for:
(a) Barium sulfate (b) Lithium sulfate

(a) Look in the periodic table for the locations of magnesium and barium. They are both in Group IIA. Since Mg and Ba are in the same group, we can predict that the ratio of $Ba^{2+}$ to $SO_4^{2-}$ in barium sulfate will be the same as the ratio of $Mg^{2+}$ to $SO_4^{2-}$ in magnesium sulfate—namely 1:1. Therefore, the formula of barium sulfate will be $BaSO_4$.

(b) Check the periodic table and locate potassium and lithium in Group IA. Since both elements are in the same group and the formula of potassium sulfate is $K_2SO_4$, the formula for lithium sulfate will be $Li_2SO_4$.

---

## 6.7 Transition Elements

transition elements

Elements in Groups IB, IIIB through VIIB, and VIII are known as the **transition elements**. There are four series of transition elements, one in each of the periods 4, 5, 6, and 7. The transition elements are characterized by an increasing number of *d* or *f* electrons in an inner shell; they all have either one or two electrons in their outer shell. In period 4, electrons enter the 3*d* sublevel. In period 5, electrons enter the 4*d* sublevel. The transition elements in period 6 include the lanthanide series (or rare earth elements, La to Lu), in which electrons are entering the 4*f* sublevel. The 5*d* sublevel also fills up in the sixth period. The seventh period of elements is an incomplete period. It includes the actinide series (Ac to Lr) in which electrons are entering the 5*f* sublevel.

In the formation of compounds of the transition metals, electrons may come from more than one energy level. For this reason these metals form multiple series of compounds.

## 6.8 New Elements

Mendeleev allowed gaps in his orderly periodic table for elements whose discovery he predicted. These were actually discovered, as were all the elements up to atomic number 92 that occur naturally on the earth. Fourteen elements beyond uranium (atomic numbers 93–106) have been discovered or synthesized since 1939. All these elements have unstable nuclei and are radioactive. Beyond element 101, the isotopes synthesized thus far have such short lives that chemical identification has not been accomplished.

Intensive research is continuing on the synthesis of still heavier elements. Extending the periodic table beyond the presently known elements, it is predicted that elements 110 to 118 will be very stable, but still radioactive. Element 114 will lie below lead (82) and should be exceptionally stable. Element 118 should be a member of the noble gas family. Elements 119 and 120 should be in Groups IA and IIA, respectively, and have electrons in the 8*s* sublevel.

## 6.9 Summary

The periodic table has been used for studying the relationships of many properties of the elements. Ionization energies, densities, melting points, atomic radii, atomic volumes, oxidation states, electrical conductance, and electronegativity are just a few of these properties. However, a detailed discussion of all these properties is not practical at this time.

The periodic table is still used as a guide in predicting the synthesis of possible new elements. It presents a very large amount of chemical information in compact form and correlates the properties and relationships of all the elements. The table is so useful that a copy hangs in nearly every chemistry lecture hall and laboratory in the world. Refer to it often.

## Questions

A. *Review the meanings of the new terms introduced in this chapter.*

1. Periodic law
2. Periodic table
3. Periods of elements
4. Groups or families of elements
5. Transition elements

B. *Answers to the following questions are to be found in tables and figures.*

1. How many elements are present in each period?
2. Write the symbols of the alkali metal family in the order of increasing size of their atoms.
3. What similarities do you observe in the elements of Group IIA?
4. Write the symbols for the elements with atomic numbers 18, 36, 54, and 86. What do these elements have in common?
5. Write the names and symbols of the halogens.
6. Write the symbols for the family of elements that have two electrons in their outer energy level.
7. Point out similarities and differences between Group IA and Group IB elements.
8. What similarities do the elements of the lanthanide series possess?
9. Where are the elements with the most metallic characteristics located in the periodic table?

C. *Review questions.*

1. Write a paragraph describing the general features of the periodic table.
2. What do you feel is the basis for Newlands' Law of Octaves?
3. How are elements in a period related to one another?
4. How are elements in a group related to one another?
5. Classify the following elements as metals, nonmetals, or metalloids:
   (a) Potassium (c) Sulfur (e) Iodine (g) Molybdenum
   (b) Plutonium (d) Antimony (f) Radium
6. What is common about the electron structures of the alkali metals?
7. Draw the electron-dot diagrams for Cs, Ba, Tl, Pb, Po, At, and Rn. How do these structures correlate with the group in which each element occurs?
8. Pick the electron structures below that represent elements in the same chemical family.
   (a) $1s^2 2s^1$
   (b) $1s^2 2s^2 2p^4$
   (c) $1s^2 2s^2 2p^2$
   (d) $1s^2 2s^2 2p^6 3s^2 3p^4$
   (e) $1s^2 2s^2 2p^6 3s^2 3p^6$
   (f) $1s^2 2s^2 2p^6 3s^2 3p^6 4s^2$
   (g) $1s^2 2s^2 2p^6 3s^2 3p^6 4s^1$
   (h) $1s^2 2s^2 2p^6 3s^2 3p^6 3d^1 4s^2$
9. In how many different principal energy levels do electrons occur in period 1, period 3, and period 5?
10. In which period and group does an electron first appear in a *d* orbital?
11. How many electrons occur in the outer shell of Groups IIIA and IIIB elements? Why are they different?

12. In which groups are transition elements located?
13. How do transition elements differ from other elements?
14. Which element in each of the following pairs has the larger atomic radius?
    (a) Na or K (b) Na or Mg (c) O or F (d) Br or I (e) Ti or Zr
15. Which element in each of Groups IA–VIIA has the smallest atomic radius?
16. Why does the atomic size increase in going down any group of the periodic table?
17. Letting E be an element in any group, the table represents the possible formulas

| Group | IA | IIA | IIIA | IVA | VA | VIA | VIIA |
|---|---|---|---|---|---|---|---|
| | $EH$ | $EH_2$ | $EH_3$ | $EH_4$ | $EH_3$ | $H_2E$ | $HE$ |
| | $E_2O$ | $EO$ | $E_2O_3$ | $EO_2$ | $E_2O_5$ | $EO_3$ | $E_2O_7$ |

    of such compounds. Following the pattern in the table, write the formulas for the hydrogen and oxygen compounds of:
    (a) Na (b) Ca (c) Al (d) Sn (e) Sb (f) Se (g) Cl
18. Group IB elements have one electron in their outer shell, as do group IA elements. Would you expect them to form compounds such as CuCl, AgCl, and AuCl? Explain.
19. The formula for lead(II) bromide is $PbBr_2$; predict formulas for tin(II) and germanium(II) bromides. (See Section 8.3b for the use of Roman numerals in naming compounds.)
20. The formula for sodium sulfate is $Na_2SO_4$. Write the names and formulas for the other alkali metal sulfates.
21. All the atoms within each Group A family of elements can be represented by the same electron-dot structure. Complete the table, expressing the electron-dot structure for each group. Use E to represent the elements.

| Group | IA | IIA | IIIA | IVA | VA | VIA | VIIA |
|---|---|---|---|---|---|---|---|
| | E· | | | | | | |

22. Oxygen and sulfur are extremely different elements in that one is a colorless gas and the other, a yellow crystalline solid. Why, then, are both located in Group VIA?
23. Why should the discovery of the existence of isotopes have bearing on the fact that the periodicity of the elements is a function of their atomic numbers and not their atomic weights?
24. Hydrogen can fit into Group VIIA as well as Group IA. Explain why it would not be wrong to place hydrogen in Group VIIA.

*Try to answer Questions 25 through 28 without referring to the periodic table.*

25. The atomic numbers of the noble gases are 10, 18, 36, 54, and 86. What are the atomic numbers for the elements with six electrons in their outer electron shells?
26. Element number 87 is in Group IA, period 7. Describe its outermost energy level. How many energy levels of electrons does it have?
27. If element 36 is a noble gas, in which group would you expect elements 35 and 37 to occur?
28. Which of the following statements about the elements in the periodic table are correct?
    (a) Properties of the elements are periodic functions of their atomic numbers.

(b) There are more nonmetallic elements than metallic elements.
(c) Metallic properties of the elements increase as you go from left to right across a period.
(d) Metallic properties of the elements increase as you go from top to bottom in a family of elements.
(e) Calcium is a member of the alkaline earth family.
(f) Group A elements do not contain any *d* or *f* electrons.
(g) An atom of element 37 has a larger volume than an atom of element 19.
(h) An atom of element 13 has a larger volume than an atom of element 11.
(i) An atom of germanium (Group IVA) has six electrons in its outer shell.
(j) If the formula for calcium iodide is $CaI_2$, then the formula for cesium iodide is $CsI_2$.
(k) Uranium is a transition element.
(l) If the electron-dot structure for barium (Group IIA) is Ba:, then the electron-dot structure for thallium (Group IIIA) is Tl⁝.
(m) The element with the electron configuration $1s^22s^22p^63s^23p^63d^{10}4s^24p^3$ belongs to Group VB.

# 7 The Formation of Compounds from Atoms

*After studying Chapter 7 you should be able to:*

1. Understand the terms listed in Question A at the end of the chapter.
2. Describe how the ionization energies of the elements vary with respect to (1) position in the periodic table and (2) the removal of successive electrons.
3. Describe (1) the formation of ions by electron transfer and (2) the nature of the chemical bond formed by electron transfer.
4. Show by means of electron-dot structures the formation of an ionic compound from atoms.
5. Describe a crystal of sodium chloride.
6. Predict the formulas of the monatomic ions of Group A metals and nonmetals.
7. Predict the relative sizes of an atom and a monatomic ion for a given element.
8. Describe the covalent bond and predict whether a given covalent bond would be polar or nonpolar.
9. Determine from its electron dot structure whether a molecule is a dipole.
10. Describe the changes in electronegativity in (1) moving across a period and (2) moving down a group in the periodic table.
11. Distinguish clearly between ionic and molecular substances.
12. Predict whether the bonding in a compound is primarily ionic or covalent.
13. Distinguish coordinate covalent from covalent bonds in an electron-dot structure.
14. Draw electron-dot structures for monatomic and simple polyatomic ions.
15. Write the formulas of compounds formed by combining the ions from Tables 7.6 and 7.7 (or from the inside front cover of this book) in the correct ratios.
16. Assign the oxidation number to each element in a compound or ion.
17. Distinguish between oxidation number and ionic charge.
18. Distinguish between oxidation and reduction.
19. Identify in an equation the element that has been oxidized and the element that has been reduced.

## 7.1 Ionization Energy

Niels Bohr's description of the atom showed that electrons can exist at various energy levels when an atom absorbs energy. If sufficient energy is applied to an atom, it is possible to remove completely (or "knock out") one or more electrons from its structure, thereby forming a positive ion.

Atom + Energy ⟶ Positive ion + Electron ($e^-$)

ionization energy

The amount of energy required to completely remove one electron from an atom is known as the **ionization energy**. This energy may be expressed in units of kilocalories per mole (kcal/mole), indicating the number of kilocalories required to remove an electron from each atom in a mole of atoms. Thus, 314 kcal are required to remove one electron from a mole of hydrogen atoms; 567 kcal are required to remove the first and 1254 kcal to remove the second electron from a mole of helium atoms. Other units are often used to express ionization energy. The unit of energy in the International System of measurements is the joule (pronounced *jool*), where 1 calorie = 4.184 joules.

Table 7.1 gives ionization energies for the removal of five electrons (where available) from several selected elements. The table shows that it requires increasingly higher amounts of energy to remove the second, third, fourth, and fifth electrons. This is a logical sequence because as electrons are removed, the nuclear charge remains the same, and thereby holds the remaining electrons more tightly. The data also show an extra large increase in the ionization energy when an electron is removed from a noble gas electron structure, indicating the high stability of this structure.

*Table 7.1.* Ionization energies for selected elements. Values are expressed in kilocalories per mole, showing energies required to remove up to five electrons per atom. Shaded areas indicate the energy needed to remove an electron from a noble gas electron structure.

| | Required amounts of energy (kcal/mole) | | | | |
|---|---|---|---|---|---|
| Element | 1st $e^-$ | 2nd $e^-$ | 3rd $e^-$ | 4th $e^-$ | 5th $e^-$ |
| H | 314 | | | | |
| He | 567 | 1254 | | | |
| Li | 124 | 1744 | 2823 | | |
| Be | 215 | 420 | 3548 | 5020 | |
| B | 191 | 580 | 874 | 5980 | 7843 |
| C | 260 | 562 | 1104 | 1487 | 9034 |
| Ne | 497 | 947 | 1500 | 2241 | 2913 |
| Na | 118 | 1091 | 1652 | 2280 | 3192 |

Ionization energies have been experimentally determined for most of the elements. First ionization energies are given in Table 7.2. Figure 7.1 is a graphic plot of the first ionization energies of the first 56 elements, H through Ba.

*Table 7.2.* First ionization energies of the elements (in kilocalories per mole).

Key: Atomic number — 1; Symbol — **H**; Ionization energy (kcal/mole) — 314

| | | | | | | | | | | | | | | | | | |
|---|---|---|---|---|---|---|---|---|---|---|---|---|---|---|---|---|---|
| 1<br>**H**<br>314 | | | | | | | | | | | | | | | | | 2<br>**He**<br>567 |
| 3<br>**Li**<br>124 | 4<br>**Be**<br>215 | | | | | | | | | | | 5<br>**B**<br>191 | 6<br>**C**<br>260 | 7<br>**N**<br>335 | 8<br>**O**<br>314 | 9<br>**F**<br>402 | 10<br>**Ne**<br>497 |
| 11<br>**Na**<br>118 | 12<br>**Mg**<br>176 | | | | | | | | | | | 13<br>**Al**<br>138 | 14<br>**Si**<br>188 | 15<br>**P**<br>242 | 16<br>**S**<br>239 | 17<br>**Cl**<br>300 | 18<br>**Ar**<br>363 |
| 19<br>**K**<br>100 | 20<br>**Ca**<br>141 | 21<br>**Sc**<br>151 | 22<br>**Ti**<br>158 | 23<br>**V**<br>155 | 24<br>**Cr**<br>156 | 25<br>**Mn**<br>171 | 26<br>**Fe**<br>182 | 27<br>**Co**<br>181 | 28<br>**Ni**<br>176 | 29<br>**Cu**<br>178 | 30<br>**Zn**<br>217 | 31<br>**Ga**<br>138 | 32<br>**Ge**<br>182 | 33<br>**As**<br>226 | 34<br>**Se**<br>225 | 35<br>**Br**<br>273 | 36<br>**Kr**<br>322 |
| 37<br>**Rb**<br>96.3 | 38<br>**Sr**<br>131 | 39<br>**Y**<br>147 | 40<br>**Zr**<br>158 | 41<br>**Nb**<br>159 | 42<br>**Mo**<br>164 | 43<br>**Tc**<br>168 | 44<br>**Ru**<br>170 | 45<br>**Rh**<br>172 | 46<br>**Pd**<br>192 | 47<br>**Ag**<br>175 | 48<br>**Cd**<br>207 | 49<br>**In**<br>133 | 50<br>**Sn**<br>169 | 51<br>**Sb**<br>199 | 52<br>**Te**<br>208 | 53<br>**I**<br>241 | 54<br>**Xe**<br>280 |
| 55<br>**Cs**<br>89.8 | 56<br>**Ba**<br>120 | 57<br>**La**<br>129 | 72<br>**Hf**<br>127 | 73<br>**Ta**<br>182 | 74<br>**W**<br>184 | 75<br>**Re**<br>181 | 76<br>**Os**<br>201 | 77<br>**Ir**<br>207 | 78<br>**Pt**<br>207 | 79<br>**Au**<br>213 | 80<br>**Hg**<br>240 | 81<br>**Tl**<br>141 | 82<br>**Pb**<br>171 | 83<br>**Bi**<br>184 | 84<br>**Po**<br>194 | 85<br>**At**<br>219 | 86<br>**Rn**<br>248 |
| 87<br>**Fr**<br>88.3 | 88<br>**Ra**<br>122 | 89<br>**Ac**<br>159 | 104<br>**Ku** | 105<br>**Ha** | 106<br>— | | | | | | | | | | | | |

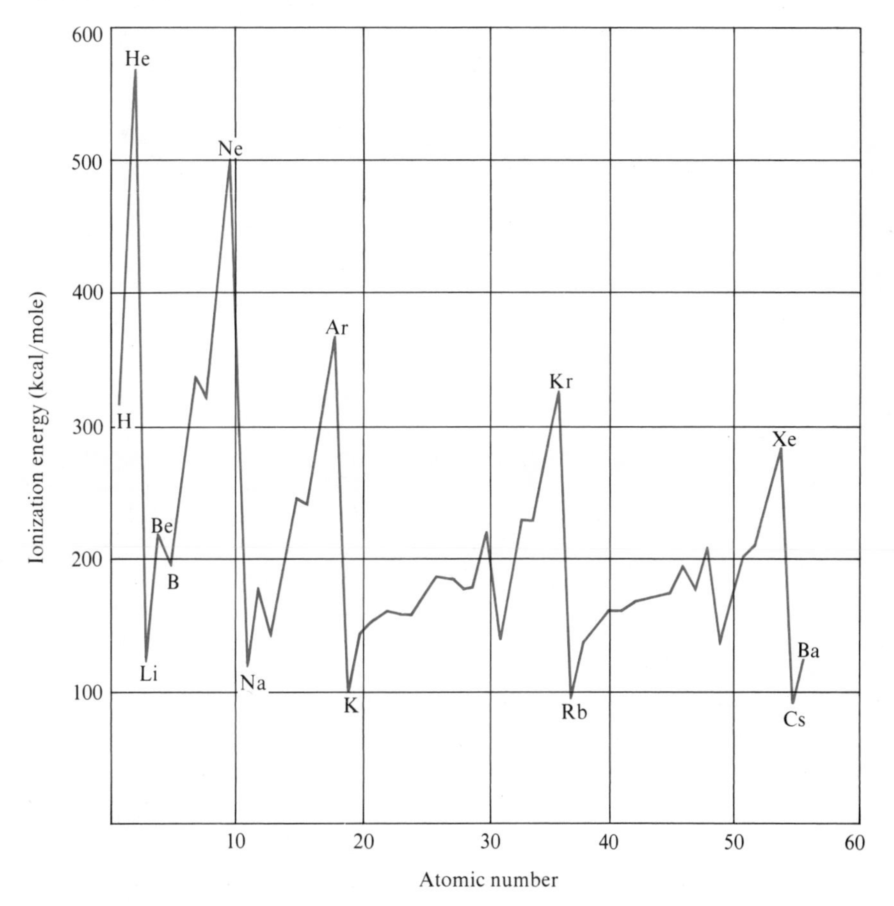


*Figure 7.1.* Periodic relationship of the first ionization energies to the atomic number of the elements.

Certain periodic relationships are noted in Table 7.2 and Figure 7.1. All the alkali metals have relatively low ionization energies, indicating that they each have one electron that is easily removed. Furthermore, the ionization energy decreases from Li to Cs, showing that Cs loses an electron more easily than the other elements. There are two main reasons for this family trend: the electron being removed (1) is farther away from its nucleus and (2) is shielded from its nucleus by more shells of electrons as one goes down the family from Li to Cs.

From left to right within a period, the ionization energy, despite some irregularities, gradually increases. The noble gases have relatively high values, confirming the nonreactivity of these elements and the stability of an electron structure containing eight electrons in the outer energy level.

## 7.2 Electrons in the Outer Shell

Two outstanding properties of the elements are their tendencies (1) to have two electrons in each atomic orbital and (2) to form a stable outer-shell electron structure. For many elements this stable outer shell contains eight electrons and is similar to the outer-shell electron structure of the noble gases. Atoms undergo electron structure rearrangements to attain a state of greater stability. These rearrangements are accomplished by losing, gaining, or sharing electrons with other atoms. For example, a hydrogen atom has a tendency to accept another electron and thus attain an electron structure like that of the stable noble gas helium; a fluorine atom can accommodate one more electron and attain a stable electron structure like neon; a sodium atom tends to lose one electron to attain a stable electron structure like neon. This process requires energy.

$$Na + Energy \longrightarrow Na^+ + 1\,e^-$$

valence electrons

The electrons in the outermost shell of an atom are responsible for most of this electron activity and are called the **valence electrons**. In electron-dot formulas of atoms the dots represent the outer-shell electrons and thus also represent the valence electrons. For example, hydrogen has one valence electron; sodium, one; aluminum, three; and oxygen, six. When a rearrangement of these electrons takes place between atoms, a chemical change occurs.

## 7.3 Transfer of Electrons from One Atom to Another

Let us look at the electron structures of sodium and chlorine to see how each element may attain a structure of 8 electrons in its outer shell. A sodium atom has 11 electrons: 2 in the first energy level, 8 in the second energy level, and 1 in the third energy level. Chlorine has 17 electrons: 2 in the first energy level, 8 in the second energy level, and 7 in the third energy level. If a sodium atom transfers or loses its 3*s* electron, its third energy level becomes vacant and it becomes a sodium ion with an electron configuration identical to that of the noble gas neon.

| 11 p, 12 n | $2e^-8e^-1e^-$ | ⟶ | 11 p, 12 n $^{1+}$ | $2e^-8e^- + 1e^-$ |
|---|---|---|---|---|

Na atom ($1s^2 2s^2 2p^6 3s^1$)    $Na^+$ ion ($1s^2 2s^2 2p^6$)

An atom that has lost or gained electrons will have a plus or minus electrical charge, depending on which charged particles, protons or electrons, are in excess. A charged atom or group of atoms is called an *ion*. A positively charged ion is called a *cation* (pronounced çat-ion); a negative ion is called an *anion* (pronounced an-ion).

By losing a negatively charged electron, the sodium atom becomes a positively charged particle known as a sodium ion. The charge, +1, occurs because the nucleus still contains 11 positively charged protons but the electron orbitals now contain only 10 negatively charged electrons. The charge is indicated by a plus sign (+) and is written as a superscript after the symbol of the element ($Na^+$).

Determination of the charge of a sodium ion:

| Na atom | 11 p | $Na^+$ ion | 11 p |
|---|---|---|---|
| | $11e^-$ | | $10e^-$ |
| Charge | 0 | Charge | +1 |

A chlorine atom with 7 electrons in the third energy level needs 1 electron to pair up with its one unpaired 3*p* electron to attain the stable outer-shell electron structure of argon. By gaining 1 electron, the chlorine atom becomes a chloride ion ($Cl^-$), a negatively charged particle containing 17 protons and 18 electrons.

| 17 p, 18 n | $2e^-8e^-7e^- + 1e^-$ | ⟶ | 17 p, 18 n $^{1-}$ | $2e^-8e^-8e^-$ |
|---|---|---|---|---|

Cl atom ($1s^2 2s^2 2p^6 3s^2 3p^5$)    $Cl^-$ ion ($1s^2 2s^2 2p^6 3s^2 3p^6$)

Determination of the charge of a chloride ion:

| Cl atom | 17 p | $Cl^-$ ion | 17 p |
|---|---|---|---|
| | $17e^-$ | | $18e^-$ |
| Charge | 0 | Charge | −1 |

Consider the case in which sodium and chlorine atoms react with each other. The 3*s* electron from the sodium atom transfers to the vacant 3*p* orbital in the chlorine atom to form a positive sodium ion and a negative chloride ion.

The compound sodium chloride results because the $Na^+$ and $Cl^-$ ions are strongly attracted to one another by their opposite electrostatic charges:

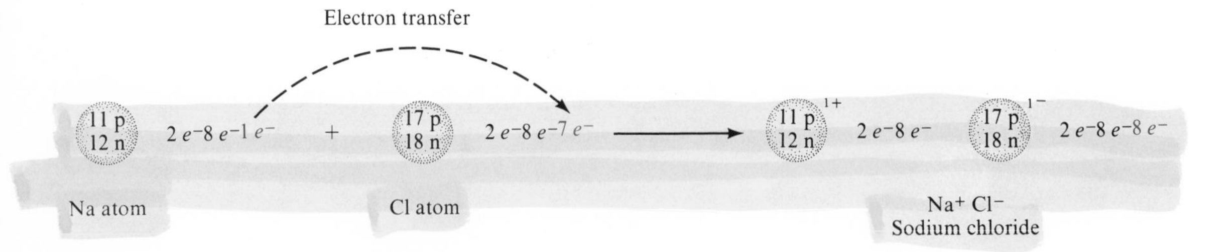


The electron-dot representation of sodium chloride formation is shown below.

$$\text{Na}\cdot + \cdot\ddot{\underset{..}{\text{Cl}}}: \longrightarrow \text{Na}^+ :\ddot{\underset{..}{\text{Cl}}}:^-$$

The chemical reaction between sodium and chlorine is a very vigorous one. It is highly *exothermic* (evolving heat), liberating 90,200 cal when 1 gram-atomic weight of sodium (23.0 g) combines with 1 gram-atomic weight of chlorine (35.5 g).

Sodium chloride is actually made up of cubic crystals, in which each sodium ion is surrounded by six chloride ions and each chloride ion by six sodium ions, except at the crystal surface. A visible crystal is a regularly arranged aggregate of millions of these ions, but the ratio of sodium to chloride ions is one-to-one. The cubic crystalline lattice arrangement of sodium chloride is shown in Figure 7.2.

Figure 7.3 contrasts the relative sizes of sodium and chlorine atoms with that of their ions. The sodium ion is smaller than the atom due primarily to

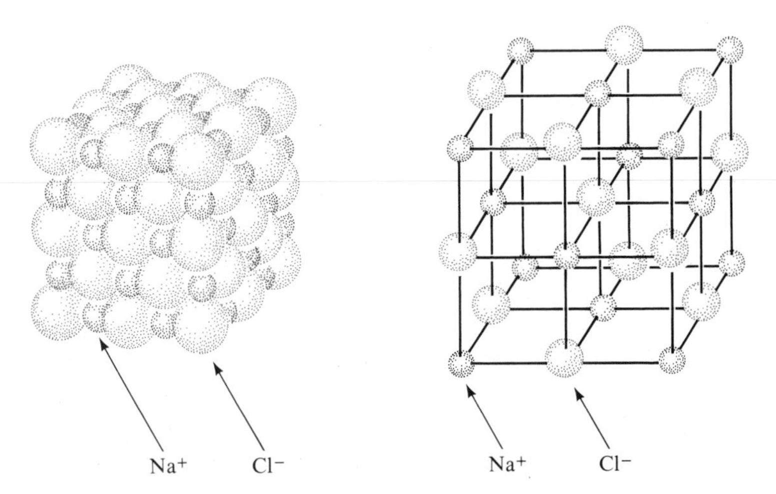


*Figure 7.2.* Sodium chloride crystal. Diagram represents a small fragment of sodium chloride, which forms cubic crystals. Each sodium ion is surrounded by six chloride ions, and each chloride ion is surrounded by six sodium ions.

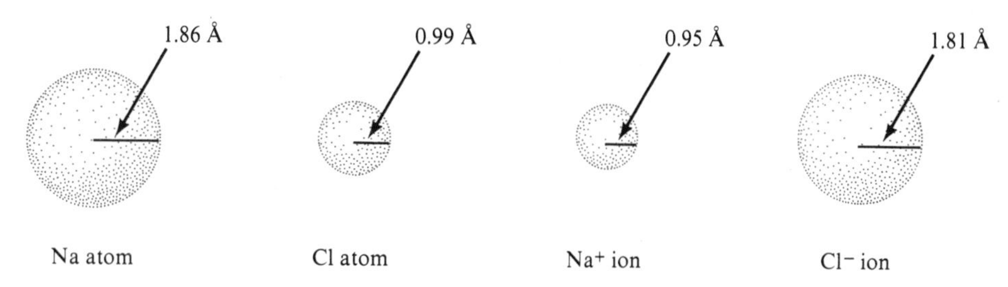


*Figure 7.3.* Relative sizes of sodium and chlorine atoms and their ions.

two factors: (1) the sodium atom has lost its outer shell, consisting of 1 electron, thereby reducing its size; and (2) the 10 remaining electrons are now attracted by 11 protons and are thus drawn closer to the nucleus. Conversely, the chloride ion is larger than the atom because it has 18 electrons but only 17 protons. The nuclear attraction on each electron is thereby decreased, allowing the chlorine atom to expand as it forms an ion.

A metal will ordinarily have one, two, or three electrons in its outer energy level. In reacting, metal atoms characteristically lose these electrons, attain the electron structure of a noble gas, and become positive ions. A nonmetal, on the other hand, lacks a small number of electrons of having a complete octet in its outer energy level and thus has a tendency to gain electrons (electron affinity). In reacting with metals, nonmetal atoms characteristically gain one, two, or three electrons, attain the electron structure of a noble gas, and become negative ions. The ions formed by loss of electrons are much smaller than the corresponding metal atoms; the ions formed by gaining electrons are larger than the corresponding nonmetal atoms. The actual dimensions of the atomic and ionic radii of several metals and nonmetals are given in Table 7.3.

*Table 7.3.* Changes in atomic size of selected metals and nonmetals. The metals shown lose electrons to become positive ions. The nonmetals gain electrons to become negative ions.

| Atomic radius (Å) | | Ionic radius (Å) | | Atomic radius (Å) | | Ionic radius (Å) | |
|---|---|---|---|---|---|---|---|
| Li | 1.52 | $Li^+$ | 0.60 | F | 0.71 | $F^-$ | 1.36 |
| Na | 1.86 | $Na^+$ | 0.95 | Cl | 0.99 | $Cl^-$ | 1.81 |
| K | 2.27 | $K^+$ | 1.33 | Br | 1.14 | $Br^-$ | 1.95 |
| Mg | 1.60 | $Mg^{2+}$ | 0.65 | O | 0.74 | $O^{2-}$ | 1.40 |
| Al | 1.43 | $Al^{3+}$ | 0.50 | S | 1.03 | $S^{2-}$ | 1.84 |

A magnesium atom of electron structure $1s^22s^22p^63s^2$ must lose two electrons or gain six electrons to reach a stable electron structure. If magnesium reacts with chlorine and each chlorine atom has room for only one electron, two chlorine atoms will be needed to accept the two electrons from one magnesium atom. The compound formed will contain one magnesium ion and two chloride ions. The magnesium ion will have a +2 charge, having lost two electrons. Each chloride ion will have a −1 charge. The transfer of electrons from a magnesium atom to two chlorine atoms is shown in the following illustration.

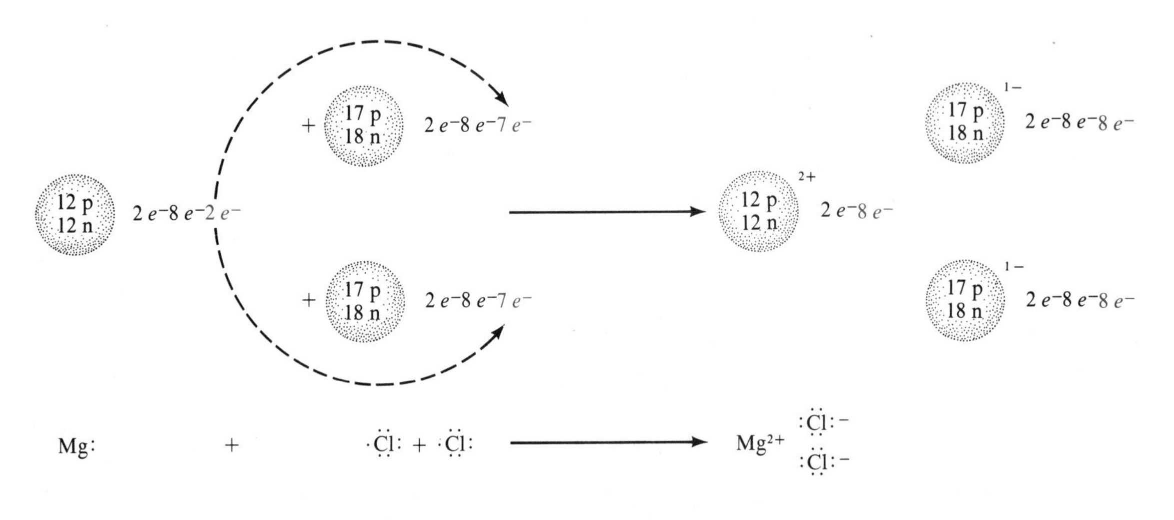


Mg atom | 2 Cl atoms | Magnesium chloride

Study the following examples. Note the loss and gain of electrons between atoms; also note that the ions in each compound have a stable noble gas electron structure.

(a) Sodium fluoride, NaF

11 p, 12 n: 2 $e^-$ 8 $e^-$ 1 $e^-$ + 9 p, 10 n: 2 $e^-$ 7 $e^-$ ⟶ 11 p, 12 n $^{1+}$: 2 $e^-$ 8 $e^-$ 9 p, 10 n $^{1-}$: 2 $e^-$ 8 $e^-$

Na· + ·F: ⟶ $Na^+$ :F:$^-$

Sodium atom | Fluorine atom | Sodium fluoride

The fluorine atom, with seven electrons in its outer shell, behaves similarly to a chlorine atom.

(b) Aluminum chloride, $AlCl_3$

Al· + ·Cl: ·Cl: ·Cl: ⟶ $Al^{3+}$ :Cl:$^-$ :Cl:$^-$ :Cl:$^-$

Aluminum atom | Chlorine atoms | Aluminum chloride

Each chlorine atom can accept only one electron. Therefore, three chlorine atoms are needed to combine with the three outer-shell electrons of one aluminum atom. The aluminum atom has lost three electrons to become an aluminum ion, $Al^{3+}$, with a +3 charge.

(c) Magnesium oxide, MgO

| | | | | | | | |
|---|---|---|---|---|---|---|---|
| (12 p, 12 n) $2e^- 8e^- 2e^-$ | + | (8 p, 8 n) $2e^- 6e^-$ | ⟶ | $(12\text{ p}, 12\text{ n})^{2+}$ $2e^- 8e^-$ | | $(8\text{ p}, 8\text{ n})^{2-}$ $2e^- 8e^-$ |
| Mg: | + | ·Ö: (with dots) | ⟶ | $Mg^{2+}$ :Ö:$^{2-}$ | | |
| Magnesium atom | | Oxygen atom | | Magnesium oxide | | |

The magnesium atom, with two electrons in the outer energy level, exactly fills the need of two electrons of one oxygen atom. The resulting compound has a ratio of one atom of magnesium to one atom of oxygen. The oxygen (oxide) ion has a −2 charge, having gained two electrons. In combining with oxygen, magnesium behaves the same way as when combining with chlorine—it loses two electrons.

(d) Sodium sulfide, $Na_2S$

| | | | | | |
|---|---|---|---|---|---|
| Na· <br> Na· | + | ·S: (with dots) | ⟶ | $Na^+$ <br> $Na^+$ | :S:$^{2-}$ (with dots) |
| Sodium atoms | | Sulfur atom | | Sodium sulfide | |

Two sodium atoms supply the electrons that one sulfur atom needs to make eight in its outer shell.

(e) Aluminum oxide, $Al_2O_3$

| | | | | | |
|---|---|---|---|---|---|
| Äl· <br> Äl· | + | ·Ö: <br> ·Ö: <br> ·Ö: | ⟶ | $Al^{3+}$ <br> $Al^{3+}$ | :Ö:$^{2-}$ <br> :Ö:$^{2-}$ <br> :Ö:$^{2-}$ |
| Aluminum atoms | | Oxygen atoms | | Aluminum oxide | |

One oxygen atom, needing two electrons, cannot accommodate the three electrons from one aluminum atom. One aluminum atom falls one electron short of the four electrons needed by two oxygen atoms. A ratio of two atoms of aluminum to three atoms of oxygen, involving the transfer of six electrons (two to each oxygen atom), gives each atom a stable electron configuration.

Note that in each of the examples above outer shells containing eight electrons were formed in all the negative ions. This formation resulted in the pairing of all the *s* and *p* electrons in these outer shells.

Chemistry would be considerably simpler if all compounds were made by the direct formation of ions as outlined in the examples just given. Unfortunately, this is only one of the two general methods of compound formation. The second general method will be outlined in the sections that follow.

## 7.4 Sharing of Electrons

The formula of chlorine gas is $Cl_2$. When the two atoms of chlorine combine to form this molecule, the electrons must interact by a method that is different from that shown in the preceding examples. Each chlorine atom would be more stable with eight electrons in its outer shell. But if an electron transfers from one chlorine atom to the other, the first chlorine atom, with only six electrons remaining in its outer shell, would be highly unstable. What actually happens when the two chlorine atoms join together is this: the unpaired $3p$ electron orbital of one chlorine atom overlaps the unpaired $3p$ electron orbital of the other atom, resulting in a pair of electrons that are mutually shared between the two atoms. Each atom furnishes one of the pair of shared electrons. Thus, each atom attains a stable structure of eight electrons by sharing an electron pair with the other atom. The pairing of the *p* electrons and formation of a chlorine molecule is illustrated in the following diagrams.

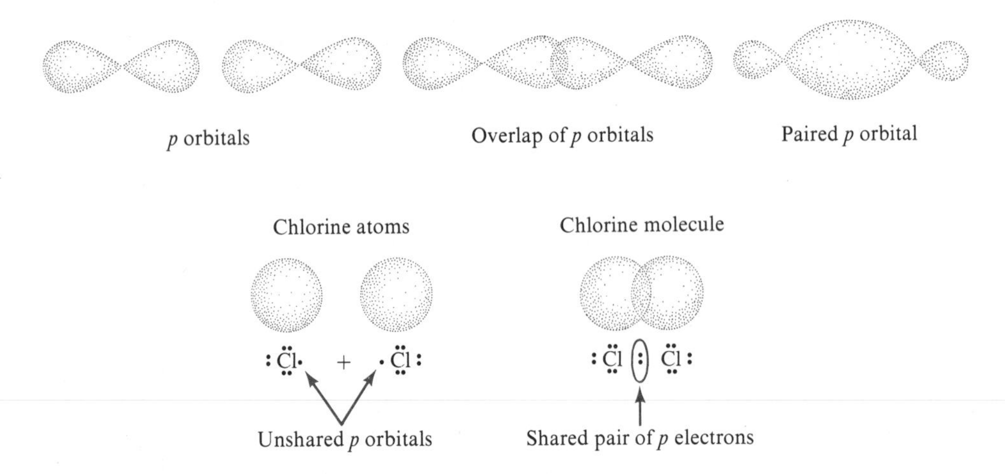


Neither chlorine atom has a positive or negative charge, since both contain the same number of protons and have equal attraction for the pair of electrons being shared. Other examples of molecules in which there is equal sharing of electrons between two atoms are hydrogen, $H_2$; oxygen, $O_2$; nitrogen, $N_2$; fluorine, $F_2$; bromine, $Br_2$; and iodine, $I_2$. Note that two or even three pairs of electrons may be shared between atoms.

| H:H | :F:F: | :Br:Br: | :I:I: | :O::O: | :N:::N: |
|---|---|---|---|---|---|
| Hydrogen | Fluorine | Bromine | Iodine | Oxygen | Nitrogen |

When two different kinds of atoms share a pair of electrons, one atom assumes a partial positive charge and the other a partial negative charge (in respect to each other). This is because the two atoms exert unequal attraction for the pair of shared electrons. The attractive force that an atom of an element has for shared electrons in a molecule is known as its **electronegativity**. Elements differ in their electronegativities. For example, both hydrogen and chlorine need one electron to form stable electron configurations. They share a pair of electrons in the substance hydrogen chloride, HCl. Chlorine is more electronegative and therefore has a greater attraction for the shared electrons than does hydrogen. As a result, the pair of electrons is displaced toward the chlorine atom, giving it a partial negative charge and leaving the hydrogen atom with a partial positive charge. It should be understood that the electron is not transferred entirely to the chlorine atom, as in the case of sodium chloride, and no ions are formed. The entire molecule, HCl, is electrically neutral.

electronegativity

Hydrogen chloride

The pair of shared electrons in HCl is closer to the more electronegative chlorine atom than to the hydrogen atom, giving chlorine a partial negative charge with respect to the hydrogen atom

The electronegativity, or ability of an atom to attract an electron, depends on several factors: (1) the charge on the nucleus, (2) the distance of the outer electrons from the nucleus, and (3) the amount of shielding of the nucleus by intervening shells of electrons between the outer-shell electrons and the nucleus. A scale of relative electronegativities, in which the most electronegative element, fluorine, is assigned a value of 4.0, was developed by Linus Pauling (1901– ) and is given in Table 7.4. This table shows that the relative electronegativity of the nonmetals is high and that of the metals is low. These electronegativities

*Table 7.4.* Relative electronegativity of the elements. The electronegativity value is given below the symbol of each element.

Key: 9 — Atomic number; F — Symbol; 4.0 — Electronegativity

| | | | | | | | | | | | | | | | | | |
|---|---|---|---|---|---|---|---|---|---|---|---|---|---|---|---|---|---|
| 1<br>**H**<br>2.1 | | | | | | | | | | | | | | | | | 2<br>**He**<br>— |
| 3<br>**Li**<br>1.0 | 4<br>**Be**<br>1.5 | | | | | | | | | | | 5<br>**B**<br>2.0 | 6<br>**C**<br>2.5 | 7<br>**N**<br>3.0 | 8<br>**O**<br>3.5 | 9<br>**F**<br>4.0 | 10<br>**Ne**<br>— |
| 11<br>**Na**<br>0.9 | 12<br>**Mg**<br>1.2 | | | | | | | | | | | 13<br>**Al**<br>1.5 | 14<br>**Si**<br>1.8 | 15<br>**P**<br>2.1 | 16<br>**S**<br>2.5 | 17<br>**Cl**<br>3.0 | 18<br>**Ar**<br>— |
| 19<br>**K**<br>0.8 | 20<br>**Ca**<br>1.0 | 21<br>**Sc**<br>1.3 | 22<br>**Ti**<br>1.4 | 23<br>**V**<br>1.6 | 24<br>**Cr**<br>1.6 | 25<br>**Mn**<br>1.5 | 26<br>**Fe**<br>1.8 | 27<br>**Co**<br>1.8 | 28<br>**Ni**<br>1.8 | 29<br>**Cu**<br>1.9 | 30<br>**Zn**<br>1.6 | 31<br>**Ga**<br>1.6 | 32<br>**Ge**<br>1.8 | 33<br>**As**<br>2.0 | 34<br>**Se**<br>2.4 | 35<br>**Br**<br>2.8 | 36<br>**Kr**<br>— |
| 37<br>**Rb**<br>0.8 | 38<br>**Sr**<br>1.0 | 39<br>**Y**<br>1.2 | 40<br>**Zr**<br>1.4 | 41<br>**Nb**<br>1.6 | 42<br>**Mo**<br>1.8 | 43<br>**Tc**<br>1.9 | 44<br>**Ru**<br>2.2 | 45<br>**Rh**<br>2.2 | 46<br>**Pd**<br>2.2 | 47<br>**Ag**<br>1.9 | 48<br>**Cd**<br>1.7 | 49<br>**In**<br>1.1 | 50<br>**Sn**<br>1.8 | 51<br>**Sb**<br>1.9 | 52<br>**Te**<br>2.1 | 53<br>**I**<br>2.5 | 54<br>**Xe**<br>— |
| 55<br>**Cs**<br>0.7 | 56<br>**Ba**<br>0.9 | 57–71<br>**La-Lu**<br>1.1–1.2 | 72<br>**Hf**<br>1.3 | 73<br>**Ta**<br>1.5 | 74<br>**W**<br>1.7 | 75<br>**Re**<br>1.9 | 76<br>**Os**<br>2.2 | 77<br>**Ir**<br>2.2 | 78<br>**Pt**<br>2.2 | 79<br>**Au**<br>2.4 | 80<br>**Hg**<br>1.9 | 81<br>**Tl**<br>1.8 | 82<br>**Pb**<br>1.8 | 83<br>**Bi**<br>1.9 | 84<br>**Po**<br>2.0 | 85<br>**At**<br>2.2 | 86<br>**Rn**<br>— |
| 87<br>**Fr**<br>0.7 | 88<br>**Ra**<br>0.9 | 89–<br>**Ac–**<br>1.1–1.7 | 104<br>**Ku** | 105<br>**Ha** | 106<br>— | | | | | | | | | | | | |

indicate that atoms of metals have a greater tendency to lose electrons than do atoms of nonmetals, and that nonmetals have a greater tendency to gain electrons. The higher the electronegativity value, the greater the attraction for electrons.

The following examples further illustrate electrons shared between atoms to form molecules. The electron-dot structure is used, showing the individual atoms and then the molecule formed from them. In each case, an electron structure is formed to give each atom a noble gas electron structure.

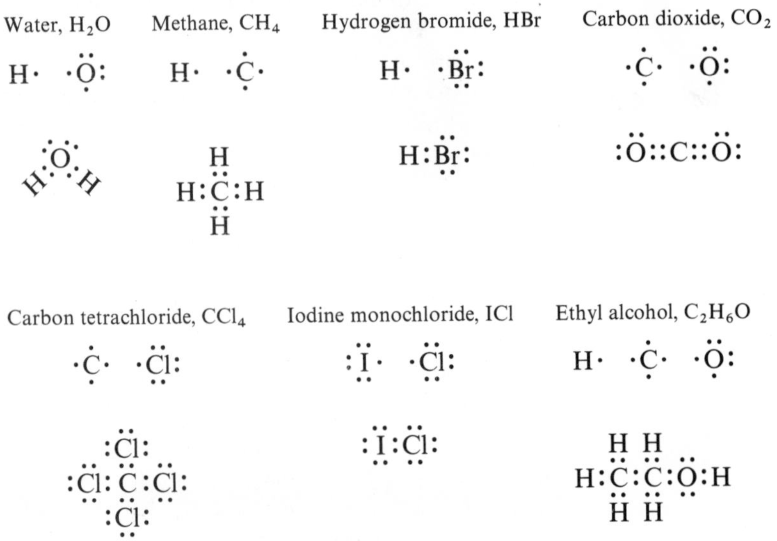

Water, $H_2O$ — H· ·Ö: → H:Ö:H

Methane, $CH_4$ — H· ·Ċ· → H:C:H with H above and below

Hydrogen bromide, HBr — H· ·Br: → H:Br:

Carbon dioxide, $CO_2$ — ·Ċ· ·Ö: → :Ö::C::Ö:

Carbon tetrachloride, $CCl_4$ — ·Ċ· ·Cl: → :Cl:C:Cl: with :Cl: above and below

Iodine monochloride, ICl — :I· ·Cl: → :I:Cl:

Ethyl alcohol, $C_2H_6O$ — H· ·Ċ· ·Ö: → H:C:C:Ö:H with H H above and below

A dash (——) written between two atoms is often used to represent the pair of shared electrons. These same structures would then appear as follows:

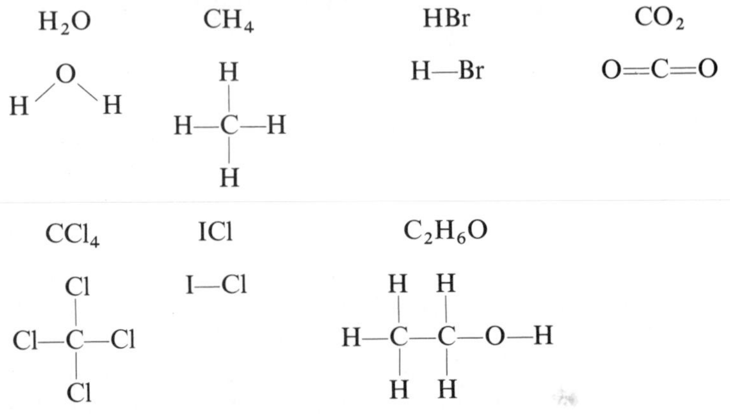

$H_2O$

```
   O
H/   \H
```

$CH_4$

```
   H
   |
H—C—H
   |
   H
```

HBr

H—Br

$CO_2$

O=C=O

$CCl_4$

```
    Cl
    |
Cl—C—Cl
    |
    Cl
```

ICl

I—Cl

$C_2H_6O$

```
  H  H
  |  |
H—C—C—O—H
  |  |
  H  H
```

One should not get the impression that these shared electrons are in a fixed position between their respective atoms. The placement of the dots in these diagrams is merely a convenient method of showing the shared pairs of electrons. Both atoms use these shared electrons in their orbitals to complete their octets of electrons.

## 7.5 Chemical Bonds

Except in very rare instances matter does not fly apart spontaneously. It is prevented from doing so by forces acting at the ionic and molecular levels. Through chemical reactions, atoms tend to attain more stable states at lower chemical potential energy levels. Atoms react chemically by losing, gaining, or sharing electrons. Forces arise from electron transferring and sharing interactions. These forces that hold oppositely charged ions together or that bind atoms together in molecules or in polyatomic ions are called **chemical bonds**. There are two principal types of bonds: the electrovalent, or ionic, bond and the covalent bond. We will study these two bond types and their modifications in more detail in the next four sections of this chapter.

chemical bond

## 7.6 The Electrovalent Bond

When sodium reacts with chlorine, each atom becomes an electrically charged ion. Sodium chloride, and indeed all ionic substances, is bonded together by the attraction existing between positive and negative charges. An **electrovalent** or **ionic bond** is the electrostatic attraction existing between oppositely charged ions.

electrovalent bond

ionic bond

Electrovalent, or ionic, bonds are formed whenever there is a complete transfer of one or more electrons from one atom to another. The metals, which have comparatively low electronegativities and little attraction for their valence electrons, tend to form ionic bonds when they combine with nonmetals. Section 7.3 gives several examples of electron transfer reactions that result in the formation of electrovalent bonds. Restudy these examples. Substances that contain polyatomic ions (see Section 7.10) are also electrostatically bonded.

It is important to recognize that electrovalently bonded substances do not exist as molecules. In sodium chloride, for example, the bond does not exist solely between a single sodium ion and a single chloride ion. Each sodium ion in the crystal attracts six near-neighbor negative chloride ions; in turn, each negative chloride ion attracts six near-neighbor positive sodium ions (see Figure 7.2).

## 7.7 The Covalent Bond

covalent bond

A pair of electrons shared between two atoms constitutes a **covalent bond**. It is the most predominant chemical bond in nature. The concept of the covalent bond was introduced in 1916 by Gilbert N. Lewis (1875–1946) of the University of California at Berkeley.

True molecules exist in compounds that are held together by covalent bonds. It is not correct to refer to "a molecule" of sodium chloride or other ionic compounds, since these compounds exist as large aggregates of positive and negative ions. But we can refer to a molecule of hydrogen, chlorine, hydrogen chloride, carbon tetrachloride, sugar, or carbon dioxide, because these compounds contain only covalent bonds and exist in molecular aggregates.

A study of the hydrogen molecule will give us a better insight into the nature of the covalent bond and its formation. The formation of a hydrogen molecule, $H_2$, involves the overlapping and pairing of 1*s* electron orbitals from two hydrogen atoms. This overlapping and pairing is shown in Figure 7.4. Each atom contributes one electron of the pair that is shared jointly by two hydrogen nuclei. The orbital of the electrons now includes both hydrogen nuclei, but probability factors show that the most likely place to find the electrons (the point of highest electron density) is between the two nuclei.

The tendency for hydrogen atoms to form a molecule is very strong. In the molecule, each electron is attracted by two positive nuclei. This attraction gives the hydrogen molecule a more stable structure than the individual hydrogen atoms had. Experimental evidence of stability is shown by the fact that 104.2 kcal are needed to break the bonds between the hydrogen atoms in one mole of hydrogen (2.0 g). The strength of a bond may be determined by the energy required to break it. The energy required to break a covalent bond is known as the *bond dissociation energy*. The following bond dissociation energies illustrate relative bond strengths. (All substances are considered to be in the gaseous state and to form neutral atoms.)

| Reaction | Bond dissociation energy (kcal/mole) |
|---|---|
| $H_2 \rightarrow 2H$ | 104.2 |
| $N_2 \rightarrow 2N$ | 226.0 |
| $O_2 \rightarrow 2O$ | 118.3 |
| $F_2 \rightarrow 2F$ | 36.6 |
| $Cl_2 \rightarrow 2Cl$ | 58.0 |
| $Br_2 \rightarrow 2Br$ | 46.1 |
| $I_2 \rightarrow 2I$ | 36.1 |

The covalent bond is designated by a dash (—) between the two atoms (for example, H–H, H–Cl, C=O). A single dash means one pair of electrons; two dashes mean two pairs, or four electrons, are shared between two atoms. All the examples in Section 7.4 contain covalent bonds between atoms. Study each

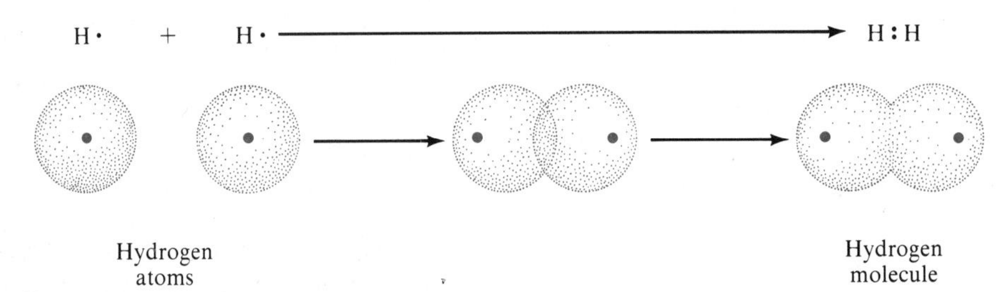


*Figure 7.4.* The formation of a hydrogen molecule from two hydrogen atoms.

example carefully, noting especially the pair of shared electrons that form each covalent bond.

## 7.8 Polar Covalent Bonds

We have considered bonds to be either covalent or electrovalent, according to whether electrons are shared between atoms or are transferred from one atom to another. In most covalent bonds, the pairs of electrons are not shared equally between the atoms. Such bonds are known as polar covalent bonds.

nonpolar covalent bond

In a **nonpolar covalent bond** the shared pair of electrons is attracted equally by the two atoms. A bond between the same kind of atoms, such as that in a hydrogen molecule, is nonpolar because the electronegativity difference between identical atoms is zero.

polar covalent bond

When a covalent bond is formed between two atoms of different electronegativities, the more electronegative atom attracts the shared electron pair toward itself. As a result, the atom with the higher electronegativity acquires a partial negative charge and the other atom, a partial positive charge. However, the overall molecule is still neutral. Due to this greater attraction of the electron pair, the bond formed between the two atoms has partial ionic character and is known as a **polar covalent bond**. The resulting molecule is said to be *polar*.

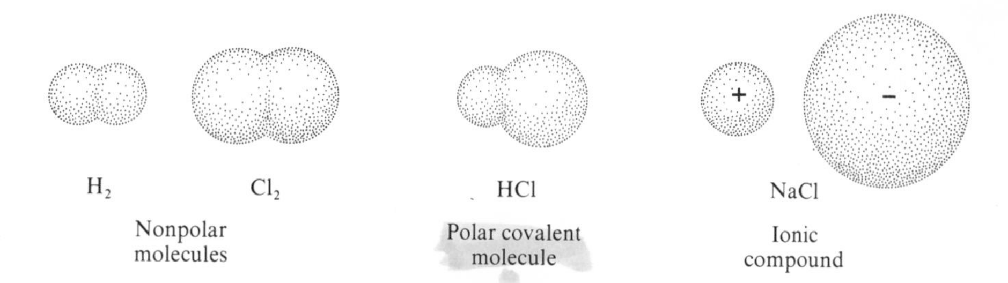


dipole

A **dipole** is a molecule that is electrically unsymmetrical, causing it to be oppositely charged at two points. A dipole is often written as (+ −). A hydrogen chloride molecule is polar and behaves as a small dipole. The HCl dipole may be written as $H \longrightarrow Cl$. The arrow points toward the negative end of the dipole. Molecules of $H_2O$, HBr, and ICl are polar; $CH_4$, $CCl_4$, and $CO_2$ are nonpolar.

$$H \longrightarrow Cl \qquad H \longrightarrow Br \qquad I \longrightarrow Cl \qquad H \nearrow O \nwarrow H$$

The greater the difference in electronegativity between two atoms, the more polar is the bond between them. When this difference is sufficiently large (greater than 1.7–1.9 electronegativity units), the bond between the two atoms will be essentially ionic (with some exceptions, of course). If the difference is less than 1.7 units, the bond will be essentially covalent. The difference in electronegativity between atoms can also give us a guide to the relative strength of covalent bonds. The greater the difference, the stronger the bond—that is, the more energy required to break the bond. For example, HF has the strongest bond in the series HF, HCl, HBr, and HI, as seen by the bond dissociation data in the following table.

| Compound | Electronegativity difference | Bond dissociation energy (kcal/mole) |
|---|---|---|
| HF | 1.9 | 134.6 |
| HCl | 0.9 | 103.2 |
| HBr | 0.7 | 87.5 |
| HI | 0.4 | 71.4 |

Care must be taken to distinguish between polar bonds and polar molecules. A covalent bond between different kinds of atoms is always polar. But a molecule containing different kinds of atoms may or may not be polar, depending on its shape or geometry. The HF, HCl, HBr, HI, and ICl molecules just mentioned are all polar because each contains a single polar bond. However, $CO_2$, $CH_4$, and $CCl_4$ are nonpolar molecules despite the fact that all three contain polar bonds. The carbon dioxide molecule, O=C=O, is nonpolar because the carbon–oxygen dipoles cancel each other by acting in opposite directions.

$$\overset{\longleftarrow\!\!+\;\;+\!\!\longrightarrow}{O=C=O}$$

Dipoles in opposite directions

Methane ($CH_4$) and carbon tetrachloride ($CCl_4$) are nonpolar because the C–H and C–Cl dipoles form tetrahedral angles (see Section 22.3) and thereby cancel one another.

## 7.9 Coordinate Covalent Bonds

In Section 7.7, we saw that a covalent bond was formed by the overlapping of electron orbitals between two atoms. The two atoms each furnish an electron to make a pair that is shared between them.

coordinate covalent bond

Covalent bonds can also be formed by a single atom furnishing both electrons that are shared between the two atoms. The bond so formed is called a **coordinate covalent**, or **semipolar**, **bond**. This bond is often designated by an arrow pointing away from the electron donor (for example, A → B). Once formed, a coordinate covalent bond has the same properties as any other covalent bond—it simply is a pair of electrons shared between two atoms.

The electron-dot structures of sulfurous and sulfuric acids show a coordinate covalent bond between the sulfur and the oxygen atoms that are not bonded to hydrogen atoms. The colored dots indicate the electrons of the sulfur atom.

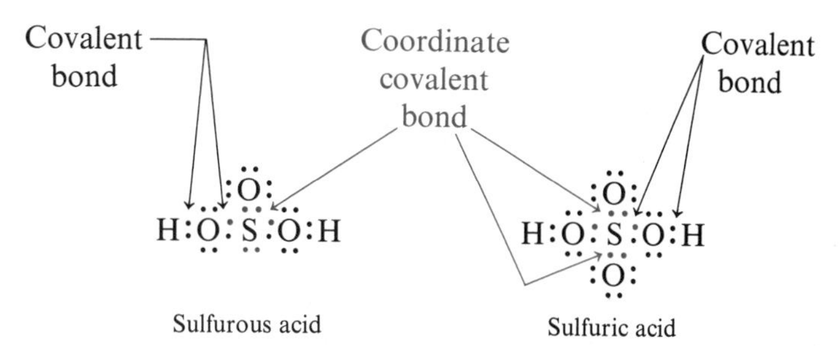


Sulfurous acid    Sulfuric acid

The open (unbonded) pair of electrons on the sulfur atom in sulfurous acid allows room for another oxygen atom with six electrons to fit perfectly into its structure and form sulfuric acid. Other atoms with six electrons in their outer shell, such as sulfur, could also fit into this pattern. The coordinate covalent bond explains the formation of many complex molecules.

## 7.10 *Polyatomic Ions*

polyatomic ion

A **polyatomic ion** is a stable group of atoms that has either a positive or a negative charge and behaves as a single unit in many chemical reactions. Sodium sulfate, $Na_2SO_4$, contains two sodium ions and a sulfate ion. The sulfate ion, $SO_4{}^{2-}$, is a polyatomic ion composed of one sulfur atom and four oxygen atoms, and has a charge of $-2$. One sulfur and four oxygen atoms have a total of 30 electrons in their outer shells. The sulfate ion contains 32 outer-shell electrons and therefore has a charge of $-2$. In this case, the two additional electrons come from the two sodium atoms, which are now sodium ions.

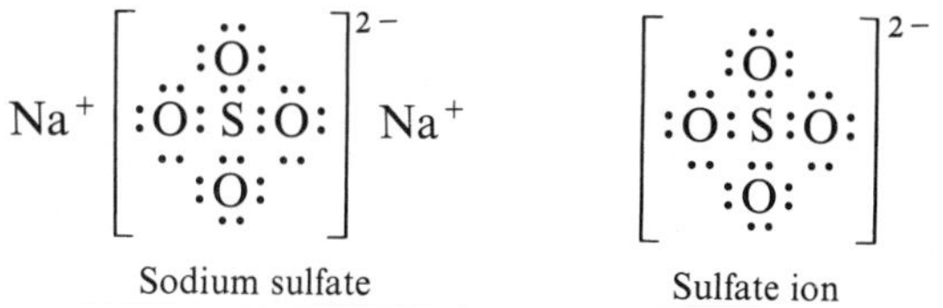

Sodium sulfate    Sulfate ion

Sodium sulfate has both ionic and covalent bonds. Ionic bonds exist between each of the sodium ions and the sulfate ion. Covalent bonds are present between the sulfur and oxygen atoms within the sulfate ion. One important difference between the ionic and covalent bonds in this compound may be demonstrated by dissolving sodium sulfate in water. It dissolves in water, forming three charged particles—two sodium ions and one sulfate ion per formula unit of sodium sulfate:

$$\underset{\text{Sodium sulfate}}{Na_2SO_4} \xrightarrow{\text{Water}} \underset{\text{Sodium ions}}{2\,Na^+} + \underset{\text{Sulfate ion}}{SO_4^{2-}}$$

The ion $SO_4^{2-}$ remains as a unit, held together by covalent bonds; whereas, where there were ionic bonds, dissociation of the ions took place. Do not think, however, that polyatomic ions are so stable that they cannot be altered. They may indeed be changed into other compounds or ions in certain chemical changes.

The electron-dot formulas for several common polyatomic ions are shown below.

```
[   H   ]+     [ :Ö:      ]-     [   :Ö:   ]3-     [:Ö:H]-
[ H:N̈:H ]      [   N::Ö:  ]      [ :Ö:P̈:Ö: ]
[   Ḧ   ]      [  :Ö:     ]      [   :Ö:   ]
```

Ammonium ion, $NH_4^+$ Nitrate ion, $NO_3^-$ Phosphate ion, $PO_4^{3-}$ Hydroxide ion, $OH^-$

## 7.11 Oxidation Numbers of Atoms

oxidation number

We have seen that atoms can combine to form compounds by losing, gaining, or sharing electrons. The **oxidation number** of an element is a number having a positive, a negative, or a zero value that may be assigned to an atom of that element in a compound. These positive and negative numbers are directly related to the positive and negative charges that result from the transfer of electrons from one atom to another in ionic compounds or from an unequal sharing of electrons between atoms forming covalent bonds. Oxidation numbers are assigned by a somewhat arbitrary system of rules. They are useful for writing formulas, naming compounds, and balancing chemical equations.

In a compound having ionic bonds, the oxidation number of an atom or group of atoms existing as an ion is the same as the *charge of the ion*. Thus, in sodium chloride, NaCl, the oxidation number of sodium is $+1$ and that of chlorine is $-1$; in magnesium oxide, MgO, the oxidation number of magnesium is $+2$ and that of oxygen is $-2$; in calcium chloride, $CaCl_2$, the oxidation number of calcium is $+2$ and that of chlorine is $-1$. The sum of the oxidation numbers of all the atoms in a compound is numerically equal to zero, since a compound is electrically neutral.

For practical purposes it is also convenient to assign oxidation numbers to the individual atoms comprising molecules and polyatomic ions. Here the electrons have not been completely transferred from one atom to another and the assignment cannot be done solely on the basis of ionic charges. However, oxidation numbers can be readily assigned to the atoms in either molecules or polyatomic ions by this general method: For each covalent bond, first assign the shared pair of electrons to the more electronegative atom. Then assign an oxidation number to each atom corresponding to its apparent net charge based (1) on the number of electrons gained or lost and (2) on the fact that the sum of the oxidation numbers must equal zero for a compound or must equal the charge on a polyatomic ion. Consider these substances, $H_2$, $H_2O$, $CH_4$, and $CCl_4$:

```
                       H          :C̈l:
H:H     H:Ö:         H:C̈:H    :C̈l:C̈:C̈l:
          H            H          :C̈l:
```

In $H_2$, the pair of electrons is shared equally between the two atoms; therefore, each H is assigned an oxidation number of zero. In $H_2O$, oxygen is the more electronegative atom and is assigned the two pairs of shared electrons. The oxygen atom now has two additional electrons over the neutral atom, and therefore is assigned an oxidation number of $-2$. Each hydrogen atom in $H_2O$ has one less electron than the neutral atom and is assigned an oxidation number of $+1$. In $CH_4$, all four shared pairs of electrons are assigned to the more electronegative carbon atom. The carbon atom then has an additional four electrons and is assigned an oxidation number of $-4$. Each hydrogen atom has one less electron than the neutral atom and is assigned an oxidation number of $+1$. In $CCl_4$, one pair of electrons is assigned to each of the four more electronegative chlorine atoms. The carbon atom therefore has four less electrons than the neutral atom and is assigned an oxidation number of $+4$. Each chlorine atom has one additional electron and is assigned an oxidation number of $-1$.

The following rules govern the assignment of oxidation numbers:

1. The oxidation number of any free element is zero, even when the atoms are combined with themselves. (*Examples*: Na, Mg, $H_2$, $O_2$)
2. Metals generally have positive oxidation numbers in compounds.
3. The oxidation number of hydrogen in a compound or an ion is $+1$ except in metal hydrides.
4. The oxidation number of oxygen in a compound or an ion is $-2$ except in peroxides.
5. The oxidation number of a monatomic ion is the same as the charge on the ion.
6. The oxidation number of an atom in a covalent compound is equal to the net apparent charge on the atom after each pair of shared electrons is assigned to the more electronegative element sharing the pair of electrons.
7. The algebraic sum of the oxidation numbers for all the atoms in a compound must equal zero.
8. The algebraic sum of the oxidation numbers for all the atoms in a polyatomic ion must equal the charge on the ion.

The oxidation numbers of many elements are predictable from their position in the periodic table. This is especially true of the Group A elements because the number of electrons in their outer shells corresponds to the group number. Remember that metals lose electrons, becoming positively charged ions. Nonmetals tend to gain electrons, and become negatively charged ions, but they can also share electrons with other atoms to assume a positive or negative oxidation number. Hydrogen can have a $+1$ or $-1$ oxidation number, depending on the relative electronegativity of the element with which it is combined.

The predictable oxidation numbers of the Group A elements are given in the following table:

| | IA | IIA | IIIA | IVA | VA | VIA | VIIA |
|---|---|---|---|---|---|---|---|
| Oxidation number | $+1$ | $+2$ | $+3$ | $+4$ to $-4$ | $-3$ to $+5$ | $-2$ to $+6$ | $-1$ to $+7$ |

*Table 7.5.* Selected binary hydrogen, oxygen, and chlorine compounds of Group A elements.

| | IA | IIA | IIIA | IVA | VA | VIA | VIIA |
|---|---|---|---|---|---|---|---|
| Hydrogen compound | $NaH$ | $CaH_2$ | $AlH_3$ | $CH_4$ | $NH_3$ | $H_2S$ | $HCl$ |
| Oxygen compound | $Na_2O$ | $CaO$ | $Al_2O_3$ | $CO_2$ | $N_2O_5$ | $SO_3$ | $Cl_2O$ |
| Chlorine compound | $NaCl$ | $CaCl_2$ | $AlCl_3$ | $CCl_4$ | $NCl_3$ | $SCl_2$ | $Cl_2$ |

Table 7.5 illustrates the use of oxidation numbers to predict formulas of binary compounds from representative members of these groups.

## 7.12 Oxidation Number Tables

The writing of formulas of compounds and chemical equations is facilitated by a knowledge of oxidation numbers and ionic charges. Table 7.6 lists the names and ionic charges of common monatomic ions. Monatomic ions of Group A elements are not given in Table 7.6 because the charges and oxidation numbers of these ions are readily determined from the periodic table. The charges and oxidation numbers of the Groups IA, IIA, and IIIA metal ions are positive and correspond to the group number (for example, $Na^+$, $Ca^{2+}$, $Al^{3+}$). The negative charges and oxidation numbers of Groups VA, VIA, and VIIA monatomic ions can be determined by subtracting eight from the group number. Sulfur, for example, is in Group VIA, and $6 - 8 = -2$. Therefore, the oxidation number of the sulfide ion ($S^{2-}$) is $-2$. All the halogens (F, Cl, Br, I) in binary compounds with metals or hydrogen have an oxidation number of $-1$.

*Table 7.6.* Names and charges of selected monatomic ions. Ions of Group A elements are not shown.

| Name | Formula | Charge | Name | Formula | Charge |
|---|---|---|---|---|---|
| Arsenic(III) | $As^{3+}$ | 3+ | Manganese(II) | $Mn^{2+}$ | 2+ |
| Cadmium | $Cd^{2+}$ | 2+ | Mercury(I) | $Hg^+$ | 1+ |
| Chromium(III) | $Cr^{3+}$ | 3+ | Mercury(II) | $Hg^{2+}$ | 2+ |
| Copper(I) | $Cu^+$ | 1+ | Nickel(II) | $Ni^{2+}$ | 2+ |
| Copper(II) | $Cu^{2+}$ | 2+ | Silver | $Ag^+$ | 1+ |
| Iron(II) | $Fe^{2+}$ | 2+ | Tin(II) | $Sn^{2+}$ | 2+ |
| Iron(III) | $Fe^{3+}$ | 3+ | Tin(IV) | $Sn^{4+}$ | 4+ |
| Lead(II) | $Pb^{2+}$ | 2+ | Zinc | $Zn^{2+}$ | 2+ |

The names, formulas, and ionic charges of some common polyatomic ions are given in Table 7.7. A more comprehensive list of both monatomic and polyatomic ions is given on the inside front cover of this book. Table 7.8 lists the principal oxidation numbers of common elements that have variable oxidation states.

*Table 7.7.* Names, formulas, and charges of some common polyatomic ions.

| Name | Formula | Charge | Name | Formula | Charge |
|---|---|---|---|---|---|
| Acetate | $C_2H_3O_2^-$ | 1− | Cyanide | $CN^-$ | 1− |
| Ammonium | $NH_4^+$ | 1+ | Dichromate | $Cr_2O_7^{2-}$ | 2− |
| Arsenate | $AsO_4^{3-}$ | 3− | Hydroxide | $OH^-$ | 1− |
| Bicarbonate | $HCO_3^-$ | 1− | Nitrate | $NO_3^-$ | 1− |
| Bisulfate | $HSO_4^-$ | 1− | Nitrite | $NO_2^-$ | 1− |
| Bromate | $BrO_3^-$ | 1− | Permanganate | $MnO_4^-$ | 1− |
| Carbonate | $CO_3^{2-}$ | 2− | Phosphate | $PO_4^{3-}$ | 3− |
| Chlorate | $ClO_3^-$ | 1− | Sulfate | $SO_4^{2-}$ | 2− |
| Chromate | $CrO_4^{2-}$ | 2− | Sulfite | $SO_3^{2-}$ | 2− |

*Table 7.8.* Principal oxidation numbers of some common elements that have variable oxidation numbers.

| Element | Oxidation number | Element | Oxidation number |
|---|---|---|---|
| Cu | +1, +2 | Cl | −1, +1, +3, +5, +7 |
| Hg | +1, +2 | Br | −1, +1, +3, +5, +7 |
| Sn | +2, +4 | I | −1, +1, +3, +5, +7 |
| Pb | +2, +4 | S | −2, +4, +6 |
| Fe | +2, +3 | N | −3, +1, +2, +3, +4, +5 |
| Au | +1, +3 | P | −3, +3, +5 |
| Ni | +2, +3 | C | −4, +4 |
| Co | +2, +3 | | |
| As | +3, +5 | | |
| Bi | +3, +5 | | |
| Cr | +2, +3, +6 | | |

## 7.13 Formulas of Electrovalent Compounds

The sum of the oxidation numbers of all the atoms in a compound is zero. This statement applies to all substances, regardless of whether they are electrovalently or covalently bonded. For electrovalently bonded compounds the sum of the charges on all the ions in the compound must also be zero. Hence the formulas of ionic (electrovalently bonded) substances can be determined and written readily. Simply combine the ions in the simplest proportion so that the sum of the ionic charges adds up to zero.

To illustrate: Sodium chloride consists of $Na^+$ and $Cl^-$ ions. Since $(1+) + (1-) = 0$, these ions combine in a one-to-one ratio, and the formula is written NaCl. Calcium fluoride is made up of $Ca^{2+}$ and $F^-$ ions; one $Ca^{2+}$ and two $F^-$ ions are needed, so the formula is $CaF_2$. Aluminum oxide is a bit more complicated, because it consists of $Al^{3+}$ and $O^{2-}$ ions. Since 6 is the lowest common multiple of 3 and 2, we have $2(3+) + 3(2-) = 0$; that is, two $Al^{3+}$ ions and three $O^{2-}$ ions are needed; therefore, the formula is $Al_2O_3$.

The foregoing compounds all are made up of monatomic ions. The same procedure is used for polyatomic ions. Consider calcium hydroxide, which is made up of $Ca^{2+}$ and $OH^-$ ions. Since $(2+) + 2(1-) = 0$, one $Ca^{2+}$ and two $OH^-$ ions are needed, the formula is $Ca(OH)_2$. The parentheses are used to indicate that the formula has two hydroxide ions. It is not correct to write $CaO_2H_2$ in place of $Ca(OH)_2$ because the identity of the compound would be lost by so doing. Note that the positive ion is written first in formulas. The following table provides examples of formula writing for ionic compounds (see Tables 7.6 and 7.7 or inside the front cover for formulas of common ions).

*The sum of the charges on the ions of an electrovalently bonded compound must equal zero.*

| Name of compound | Ions | Lowest common multiple | Sum of charges on ions | Formula |
|---|---|---|---|---|
| Sodium bromide | $Na^+$, $Br^-$ | 1 | $(1+) + (1-) = 0$ | $NaBr$ |
| Potassium sulfide | $K^+$, $S^{2-}$ | 2 | $2(1+) + (2-) = 0$ | $K_2S$ |
| Zinc sulfate | $Zn^{2+}$, $SO_4^{2-}$ | 2 | $(2+) + (2-) = 0$ | $ZnSO_4$ |
| Ammonium phosphate | $NH_4^+$, $PO_4^{3-}$ | 3 | $3(1+) + (3-) = 0$ | $(NH_4)_3PO_4$ |
| Aluminum chromate | $Al^{3+}$, $CrO_4^{2-}$ | 6 | $2(3+) + 3(2-) = 0$ | $Al_2(CrO_4)_3$ |

It is not always easy to distinguish between the terms *oxidation number* and *charge*. Oxidation numbers are assigned to atoms according to a set of rules. The charge on an ion is the actual electron excess or deficiency when compared to the neutral atom—or group of atoms in the case of a polyatomic ion. Oxidation numbers may be assigned to all the atoms, including monatomic ions, in any compound; but only ions have charges. There is no problem with covalently bonded molecules such as methane ($CH_4$) because they contain no ions. The oxidation number of the hydrogen is $+1$, and that of carbon is $-4$. Electrovalently bonded compounds are apt to be troublesome because the oxidation number and the charge on a monatomic ion have the same numerical value. In sodium chloride (NaCl), the oxidation number of the sodium ion is $+1$ and the charge is $1+$; the oxidation number of the chloride ion is $-1$ and the charge is $1-$.

In a compound composed of monatomic ions, such as sodium chloride, there is no practical difference in writing the formula regardless of whether oxidation numbers or ionic charges are used. But there is a difference with compounds containing polyatomic ions. Sodium sulfate ($Na_2SO_4$), for example, consists of two sodium ions ($Na^+$) and a polyatomic sulfate ion ($SO_4^{2-}$). The sum of the ionic charges is zero: $2(1+) + (2-) = 0$; and it is convenient to make use of this fact in writing the formula. The charge on the polyatomic sulfate ion is not an oxidation number. However, the sum of the oxidation numbers of all the atoms in sodium sulfate is zero: $2(Na^+) + (S^{6+}) + 4(O^{2-}) = 0$; and the sum of the oxidation numbers of all the atoms in the sulfate ion is $-2$: $(S^{6+}) + 4(O^{2-}) = -2$ (see Section 7.14).

*Problem 7.1* Write formulas for (a) calcium chloride; (b) iron(III) sulfide; (c) aluminum sulfate. Refer to Tables 7.6 and 7.7 as needed.

(a) *Step 1.* From the name we know that calcium chloride is composed of calcium and chloride ions. First write down the formulas of these ions.

$Ca^{2+}$ and $Cl^-$

*Step 2.* To write the formula of the compound, combine the smallest numbers of $Ca^{2+}$ and $Cl^-$ ions to give a charge sum equal to zero. In this case, the lowest common multiple of the charges is 2:

$$(Ca^{2+}) + 2(Cl^-) = 0$$
$$(2+) + 2(1-) = 0$$

Therefore, the formula is $CaCl_2$.

(b) Use the same procedure for iron(III) sulfide.

*Step 1.* Write down the formulas for the iron(III) and sulfide ions.

$Fe^{3+}$ and $S^{2-}$

*Step 2.* Use the smallest numbers of these ions required to give a charge sum equal to zero. The lowest common multiple of the charges is 6:

$$2(Fe^{3+}) + 3(S^{2-}) = 0$$
$$2(3+) + 3(2-) = 0$$

Therefore, the formula is $Fe_2S_3$.

(c) Use the same procedure for aluminum sulfate.

*Step 1.* Write down the formulas for the aluminum and sulfate ions.

$Al^{3+}$ and $SO_4^{2-}$

*Step 2.* Use the smallest numbers of these ions required to given a charge sum equal to zero. The lowest common multiple of the charges is 6:

$$2(Al^{3+}) + 3(SO_4^{2-}) = 0$$
$$2(3+) + 3(2-) = 0$$

Therefore, the formula is $Al_2(SO_4)_3$. Note the use of parentheses around the $SO_4^{2-}$ ion.

---

## 7.14 Determining Oxidation Numbers and Ionic Charges from a Formula

If the formula of a compound is known, the oxidation number of an element or the charge on a polyatomic ion in the formula can often be determined by algebraic difference. To begin, you must know the oxidation numbers of a few elements. Excellent ones with which to work are hydrogen, $H^+$, always a $+1$ except in hydrides (a hydride is a compound of hydrogen combined with a metal); oxygen, $O^{2-}$, always a $-2$ except in peroxides; and sodium, $Na^+$, always a $+1$. Using the compound sulfuric acid, $H_2SO_4$, as an example, let us determine the charge of the sulfate ion and oxidation number of the sulfur atom. The sulfate ion is combined with two hydrogen atoms, each with a +1 oxidation number. The sulfate ion must then have a −2 charge in order for the net charge in the compound to be zero:

| | |
|---|---|
| $H^+$ | +1 |
| $H^+$ | +1 |
| $SO_4^{2-}$ | −2 |
| | 0 |

To find the oxidation number of sulfur, we proceed as follows:

*Step 1.* Write the oxidation number of a single atom of hydrogen and a single atom of oxygen below the atoms in the formula.

*Step 2.* Below this, write the sums of the oxidation numbers of all the H and O atoms: $2(+1) = +2$ and $4(-2) = -8$.

*Step 3.* Then, add together the total oxidation numbers of all the atoms, including the sulfur atom, and set them equal to zero: $+2 + S + (-8) = 0$. Solving the equation for S, we determine that the oxidation number of sulfur is +6, the value needed to give the sum of zero.

| | $H_2$ | S | $O_4$ | |
|---|---|---|---|---|
| *Step 1.* | +1 | | −2 | (1) |
| *Step 2.* | $2(+1) = +2$ | | $4(-2) = -8$ | (2) |
| *Step 3.* | $+2 + S + (-8) = 0$ | | | (3) |
| | | $S = +6$ | | |

The oxidation number of sulfur in $H_2SO_4$ is +6.

What is the oxidation number of chromium in sodium dichromate, $Na_2Cr_2O_7$? Using the same method as for $H_2SO_4$, we have

| | $Na_2$ | $Cr_2$ | $O_7$ |
|---|---|---|---|
| *Step 1.* | +1 | | −2 |
| *Step 2.* | $2(+1) = +2$ | | $7(-2) = -14$ |
| *Step 3.* | $+2 + 2Cr + (-14) = 0$ | | |
| | | $2Cr = +12$ | |
| | | $Cr = +6$ | |

The oxidation number of chromium in $Na_2Cr_2O_7$ is +6.

The formula of radium chloride is $RaCl_2$. What is the oxidation number of radium? If you remember that the oxidation number of chloride is −1, then the value for radium is +2, since one radium ion is combined with two $Cl^-$ ions. If you do not remember the oxidation number of chloride, then you should try to recall the formula of another chloride. One that might come to mind is sodium chloride, NaCl, in which the chloride is −1 because of its combination with one sodium ion of +1. This recollection establishes the oxidation number of chloride, which then enables you to calculate the value for radium.

What is the oxidation number of phosphorus in the phosphate ion, $PO_4^{3-}$? First of all, note that this is a polyatomic ion with a charge of −3. The sum of the oxidation numbers of phosphorus and oxygen must equal −3 and not zero.

Four oxygen atoms, each with a $-2$, give a total of $-8$. The oxidation number of the phosphorus atom must then be $+5$:

$$\mathrm{P\ O_4^{3-}}$$
$$-2$$
$$P + 4(-2) = -3$$
$$P = +5$$

*The sum of the oxidation numbers of the atoms in a polyatomic must equal the charge of the polyatomic ion.*

## 7.15 Oxidation–Reduction

Magnesium burns brilliantly in air, forming magnesium oxide:

$$2\ Mg^0 + O_2^0 \longrightarrow 2\ Mg^{2+}O^{2-}$$

In this reaction, two electrons are transferred from each magnesium atom to the oxygen atoms, resulting in an increase in the oxidation number of magnesium from 0 to $+2$. At the same time, the oxidation number of oxygen has decreased from 0 to $-2$. Oxidation and reduction have occurred in this reaction. The $Mg^0$ was oxidized and the $O_2^0$ was reduced. **Oxidation** is defined as an increase in the oxidation number or oxidation state of an element. **Reduction** is defined as a decrease in the oxidation number or oxidation state of an element. *Oxidation–reduction* occurs as a result of a loss and gain of electrons. Oxidation and reduction occur simultaneously. The element that loses electrons (increases in oxidation number) is oxidized, and the element that gains electrons (decreases in oxidation number) is reduced.

oxidation

reduction

$$2\ Hg^{2+}O^{2-} \xrightarrow{\Delta} 2\ Hg^0 + O_2^0$$

In the decomposition of mercury(II) oxide, the oxidation number of mercury(II) changes from $+2$ to 0; the oxidation number of oxygen changes from $-2$ to 0. Therefore, oxidation–reduction occurs; oxygen ($O^{2-}$) is oxidized and mercury ($Hg^{2+}$) is reduced.

The process of oxidation and reduction involves the transfer of electrons and results in changes of oxidation numbers. The element oxygen is not necessarily involved in this process. For example, in the chemical reaction between sodium and chlorine to form sodium chloride, we saw that electrons were transferred from sodium atoms to chlorine atoms:

$$2\ Na^0 + Cl_2^0 \longrightarrow 2\ Na^+Cl^-$$

In this reaction, the oxidation number of sodium increases from 0 to $+1$, and therefore sodium is oxidized. The oxidation number of chlorine decreases from 0 to $-1$, and consequently, chlorine is reduced.

A more detailed discussion of oxidation–reduction is given in Chapter 17.

## Questions

*A. Review the meanings of the new terms introduced in this chapter.*

1. Ionization energy
2. Valence electrons
3. Electronegativity
4. Chemical bond
5. Electrovalent bond
6. Ionic bond
7. Covalent bond
8. Nonpolar covalent bond
9. Polar covalent bond
10. Dipole
11. Coordinate covalent bond
12. Polyatomic ion
13. Oxidation number
14. Ionic charge
15. Oxidation
16. Reduction

*B. Answers to the following questions will be found in tables and figures.*

1. Explain the large increase in ionization energy required to remove the second electron from a lithium atom compared to the first electron removed. (See Table 7.1.)
2. Arrange the following elements in the order of increasing attraction by which their valence electrons are held in the atom: aluminum, sulfur, silicon, magnesium, chlorine, phosphorus, argon, and sodium.
3. In which general areas of the periodic table are the elements with the lowest and the highest ionization energies and electronegativities located?
4. Using the table of electronegativities (Table 7.4), indicate which element is positive and which is negative in the following compounds:
   (a) $MgH_2$
   (b) NaCl
   (c) $CO_2$
   (d) $PCl_3$
   (e) $NH_3$
   (f) IC1
   (g) $Br_2$
   (h) $Cl_2O$
   (i) $OF_2$
   (j) $NO_2$
   (k) $CF_4$
   (l) KBr
   (m) CuS
5. Classify the bond between the following pairs of atoms as either principally ionic or covalent (use Table 7.4):
   (a) Phosphorus and hydrogen
   (b) Sodium and fluorine
   (c) Chlorine and carbon
   (d) Hydrogen and chlorine
   (e) Magnesium and oxygen
   (f) Hydrogen and sulfur
6. Using the principle employed in Table 7.5, write formulas for:
   (a) The hydrogen compounds of Li, Ca, Sb, Br, and S
   (b) The oxygen compounds of K, Si, N, Sr, and Ga
   (c) The bromine compounds of Na, Mg, Al, C, P, and S
7. Use the oxidation number tables and determine the formulas for compounds composed of the following ions:
   (a) Hydrogen and cyanide
   (b) Silver and nitrate
   (c) Potassium and carbonate
   (d) Mercury(I) and oxide
   (e) Aluminum and acetate
   (f) Iron(III) and sulfate

*C. Review questions.*

1. Write an equation representing the change of a fluorine atom to a fluoride ion ($F^-$).
2. Write an equation representing the change of a calcium atom to a calcium ion ($Ca^{2+}$).
3. Why does barium (Ba) have a lower ionization energy than beryllium (Be)?
4. Why is there such a large increase in the ionization energy required to remove the second electron from a sodium atom?

5. How many electrons must be gained or lost for each of the following to achieve a noble gas electron structure?
   (a) A calcium atom (d) A chloride ion
   (b) A sulfur atom (e) A nitrogen atom
   (c) A helium atom
6. Explain why potassium forms a $K^+$ ion but not a $K^{2+}$ ion.
7. What portion of an atom is represented by the kernel?
8. Why does an aluminum ion have a +3 charge?
9. Which would be larger, a potassium ion or a potassium atom? Explain.
10. Which would be smaller, a bromine atom or a bromide ion? Explain.
11. Which would be larger, a magnesium ion or an aluminum ion? Explain.
12. What causes a bond to be polar?
13. How does a coordinate covalent bond differ from an ordinary covalent bond?
14. How does a covalent bond differ from an electrovalent bond?
15. Draw electron-dot structures for:
    (a) Mg (c) $H_2O$ (e) $CO_2$ (g) $H_2S$ (i) $CaF_2$
    (b) $F_2$ (d) $NH_3$ (f) HCl (h) MgO (j) $ZnI_2$
16. Draw electron-dot structures for:
    (a) $Mg^{2+}$ (c) $Cl^-$ (e) $SO_4^{2-}$ (g) $CN^-$ (i) $ClO_3^-$
    (b) $Al^{3+}$ (d) $S^{2-}$ (f) $SO_3^{2-}$ (h) $CO_3^{2-}$ (j) $NO_3$
17. The electron-dot structure for chloric acid is

```
  ..  ..  ..
H:O:Cl:O:
  ..  ..  ..
     :O:
      ..
```

    Point out the covalent and coordinate covalent bonds in this structure.
18. Draw the electron-dot structure for ammonia ($NH_3$). What type of bonds are present? Can this molecule form coordinate covalent bonds? How many?
19. Classify the following molecules as polar or nonpolar:
    (a) $NH_3$ (b) HBr (c) $CF_4$ (d) $F_2$ (e) $CO_2$
20. Is it possible for a molecule to be nonpolar even though it contains polar covalent bonds? Explain.
21. Determine the oxidation number of each element in the following:
    (a) HBr (c) MgO (e) $F_2$ (g) $H_2SO_4$ (i) $K_2CrO_4$
    (b) $NH_3$ (d) $FeCl_2$ (f) $SiH_4$ (h) $KNO_3$ (j) $C_2H_6$
22. Determine the oxidation number of the element underlined in each formula:
    (a) $Ba\underline{S}O_4$ (c) $Li\underline{H}$ (e) $\underline{As}_2O_5$ (g) $\underline{Cu}Cl_2$ (i) $\underline{C}O$
    (b) $\underline{Cl}_2O_7$ (d) $H\underline{N}O_3$ (f) $\underline{Na}_2O$ (h) $Na_2\underline{Cr}_2O_7$ (j) $\underline{HC}_2H_3O_2$
23. Write the formula of the compound that would be formed between the given elements:
    (a) Na and I (d) H and N (g) Ca and P (i) C and F
    (b) Ba and Br (e) Rb and Cl (h) Li and Se (j) I and Cl
    (c) H and S (f) Al and S
24. Write the formula of the compound formed from the given ions:
    (a) $Na^+$ and $F^-$ (f) $PO_4^{3-}$ and $Fe^{3+}$
    (b) $Mg^{2+}$ and $O^{2-}$ (g) $NH_4^+$ and $S^{2-}$
    (c) $Cl^-$ and $Al^{3+}$ (h) $Zn^{2+}$ and $AsO_4^{3-}$
    (d) $Fe^{2+}$ and $OH^-$ (i) $NO_3^-$ and $Ni^{2+}$
    (e) $SO_4^{2-}$ and $K^+$ (j) $NH_4^+$ and $CrO_4^{2-}$
25. Which of these statements are correct? (Try to answer this question without referring to your book.)

(a) A chlorine atom has less electrons than a chloride ion.
(b) The noble gases have a tendency to lose one electron to become a positively charged ion.
(c) The chemical bonds in a water molecule are ionic.
(d) The chemical bonds in a water molecule are polar.
(e) Valence electrons are those electrons in the highest occupied energy level, **n**, of an atom.
(f) An atom with eight electrons in its outer shell has all its *s* and *p* orbitals filled.
(g) Fluorine has the lowest electronegativity of all the elements.
(h) A neutral atom with eight electrons in its valence shell must be an atom of a noble gas.
(i) A nitrogen atom has four valence electrons.
(j) An aluminum atom must lose three electrons to become an aluminum ion, $Al^{3+}$.
(k) In an ethylene molecule, $C_2H_4$,

```
H       H
 \     /
  C=C
 /     \
H       H
```

two pairs of electrons are shared between the carbon atoms.
(l) The octet rule is mainly useful for atoms where only *s* and *p* electrons enter into bonding.
(m) When electrons are transferred from one atom to another, the resulting compound contains ionic bonds.
(n) A phosphorus atom, $\cdot\ddot{\text{P}}\cdot$ (with one dot below), needs three additional electrons to attain a stable octet of electrons.
(o) The simplest compound between oxygen, $\cdot\ddot{\underset{\cdot\cdot}{\text{O}}}\cdot$, and fluorine, $:\ddot{\underset{\cdot\cdot}{\text{F}}}\cdot$, atoms is $FO_2$.
(p) In the molecule $\text{H}:\ddot{\underset{\cdot\cdot}{\text{Cl}}}:$, there are three unshared pairs of electrons.
(q) The bonds in a water molecule are formed by overlapping of *s* and *p* electron orbitals.
(r) The smaller the difference in electronegativity between two atoms, the more ionic will be the bond between them.
(s) In the reaction $2\ H_2O \rightarrow 2\ H_2 + O_2$, hydrogen is oxidized and oxygen is reduced.
(t) Oxidation occurs when an atom loses electrons.
(u) In the oxide $WO_3$, the oxidation number of tungsten is $+6$ and that of oxygen is $-2$.

*D. Review exercises.*

1. (a) In terms of electron structure, why is the oxidation number of nitrogen never higher than $+5$ or lower than $-3$?
   (b) What are the highest and lowest possible oxidation states for bromine?
2. Why do chemical bonds form?

# 8 Nomenclature of Inorganic Compounds

*After studying Chapter 8 you should be able to:*

1. Give the name or formula for inorganic binary compounds in which the metal has only one common oxidation state.
2. Give the name or formula for inorganic binary compounds that contain metals of variable oxidation state, using either the Stock System or classical nomenclature.
3. Give the name or formula for inorganic binary compounds that contain two nonmetals.
4. Give the name or formula for binary acids.
5. Give the name or formula for ternary inorganic acids.
6. Give the name or formula for ternary salts.
7. Given the formula of a salt, write the name and formula of the acid from which the salt may be derived.
8. Give the name or formula for salts that contain more than one positive ion.
9. Give the name or formula for inorganic bases.

## 8.1 Common, or Trivial, Names

Chemical nomenclature is the system of names that chemists use to identify compounds. When a new substance is formulated, it must be named in order to distinguish it from all other substances. Before chemistry was systematized, a substance was given a name that generally associated it with one of its outstanding physical or chemical properties. For example, *quicksilver* is a common name for mercury; it describes two properties of mercury—a silvery appearance and quick, liquidlike movement. Nitrous oxide, $N_2O$, used as an anesthetic in dentistry, has been called *laughing gas*, because it induces laughter when inhaled. The name *nitrous oxide* is now giving way to the more systematic name *dinitrogen oxide*. Nonsystematic names are called *common*, or *trivial*, names.

Common names for chemicals are widely used in many industries, since the systematic name frequently is too long or too technical for everyday use.

For example, CaO is called *lime*, not *calcium oxide*, by plasterers; photographers refer to $Na_2S_2O_3$ as *hypo*, rather than *sodium thiosulfate*; gardeners refer to $CCl_3CH(C_6H_4Cl)_2$ by the abbreviation *DDT*, not as *dichlorodiphenyltrichloroethane*. These common names are chemical nicknames, and, as the DDT example shows, there is a practical need for short, common names. Table 8.1 lists the common names, formulas, and chemical names of some familiar substances.

*Table 8.1.* Common names, formulas, and chemical names of some familiar substances.

| Common name | Formula | Chemical name |
|---|---|---|
| Acetylene | $C_2H_2$ | Ethyne |
| Lime | $CaO$ | Calcium oxide |
| Slaked lime | $Ca(OH)_2$ | Calcium hydroxide |
| Water | $H_2O$ | Water |
| Galena | $PbS$ | Lead(II) sulfide |
| Alumina | $Al_2O_3$ | Aluminum oxide |
| Baking soda | $NaHCO_3$ | Sodium hydrogen carbonate |
| Cane or beet sugar | $C_{12}H_{22}O_{11}$ | Sucrose |
| Blue stone, blue vitriol | $CuSO_4 \cdot 5H_2O$ | Copper(II) sulfate pentahydrate |
| Borax | $Na_2B_4O_7 \cdot 10H_2O$ | Sodium tetraborate decahydrate |
| Brimstone | $S$ | Sulfur |
| Calcite, marble, limestone | $CaCO_3$ | Calcium carbonate |
| Cream of tartar | $KHC_4H_4O_6$ | Potassium hydrogen tartrate |
| Epsom salts | $MgSO_4 \cdot 7H_2O$ | Magnesium sulfate heptahydrate |
| Gypsum | $CaSO_4 \cdot 2H_2O$ | Calcium sulfate dihydrate |
| Grain alcohol | $C_2H_5OH$ | Ethyl alcohol, ethanol |
| Hypo | $Na_2S_2O_3$ | Sodium thiosulfate |
| Laughing gas | $N_2O$ | Dinitrogen oxide |
| Litharge | $PbO$ | Lead(II) oxide |
| Lye, caustic soda | $NaOH$ | Sodium hydroxide |
| Milk of magnesia | $Mg(OH)_2$ | Magnesium hydroxide |
| Muriatic acid | $HCl$ | Hydrochloric acid |
| Oil of vitriol | $H_2SO_4$ | Sulfuric acid |
| Plaster of paris | $CaSO_4 \cdot \frac{1}{2}H_2O$ | Calcium sulfate hemihydrate |
| Potash | $K_2CO_3$ | Potassium carbonate |
| Pyrites (fool's gold) | $FeS_2$ | Iron disulfide |
| Quicksilver | $Hg$ | Mercury |
| Sal ammoniac | $NH_4Cl$ | Ammonium chloride |
| Saltpeter | $NaNO_3$ | Sodium nitrate |
| Table salt | $NaCl$ | Sodium chloride |
| Washing soda | $Na_2CO_3 \cdot 10H_2O$ | Sodium carbonate decahydrate |
| Wood alcohol | $CH_3OH$ | Methyl alcohol, methanol |

## 8.2 Systematic Chemical Nomenclature

The trivial name is not entirely satisfactory to the chemist, who requires a name that will identify precisely the composition of each substance. Therefore, as the number of known compounds increased, it became more and more necessary to develop a scientific, systematic method of identifying compounds by name. The systematic method of naming inorganic compounds considers the compound to be composed of two parts, one positive and one negative. The positive part, which is either a metal, hydrogen or other positively charged group, is named and written first. The negative part, generally nonmetallic, follows. The names of the elements are modified with suffixes and prefixes to identify the different types or classes of compounds. Thus, the compound composed of sodium ions and chloride ions is named sodium chloride; the compound composed of calcium ions and bromide ions is named calcium bromide; the compound composed of iron(II) ions and chloride ions is named iron(II) chloride (read as "iron-two chloride").

We will consider the naming of acids, bases, salts, and oxides. Refer to Tables 7.6, 7.7, and 7.8 for the names, formulas, and oxidation numbers of ions. For handy, quick reference, the names and formulas of some common ions are given on the inside front cover of this book.

## 8.3 Binary Compounds

Binary compounds contain only two different elements. Their names consist of two parts: the name of the more electropositive element followed by the name of the electronegative element, which is modified to end in *ide*. [The names of nonbinary compounds that use the *ide* ending but are exceptions to the rule are discussed in part (d) of this section.]

(a) Binary compounds in which the electropositive element has a fixed oxidation state. The majority of these compounds contain a metal and a nonmetal. The chemical name is composed of the name of the metal, which is written first, followed by the name of the nonmetal, which has been modified to an identifying stem plus the suffix *ide*. For example, sodium chloride, NaCl, is composed of one atom each of sodium and chlorine. The name of the metal, sodium, is written first and is not modified. The second part of the name is derived from the nonmetal, chlorine, by using the stem *chlor* and adding the ending *ide*; it is named *chloride*. The compound name is sodium chloride.

**NaCl**

| | |
|---|---|
| Elements: | Sodium (metal) |
| | Chlorine (nonmetal) |
| | name modified to the stem *chlor* + *ide* |
| Name of compound: | Sodium chloride |

Stems of the more common negative-ion forming elements are shown in the following table.

| Symbol | Element | Stem | Binary name ending |
|---|---|---|---|
| B | Boron | Bor | Boride |
| Br | Bromine | Brom | Bromide |
| Cl | Chlorine | Chlor | Chloride |
| F | Fluorine | Fluor | Fluoride |
| H | Hydrogen | Hydr | Hydride |
| I | Iodine | Iod | Iodide |
| N | Nitrogen | Nitr | Nitride |
| O | Oxygen | Ox | Oxide |
| P | Phosphorus | Phosph | Phosphide |
| S | Sulfur | Sulf | Sulfide |

Compounds may contain more than one atom of the same element, but as long as they contain only two different elements and if only one compound of these two elements exists, the name follows the rule for binary compounds:

*Examples:* $CaBr_2$ Calcium bromide; $Mg_3N_2$ Magnesium nitride; KI Potassium iodide

Table 8.2 shows more examples of compounds with names ending in *ide.*

*Table 8.2.* Examples of compounds with names ending in *ide*.

| Formula | Name |
|---|---|
| $MgBr_2$ | Magnesium bromide |
| $Na_2O$ | Sodium oxide |
| NaH | Sodium hydride |
| HCl | Hydrogen chloride |
| HI | Hydrogen iodide |
| $CaC_2$ | Calcium carbide |
| $AlCl_3$ | Aluminum chloride |
| PbS | Lead(II) sulfide |
| LiI | Lithium iodide |
| $Al_2O_3$ | Aluminum oxide |

(b) Binary compounds containing metals of variable oxidation numbers. Two systems are commonly used for compounds in this category. The official system, designated by the International Union of Pure and Applied Chemistry (IUPAC), is known as the *Stock System*. In the Stock System, when a compound contains a metal that can have more than one oxidation number, the oxidation number of the metal in the compound is designated by a roman numeral written immediately after the name of the metal. The negative element is treated in the usual manner for binary compounds.

*Examples:*

| | | |
|---|---|---|
| $FeCl_2$ | Iron(II) chloride | $Fe^{2+}$ |
| $FeCl_3$ | Iron(III) chloride | $Fe^{3+}$ |
| CuCl | Copper(I) chloride | $Cu^{+}$ |
| $CuCl_2$ | Copper(II) chloride | $Cu^{2+}$ |

When a metal has only one possible oxidation state, there is no need to distinguish one oxidation state from another, so roman numerals are not needed. Thus, we do not say calcium(II) chloride for $CaCl_2$, but rather calcium chloride, since the oxidation number of calcium is understood to be $+2$.

In classical nomenclature, when the metallic ion has only two oxidation numbers, the name of the metal is modified with the suffixes *ous* and *ic* to distinguish between the two. The lower oxidation state is given the *ous* ending and the higher one, the *ic* ending.

*Examples:*

| | | |
|---|---|---|
| $FeCl_2$ | Ferrous chloride | $Fe^{2+}$ |
| $FeCl_3$ | Ferric chloride | $Fe^{3+}$ |
| CuCl | Cuprous chloride | $Cu^{+}$ |
| $CuCl_2$ | Cupric chloride | $Cu^{2+}$ |

Table 8.3 lists some common metals with more than one oxidation number.

*Table 8.3.* Names and oxidation numbers of some common metal ions that have more than one oxidation number.

| Formula | Stock System name | Classical name |
|---|---|---|
| $Cu^{1+}(Cu_2)^{2+}$ | Copper(I) | Cuprous |
| $Cu^{2+}$ | Copper(II) | Cupric |
| $Hg^{1+}(Hg_2)^{2+}$ | Mercury(I) | Mercurous |
| $Hg^{2+}$ | Mercury(II) | Mercuric |
| $Fe^{2+}$ | Iron(II) | Ferrous |
| $Fe^{3+}$ | Iron(III) | Ferric |
| $Sn^{2+}$ | Tin(II) | Stannous |
| $Sn^{4+}$ | Tin(IV) | Stannic |
| $As^{3+}$ | Arsenic(III) | Arsenous |
| $As^{5+}$ | Arsenic(V) | Arsenic |
| $Sb^{3+}$ | Antimony(III) | Stibnous |
| $Sb^{5+}$ | Antimony(V) | Stibnic |
| $Ti^{3+}$ | Titanium(III) | Titanous |
| $Ti^{4+}$ | Titanium(IV) | Titanic |

Notice that the *ous–ic* naming system does not give the oxidation state of an element but merely indicates that at least two oxidation states exist. The Stock System avoids any possible uncertainty by clearly stating the oxidation number.

(c) Binary compounds containing two nonmetals. The chemical bond that exists between two nonmetals is predominantly covalent. In a covalent compound, positive and negative oxidation numbers are assigned to the elements according to their electronegativities. The most electropositive element is named first. In a compound between two nonmetals, the element that occurs earlier in the following sequence is written and named first.

$B_2$, Si, C, P, N, H, S, I, Br, Cl, O, F.

A Latin or Greek prefix is attached to each element to indicate the number of atoms of that element in the molecule. The second element still retains the modified binary ending. The prefix *mono* is generally omitted except when needed to distinguish between two or more compounds, such as carbon monoxide, CO, and carbon dioxide, $CO_2$. Some common prefixes and their numerical equivalences are the following:

| | |
|---|---|
| Mono = 1 | Hexa = 6 |
| Di = 2 | Hepta = 7 |
| Tri = 3 | Octa = 8 |
| Tetra = 4 | Nona = 9 |
| Penta = 5 | Deca = 10 |

Here are some examples of compounds that illustrate this system:

| | |
|---|---|
| CO | Carbon monoxide |
| $CO_2$ | Carbon dioxide |
| $PCl_3$ | Phosphorus trichloride |
| $PCl_5$ | Phosphorus pentachloride |
| $P_2O_5$ | Diphosphorus pentoxide |
| $CCl_4$ | Carbon tetrachloride |
| $N_2O$ | Dinitrogen oxide |
| $S_2Cl_2$ | Disulfur dichloride |
| $N_2O_4$ | Dinitrogen tetroxide |
| NO | Nitrogen oxide |
| $N_2O_3$ | Dinitrogen trioxide |

(d) Exceptions that use *ide* endings. Three notable exceptions that use the *ide* ending are hydroxides ($OH^-$), cyanides ($CN^-$), and ammonium ($NH_4^+$) compounds. These polyatomic ions, when combined with another element, take the ending *ide*, even though more than two elements are present in the compound.

| | |
|---|---|
| $NH_4I$ | Ammonium iodide |
| $Ca(OH)_2$ | Calcium hydroxide |
| KCN | Potassium cyanide |

(e) Acids derived from binary compounds. Certain binary hydrogen compounds, when dissolved in water, form solutions that have **acid** properties. Because of this property, these compounds are given acid names in addition to their regular *ide* names. For example, HCl is a gas and is called *hydrogen chloride*, but its water solution is known as *hydrochloric acid*. Binary acids are composed of hydrogen and one other nonmetallic element. However, not all binary hydrogen compounds are acids. To express the formula of a binary acid, it is customary to write the symbol of hydrogen first, followed by the symbol of the second element (for example, HCl, HBr, $H_2S$). When we see formulas such as $CH_4$ or $NH_3$, we understand that these compounds are not normally considered to be acids.

To name a binary acid, place the prefix *hydro* in front of, and the suffix *ic* after, the stem of the nonmetal. Then add the word *acid*.

| | HCl | $H_2S$ |
|---|---|---|
| *Examples:* | *Hydro* chlor/ic acid<br>(hydrochloric acid) | *Hydro* sulfur/ic acid<br>(hydrosulfuric acid) |

Acids are hydrogen-containing substances that liberate hydrogen ions when dissolved in water. The same formula is often used to express binary hydrogen compounds such as HCl, regardless of whether they are dissolved in water. Table 8.4 shows examples of other binary acids.

*Table 8.4.* Names and formulas of selected binary acids.

| Formula | Acid name |
|---|---|
| HF | Hydrofluoric acid |
| HCl | Hydrochloric acid |
| HBr | Hydrobromic acid |
| HI | Hydriodic acid |
| $H_2S$ | Hydrosulfuric acid |
| $H_2Se$ | Hydroselenic acid |

## 8.4 Ternary Compounds

Ternary compounds contain three elements: an electropositive group, which is either a metal or hydrogen, combined with a polyatomic negative ion. We will consider the naming of compounds in which one of the three elements is oxygen.

In general, in naming ternary compounds the positive group is given first, followed by the name of the negative ion. Rules for naming the positive groups are identical to those used in naming binary compounds. The negative group usually contains two elements: oxygen and a metal or a nonmetal. To name the polyatomic negative ion, add the endings *ite* or *ate* to the stem of the element other than oxygen. Thus, $SO_4^{2-}$ is called *sulfate*, and $SO_3^{2-}$ is called *sulfite*. Note that oxygen is not specifically included in the name, but is understood to be present when the endings *ite* and *ate* are used. The suffixes *ite* and *ate* repre-

| *ite* | *ate* |
|---|---|
| +4<br>$CaSO_3$<br>Calcium sulfite<br>*ite* ending indicates lower oxidation state of sulfur | +6<br>$CaSO_4$<br>Calcium sulfate<br>*ate* ending indicates higher oxidation state of sulfur |

sent different oxidation states of the element other than oxygen in the polyatomic ion. The *ite* ending represents the lower and the *ate* the higher oxidation state. When an element has only one oxidation state, such as C in carbonate,

*Table 8.5.* Names and formulas of selected ternary compounds.

| Formula | Name |
|---|---|
| $Na_2SO_3$ | Sodium sulfite |
| $Na_2SO_4$ | Sodium sulfate |
| $K_2CO_3$ | Potassium carbonate |
| $CaCO_3$ | Calcium carbonate |
| $Al_2(SO_4)_3$ | Aluminum sulfate |
| $KClO_3$ | Potassium chlorate |
| $AlPO_4$ | Aluminum phosphate |
| $FeSO_4$ | Iron(II) sulfate or ferrous sulfate |
| $Fe_2(SO_4)_3$ | Iron(III) sulfate or ferric sulfate |
| $PbCrO_4$ | Lead(II) chromate |
| $H_2SO_4$ | Hydrogen sulfate |
| $HNO_3$ | Hydrogen nitrate |
| $Cu_2SO_4$ | Copper(I) sulfate or cuprous sulfate |
| $CuSO_4$ | Copper(II) sulfate or cupric sulfate |
| $Li_3AsO_4$ | Lithium arsenate |
| $NaNO_2$ | Sodium nitrite |
| $ZnMoO_4$ | Zinc molybdate |

the *ate* ending is used. Examples of ternary compounds and their names are given in Table 8.5.

Ternary oxy-acids. Inorganic ternary compounds containing hydrogen, oxygen, and one other element are called *oxy-acids*. The element other than hydrogen or oxygen in these acids is usually a nonmetal, but in some cases it can be a metal. The *ous–ic* system is used in naming ternary acids. The suffixes *ous* and *ic* are used to indicate different oxidation states of the element other than hydrogen and oxygen. The *ous* ending again indicates the lower oxidation state and the *ic* ending, the higher oxidation state.

To name these acids, we place the ending *ic* or *ous* after the stem of the element other than hydrogen and oxygen, and add the word *acid*. If an element has only one usual oxidation state, the *ic* ending is used. Hydrogen in a ternary oxy-acid is not specifically designated in the acid name but its presence is implied by use of the word *acid*.

*Examples:* $H_2SO_3$ — Sulfur/*ous acid*; $H_2SO_4$ — Sulfur/*ic acid*

Once again, the acid name is associated with the water solution of the pure compound. In the pure state, the usual ternary name may be used. Thus, $H_2SO_4$ is called both hydrogen sulfate and sulfuric acid.

In cases where there are more than two oxy-acids in a series, the *ous–ic* names are further modified with the prefixes *per* and *hypo*. *Per* is placed before the stem of the element other than hydrogen and oxygen when the element has a higher oxidation number than in the *ic* acid. *Hypo* is used as a prefix before the stem when the element has a lower oxidation number than in the *ous* acid. The use of *per* and *hypo* is illustrated in the oxy-acids of chlorine.

| Formula | Name | Oxidation number of chlorine |
|---|---|---|
| $HClO$ | Hypochlorous acid | +1 |
| $HClO_2$ | Chlorous acid | +3 |
| $HClO_3$ | Chloric acid | +5 |
| $HClO_4$ | Perchloric acid | +7 |

The electron-dot structures of the oxy-acids of chlorine are

```
                                                              ..
                                                             :O:
                          ..            ..             ..  ..  ..
                         :O:           :O:           H:O:Cl:O:
     .. ..              .. ..         .. ..  ..        ..  ..  ..
   H:O:Cl:            H:O:Cl:       H:O:Cl:O:            :O:
     .. ..              .. ..         .. ..  ..           ..
Hypochlorous acid   Chlorous acid   Chloric acid   Perchloric acid
```

Check the oxidation number of chlorine in each of these oxy-acids using the method for assigning oxidation numbers.

Examples of other ternary oxy-acids and their names are shown in Table 8.6.

*Table 8.6.* Names and formulas of selected ternary oxy-acids.

| Formula | Acid name | Formula | Acid name |
|---|---|---|---|
| $H_2SO_3$ | Sulfurous acid | $HNO_2$ | Nitrous acid |
| $H_2SO_4$ | Sulfuric acid | $HNO_3$ | Nitric acid |
| $H_3PO_2$ | Hypophosphorous acid | $HBrO_3$ | Bromic acid |
| $H_3PO_3$ | Phosphorous acid | $HIO_3$ | Iodic acid |
| $H_3PO_4$ | Phosphoric acid | $H_3BO_3$ | Boric acid |
| $HClO$ | Hypochlorous acid | $H_2C_2O_4$ | Oxalic acid |
| $HClO_2$ | Chlorous acid | $HC_2H_3O_2$ | Acetic acid |
| $HClO_3$ | Chloric acid | $H_2CO_3$ | Carbonic acid |
| $HClO_4$ | Perchloric acid | | |

The endings *ous*, *ic*, *ite*, and *ate* are part of classical nomenclature; they are not used in the Stock System to indicate different oxidation states of the elements. These endings are still used, however, in naming many common compounds. The Stock name for $H_2SO_4$ is tetraoxosulfuric(VI) acid, and that for $H_2SO_3$ is trioxosulfuric(IV) acid. These Stock names are awkward and are not commonly used.

## 8.5 Salts

When the hydrogen of an acid is replaced by a metal ion or an ammonium ($NH_4^+$) ion, the compound formed is classified as a *salt*. Therefore, we can have a series of metal chlorides, bromides, sulfides, sulfates, sulfites, nitrates, phosphites, borates, and so on.

We have already considered the rules for naming salts, but a comparison of the salt and acid names will reveal definite patterns that are used. The same rules given above are used in naming the positive part of a salt. In binary compounds, the usual *ide* ending is given to the negative part of the salt name. In ternary compounds, the *ous* and *ic* endings of the acids become *ite* and *ate*, respectively, in the salt names, but the names of the stems remain the same.

| *Ternary oxy-acid* | | *Ternary oxy-salt* |
|---|---|---|
| *ous* ending of acid | becomes | *ite* ending in salt |
| *ic* ending of acid | becomes | *ate* ending in salt |

| *Acid* | | *Salt* | |
|---|---|---|---|
| $H_2SO_4$ | Sulfur/*ic* acid | $Na_2SO_4$ | Sodium sulf/*ate* |
| $H_2SO_3$ | Sulfur/*ous* acid | $CaSO_3$ | Calcium sulf/*ite* |
| $HClO$ | Hypochlor/*ous* acid | $LiClO$ | Lithium hypochlor/*ite* |
| $HClO_4$ | Perchlor/*ic* acid | $NaClO_4$ | Sodium perchlor/*ate* |

Other examples of ternary acids and salts are given in Table 8.7.

*Table 8.7.* Comparison of acid and salt names in ternary oxy-compounds.

| Acid | Salt | Name of salt |
|---|---|---|
| $H_2SO_4$ | $Na_2SO_4$ | Sodium sulfate |
| Sulfuric acid | $CuSO_4$ | Copper(II) sulfate or cupric sulfate |
| | $CaSO_4$ | Calcium sulfate |
| | $Fe_2(SO_4)_3$ | Iron(III) sulfate or ferric sulfate |
| $HNO_3$ | $KNO_3$ | Potassium nitrate |
| Nitric acid | $HgNO_3$ | Mercury(I) nitrate or mercurous nitrate |
| | $Hg(NO_3)_2$ | Mercury(II) nitrate or mercuric nitrate |
| | $Fe(NO_3)_2$ | Iron(II) nitrate or ferrous nitrate |
| | $Al(NO_3)_3$ | Aluminum nitrate |
| $HNO_2$ | $KNO_2$ | Potassium nitrite |
| Nitrous acid | $Co(NO_2)_2$ | Cobalt(II) nitrite or cobaltous nitrite |
| | $Mg(NO_2)_2$ | Magnesium nitrite |
| $HClO$ | $NaClO$ | Sodium hypochlorite |
| $HClO_2$ | $NaClO_2$ | Sodium chlorite |
| $HClO_3$ | $NaClO_3$ | Sodium chlorate |
| $HClO_4$ | $NaClO_4$ | Sodium perchlorate |
| $H_2CO_3$ | $Li_2CO_3$ | Lithium carbonate |
| Carbonic acid | $CaCO_3$ | Calcium carbonate |

## 8.6 Salts with More than one Positive Ion

Salts may be formed from acids that contain two or more acid hydrogen atoms by replacing only one of the hydrogen atoms with a metal or by replacing both hydrogen atoms with different metals. Each positive group is named first and then the appropriate salt ending is added.

| Acid | Salt | Name of salt |
|---|---|---|
| $H_2CO_3$ | $NaHCO_3$ | Sodium hydrogen carbonate or sodium bicarbonate |
| $H_2S$ | $NaHS$ | Sodium hydrogen sulfide or sodium bisulfide |
| $H_3PO_4$ | $MgNH_4PO_4$ | Magnesium ammonium phosphate |
| $H_2SO_4$ | $NaKSO_4$ | Sodium potassium sulfate |

Note the name *sodium bicarbonate* given in the table. The prefix *bi* is commonly used to indicate a compound in which one of two acid hydrogen atoms has been replaced by a metal. Another example is sodium bisulfate, which has the formula $NaHSO_4$. Table 8.8 shows examples of other salts that contain more than one positive ion.

*Table 8.8.* Names of selected salts that contain more than one positive ion.

| Acid | Salt | Name of salt |
|---|---|---|
| $H_2SO_4$ | $KHSO_4$ | Potassium hydrogen sulfate or potassium bisulfate |
| $H_2SO_3$ | $Ca(HSO_3)_2$ | Calcium hydrogen sulfite or calcium bisulfite |
| $H_2S$ | $NH_4HS$ | Ammonium hydrogen sulfide or ammonium bisulfide |
| $H_3PO_4$ | $MgNH_4PO_4$ | Magnesium ammonium phosphate |
| $H_3PO_4$ | $NaH_2PO_4$ | Sodium dihydrogen phosphate |
| $H_3PO_4$ | $Na_2HPO_4$ | Disodium hydrogen phosphate |
| $H_2C_2O_4$ | $KHC_2O_4$ | Potassium hydrogen oxalate or potassium binoxalate |
| $H_2SO_4$ | $KAl(SO_4)_2$ | Potassium aluminum sulfate |
| $H_2CO_3$ | $Al(HCO_3)_3$ | Aluminum hydrogen carbonate or aluminum bicarbonate |

Note that prefixes are also used in chemical nomenclature to give special clarity or emphasis to certain compounds as well as to distinguish between two or more compounds.

*Examples:* $Na_3PO_4$ Trisodium phosphate
$Na_2HPO_4$ Disodium hydrogen phosphate
$NaH_2PO_4$ Sodium dihydrogen phosphate

## 8.7 Bases

Inorganic bases contain the hydroxyl group, $OH^-$, in chemical combination with a metal ion. These compounds are called *hydroxides*. The $OH^-$ group is named as a single ion and is given the ending *ide*. Several common bases are listed below.

| | |
|---|---|
| $NaOH$ | Sodium hydroxide |
| $KOH$ | Potassium hydroxide |
| $NH_4OH$ | Ammonium hydroxide |
| $Ca(OH)_2$ | Calcium hydroxide |
| $Ba(OH)_2$ | Barium hydroxide |

We have now looked at methods of naming inorganic acids, bases, salts, and oxides. These four classes are just a handful of the classified chemical compounds. Most other classes fall under the broad field of organic chemistry. A few of these are alcohols, hydrocarbons, ethers, aldehydes, ketones, phenols, and carboxylic acids.

*Problem 8.1*

Name the compound CaS.

*Step 1.* From the formula, it is a two-element compound and follows the rules for binary compounds.

*Step 2.* The compound is composed of Ca, a metal, and S, a nonmetal. From oxidation number tables, determine whether the metal has a single or a variable oxidation number. We find that Ca has only one oxidation state; therefore, we name the positive part of the compound *calcium*.

*Step 3.* Modify the name of the second element to the identifying stem *sulf* and add the binary ending *ide* to form the name of the negative part, *sulfide*.

*Step 4.* The name of the compound, therefore, is *calcium sulfide*.

*Problem 8.2*

Name the compound FeS.

*Step 1.* This compound follows the rules for a binary compound and, like CaS, must be a sulfide.

*Step 2.* It is a compound of Fe, a metal, and S, a nonmetal. From the oxidation number tables, we see that Fe has variable oxidation numbers. In sulfides, the oxidation number of S is $-2$. Therefore, the oxidation number of Fe must be $+2$. Thus, the name of the positive part of the compound is *iron(II)* or *ferrous*.

*Step 3.* We have already determined that the name of the negative part of the compound will be *sulfide*.

*Step 4.* The name of FeS is *iron(II) sulfide* or *ferrous sulfide*.

*Problem 8.3*

(a) Name the salt $KNO_3$; and (b) name the acid $HNO_3$, from which this salt can be derived.

(a) *Step 1.* From the formula, the compound contains three elements and follows the rules for ternary compounds.

*Step 2.* The salt is composed of a $K^+$ ion and a $NO_3^-$ ion. The name of the positive part of the compound is *potassium.*

*Step 3.* Since it is a ternary salt, the name will end in *ite* or *ate.* From the oxidation number tables, we see that the name of the $NO_3^-$ ion is *nitrate.*

*Step 4.* The name of the compound is *potassium nitrate.*

(b) The name of the acid follows the rules for ternary oxy-acids. Since the name of the salt $KNO_3$ ends in *ate*, the name of the corresponding acid will end in *ic acid.* Change the *ate* ending of nitrate to *ic.* Thus, *nitrate* becomes *nitric*, and the name of the acid is *nitric acid.*

---

## Questions

*In naming compounds, be careful to use correct spelling. For additional assistance in naming compounds refer to Tables 7.6, 7.7, and 7.8.*

1. Write formulas for the following cations (do not forget to include the charges): sodium, magnesium, aluminum, copper(II), iron(II), ferric, lead(II), silver, barium, hydrogen, mercury(II), tin(II), chromium(III).
2. Write formulas for the following anions (do not forget to include the charges): chloride, bromide, fluoride, iodide, cyanide, oxide, hydroxide, sulfide, sulfate, bisulfate, bisulfite, chromate, carbonate, bicarbonate, acetate, chlorate, permanganate, oxalate.
3. Complete the table, filling in each box with the proper formula.

| | $Br^-$ | $CO_3^{2-}$ | $NO_3^-$ | $PO_4^{3-}$ | $OH^-$ | $C_2H_3O_2^-$ |
|---|---|---|---|---|---|---|
| $K^+$ | KBr | $K_2CO_3$ | | | | |
| $Mg^{2+}$ | | | | | | |
| $Al^{3+}$ | | | | | | |
| $Fe^{3+}$ | | | | | | |
| $Zn^{2+}$ | | | | | | |
| $Ag^+$ | | | | | | |

4. State how each of the following terms are used in naming inorganic compounds: ide, ous, ic, hypo, per, ite, ate, roman numerals.
5. Name the following binary compounds, all of which are composed of nonmetals:
   (a) CO (c) $NF_3$ (e) $SO_2$ (g) $CCl_4$ (i) $N_2O_5$ (k) $NH_3$
   (b) $CO_2$ (d) $PBr_5$ (f) $SO_3$ (h) $Cl_2O_7$ (j) $H_2S$ (l) $OF_2$
6. Name each compound listed by (1) the Stock (IUPAC) System and (2) the *ous–ic* system:
   (a) $CuCl_2$ (c) $TiCl_4$ (e) $Fe(OH)_3$ (g) $SnCl_4$ (i) $Hg_2S$
   (b) $SnF_2$ (d) $Ti_2O_3$ (f) FeO (h) $MnCO_3$ (j) $HgSO_4$
7. Provide two names for each compound listed: (1) the name of the pure substance and (2) the name of the substance when dissolved in water to form an acid or a base.
   (a) HCl (d) $NH_3$ (g) $H_2Se$ (j) HBr (l) $H_2SO_3$
   (b) $H_2SO_4$ (e) $H_2S$ (h) NaOH (k) $H_3PO_4$ (m) $Ba(OH)_2$
   (c) $HNO_3$ (f) $HC_2H_3O_2$ (i) $HClO_4$

8. Name each compound.
   (a) $AgCl$
   (b) $Na_2SO_3$
   (c) $CuI_2$
   (d) $MgBr_2$
   (e) $Cu_2S$
   (f) $Fe(NO_3)_2$
   (g) $AlPO_4$
   (h) $KHCO_3$
   (i) $KOCl$
   (j) $Na_2C_2O_4$
   (k) $NH_4SCN$
   (l) $KAl(SO_4)_2$
   (m) $BaCr_2O_7$
   (n) $Cd(CN)_2$
   (o) $NaMnO_4$
   (p) $HgF_2$
   (q) $ZnCO_3$
   (r) $PbCrO_4$
   (s) $Na_2O_2$
   (t) $CO$
   (u) $Co(NO_2)_2$
   (v) $SO_3$
   (w) $K_2SiO_3$
   (x) $LiH$
   (y) $Mn(C_2H_3O_2)_2$
   (z) $Si_3N_4$
9. Name each compound.
   (a) $NH_4I$
   (b) $PF_3$
   (c) $KOH$
   (d) $Ca(ClO_3)_2$
   (e) $NaC_2H_3O_2$
   (f) $CaF_2$
   (g) $Sn_3(PO_4)_2$
   (h) $SbCl_5$
   (i) $MnSO_4$
   (j) $Hg_2O$
   (k) $Co(HCO_3)_2$
   (l) $BaSO_3$
   (m) $HOCl$
   (n) $N_2O_5$
   (o) $NH_4NO_2$
   (p) $Na_2Cr_2O_7$
   (q) $Li_2CO_3$
   (r) $NiCl_2$
   (s) $Na_2SiO_3$
   (t) $CaC_2O_4$
   (u) $BF_3$
   (v) $CaH_2$
   (w) $HgBr_2$
   (x) $Sn(CN)_4$
   (y) $KSCN$
   (z) $BaO_2$
10. Write formulas for each substance.
   (a) Ammonium iodide
   (b) Silver oxide
   (c) Sulfurous acid
   (d) Chlorous acid
   (e) Stannous fluoride
   (f) Copper(I) oxide
   (g) Beryllium nitride
   (h) Potassium chlorate
   (i) Calcium carbonate
   (j) Magnesium oxalate
   (k) Bismuth(III) sulfide
   (l) Aluminum carbide
   (m) Silicon carbide
   (n) Sodium permanganate
   (o) Zinc hypochlorite
   (p) Carbonic acid
   (q) Zinc phosphate
   (r) Hydriodic acid
   (s) Mercuric oxide
   (t) Nickel(II) carbonate
   (u) Ammonium hydroxide
   (v) Strontium chloride
   (w) Cadmium nitrate
   (x) Lead(II) nitrate
   (y) Iodine
   (z) Acetic acid
   (aa) Iron(III) sulfate
   (bb) Zinc oxide
   (cc) Ferric acetate
   (dd) Oxalic acid
11. Write the name of each salt and the formula and name of the acid from which the salt may be derived.
   (a) $Mg(NO_2)_2$
   (b) $NiCl_2$
   (c) $KNO_3$
   (d) $CaCO_3$
   (e) $KMnO_4$
   (f) $NaNO_3$
   (g) $NaC_2H_3O_2$
   (h) $KNO_2$
   (i) $HgI_2$
   (j) $KHSO_4$
   (k) $Na_3BO_3$
   (l) $KOCl$
   (m) $NaF$
   (n) $BaBr_2$
   (o) $Fe_2(SO_4)_3$
   (p) $Na_2HPO_4$
   (q) $Zn(HSO_3)_2$
   (r) $RaCl_2$
   (s) $SnS_2$
   (t) $Mn(ClO_3)_2$
   (u) $Na_2C_2O_4$
   (v) $LiBrO_3$
   (w) $Fe(ClO_2)_2$
   (x) $Ca(ClO_4)_2$
12. Refer to an outside reference (chemical handbook, encyclopedia, dictionary) and write the chemical formulas for each of the following substances.
   (a) Baking soda
   (b) Calomel
   (c) Carbolic acid
   (d) Epsom salts
   (e) Fool's gold
   (f) Glauber's salt
   (g) Litharge
   (h) Lunar caustic
   (i) Muriatic acid
   (j) Prussian blue
   (k) Prussic acid
   (l) Quicksilver
   (m) Sal soda
   (n) Vinegar
   (o) Vitriolic acid

# 9 Quantitative Composition of Compounds

*After studying Chapter 9 you should be able to:*

1. Understand the new terms listed in Question A at the end of the chapter.
2. Determine the formula weight or molecular weight of a compound from the formula.
3. Convert moles (gram-molecular weights or gram-formula weights) to grams, to molecules, or to formula units, and vice versa.
4. Calculate the percentage composition by weight of a compound from its formula.
5. Explain the relationship between an empirical formula and a molecular formula.
6. Determine the empirical formula of a compound from its percentage composition.
7. Calculate the molecular formula of a compound from its percentage composition and molecular weight.

## 9.1 Formula Weight or Molecular Weight

The quantitative composition of a compound can be determined from its formula; and the formula of a compound can be determined from its quantitative composition. In order to make these determinations, chemists have established a scale of relative masses for atoms known as *atomic weights*. This scale is based on the carbon-12 isotope having a mass of exactly 12 amu (see Section 5.18).

formula weight

molecular weight

Because compounds are composed of atoms, they may be represented by a mass known as the formula weight or the molecular weight. The **formula weight** of a substance is the total mass of all the atoms in the chemical formula of that substance. The **molecular weight** of a substance is the total mass of all the atoms in a molecule of that substance. Formula weight and molecular weight are used interchangeably. However, the term *formula weight* is more inclusive, since it includes both molecular and ionic substances.

## 9.2 Determination of Molecular Weights from Formulas

If the formula of a substance is known, its formula weight or molecular weight may be determined by adding together the atomic weights of all the atoms in the formula. If more than one atom of any element is present, it must be added as many times as it is used in the formula.

*Problem 9.1*

The molecular formula for water is $H_2O$. What is its molecular weight?
Proceed by looking up the atomic weights of H (1.008 amu) and O (15.999 amu) and adding together the masses of all the atoms in the formula unit. Water contains two atoms of H and one atom of O. Thus,

$$\begin{aligned} 2\,\text{H atoms} &= 2 \times 1.008 = 2.016\ \text{amu} \\ 1\,\text{O atom} &= 1 \times 15.999 = 15.999\ \text{amu} \\ & \qquad\qquad 18.015\ \text{amu} = \text{Molecular or formula weight} \end{aligned}$$

---

*Problem 9.2*

Calculate the formula weight of calcium hydroxide, $Ca(OH)_2$.
The formula of this substance contains one atom of Ca and two atoms each of O and H. Thus,

$$\begin{aligned} 1\,\text{Ca atom} &= 1 \times 40.08 = 40.08\ \text{amu} \\ 2\,\text{O atoms} &= 2 \times 15.999 = 31.998\ \text{amu} \\ 2\,\text{H atoms} &= 2 \times 1.008 = 2.016\ \text{amu} \\ & \qquad\qquad 74.094\ \text{amu} = \text{Formula weight} \end{aligned}$$

The atomic weights of elements are often rounded off to one decimal place to simplify calculations. (However, this simplification cannot be made in the most exacting chemical work.) If we calculate the formula weight of $Ca(OH)_2$ on the basis of one decimal place, we find the value to be 74.1 amu instead of 74.094 amu. The formula weight of $Ca(OH)_2$ would then be calculated as follows:

$$\begin{aligned} 1\,\text{Ca} &= 1 \times 40.1 = 40.1\ \text{amu} \\ 2\,\text{O} &= 2 \times 16.0 = 32.0\ \text{amu} \\ 2\,\text{H} &= 2 \times 1.0 = 2.0\ \text{amu} \\ & \qquad\qquad 74.1\ \text{amu} = \text{Formula weight} \end{aligned}$$

---

*Problem 9.3*

The formula for barium chloride dihydrate is $BaCl_2 \cdot 2H_2O$. What is its formula weight?
This formula contains one atom of Ba, two atoms of Cl, and two molecules of $H_2O$. Thus,

$$\begin{aligned} 1\,\text{Ba} &= 1 \times 137.3 = 137.3\ \text{amu} \\ 2\,\text{Cl} &= 2 \times 35.5 = 71.0\ \text{amu} \\ 2\,\text{H}_2\text{O} &= 2 \times 18.0 = 36.0\ \text{amu} \\ & \qquad\qquad 244.3\ \text{amu} = \text{Formula weight} \end{aligned}$$

---

## 9.3 *Gram-Molecular Weight; Gram-Formula Weight; The Mole*

gram-formula weight

gram-molecular weight

The quantity of any substance having a mass in grams that is numerically equal to its formula weight is the **gram-formula weight (g-form. wt)** or **gram-molecular weight (g-mol. wt)** of that substance. This quantity represents the weight of 1 mole ($6.02 \times 10^{23}$ formula units or molecules) of the substance. As an illustration, consider the compound hydrogen chloride, HCl. When 1 gram-atomic weight of H (1.00 gram representing $6.02 \times 10^{23}$, or 1 mole of, H atoms) and 1 gram-atomic weight of Cl (35.5 grams representing $6.02 \times 10^{23}$, or 1 mole of, Cl atoms) combine, they produce 1 gram-molecular weight of HCl (36.5 grams representing $6.02 \times 10^{23}$, or 1 mole of, HCl molecules). Since 36.5 grams of HCl contains $6.02 \times 10^{23}$ molecules, we may refer to this quantity as a gram-molecular weight, a gram-formula weight, or simply as a mole of HCl. These relationships are summarized in tabular form for hydrogen chloride.

| H | Cl | HCl |
|---|---|---|
| $6.02 \times 10^{23}$ H *atoms* | $6.02 \times 10^{23}$ Cl *atoms* | $6.02 \times 10^{23}$ HCl *molecules* |
| 1 mole H *atoms* | 1 mole Cl *atoms* | 1 mole HCl *molecules* |
| 1.00 g H | 35.5 g Cl | 36.5 g HCl |
| 1 g-at. wt H | 1 g-at. wt Cl | 1 g-mol. wt HCl or |
| | | 1 g-form. wt HCl |

In dealing with diatomic elements ($H_2$, $O_2$, $N_2$, $F_2$, $Cl_2$, $Br_2$, $I_2$), special care must be taken to distinguish between a mole of atoms (gram-atomic weight) and a mole of molecules (gram-molecular weight). For example, consider *one* mole of oxygen molecules, which weighs 32.0 g. This quantity is equal to *two* gram-atomic weights of the element oxygen and thus represents *two* moles of oxygen atoms. It is important, therefore, in the case of diatomic elements, to be certain which form we are considering, the atom or the molecule.

*1 mole = $6.02 \times 10^{23}$ molecules or formula units*

*1 mole = 1 gram-molecular weight of a compound*

*1 mole = 1 gram-atomic weight of a monatomic element*

The conversion factors for changing grams of a compound to moles and vice versa are

$$\text{Grams to moles:} \quad \frac{\text{1 mole of a compound}}{\text{1 g-mol. wt of the compound}}$$

$$\text{Moles to grams:} \quad \frac{\text{1 g-mol. wt of a compound}}{\text{1 mole of the compound}}$$

*Problem 9.4*

What is the weight of 1 mole (gram-molecular weight) of sulfuric acid, $H_2SO_4$? This problem is solved in a similar manner to Problems 9.1 through 9.3, using atomic weights of the elements in gram units instead of amu. Look up the atomic weights of H, S, and O, and solve.

$$\begin{aligned} 2\,H &= 2 \times 1.0 = 2.0 \text{ g} \\ 1\,S &= 1 \times 32.1 = 32.1 \text{ g} \\ 4\,O &= 4 \times 16.0 = \underline{64.0 \text{ g}} \\ & \qquad 98.1 \text{ g} = \text{Weight of 1 mole (1 g-mol. wt) of } H_2SO_4 \end{aligned}$$

*Problem 9.5*

How many moles of NaOH are there in 1 kg of sodium hydroxide?

First, we know that

1 mole = 1 g-mol. wt = 40.0 g (23.0 + 16.0 + 1.0 g) NaOH
1 kg = 1000 g

To convert grams to moles we use the conversion factor

$$\frac{1 \text{ mole}}{1 \text{ g-mol. wt}} = \frac{1 \text{ mole NaOH}}{40.0 \text{ g NaOH}}$$

The calculation is

$$1000 \not{\text{g NaOH}} \times \frac{1 \text{ mole NaOH}}{40.0 \not{\text{g NaOH}}} = 25.0 \text{ moles NaOH}$$

$$1 \text{ kg NaOH} = 25.0 \text{ moles NaOH}$$

*Problem 9.6*

What is the weight in grams of 5.00 moles of water?

First, we know that

1 mole $H_2O$ = 18.0 g (Problem 9.1)

To convert moles to grams, use the conversion factor

$$\frac{1 \text{ g-mol. wt } H_2O}{1 \text{ mole } H_2O} = \frac{18.0 \text{ g } H_2O}{1 \text{ mole } H_2O}$$

The calculation is

$$5.00 \text{ moles } H_2O \times \frac{18.0 \text{ g } H_2O}{1 \text{ mole } H_2O} = 90.0 \text{ g } H_2O \quad \text{(Answer)}$$

*Problem 9.7*

How many molecules of HCl are there in 25.0 g of hydrogen chloride?

From the formula we find that the gram-molecular weight (1 mole) of HCl is 36.5 g. The sequence of conversions is from

$$\text{grams HCl} \longrightarrow \text{moles HCl} \longrightarrow \text{molecules HCl}$$

using the conversion factors

$$\frac{1 \text{ mole HCl}}{36.5 \text{ g HCl}} \quad \text{and} \quad \frac{6.02 \times 10^{23} \text{ molecules HCl}}{1 \text{ mole HCl}}$$

$$25.0 \text{ g HCl} \times \frac{1 \text{ mole HCl}}{36.5 \text{ g HCl}} \times \frac{6.02 \times 10^{23} \text{ molecules HCl}}{1 \text{ mole HCl}}$$

$$= 4.12 \times 10^{23} \text{ molecules HCl} \quad \text{(Answer)}$$

## 9.4 Percentage Composition of Compounds

Just as each piece of pie represents a percentage of the whole pie, so the mass of each element in a compound represents a percentage of the total mass of that compound. The formula weight may be used to represent the total mass, or 100%, of a compound. If the *weight-percent* of each element in a compound is known, we have the **percentage composition of the compound**. The composition of water, $H_2O$, is two atoms of H and one atom of O per molecule, or 11.1% H and 88.9% O by weight.

percentage composition of a compound

The percentage composition of a compound can be determined if its formula is known or if the weights of two or more elements that have combined with each other are known or are experimentally determined.

If the formula is known, it is essentially a two-step process to determine the percentage composition.

*Step 1.* Determine the gram-formula weight.

*Step 2.* Determine the total weight of each element in the gram-formula weight. Divide each of these weights by the gram-formula weight and multiply by 100%. This gives the percentage.

*Problem 9.8*

Calculate the percentage composition of sodium chloride, NaCl.

*Step 1.* Gram-formula weight of NaCl:

$$\begin{aligned} 1\ \text{Na} &= 23.0\ \text{g} \\ 1\ \text{Cl} &= \underline{35.5\ \text{g}} \\ & \phantom{=}\ 58.5\ \text{g} \end{aligned}$$

*Step 2.*

$$\text{Na} = \frac{23.0\ \text{g}}{58.5\ \text{g}} \times 100\% = 39.3\%\ \text{Na}$$

$$\text{Cl} = \frac{35.5\ \text{g}}{58.5\ \text{g}} \times 100\% = \underline{60.7\%\ \text{Cl}}$$

$$100.0\%\ \text{Total}$$

In any two-component system, if one percentage is known, the other is automatically defined by difference; that is, if Na = 39.3%, then Cl = 100 − 39.3 = 60.7%. However, the calculation of the percentage of each component should be carried out, since this provides a check against possible error. The percentage composition data should add up to 100 ± 0.5%.

---

*Problem 9.9*

Calculate the percentage composition of potassium sulfate, $K_2SO_4$.

*Step 1.* Gram-formula weight of $K_2SO_4$:

$$\begin{aligned} 2\ \text{K} &= 2 \times 39.1 = 78.2\ \text{g} \\ 1\ \text{S} &= 1 \times 32.1 = 32.1\ \text{g} \\ 4\ \text{O} &= 4 \times 16.0 = \underline{64.0\ \text{g}} \\ & \phantom{= 4 \times 16.0 =}\ 174.3\ \text{g} \end{aligned}$$

*Step 2.* $K = \frac{78.2\text{ g}}{174.3\text{ g}} \times 100\% = 44.9\%\ K$

$S = \frac{32.1\text{ g}}{174.3\text{ g}} \times 100\% = 18.4\%\ S$

$O = \frac{64.0\text{ g}}{174.3\text{ g}} \times 100\% = 36.7\%\ O$

$100.0\%$ Total

*Problem 9.10*

When heated in the air, 1.63 g of zinc, Zn, combine with 0.40 g of oxygen, $O_2$, to form an oxide of zinc. Calculate the percentage composition of the compound formed.

The percentage composition may be calculated on the basis of the individual elements as parts or percentages of the total weight of the compound formed. First, calculate the total weight of the compound formed.

1.63 g Zn
0.40 g O
2.03 g = Total weight of product

Then divide the weight of each element by the total weight (Step 1) and multiply by 100%.

$\frac{1.63\text{ g}}{2.03\text{ g}} \times 100\% = 80.3\%\ Zn$

$\frac{0.40\text{ g}}{2.03\text{ g}} \times 100\% = 19.7\%\ O$

$100.0\%$ Total

The compound formed contains 80.3% Zn and 19.7% O.

## 9.5 Empirical Formula versus Molecular Formula

empirical formula

The **empirical formula**, or **simplest formula**, gives the smallest ratio of the atoms that are present in a compound. This formula gives the relative number of atoms of each element in the compound. The empirical formula contains the smallest whole-number ratio that can be derived from the percentages of the different elements in the compound.

molecular formula

The **molecular formula** is the true formula, representing the total number of atoms of each element present in one molecule of a compound. It is entirely possible that two or more substances will have the same percentage composition, yet be distinctly different compounds. For example, acetylene, $C_2H_2$, is a common gas used in welding; benzene, $C_6H_6$, is an important solvent obtained from coal tar and is used in the synthesis of styrene and nylon. Both acetylene and benzene contain 92.3% C and 7.7% H. The smallest ratio of C and H corresponding to these percentages is CH (1:1). Therefore, the empirical formula for both acetylene and benzene is CH—even though it is known that the molecular formulas are $C_2H_2$ and $C_6H_6$, respectively. It is not uncommon for the molecular formula to be the same as the empirical formula. If the molecular formula is not the same, it will be an integral multiple of the empirical formula.

$$CH = \text{Empirical formula}$$
$$(CH)_2 = C_2H_2 = \text{Acetylene (molecular formula)}$$
$$(CH)_6 = C_6H_6 = \text{Benzene (molecular formula)}$$

Table 9.1 summarizes the data concerning these CH formulas. Table 9.2 shows empirical and molecular formula relationships of other compounds.

*Table 9.1.* Molecular formulas of two compounds having an empirical formula with a 1:1 ratio of carbon and hydrogen atoms.

| Formula | Composition %C | %H | Molecular weight |
|---|---|---|---|
| CH (empirical) | 92.3 | 7.7 | 13.0 (empirical) |
| $C_2H_2$ (acetylene) | 92.3 | 7.7 | 26.0 (2 × 13.0) |
| $C_6H_6$ (benzene) | 92.3 | 7.7 | 78.0 (6 × 13.0) |

*Table 9.2.* Some empirical and molecular formulas.

| Compound | Empirical formula | Molecular formula | Compound | Empirical formula | Molecular formula |
|---|---|---|---|---|---|
| Acetylene | $CH$ | $C_2H_2$ | Diborane | $BH_3$ | $B_2H_6$ |
| Benzene | $CH$ | $C_6H_6$ | Hydrazine | $NH_2$ | $N_2H_4$ |
| Ethylene | $CH_2$ | $C_2H_4$ | Hydrogen | $H$ | $H_2$ |
| Formaldehyde | $CH_2O$ | $CH_2O$ | Chlorine | $Cl$ | $Cl_2$ |
| Acetic acid | $CH_2O$ | $C_2H_4O_2$ | Bromine | $Br$ | $Br_2$ |
| Dextrose | $CH_2O$ | $C_6H_{12}O_6$ | Oxygen | $O$ | $O_2$ |
| Hydrogen chloride | $HCl$ | $HCl$ | Nitrogen | $N$ | $N_2$ |
| Carbon dioxide | $CO_2$ | $CO_2$ | | | |

## 9.6 Calculation of Empirical Formula

It is possible to establish an empirical formula because (1) the individual atoms in a compound are combined in whole-number ratios and (2) each element has a specific atomic weight.

In order to calculate the empirical formula, we need to know (1) the elements that are combined, (2) their atomic weights, and (3) the ratio by weight or percentage in which they are combined. If elements A and B form a compound, we may represent the empirical formula as $A_xB_y$, where $x$ and $y$ are small whole numbers that represent the number of atoms of A and B. To write the empirical formula, we must determine $x$ and $y$.

The solution to this problem requires three or four steps.

*Step 1.* Assume a definite quantity (usually 100 g) of the compound, if not given, and express the weight of each element in grams.

*Step 2.* Multiply the weight (grams) of each element by the factor 1 mole/1 g-at. wt to convert grams to moles. This conversion gives the number of moles of atoms of each element in the quantity assumed. At this point, these numbers will usually not be whole numbers.

*Step 3.* Divide each of the values obtained in Step 2 by the smallest of these values. If the numbers obtained by this procedure are whole numbers, use them as subscripts in writing the empirical formula. If the numbers obtained are not whole numbers, go on to Step 4.

*Step 4.* Multiply the values obtained in Step 3 by the smallest number that will convert them to whole numbers. Use these whole numbers as the subscripts in the empirical formula. For example, if the ratio of A to B is 1.0:1.5, multiply both numbers by 2 to obtain a ratio of 2:3. The empirical formula would then be $A_2B_3$.

*Problem 9.11*

Calculate the empirical formula of a compound containing 11.19% hydrogen, H, and 88.89% oxygen, O.

*Step 1.* Express each element in grams; if we assume that there are 100 g of material, then the percentage of each element is equal to the grams of each element in 100 g.

$$\text{H} = 11.19\% = \frac{11.19\text{ g}}{100\text{ g}}$$

$$\text{O} = 88.89\% = \frac{88.89\text{ g}}{100\text{ g}}$$

*Step 2.* Multiply the grams of each element by the proper factor to obtain the relative number of moles of atoms:

$$\text{H:}\quad 11.19\text{ g H} \times \frac{1\text{ mole H atoms}}{1.01\text{ g H}} = 11.1\text{ moles H atoms}$$

$$\text{O:}\quad 88.89\text{ g O} \times \frac{1\text{ mole O atoms}}{16.0\text{ g O}} = 5.55\text{ moles O atoms}$$

The formula could be expressed as $H_{11.1}O_{5.55}$. However, it is customary to use the smallest whole-number ratio of atoms. This ratio is calculated in Step 3.

*Step 3.* Change these numbers to whole numbers by dividing each by the smallest.

$$\text{H} = \frac{11.1\text{ moles}}{5.55\text{ moles}} = 2 \qquad \text{O} = \frac{5.55\text{ moles}}{5.55\text{ moles}} = 1$$

In this step, the ratio of atoms has not changed, because we divided the number of moles of each element by the same number.

The simplest ratio of H to O is 2:1.

Empirical formula = $H_2O$

---

*Problem 9.12*

The analysis of a salt showed that it contained 56.58% potassium, K, 8.68% carbon, C, and 34.73% oxygen, O. Calculate the empirical formula for this substance.

*Steps 1 and 2.* After changing the percentage of each element to grams, find the relative number of moles of each element by multiplying by the proper mole/g-at. wt factor.

$$K: \quad 56.58 \text{ g K} \times \frac{1 \text{ mole K atoms}}{39.1 \text{ g K}} = 1.45 \text{ moles K atoms}$$

$$C: \quad 8.68 \text{ g C} \times \frac{1 \text{ mole C atoms}}{12.0 \text{ g C}} = 0.720 \text{ mole C atoms}$$

$$O: \quad 34.73 \text{ g O} \times \frac{1 \text{ mole O atoms}}{16.0 \text{ g O}} = 2.17 \text{ moles O atoms}$$

*Step 3.* Divide each number of moles by the smallest.

$$K = \frac{1.45 \text{ moles}}{0.720 \text{ mole}} = 2.01$$

$$C = \frac{0.720 \text{ mole}}{0.720 \text{ mole}} = 1.00$$

$$O = \frac{2.17 \text{ moles}}{0.720 \text{ mole}} = 3.01$$

The simplest ratio of K:C:O is therefore 2:1:3.

Empirical formula = $K_2CO_3$

---

*Problem 9.13*

A sulfide of iron was formed by combining 2.233 g of iron, Fe, with 1.926 g of sulfur, S. What is the empirical formula of the compound?

*Steps 1 and 2.* Find the relative number of moles of each element by multiplying grams of each element by the proper mole/g-at. wt factor.

$$Fe: \quad 2.233 \text{ g Fe} \times \frac{1 \text{ mole Fe atoms}}{55.8 \text{ g Fe}} = 0.0400 \text{ mole Fe atoms}$$

$$S: \quad 1.926 \text{ g S} \times \frac{1 \text{ mole S atoms}}{32.1 \text{ g S}} = 0.0600 \text{ mole S atoms}$$

*Step 3.* Divide each number of moles by the smaller of the two numbers.

$$Fe = \frac{0.0400 \text{ mole}}{0.0400 \text{ mole}} = 1.00$$

$$S = \frac{0.0600 \text{ mole}}{0.0400 \text{ mole}} = 1.50$$

*Step 4.* Since we still have not reached a ratio that will give a formula containing whole numbers of atoms, we must double each value to obtain a ratio of 2.00 atoms of Fe to 3.00 atoms of S. Doubling both values does not change the ratio of Fe and S atoms.

Fe: $1.00 \times 2 = 2.00$

S: $1.50 \times 2 = 3.00$

Empirical formula = $Fe_2S_3$

In many of these calculations, results may vary somewhat from an exact whole number. This can be due to experimental errors in obtaining the data.

Calculations that vary by no more than $\pm 0.1$ from a whole number can usually be rounded off to the nearest whole number. Deviations greater than about 0.1 unit usually mean that the calculated ratios need to be multiplied by a factor to make them all whole numbers.

## 9.7 Calculation of the Molecular Formula from the Empirical Formula

The molecular formula can be calculated from the empirical formula if the molecular weight, in addition to data for calculating the empirical formula, is known. The molecular formula, as stated in Section 9.5, will be equal to, or some multiple of, the empirical formula. For example, if the empirical formula of a compound between hydrogen and fluorine is HF, the molecular formula can be expressed as $(HF)n$, where $n = 1, 2, 3, 4, \ldots$. This $n$ means that the molecular formula could be HF, $H_2F_2$, $H_3F_3$, $H_4F_4$, and so on. To determine the molecular formula, we must evaluate $n$.

$$n = \frac{\text{Molecular weight}}{\text{Empirical formula weight}} = \text{Number of empirical formula units}$$

*Problem 9.14*

A compound of nitrogen and oxygen with a molecular weight of 92.0 was found to have an empirical formula of $NO_2$. What is its molecular formula?

*Step 1.* Let $n$ be the number of $(NO_2)$ units in a molecule; then the molecular formula is $(NO_2)n$.

*Step 2.* Each $(NO_2)$ unit weighs 46.0 g $[14 + (2)(16)]$. The gram-molecular weight of $(NO_2)n = 92.0$ g and the number of (46.0) units in 92.0 is 2.

$$n = \frac{92.0\ g}{46.0\ g} = 2 \quad \text{(Empirical formula units)}$$

*Step 3.* The molecular formula is $(NO_2)_2$, or $N_2O_4$.

*Problem 9.15*

The hydrocarbon propylene has a gram-molecular weight of 42.0 g/mole and contains 14.3% H and 85.7% C. What is its molecular formula?

*Step 1.* First find the empirical formula:

$$\text{C:}\quad 85.7\text{ g C} \times \frac{1\text{ mole C atoms}}{12.0\text{ g C}} = 7.14\text{ moles C atoms}$$

$$\text{H:}\quad 14.3\text{ g H} \times \frac{1\text{ mole H atoms}}{1.0\text{ g H}} = 14.3\text{ moles H atoms}$$

Divide each value by the smallest number of moles.

$$C = \frac{7.14\text{ moles}}{7.14\text{ moles}} = 1.0$$

$$H = \frac{14.3\text{ moles}}{7.14\text{ moles}} = 2.0$$

Empirical formula $= CH_2$

*Step 2.* Determine the molecular formula from the empirical formula and molecular weight.

Molecular formula = $(CH_2)_n$
Molecular weight = 42.0

Each $CH_2$ unit weighs 14.0 (12 + 2). The number of $CH_2$ units in 42.0 is 3.

$$n = \frac{42.0}{14.0} = 3 \text{ (Empirical formula units)}$$

The molecular formula is $(CH_2)_3$, or $C_3H_6$.

## Questions

A. *Review the meanings of the new terms introduced in this chapter.*

1. Formula weight
2. Molecular weight
3. Gram-formula weight
4. Gram-molecular weight
5. Percentage composition of a compound
6. Empirical formula
7. Molecular formula

B. *Review questions.*

1. How are formula weight and molecular weight related to each other? In what respects are they different?
2. How many molecules are present in 1 g-mol. wt of nitric acid, $HNO_3$? How many atoms are present?
3. What is the relationship between the following?
   (a) Mole and molecular weight (b) Mole and formula weight
4. Why is it correct to refer to the weight of 1 mole of sodium chloride, but incorrect to refer to a molecular weight of sodium chloride?
5. In calculating the empirical formula of a compound from its percentage composition, why do we choose to start with 100 g of the compound?
6. Which of the following statements are correct?
   (a) A mole of sodium and a mole of sodium chloride contain the same number of sodium atoms.
   (b) A compound such as NaCl has a formula weight but no true molecular weight.
   (c) One mole of nitrogen gas weighs 14.0 g.
   (d) The percentage of oxygen is higher in $K_2CrO_4$ than it is in $Na_2CrO_4$.
   (e) The number of Cr atoms is the same in a mole of $K_2CrO_4$ as it is in a mole of $Na_2CrO_4$.
   (f) Both $K_2CrO_4$ and $Na_2CrO_4$ contain the same percentage by weight of Cr.
   (g) A gram-molecular weight of sucrose, $C_{12}H_{22}O_{11}$, contains 1 mole of sucrose molecules.
   (h) Two moles of sulfuric acid, $H_2SO_4$, contain 8 moles of oxygen atoms.
   (i) The empirical formula of sucrose, $C_{12}H_{22}O_{11}$, is $CH_2O$.
   (j) A hydrocarbon that has a molecular weight of 280 and an empirical formula of $CH_2$ has a molecular formula of $C_{22}H_{44}$.

C. *Problems.*

1. Determine the molecular weight of the following compounds:
   (a) NaI (d) $Mn_3O_4$ (g) $C_6H_{12}O_6$
   (b) $Fe(OH)_3$ (e) $Al_2(SO_4)_3$ (h) $Br_2$
   (c) $K_2SO_4$ (f) $C_2H_5Cl$ (i) $KH_2PO_4$

2. Determine the gram-molecular weight of the following compounds:
(a) $HC_2H_3O_2$ (d) $HNO_3$ (g) $UF_6$
(b) $Pb(NO_3)_2$ (e) $Ca(HCO_3)_2$ (h) $K_3Fe(CN)_6$
(c) $C_6H_5NO_2$ (f) $ZnCl_2$ (i) $CoCl_2 \cdot 6\,H_2O$
3. How many moles are contained in the following?
(a) 32.0 g NaOH (d) 40.0 g $Ca(NO_3)_2$
(b) 18.0 g $N_2$ (e) 0.953 g $MgCl_2$
(c) 50.0 g $CH_3OH$ (f) 1.0 lb KCl
4. How many moles of atoms are contained in each of the following?
(a) 7.2 g Mg (c) 12.0 g $Cl_2$
(b) 39.9 g Ar (d) $3.01 \times 10^{23}$ atoms F
5. Calculate the number of grams contained in each of the following:
(a) 50.0 moles $H_2O$ (d) 0.500 mole $O_2$
(b) 0.100 mole $SnCl_2$ (e) $4.25 \times 10^{-4}$ mole $NH_4Br$
(c) 1.21 moles $H_3PO_4$ (f) 4.50 moles $CH_4$
6. How many molecules are contained in each of the following?
(a) 1.0 mole $F_2$ (c) 3.0 g $C_2H_6$
(b) 0.35 mole $N_2$ (d) 50.0 g $SO_3$
How many atoms are present in each of these amounts?
7. What is the weight in grams of each of the following?
(a) 1 atom of mercury (c) 1 atom of helium
(b) 1 molecule of water (d) 1 molecule of $CO_2$
8. How many moles are contained in each of the following?
(a) 1000 molecules $C_6H_6$ (d) 6000 molecules $NO_2$
(b) $1 \times 10^{12}$ atoms Zn (e) 1 atom Mg
(c) 1000 molecules $CH_4$ (f) $2 \times 10^6$ molecules $H_2O$
9. How many atoms of carbon are contained in each of the following?
(a) 1.00 mole $C_4H_{10}$ (d) 2.00 moles $C_2H_5Br$
(b) 11.5 g $CaCO_3$ (e) 5.5 g $CO_2$
(c) $6.00 \times 10^{20}$ molecules $C_3H_8O_3$ (f) $3.00 \times 10^{10}$ molecules $CH_4$
10. Calculate the number of:
(a) Grams of silver in 40.0 g AgCl
(b) Grams of bromine in 10.0 g $CaBr_2$
(c) Grams of sulfur in 500 g $Na_2S_2O_7$
(d) Grams of chromium in 25.0 g $K_2CrO_4$
11. A solution was made by dissolving 12.0 g of potassium chromate, $K_2CrO_4$, in 200 g of water. How many moles of each compound were used?
12. A sulfuric acid solution contains 65.0% $H_2SO_4$ by weight and has a density of 1.55 g/ml. How many moles of the acid are present in 1.00 litre of the solution?
13. A nitric acid solution containing 72.0% $HNO_3$ by weight has a density of 1.42 g/ml. How many moles of $HNO_3$ are present in 100 ml of the solution?
14. Calculate the percentage composition of the following compounds:
(a) MgO (c) $CaSO_4$ (e) AlN (g) $AgNO_3$
(b) $CCl_4$ (d) $KNO_3$ (f) HCl (h) $Fe(OH)_3$
15. What is the oxidation number of the first element in each of the compounds in Problem 14?
16. Calculate the percentage of iron, Fe, in the following compounds:
(a) FeO (b) $Fe_2O_3$ (c) $Fe_3O_4$ (d) $K_4Fe(CN)_6$
17. Which one of the following oxides has the highest and which has the lowest percentage of oxygen, O, by weight, in its formula?
(a) $Li_2O$ (b) MgO (c) $Bi_2O_3$ (d) $TiO_2$

18. A 6.23 g sample of silver, Ag, was converted to the oxide, which was found to weigh 6.69 g. Calculate the percentage composition of the compound formed.
19. Calculate the empirical formula of each compound from the percentage compositions given below.
    (a) 66.4% Cu, 33.6% S
    (b) 79.8% Cu, 20.2% S
    (c) 62.6% Ca, 37.4% C
    (d) 36.8% N, 63.2% O
    (e) 38.9% Cl, 61.2% O
    (f) 39.8% K, 27.8% Mn, 32.5% O
    (g) 32.4% Na, 22.6% S, 45.0% O
    (h) 52.0% Zn, 9.60% C, 38.4% O
    (i) 1.90% H, 67.6% Cl, 30.5% O
    (j) 60.0% C, 13.3% H, 26.7% O
20. Calculate the percentage of:
    (a) Cadmium in $CdCO_3$
    (b) Carbon in $C_5H_{11}Cl$
    (c) Manganese in $KMnO_4$
    (d) Nitrogen in $NH_4NO_3$
21. Answer the following by consideration of the formulas. Check your answers by calculation if you wish. Which compound has the:
    (a) Higher percent by weight of hydrogen, $H_2O$ or $H_2O_2$?
    (b) Lower percent by weight of manganese, $NaMnO_4$ or $Na_2MnO_4$?
    (c) Higher percent by weight of chromium, $K_2CrO_4$ or $K_2Cr_2O_7$?
    (d) Higher percent by weight of nitrogen, $NO_2$ or $N_2O_4$?
    (e) Lower percent by weight of sulfur, $NaHSO_4$ or $Na_2SO_4$?
22. A 7.615 g sample of gallium, Ga, was found to react with 2.622 g of oxygen, $O_2$. What is the empirical formula of the compound formed?
23. Magnesium reacts with nitrogen to form the compound $Mg_3N_2$. Will a mixture of 20.0 g Mg and 10.0 g $N_2$ have sufficient magnesium atoms to react with all the nitrogen atoms? Show evidence for your answer.
24. Hydroquinone is an organic compound commonly used as a photographic developer. It has a molecular weight of 110 g/mole and a composition of 65.45% C, 5.45% H, and 29.09% O. Calculate the molecular formula of hydroquinone.
25. Fructose is a very sweet natural sugar that is present in honey, fruits, and fruit juices. It has a molecular weight of 180 g/mole and a composition of 40.0% C, 6.7% H, and 53.3% O. Calculate the molecular formula of fructose.
26. Listed below are the compositions of four different compounds of carbon, C, and chlorine, Cl. Derive the empirical and molecular formulas for each.

| | Percent C | Percent Cl | Molecular weight |
|---|---|---|---|
| (a) | 7.79 | 92.21 | 154 |
| (b) | 10.13 | 89.87 | 237 |
| (c) | 25.26 | 74.74 | 285 |
| (d) | 11.25 | 88.75 | 320 |

# 10 Chemical Equations

*After studying Chapter 10 you should be able to:*

1. Understand the terms listed in Question A at the end of the chapter.
2. Know the format used in setting up chemical equations.
3. Recognize the various symbols commonly used in writing chemical equations.
4. Balance simple chemical equations.
5. Interpret a balanced equation in terms of the relative numbers or amounts of molecules, atoms, grams, or moles of each substance represented.
6. Classify equations as representing combination, decomposition, single-replacement, or double-replacement reactions.
7. Complete and balance equations for simple combination, decomposition, single-replacement, and double-replacement reactions when given the reactants.
8. Distinguish between exothermic and endothermic reactions, and relate the quantity of heat to the amounts of substances involved in the reaction.

## 10.1 The Chemical Equation

chemical equation

word equation

A **chemical equation** is a shorthand expression for a chemical change or reaction. It shows, among other things, the rearrangement of atoms that are involved in the reaction. A **word equation** states in words, in equation form, the substances involved in a chemical reaction. For example, when mercuric oxide is heated, it decomposes to form mercury and oxygen. The word equation for this decomposition is

$$\text{Mercuric oxide} + \text{Heat} \longrightarrow \text{Mercury} + \text{Oxygen}$$

From the chemist's point of view, this method of describing a chemical change is very inadequate. It is bulky and cumbersome to use and does not give quantitative information. The chemical equation, using symbols and formulas, is a far better way to describe the decomposition of mercuric oxide:

$$2\,HgO \xrightarrow{\Delta} 2\,Hg + O_2\uparrow$$

This equation gives all the information from the word equation plus formulas, composition, reactive amounts of all the substances involved in the reaction, and much additional information (see Section 10.4). Even though a chemical equation provides much quantitative information, it is still not a complete description; it does not tell us how much heat is needed to cause decomposition, what we observe during the reaction, or anything about the rate of reaction. This information must be obtained from other sources or from experimentation.

## 10.2 Format for Writing Chemical Equations

reactant

product

The **reactants** are the substances that enter into a chemical change or reaction. The **products** are the substances produced by the reaction. A chemical equation uses the chemical symbols and formulas of the reactants and products and other symbolic terms to represent a chemical reaction. Equations are written according to this general format:

1. The reactants and products are separated by an arrow or other sign indicating equality between reactants and products ($\longrightarrow$, $=$, $\rightleftarrows$).
2. The reactants are placed to the left and the products to the right of the arrow or equality sign. A plus sign (+) is placed between reactants and between products when needed.
3. Conditions required to carry out the reaction may, if desired, be placed above or below the arrow or equality sign. For example, a delta sign placed over the arrow ($\xrightarrow{\Delta}$) indicates that heat is supplied to the reaction.
4. Small integral numbers in front of substances (for example, 2 $H_2O$) are used to balance the equation and to indicate the number of formula units (atoms, molecules, moles, ions) of each substance reacting or being produced. When no number is shown, it is understood that one formula unit of the substance is indicated.

Symbols commonly used in equations are given in Table 10.1.

*Table 10.1.* Symbols commonly used in chemical equations.

| Symbol | Meaning |
|---|---|
| $\rightarrow$ | Yields, produces (points to products) |
| $=$ | Equals; equilibrium between reactants and products |
| $\rightleftarrows$ | Reversible reaction; equilibrium between reactants and products |
| $\uparrow$ | Gas evolved (written after a substance) |
| $\downarrow$ | Solid or precipitate formed (written after a substance) |
| $(s)$ | Solid (written after a substance) |
| $(l)$ | Liquid (written after a substance) |
| $(g)$ | Gas (written after a substance) |
| $\Delta$ | Heat |
| $+$ | Plus or added to |
| $(aq)$ | Aqueous solution (substance dissolved in water) |

## 10.3 Writing and Balancing Equations

balanced equation

To represent the quantitative relationships of a reaction, the chemical equation must be balanced. A **balanced equation** is one that contains the same number of each kind of atom on each side of the equation. The balanced equation, therefore, obeys the Law of Conservation of Mass.

The ability to balance equations must be acquired by every chemistry student. Simple equations are easy to balance, but some care and attention to detail are required. Clearly, the way to balance an equation is to adjust the number of atoms of each element so that it is the same on each side of the equation. But we must not change a correct formula in order to achieve a balanced equation. Each equation must be treated on its own merits; there is no simple "plug in" formula for balancing equations. The following outline gives a general procedure for balancing equations. Study this outline and refer to it as needed when working examples. There is no substitute for practice in learning to write and balance equations.

1. Identify the reaction for which the equation is to be written. Formulate a description or word equation for this reaction (for example, mercuric oxide decomposes yielding mercury and oxygen). This, of course, need not be done when the reactants and products are identified and their formulas are given.
2. Write the unbalanced, or skeleton, equation. Make sure that the formula for each substance is correct and that the reactants are written to the left and the products to the right of the arrow (for example, $HgO \rightarrow Hg + O_2$). The correct formulas must be known or ascertained from the periodic table, oxidation numbers, lists of ions, or experimental data.
3. Balance the equation. Use the following steps as necessary:
   (a) Count and compare the number of atoms of each element on each side of the equation and determine those that must be balanced.
   (b) Balance each element, one at a time, by placing small whole numbers (coefficients) in front of the formulas containing the unbalanced element. It is usually best to balance first metals, then nonmetals, then hydrogen and oxygen. Select the smallest coefficients that will give the same number of atoms of the element on each side. A coefficient placed before a formula multiplies every atom in the formula by that number (for example, $2\,H_2SO_4$ means two molecules of sulfuric acid and also means four H atoms, two S atoms, and eight O atoms).
   (c) Check all other elements after each individual element is balanced to see if, in balancing one, other elements have become unbalanced. Make adjustments as needed.
   (d) Balance polyatomic ions such as $SO_4^{2-}$, which remain unchanged from one side of the equation to the other, in the same way as individual atoms.
   (e) Do a final check, making sure that each element and/or polyatomic ion is balanced and that the smallest possible set of whole-number coefficients has been used.

$$4\,HgO \longrightarrow 4\,Hg + 2\,O_2 \quad \text{(Incorrect form)}$$
$$2\,HgO \longrightarrow 2\,Hg + O_2 \quad \text{(Correct form)}$$

The following examples show stepwise sequences leading to balanced equations. Study each example carefully.

*Example 10.1*

Write the balanced equation for the reaction that takes place when magnesium metal is burned in air to produce magnesium oxide.

1. *Word equation*:

$$\text{Magnesium} + \text{Oxygen} \longrightarrow \text{Magnesium oxide}$$

2. *Skeleton equation*:

$$Mg + O_2 \longrightarrow MgO \quad \text{(Unbalanced)}$$

3. *Balance*:
   (a) Oxygen is not balanced. There are two O atoms on the left side and one on the right side.
   (b) Place the coefficient 2 before MgO.

$$Mg + O_2 \longrightarrow 2\,MgO \quad \text{(Unbalanced)}$$

   (c) Now Mg is not balanced. There is one Mg atom on the left side and two on the right side. Place a 2 before Mg.

$$2\,Mg + O_2 \longrightarrow 2\,MgO \quad \text{(Balanced)}$$

   (d) *Check*: Each side has two Mg and two O atoms.

---

*Example 10.2*

Methane, $CH_4$, undergoes complete combustion to produce carbon dioxide and water. Write the balanced equation for this reaction.

1. *Word equation*:

$$\text{Methane} + \text{Oxygen} \longrightarrow \text{Carbon dioxide} + \text{Water}$$

2. *Skeleton equation*:

$$CH_4 + O_2 \longrightarrow CO_2 + H_2O \quad \text{(Unbalanced)}$$

3. *Balance*:
   (a) Hydrogen and oxygen are not balanced.
   (b) Balance H atoms by placing a 2 before $H_2O$

$$CH_4 + O_2 \longrightarrow CO_2 + 2\,H_2O \quad \text{(Unbalanced)}$$

   Each side of the equation has four H atoms; oxygen is still not balanced. Place a 2 before $O_2$ to balance the oxygen atoms.

$$CH_4 + 2\,O_2 \longrightarrow CO_2 + 2\,H_2O \quad \text{(Balanced)}$$

   (c) *Check*: The equation is correctly balanced; it has one C atom, four O atoms, and four H atoms on each side.

---

*Example 10.3*

Oxygen is prepared by heating potassium chlorate.

1. *Word equation*:

$$\text{Potassium chlorate} \xrightarrow{\Delta} \text{Potassium chloride} + \text{Oxygen}$$

2. *Skeleton equation*:

$$KClO_3 \xrightarrow{\Delta} KCl + O_2 \quad \text{(Unbalanced)}$$

3. *Balance*:
   (a) Oxygen is unbalanced (three O atoms on the left and two on the right side).
   (b) Balance by placing a 2 before $KClO_3$ and a 3 before $O_2$ to give six O atoms on each side.

$$2\,KClO_3 \xrightarrow{\Delta} KCl + 3\,O_2 \quad \text{(Unbalanced)}$$

Now K and Cl are not balanced. Place a 2 before KCl, which balances both K and Cl at the same time.

$$2\,KClO_3 \xrightarrow{\Delta} 2\,KCl + 3\,O_2 \quad \text{(Balanced)}$$

(c) *Check*: Each side contains two K, two Cl, and six O atoms.

---

*Example 10.4*

Balance by starting with the word equation given.

1. *Word equation*:

   Silver nitrate + Hydrogen sulfide ⟶ Silver sulfide + Nitric acid

2. *Skeleton equation*:

   $$AgNO_3 + H_2S \longrightarrow Ag_2S + HNO_3 \quad \text{(Unbalanced)}$$

3. *Balance*:

   (a) Ag and H are unbalanced.

   (b) Place a 2 in front of $AgNO_3$.

   $$2\,AgNO_3 + H_2S \longrightarrow Ag_2S + HNO_3 \quad \text{(Unbalanced)}$$

   (c) This leaves H and $NO_3^-$ unbalanced. Balance by placing a 2 in front of $HNO_3$.

   $$2\,AgNO_3 + H_2S \longrightarrow Ag_2S + 2\,HNO_3 \quad \text{(Balanced)}$$

   (d) In this example, N and O atoms are balanced by balancing the $NO_3^-$ ion as a unit.

   (e) *Check*: Each side has two Ag, two H, and one S atom. Also, each side has two $NO_3^-$ ions.

---

*Example 10.5*

Balance by starting with the word equation given.

1. *Word equation*:

   Aluminum hydroxide + Sulfuric acid ⟶ Aluminum sulfate + Water

2. *Skeleton equation*:

   $$Al(OH_3) + H_2SO_4 \longrightarrow Al_2(SO_4)_3 + HOH \quad \text{(Unbalanced)}$$

3. *Balance*:

   (a) All elements are unbalanced.

   (b) Balance Al by placing a 2 in front of $Al(OH)_3$. Treat the unbalanced $SO_4^{2-}$ ion as a unit and balance by placing a 3 before $H_2SO_4$. Note that Step (d) may sometimes be combined with Step (b).

   $$2\,Al(OH)_3 + 3\,H_2SO_4 \longrightarrow Al_2(SO_4)_3 + HOH \quad \text{(Unbalanced)}$$

   Balance the unbalanced H and O by placing a 6 in front of HOH.

   $$2\,Al(OH)_3 + 3\,H_2SO_4 \longrightarrow Al_2(SO_4)_3 + 6\,HOH \quad \text{(Balanced)}$$

   (c) *Check*: Each side has two Al, eighteen H, and six O atoms; and also three $SO_4^{2-}$ ions.

---

*Example 10.6*

The fuel in a butane gas stove undergoes complete combustion to carbon dioxide and water.

1. *Word equation*:

   Butane + Oxygen ⟶ Carbon dioxide + Water

2. *Skeleton equation*:

$$C_4H_{10} + O_2 \longrightarrow CO_2 + H_2O \quad \text{(Unbalanced)}$$

3. *Balance*:
   (a) All elements are unbalanced.
   (b) Balance C by placing a 4 in front of $CO_2$.

$$C_4H_{10} + O_2 \longrightarrow 4\ CO_2 + H_2O \quad \text{(Unbalanced)}$$

Balance H by placing a 5 in front of $H_2O$.

$$C_4H_{10} + O_2 \longrightarrow 4\ CO_2 + 5\ H_2O \quad \text{(Unbalanced)}$$

Oxygen remains unbalanced. When we try to balance the O atoms, we find that there is no integer (whole number) that can be placed in front of $O_2$ to bring about a balance. The equation can be balanced if we use $6\frac{1}{2}\ O_2$ and then double the coefficients of each substance, including the $6\frac{1}{2}\ O_2$, to obtain the balanced equation.

$$2\ C_4H_{10} + 13\ O_2 \longrightarrow 8\ CO_2 + 10\ H_2O \quad \text{(Balanced)}$$

An alternate procedure is to rewrite the last unbalanced equation, doubling all the coefficients except that of $O_2$:

$$2\ C_4H_{10} + O_2 \longrightarrow 8\ CO_2 + 10\ H_2O$$

Now balance the $O_2$—the result is the same balanced equation as above.

   (c) *Check*: Each side now has eight C, twenty H, and twenty-six O atoms.

## 10.4 What Information Does an Equation Tell Us?

Interpreting the information given in an equation is important if we are to gain the full benefit of its use in evaluating a chemical reaction. The balanced equation tells us

1. what the reactants are and what the products are,
2. the formulas of the reactants and products,
3. the number of molecules or formula units of reactants and products in the reaction,
4. the number of atoms of each element involved in the reaction,
5. the number of molecular weights or formula weights of each substance used or produced,
6. the number of moles of each substance,
7. the number of gram-molecular weights or gram-formula weights of each substance used or produced,
8. the number of grams of each substance used or produced.

Consider the equation

$$H_2(g) + Cl_2(g) \longrightarrow 2\ HCl(g)$$

This equation states that hydrogen gas reacts with chlorine gas to produce hydrogen chloride, also a gas. Let us summarize all the information relating to the equation. The information that can be stated about the relative amount of each substance, with respect to all other substances in the balanced equation, is written below its formula in the following equation:

| $H_2(g)$ | + $Cl_2(g)$ | $\longrightarrow$ $2\ HCl(g)$ |
|---|---|---|
| Hydrogen | Chlorine | Hydrogen chloride |
| 1 molecule | 1 molecule | 2 molecules |
| 2 atoms | 2 atoms | 2 atoms H + 2 atoms Cl |
| 1 mol. wt | 1 mol. wt | 2 mol. wt |
| 1 mole | 1 mole | 2 moles |
| 1 form. wt | 1 form. wt | 2 form. wt |
| 1 g-mol. wt | 1 g-mol. wt | 2 g-mol. wt |
| 2.0 g | 71.0 g | 2 × 36.5 g (73.0 g) |

These data are very useful in calculating quantitative relationships that exist among substances in a chemical reaction. For example, if we react 2 moles of hydrogen (twice as much as is indicated by the equation) with 2 moles of chlorine, we can expect to obtain 4 moles, or 146 g of hydrogen chloride, as a product. We will study this phase of using equations in more detail in the next chapter.

## 10.5 Types of Chemical Equations

Chemical equations represent chemical changes or reactions. To be of any significance, an equation must represent an actual or possible reaction. Part of the problem of writing equations is determining the products formed. There is no sure method of predicting products, nor do we have time to carry out experimentally all the reactions we may wish to consider. Therefore, we must use data reported in the writings of other workers, certain rules to aid in our predictions, and the atomic structure and combining capacities of the elements to help us predict the formulas of the products of a chemical reaction. The final proof of the existence of any reaction, of course, is in the actual observation of the reaction in the laboratory (or elsewhere).

Reactions are classified into types to assist in writing equations and to aid in predicting other reactions. Many chemical reactions fit one or another of the four principal reaction types that are discussed in the following paragraphs. Reactions are also classified as oxidation–reduction. Special methods are used to balance complex oxidation–reduction equations (see Chapter 17).

combination or synthesis reaction

**1. Combination or synthesis reaction.** In this type of reaction, direct union or combination of two substances produces one new substance. Oxidation–reduction is involved in some, but not all, combination reactions. The general form of the equation is

$$A + B \longrightarrow AB$$

*Examples:*

$$S(s) + O_2(g) \longrightarrow SO_2(g)$$
$$2\,Mg(s) + O_2(g) \longrightarrow 2\,MgO(s)$$
$$2\,Na(s) + Cl_2(g) \longrightarrow 2\,NaCl(s)$$
$$CaO(s) + H_2O \longrightarrow Ca(OH)_2(aq)$$
$$SO_3(g) + H_2O \longrightarrow H_2SO_4(aq)$$

decomposition reaction

**2. Decomposition reaction.** In this type of reaction a single substance is decomposed or broken down into two or more different substances. The reaction may be considered the reverse of combination. The starting material must be a compound, and the products may be elements or compounds. Oxidation–reduction is involved in some, but not all, decomposition reactions. The general form of the equation is

$$AB \longrightarrow A + B$$

*Examples:*

$$2\,HgO(s) \longrightarrow 2\,Hg(l) + O_2(g)$$
$$2\,H_2O \longrightarrow 2\,H_2(g) + O_2(g)$$
$$CaCO_3(s) \longrightarrow CaO(s) + CO_2(g)$$
$$2\,KClO_3(s) \longrightarrow 2\,KCl(s) + 3\,O_2(g)$$

single-replacement or substitution reaction

**3. Single-replacement or substitution reaction.** In this type of reaction one simple substance (element) reacts with a compound substance to form a new simple substance and a new compound, one element displacing another in a compound. Oxidation–reduction is always present in single-replacement reactions. The general form of the equation is

$$A + BC \longrightarrow B + AC$$

If A is a metal, it will replace B to form AC; if A is a nonmetal, it will replace C to form BA. Some reactions that fall into this category are the following:

(a) Active metal + Acid ⟶ Hydrogen + Salt

$$Zn(s) + 2\,HCl(aq) \longrightarrow H_2(g) + ZnCl_2(aq)$$
$$2\,Al(s) + 3\,H_2SO_4(aq) \longrightarrow 3\,H_2(g) + Al_2(SO_4)_3(aq)$$

(b) Active metal + Water ⟶ Hydrogen + Metal hydroxide or Metal oxide

$$2\,Na(s) + 2\,H_2O \longrightarrow H_2(g) + 2\,NaOH(aq)$$
$$Ca(s) + 2\,H_2O \longrightarrow H_2(g) + Ca(OH)_2(aq)$$
$$3\,Fe(s) + 4\,H_2O(g) \xrightarrow[\text{steam}]{} 4\,H_2(g) + Fe_3O_4(s)$$

(c) Metal + Salt ⟶ Metal + Salt

$$Fe(s) + CuSO_4(aq) \longrightarrow Cu(s) + FeSO_4(aq)$$
$$Cu(s) + 2\,AgNO_3(aq) \longrightarrow 2\,Ag(s) + Cu(NO_3)_2(aq)$$

In the reactions in (c), the starting free metal must be more reactive than the metal ion in the salt that it displaces.

(d) Nonmetal + Salt ⟶ Nonmetal + Salt

$$Cl_2(g) + 2\,NaBr(aq) \longrightarrow Br_2(l) + 2\,NaCl(aq)$$
$$Cl_2(g) + 2\,KI(aq) \longrightarrow I_2(s) + 2\,KCl(aq)$$

In the reactions in (d), the starting free nonmetal must be more reactive than the nonmetal ion in the salt that it displaces.

double-replacement or metathesis reaction

**4. Double-replacement or metathesis reaction.** In this type of reaction, two compounds react with each other to produce two different compounds. Oxidation–reduction does not occur in double-replacement reactions. The general form of the equation is

$$AB + CD \longrightarrow AD + CB$$

This reaction may be thought of as an exchange of positive and negative groups, where the positive group (A) of the first reactant combines with the negative group (D) of the second reactant, and the positive group (C) of the second reactant combines with the negative group (B) of the first reactant. In writing the formulas of the products, we must take into account the oxidation numbers of the combining groups. Some reactions that fall into this category are the following:

(a) Neutralization of an acid and a base

$$\text{Acid} + \text{Base} \longrightarrow \text{Salt} + \text{Water}$$

$$HCl(aq) + NaOH(aq) \longrightarrow NaCl(aq) + H_2O$$

$$H_2SO_4(aq) + Ba(OH)_2(aq) \longrightarrow BaSO_4\downarrow + 2\ H_2O$$

(b) Formation of an insoluble precipitate

$$BaCl_2(aq) + 2\ AgNO_3(aq) \longrightarrow 2\ AgCl\downarrow + Ba(NO_3)_2(aq)$$

$$FeCl_3(aq) + 3\ NH_4OH(aq) \longrightarrow Fe(OH)_3\downarrow + 3\ NH_4Cl(aq)$$

(c) Metal oxide + Acid $\longrightarrow$ Salt + Water

$$CuO(s) + 2\ HNO_3(aq) \longrightarrow Cu(NO_3)_2(aq) + H_2O$$

$$CaO(s) + 2\ HCl(aq) \longrightarrow CaCl_2(aq) + H_2O$$

(d) Formation of a gas

$$H_2SO_4(l) + NaCl(s) \longrightarrow NaHSO_4(s) + HCl\uparrow$$

$$2\ HCl(aq) + ZnS(s) \longrightarrow ZnCl_2(aq) + H_2S\uparrow$$

$$2\ HCl(aq) + Na_2CO_3(s) \longrightarrow 2\ NaCl(aq) + H_2O(l) + CO_2\uparrow$$

All substances that we attempt to react may not react, or the conditions under which they react may not be present. For example, mercuric oxide does not decompose until it is heated; magnesium does not burn in air or oxygen until the temperature is raised to the point at which it begins to react. When silver is placed in a solution of copper(II) sulfate, no reaction takes place; however, when a strip of copper is placed in a solution of silver nitrate, the single replacement reaction as given in 3 (c) above takes place because copper is a more reactive metal than silver. The successful prediction of the products of a reaction is not always easy. The ability to predict products correctly comes with knowledge and experience. Although you may not be able to predict many reactions at this point, as you continue, you will find that reactions can be categorized, as above, and that prediction of the products thereby becomes easier—if not always certain.

Consider the reaction between aqueous solutions of potassium hydroxide and hydrobromic acid. First, write the formula for each reactant, KOH and HBr. Then, after examining these formulas, ascertain that one is a base and the other is an acid. From the type reaction (Acid + Base → Salt + Water), begin to write the equation by putting down the formulas for the known substances.

$$HBr(aq) + KOH(aq) \longrightarrow \text{Salt} + H_2O$$

In this double-replacement reaction, the $H^+$ from the acid combines with the $OH^-$ from the base to form water. The salt must be composed of the other two ions, $K^+$ and $Br^-$. We determine the formula of the salt to be KBr from the fact that K is a + ion and Br is a − ion. The final balanced equation is

$$HBr(aq) + KOH(aq) \longrightarrow KBr(aq) + H_2O$$

There is a great deal yet to learn about which substances react with each other, how they react, and what conditions are necessary to bring about their reaction. It is possible to make accurate predictions concerning the occurrence of proposed reactions. Such predictions require, in addition to appropriate data, a good knowledge of thermodynamics. But the study of this subject is usually reserved for advanced courses in chemistry and physics. Even without the formal use of thermodynamics, your knowledge of such generalities as the four reaction types just cited, the periodic table, atomic structure, oxidation numbers, and so on, can be put to good use in predicting reactions and in writing equations. Indeed, such applications serve to make chemistry an interesting and fascinating study.

## 10.6 Heat in Chemical Reactions

Energy changes always accompany chemical reactions. One reason why reactions may occur is that the products attain a lower, more stable energy state than the reactants. For the products to attain this more stable state, energy must be liberated and given off to the surroundings as heat (or as heat and work). When a solution of a base is neutralized by the addition of an acid, the liberation of heat energy is signaled by an immediate rise in the temperature of the solution. When an automobile engine burns gasoline, heat is certainly liberated, and, at the same time, part of the liberated energy does the work of moving the automobile.

exothermic reaction

endothermic reaction

Reactions are either exothermic or endothermic. **Exothermic reactions** liberate heat; **endothermic reactions** absorb heat. In an exothermic reaction, heat is a product and may be written on the right side of the equation for the reaction. In an endothermic reaction, heat can be regarded as a reactant and is written on the left side of the equation. Examples indicating heat in an exothermic and an endothermic reaction are shown below.

$$H_2(g) + Cl_2(g) \longrightarrow 2\,HCl(g) + 44.2\text{ kcal} \quad \text{(Exothermic)}$$

$$N_2(g) + O_2(g) + 43.2\text{ kcal} \longrightarrow 2\,NO(g) \quad \text{(Endothermic)}$$

heat of reaction

The quantity of heat produced by a reaction is known as the **heat of reaction**. The units used are calories or kilocalories. Consider the reaction represented by this equation:

$$C(s) + O_2(g) \longrightarrow CO_2(g) + 94.0 \text{ kcal}$$

When the heat liberated is expressed as part of the equation, the substances are expressed in units of moles. Thus, when 1 mole (12.0 g) of C combines with 1 mole (32.0 g) of $O_2$, 1 mole (44.0 g) of $CO_2$ is formed and 94.0 kcal of heat are liberated. Assuming that coal is 90% C and the combustion product is only $CO_2$, $6.4 \times 10^9$ cal would be released when 1 ton of coal is burned. In this reaction, as in many others, the heat or energy is more useful than the chemical products. At 1 cal per gram per degree, $6.4 \times 10^9$ cal is sufficient energy to heat about 20,000 gal

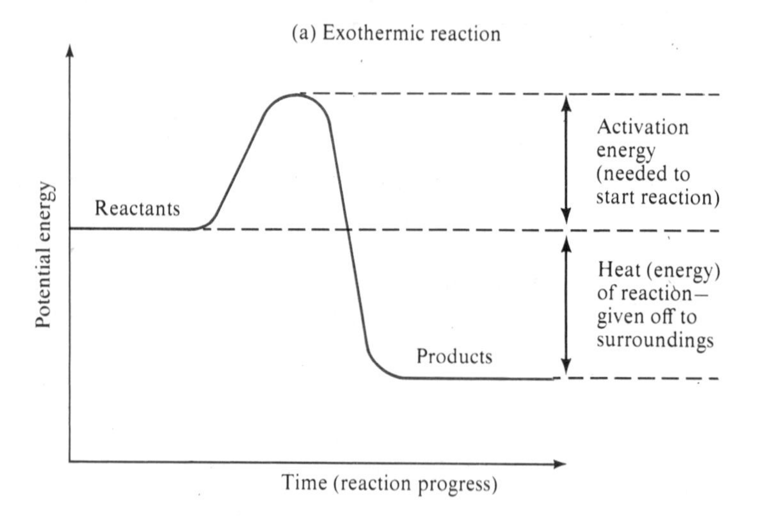


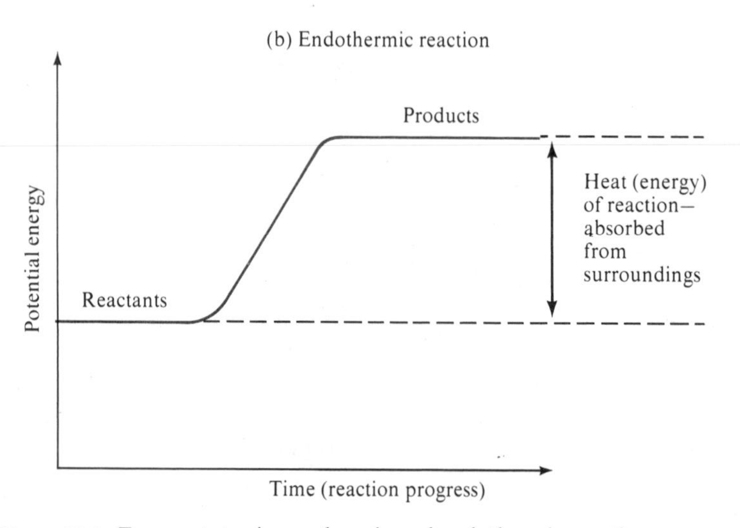


*Figure 10.1.* Energy states in exothermic and endothermic reactions.

of water from room temperature (20°C) to 100°C.

As another example, aluminum metal is produced by electrolyzing aluminum oxide.

$$2\,Al_2O_3 + 779\,\text{kcal} \longrightarrow 4\,AL + 3\,O_2$$

For each mole of $Al_2O_3$ the equivalent of 389.5 kcal must be supplied as electrical energy. Since only 54 g of aluminum are obtained from 1 mole of $Al_2O_3$, this means that each ton of aluminum produced requires more than $6.56 \times 10^9$ kcal of energy.

Be careful not to confuse an exothermic reaction that merely requires heat (activation energy) to get it started with a truly endothermic process. The combustion of magnesium is highly exothermic, yet magnesium must be heated to a fairly high temperature in air before combustion begins. Once started, however, the combustion reaction goes very vigorously until either the magnesium or the available supply of oxygen is exhausted. The electrolytic decomposition of water to hydrogen and oxygen is highly endothermic. If the electric current is shut off when this process is going on, the reaction stops instantly. The relative energy levels of reactants and products in exothermic and in endothermic processes are presented graphically in Figure 10.1.

In reaction (a) of Figure 10.1, the products are at a lower energy level than the reactants. Energy (heat) was given off to the surroundings and the reaction is exothermic. In reaction (b), the products are at a higher energy level than the reactants. Energy has therefore been absorbed and the reaction is endothermic.

## Questions

A. *Review the meanings of the new terms introduced in this chapter.*

1. Chemical equation
2. Word equation
3. Reactant
4. Product
5. Balanced equation
6. Combination reaction
7. Decomposition reaction
8. Single-replacement reaction
9. Double-replacement or metathesis reaction
10. Exothermic reaction
11. Endothermic reaction
12. Heat of reaction

B. *Study Table 10.1 so that you will be familiar with the more common symbols used in equations.*

C. *Review questions.*

1. Balance the following equations:
   (a) $H_2 + I_2 \rightarrow HI$
   (b) $Mg + N_2 \rightarrow Mg_3N_2$
   (c) $NH_4NO_2 \xrightarrow{\Delta} N_2\uparrow + H_2O$
   (d) $Ca(ClO_3)_2 \xrightarrow{\Delta} CaCl_2 + O_2\uparrow$
   (e) $Ca + HCl \rightarrow CaCl_2 + H_2\uparrow$
   (f) $K + H_2O \rightarrow KOH + H_2\uparrow$
   (g) $HCl + Ba(OH)_2 \rightarrow BaCl_2 + H_2O$
   (h) $Pb(NO_3)_2 + NaCl \rightarrow PbCl_2\downarrow + NaNO_3$
   (i) $Cl_2 + KI \rightarrow KCl + I_2$
   (j) $As_2S_3 + HCl \rightarrow AsCl_3 + H_2S\uparrow$
2. Balance the following equations:
   (a) $SO_2 + O_2 \rightarrow SO_3$

(b) $Al + Br_2 \rightarrow AlBr_3$
(c) $NH_4NO_3 \xrightarrow{\Delta} N_2O + H_2O$
(d) $CuSO_4 \cdot 5\,H_2O \rightarrow CuSO_4 + H_2O$
(e) $Al + H_2SO_4 \rightarrow Al_2(SO_4)_3 + H_2\uparrow$
(f) $Zn + HC_2H_3O_2 \rightarrow Zn(C_2H_3O_2)_2 + H_2\uparrow$
(g) $C_5H_{12} + O_2 \rightarrow CO_2 + H_2O$
(h) $C_6H_{14} + O_2 \rightarrow CO_2 + H_2O$
(i) $Fe_2O_3 + HCl \rightarrow FeCl_3 + H_2O$
(j) $BaCl_2 + (NH_4)_2CO_3 \rightarrow BaCO_3\downarrow + NH_4Cl$

3. Balance the following equations:
(a) $Bi + O_2 \rightarrow Bi_2O_3$
(b) $LiAlH_4 \xrightarrow{\Delta} LiH + Al + H_2$
(c) $Mg + B_2O_3 \xrightarrow{\Delta} MgO + B$
(d) $Al + MnO_2 \xrightarrow{\Delta} Mn + Al_2O_3$
(e) $Na_2CO_3 + HCl \rightarrow NaCl + CO_2\uparrow + H_2O$
(f) $Al_2S_3 + H_2O \rightarrow Al(OH)_3 + H_2S\uparrow$
(g) $B_2O_3 + C \rightarrow B_4C + CO$
(h) $C_5H_{11}OH + O_2 \rightarrow CO_2 + H_2O$
(i) $Ca_3(PO_4)_2 + H_3PO_4 \rightarrow Ca(H_2PO_4)_2$
(j) $C_3H_5(NO_3)_3 \xrightarrow{\Delta} N_2 + O_2 + CO_2 + H_2O$
Nitroglycerin

4. Change the following word equations into formula equations and balance them:
(a) Copper + Sulfur → Copper(I) sulfide
(b) Acetic acid + Sodium hydroxide → Sodium acetate + Water
(c) Iron(III) chloride + Potassium hydroxide → Iron(III) hydroxide + Potassium chloride
(d) Aluminum + Copper(II) sulfate → Copper + Aluminum sulfate
(e) Calcium hydroxide + Phosphoric acid → Calcium phosphate + Water
(f) Iron(II) sulfide + Hydrobromic acid → Iron(II) bromide + Hydrogen sulfide
(g) Ammonium sulfate + Barium chloride → Ammonium chloride + Barium sulfate
(h) Sodium hydroxide + Sulfuric acid → Sodium hydrogen sulfate + Water
(i) Silver nitrate + Aluminum chloride → Silver chloride + Aluminum nitrate
(j) Sulfur tetrafluoride + Water → Sulfur dioxide + Hydrogen fluoride

5. Complete and balance the following equations: (Combination, a–d; decomposition, e–h; single replacement, i–m; double replacement, n–s.)
(a) $Li + O_2 \rightarrow$
(b) $Al + Cl_2 \rightarrow$
(c) $SO_2 + H_2O \rightarrow$
(d) $MgO + H_2O \rightarrow$
(e) $CuSO_4 \xrightarrow{\Delta} SO_3 +$
(f) $H_2O_2 \xrightarrow{\Delta}$
(g) $Ca(HCO_3)_2 \xrightarrow{\Delta} CaO +$
(h) $MnO_2 \xrightarrow{\Delta} Mn_3O_4 +$
(i) $Al + Fe_2O_3 \rightarrow$
(j) $Pb + AgNO_3 \rightarrow$
(k) $Al + SnCl_2 \rightarrow$
(l) $Br_2 + NaI \rightarrow$
(m) $Ca + H_2O \rightarrow$
(n) $Ba(OH)_2 + HNO_3 \rightarrow$
(o) $(NH_4)_2S + HCl \rightarrow$
(p) $Na_2O + H_2O \rightarrow$
(q) $FeBr_2 + NH_4OH \rightarrow$
(r) $Bi(NO_3)_3 + H_2S \rightarrow$
(s) $CdO + HCl \rightarrow$

6. What is the purpose of balancing equations?
7. What is represented by the numbers that are placed in front of the formulas in a balanced equation?
8. Interpret the meaning of each of the following equations in terms of a chemical

reaction. Give the relative number of moles of each substance involved and indicate whether the reaction is exothermic or endothermic.

(a) $2\ Na + Cl_2 \rightarrow 2\ NaCl + 196.4$ kcal

(b) $PCl_5 + 22.2$ kcal $\rightarrow PCl_3 + Cl_2$

9. Why is an equation balanced when we have the same number of atoms of each kind of element on both sides of the equation?
10. Interpret the following chemical reactions from the point of view of the number of moles of reactants and products:

(a) $HBr + KOH \rightarrow KBr + H_2O$

(b) $MgBr_2 + 2\ AgNO_3 \rightarrow Mg(NO_3)_2 + 2\ AgBr\downarrow$

(c) $N_2 + 3\ H_2 \rightarrow 2\ NH_3$

(d) $3\ Ag + 4\ HNO_3 \rightarrow 3\ AgNO_3 + NO + 2\ H_2O$

(e) $2\ CH_3CH(OH)CH_3 + 9\ O_2 \rightarrow 6\ CO_2 + 8\ H_2O$
Isopropyl alcohol

11. In the reaction $H_2(g) + Br_2(g) \rightarrow 2\ HBr(g) + 24.7$ kcal, the net heat liberated is a result of breaking the bonds of $H_2$ and $Br_2$ molecules and forming the bonds of 2 HBr molecules. Energy is absorbed in breaking bonds and energy is released in forming bonds. Use the bond dissociation energy data give in Section 7.7 and 7.8 to verify that this reaction is exothermic by 24.7 kcal.
12. Which of the following statements are correct?

(a) The coefficients in front of the formulas in a balanced chemical equation give the relative number of moles of the reactants and products in the reaction.

(b) A balanced chemical equation is one that has the same number of moles on each side of the equation.

(c) In a chemical equation, the symbol $\xrightarrow{\Delta}$ indicates that the reaction is exothermic.

(d) A chemical change that absorbs heat energy is said to be endothermic.

(e) In the reaction $H_2 + Cl_2 \rightarrow 2\ HCl$, 100 molecules of HCl are produced for every 50 molecules of $H_2$ reacted.

(f) The symbol (*aq*) after a substance in an equation means that the substance is in a water solution.

(g) The equation $H_2O \rightarrow H_2 + O_2$ can be balanced by placing a 2 in front of $H_2O$.

(h) In the equation $3\ H_2 + N_2 \rightarrow 2\ NH_3$ there are fewer moles of product than there are moles of reactants.

(i) The total number of moles of reactants and products represented by this equation is 5 moles.

$$Mg + 2\ HCl \rightarrow MgCl_2 + H_2$$

(j) One mole of glucose, $C_6H_{12}O_6$, contains 6 moles of carbon atoms.

# 11 Calculations from Chemical Equations

*After studying Chapter 11 you should be able to:*

1. Understand the new terms listed in Question A at the end of the chapter.
2. Give mole ratios for any two substances involved in a chemical reaction.
3. Outline the mole or mole-ratio method for making stoichiometric calculations.
4. Calculate the number of moles of a desired substance obtainable from a given number of moles of a starting substance in a chemical reaction (mole to mole calculations).
5. Calculate the weight of a desired substance obtainable from a given number of moles of a starting substance in a chemical reaction and vice versa (mole to weight and weight to mole calculations).
6. Calculate the weight of a desired substance involved in a chemical reaction from a given weight of a starting substance (weight to weight calculations).
7. Deduce the limiting reagent or reactant when given the amounts of starting substances and then calculate the moles or weight of desired substance obtainable from a given chemical reaction (limiting reagent calculations).
8. Apply theoretical yield or actual yield to any of the foregoing types of problems, or calculate theoretical and actual yields of a chemical reaction.

This chapter shows the quantitative relationship between reactants and products in chemical reactions and also reviews and correlates such relationships as molecular weight, the molecule, the mole concept, and balancing equations.

## 11.1 A Short Review

(a) Molecular weight or formula weight. The molecular weight is the sum of the atomic weights of all the atoms in a molecule. The formula weight is the sum of the atomic weights of all the atoms in a given formula of a compound or an ion. The terms *molecular weight* and *formula weight* are commonly used interchangeably.

(b) Relationship between molecule and mole. A molecule is the smallest unit of a molecular substance (e.g., $Cl_2$) and a mole is Avogadro's Number,

$6.02 \times 10^{23}$, of molecules of that substance. A mole of chlorine ($Cl_2$) has the same number of molecules as a mole of carbon dioxide, a mole of water, or a mole of any other molecular substance. When we relate molecules to gram-molecular weight, 1 g-mol. wt = 1 mole, or $6.02 \times 10^{23}$ molecules.

In addition to molecular substances, the term *mole* may refer to any chemical species. It represents a quantity in grams equal to the formula weight (1 g-form. wt) and may be applied to atoms, ions, electrons, and formula units of nonmolecular substances. For example, a mole of water consists of 18.0 grams of water, or $6.02 \times 10^{23}$ molecules; a mole of sodium is 23.0 grams of sodium, or $6.02 \times 10^{23}$ atoms; a mole of chloride ions is 35.5 grams of chloride ions, or $6.02 \times 10^{23}$ of these ions; a mole of electrons is $5.48 \times 10^{-4}$ grams of electrons, or $6.02 \times 10^{23}$ electrons, and a mole of sodium chloride (a nonmolecular substance) is 58.5 grams of sodium chloride, or $6.02 \times 10^{23}$ formula units.

$$1 \text{ mole} = \begin{cases} 1 \text{ g-mol. wt} = 6.02 \times 10^{23} \text{ molecules} \\ 1 \text{ g-form. wt} = 6.02 \times 10^{23} \text{ formula units} \\ 1 \text{ g-atomic wt} = 6.02 \times 10^{23} \text{ atoms} \\ 1 \text{ g-ionic wt} = 6.02 \times 10^{23} \text{ ions} \end{cases}$$

Other useful mole relationships are

$$\text{Number of moles} = \frac{\text{Grams of a substance}}{\text{Gram-molecular weight of the substance}}$$

$$\text{Number of moles} = \frac{\text{Grams of a monatomic element}}{\text{Gram-atomic weight of the element}}$$

$$\text{Number of moles} = \frac{\text{Number of molecules}}{6.02 \times 10^{23} \text{ molecules/mole}}$$

Two other useful equalities may be derived algebraically from each of these mole relationships. What are they?

(c) Balanced equations. In using equations for calculations of mole–weight–volume relationships between reactants and products, the equations must be balanced. Remember that the number in front of a formula in a balanced equation represents the number of moles of that substance reacting in the chemical change.

## 11.2 Calculations from Chemical Equations: The Mole Method

In chemical work it is often necessary to calculate the amount of a substance that is produced from, or needed to react with, a given quantity of another substance. The area of chemistry that deals with the quantitative relationships among reactants and products is known as **stoichiometry**. Although

stoichiometry

several methods are known, we believe that the *mole* or *mole-ratio* method is best for solving problems in stoichiometry. This method is straightforward and, in our opinion, makes it easy to see and understand the relationships of the reacting species.

mole ratio

A **mole ratio** is a ratio between the number of moles of any two species involved in a chemical reaction. For example in the reaction

$$\underset{2\ \text{moles}}{2H_2} + \underset{1\ \text{mole}}{O_2} \longrightarrow \underset{2\ \text{moles}}{2H_2O}$$

there are six mole ratios that apply only to this reaction:

$$\frac{2\ \text{moles}\ H_2}{1\ \text{mole}\ O_2};\quad \frac{2\ \text{moles}\ H_2}{2\ \text{moles}\ H_2O};\quad \frac{1\ \text{mole}\ O_2}{2\ \text{moles}\ H_2};$$

$$\frac{1\ \text{mole}\ O_2}{2\ \text{moles}\ H_2O};\quad \frac{2\ \text{moles}\ H_2O}{2\ \text{moles}\ H_2};\quad \frac{2\ \text{moles}\ H_2O}{1\ \text{mole}\ O_2}$$

Since stoichiometric problems are encountered throughout the entire field of chemistry, it is profitable to master this general method for their solution. The mole method makes use of three simple basic operations.

A. Conversion of the starting substance to moles (if it is not given in moles).
B. Calculation of the moles of desired substance obtainable from the available moles of starting substance. This is done by multiplying the moles of starting substance (from A) by the mole ratio. This mole ratio is taken from the balanced equation and is the number of moles of the desired substance over the number of moles of the starting substance.
C. Conversion of the moles of desired substance (from B) to whatever units are required.

Like learning to balance chemical equations, learning to make stoichiometric calculations requires some practice. A detailed step-by-step description of the general method, together with a variety of worked examples, is given in the following paragraphs. Study this material and apply the method to the problems at the end of this chapter.

*Step 1.* Use a balanced equation. Write a balanced equation for the chemical reaction in question or check to see that the equation given is balanced.

*Step 2.* Determine the number of moles of starting substance. Identify the starting substance from the data given in the statement of the problem. When the starting substance is given in moles, use it in that form; if it is not in moles, convert the quantity of the starting substance to moles.

*Step 3.* Determine the mole ratio of the desired substance to the starting substance. The number of moles of each substance in the balanced equation is indicated by the coefficient in front of each substance. Use these coefficients to set up the mole ratio:

$$\text{Mole ratio} = \frac{\text{Moles of desired substance in the equation}}{\text{Moles of starting substance in the equation}}$$

*Step 4.* Calculate the number of moles of the desired substance. Multiply the number of moles of starting substance (from Step 2) by the mole ratio (from Step 3) to obtain the number of moles of desired substance:

$$\begin{pmatrix}\text{Moles of desired}\\ \text{substance}\end{pmatrix} = \underset{\text{From Step 2}}{\begin{pmatrix}\text{Moles of starting}\\ \text{substance}\end{pmatrix}} \times \underset{\text{Mole ratio from Step 3}}{\frac{\begin{pmatrix}\text{Moles of desired}\\ \text{substance in}\\ \text{the equation}\end{pmatrix}}{\begin{pmatrix}\text{Moles of starting}\\ \text{substance in}\\ \text{the equation}\end{pmatrix}}}$$

Note that the units of moles of starting substance cancel out in the numerator and the denominator.

*Step 5.* Calculate the desired substance in the units asked for in the problem. If the answer is requested in moles, the problem is finished in Step 4. If units other than moles are wanted, multiply the moles of the desired substance from Step 4 by the appropriate factor to convert moles to the units required.

For example, if grams of the desired substance are wanted,

$$\begin{pmatrix}\text{Grams of desired}\\ \text{substance}\end{pmatrix} = \underset{\text{From Step 4}}{\begin{pmatrix}\text{Moles of desired}\\ \text{substance}\end{pmatrix}} \times \begin{pmatrix}\text{Gram-molecular weight}\\ \text{of desired substance}\end{pmatrix}$$

These steps are summarized in Figure 11.1.

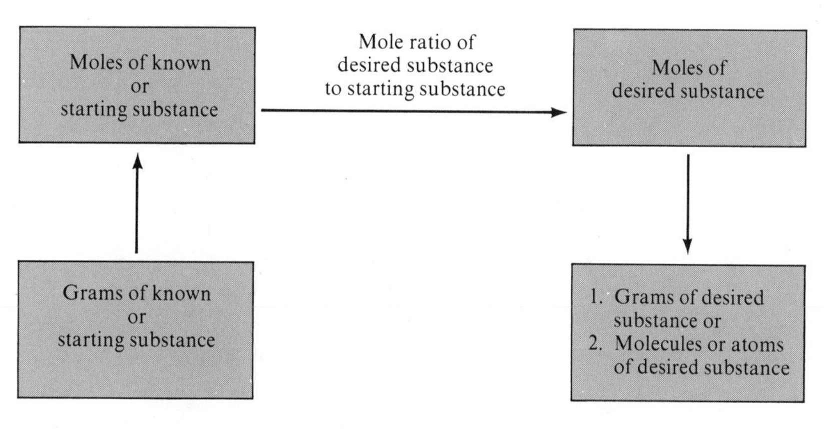


*Figure 11.1.* Basic steps in using the mole ratio to convert moles of one substance to moles of another substance in a chemical reaction.

## 11.3 Mole–Mole Calculations

The object of this type of problem is to calculate the moles of one substance that react with, or are produced from, a given number of moles of another substance. Illustrative problems follow.

*Problem 11.1*

How many moles of water will be produced from 3.5 moles of magnesium hydroxide, $Mg(OH)_2$, according to the following reaction?

$$\underset{1\text{ mole}}{Mg(OH)_2} + \underset{1\text{ mole}}{H_2CO_3} \longrightarrow \underset{1\text{ mole}}{MgCO_3} + \underset{2\text{ moles}}{2\,H_2O}$$

When the equation is balanced, it states that 2 moles of water will be produced from 1 mole of $Mg(OH)_2$. Even though we can quickly ascertain that 7.0 moles of water will be formed from 3.5 moles of $Mg(OH)_2$, the mole method for solving the problem is shown below. This method of working with mole ratios will be very helpful in solving later problems.

*Step 1.* The equation given is balanced.
*Step 2.* The moles of starting substance are 3.5 moles of $Mg(OH)_2$.
*Step 3.* From the balanced equation, set up the mole ratio between the two substances in question, placing the moles of the substance being sought in the numerator and the moles of the starting substance in the denominator. The number of moles, in each case, is the same as the coefficient in front of the substance in the balanced equation.

$$\text{Mole ratio} = \frac{2\text{ moles }H_2O}{1\text{ mole }Mg(OH)_2} \quad \text{(From equation)}$$

*Step 4.* Multiply the 3.5 moles of $Mg(OH)_2$ given in the problem by this mole ratio.

$$3.5\,\cancel{\text{moles }Mg(OH)_2} \times \frac{2\text{ moles }H_2O}{1\,\cancel{\text{mole }Mg(OH)_2}} = 7.0\text{ moles }H_2O \quad \text{(Answer)}$$

Again note the use of units. The moles of $Mg(OH)_2$ cancel, leaving the answer in units of moles of $H_2O$.

---

*Problem 11.2*

How many moles of ammonia can be produced from 8.00 moles of hydrogen reacting with nitrogen?

*Step 1.* First, we need the balanced equation

$$3\,H_2 + N_2 \longrightarrow 2\,NH_3$$

*Step 2.* The moles of starting substance are 8.00 moles of hydrogen.
*Step 3.* The balanced equation states that we get 2 moles of $NH_3$ for every 3 moles of $H_2$ that react. Set up the mole ratio of desired substance ($NH_3$) to starting substance ($H_2$):

$$\text{Mole ratio} = \frac{2\text{ moles }NH_3}{3\text{ moles }H_2}$$

*Step 4.* Multiplying the 8.00 moles of starting $H_2$ by this mole ratio, we get

$$8.00\text{ moles }H_2 \times \frac{2\text{ moles }NH_3}{3\text{ moles }H_2} = 5.33\text{ moles }NH_3 \quad \text{(Answer)}$$

---

*Problem 11.3*

Given the balanced equation

$$\underset{1\text{ mole}}{K_2Cr_2O_7} + \underset{6\text{ moles}}{6\,KI} + 7\,H_2SO_4 \longrightarrow Cr_2(SO_4)_3 + 4\,K_2SO_4 + \underset{3\text{ moles}}{3\,I_2} + 7\,H_2O$$

Calculate (a) the number of moles of potassium dichromate ($K_2Cr_2O_7$) that will react with 2.0 moles of potassium iodide (KI); (b) the number of moles of iodine ($I_2$) that will be produced from 2.0 moles of potassium iodide.

After the equation is balanced, we are concerned only with $K_2Cr_2O_7$, KI, and $I_2$, and we can ignore all the other substances. The equation states that 1 mole of $K_2Cr_2O_7$ will react with 6 moles of KI to produce 3 moles of $I_2$.

(a) Calculate the moles of $K_2Cr_2O_7$.

*Step 1.* The equation given is balanced.

*Step 2.* The moles of starting substance are 2.0 moles of KI.

*Step 3.* Set up the mole ratio of desired substance to starting substance:

$$\text{Mole ratio} = \frac{1 \text{ mole } K_2Cr_2O_7}{6 \text{ moles KI}} \quad \text{(From equation)}$$

*Step 4.* Multiply the moles of starting material by this ratio to obtain the answer.

$$2.0 \text{ moles KI} \times \frac{1 \text{ mole } K_2Cr_2O_7}{6 \text{ moles KI}} = 0.33 \text{ mole } K_2Cr_2O_7 \quad \text{(Answer)}$$

(b) Calculate the moles of $I_2$.

*Steps 1 and 2.* The equation given is balanced and the moles of starting substance are 2.0 moles KI as in part (a).

*Step 3.* Set up the mole ratio of desired substance to starting substance:

$$\text{Mole ratio} = \frac{3 \text{ moles } I_2}{6 \text{ moles KI}} \quad \text{(From equation)}$$

*Step 4.* Multiply the moles of starting material by this ratio to obtain the answer.

$$2.0 \text{ moles KI} \times \frac{3 \text{ moles } I_2}{6 \text{ moles KI}} = 1.0 \text{ mole } I_2 \quad \text{(Answer)}$$

Thus, 2.0 moles KI will react with 0.33 moles $K_2Cr_2O_7$ to produce 1.0 mole $I_2$.

---

*Problem 11.4*

How many molecules of water can be produced by reacting 0.010 mole of oxygen with hydrogen?

The sequence of conversions needed in the calculation is

$$\text{Moles } O_2 \longrightarrow \text{Moles } H_2O \longrightarrow \text{Molecules } H_2O$$

*Step 1.* First, we write the balanced equation:

$$\underset{\phantom{1 \text{ mole}}}{2H_2} + \underset{1 \text{ mole}}{O_2} \longrightarrow \underset{2 \text{ moles}}{2H_2O}$$

*Step 2.* The moles of starting substance is 0.010 mole $O_2$.

*Step 3.* Set up the mole ratio of desired substance to starting substance:

$$\text{Mole ratio} = \frac{2 \text{ moles } H_2O}{1 \text{ mole } O_2} \quad \text{(From equation)}$$

*Step 4.* Multiplying the 0.010 mole of oxygen by this ratio, we obtain

$$0.010 \text{ mole } O_2 \times \frac{2 \text{ moles } H_2O}{1 \text{ mole } O_2} = 0.020 \text{ mole } H_2O$$

*Step 5.* Since the problem asks for molecules instead of moles of $H_2O$, we must convert moles to molecules. Use the conversion factor ($6.02 \times 10^{23}$ molecules)/mole.

$$0.020 \text{ mole } H_2O \times \frac{6.02 \times 10^{23} \text{ molecules}}{\text{mole}} = 1.2 \times 10^{22} \text{ molecules } H_2O$$

We should note that 0.020 mole is still quite a large number of water molecules

## 11.4 Mole–Weight and Weight–Weight Calculations

The object of these types of problems is to calculate the weight of one substance that reacts with, or is produced from, a given number of moles or a given weight of another substance in a chemical reaction. The mole ratio is used to convert from moles of starting substance to moles of desired substance.

*Problem 11.5*

What weight of hydrogen can be produced by reacting 6.0 moles of aluminum with hydrochloric acid?

First calculate the moles of hydrogen produced, using the mole-ratio method, and then calculate the weight of hydrogen by multiplying the moles of hydrogen by its weight per mole. The sequence of conversions in the calculation is

$$\text{Moles Al} \longrightarrow \text{Moles } H_2 \longrightarrow \text{Grams } H_2$$

*Step 1.* The balanced equation is

$$\underset{\text{2 moles}}{2\,Al(s)} + 6\,HCl(aq) \longrightarrow 2\,AlCl_3(aq) + \underset{\text{3 moles}}{3\,H_2(g)}$$

*Step 2.* The moles of starting substance are 6.0 moles of aluminum.
*Steps 3 and 4.* Calculate moles of $H_2$.

$$6.0 \text{ moles Al} \times \frac{3 \text{ moles } H_2}{2 \text{ moles Al}} = 9.0 \text{ moles } H_2$$

*Step 5.* Convert moles of $H_2$ to grams [g = moles × (g/mole)]:

$$9.0 \text{ moles } H_2 \times \frac{2.0 \text{ g } H_2}{1 \text{ mole } H_2} = 18 \text{ g } H_2$$

We see that 18 g of $H_2$ can be produced by reacting 6.0 moles of Al with HCl. The following setup combines all the above steps into one continuous calculation:

$$6.0 \text{ moles Al} \times \frac{3 \text{ moles } H_2}{2 \text{ moles Al}} \times \frac{2.0 \text{ g } H_2}{1 \text{ mole } H_2} = 18 \text{ g } H_2$$

*Problem 11.6*

What weight of carbon dioxide is produced by the complete combustion of 100 g of propane gas, $C_3H_8$?

The sequence of conversions in the calculation is

$$\text{Grams } C_3H_8 \longrightarrow \text{Moles } C_3H_8 \longrightarrow \text{Moles } CO_2 \longrightarrow \text{Grams } CO_2$$

*Step 1.* The balanced equation is

$$\underset{\text{1 mole}}{C_3H_8} + 5\,O_2 \rightarrow \underset{\text{3 moles}}{3\,CO_2} + 4\,H_2O$$

*Step 2.* The starting substance is 100 g of $C_3H_8$. Convert 100 g of $C_3H_8$ to moles [moles = g × (moles/g)]:

$$100 \text{ g } C_3H_8 \times \frac{1 \text{ mole } C_3H_8}{44.0 \text{ g } C_3H_8} = 2.27 \text{ moles } C_3H_8$$

*Steps 3 and 4.* Calculate the moles of $CO_2$ by the mole-ratio method.

$$2.27 \text{ moles } C_3H_8 \times \frac{3 \text{ moles } CO_2}{1 \text{ mole } C_3H_8} = 6.81 \text{ moles } CO_2$$

*Step 5.* Convert the moles of $CO_2$ to grams.

Moles $CO_2$ × Gram-molecular weight $CO_2$ = Grams $CO_2$

$$6.81 \text{ moles } CO_2 \times \frac{44.0 \text{ g } CO_2}{1 \text{ mole } CO_2} = 300 \text{ g } CO_2 \quad \text{(Answer)}$$

We see that 300 g of $CO_2$ are produced from the complete combustion of 100 g of $C_3H_8$. The calculation in a continuous setup is

$$\boxed{100 \text{ g } C_3H_8} \times \boxed{\frac{1 \text{ mole } C_3H_8}{44.0 \text{ g } C_3H_8}} \times \boxed{\frac{3 \text{ moles } CO_2}{1 \text{ mole } C_3H_8}} \times \boxed{\frac{44.0 \text{ g } CO_2}{1 \text{ mole } CO_2}} = 300 \text{ g } CO_2$$

Grams $C_3H_8$ → Moles $C_3H_8$ → Moles $CO_2$ → Grams $CO_2$

Note that in the continuous setup, since both $C_3H_8$ and $CO_2$ have the same molecular weight, the numbers cancel, saving two steps in the calculation.

---

## 11.5 Limiting Reagent and Yield Calculations

In many chemical processes the quantities of the reactants used are such that the moles of one reactant are in excess of the moles of a second reactant in the reaction. The amount of the product(s) formed in such a case will be dependent on the reactant that is not in excess. Thus, the reactant that is not in excess is known as the **limiting reagent**, since it limits the amount of product that can be formed.

limiting reagent

As an example, consider the case where solutions containing 1.0 mole of sodium hydroxide and 1.5 moles of hydrochloric acid are mixed:

$$\underset{1 \text{ mole}}{NaOH} + \underset{1 \text{ mole}}{HCl} \longrightarrow \underset{1 \text{ mole}}{NaCl} + \underset{1 \text{ mole}}{H_2O}$$

According to the equation it is possible to obtain 1.0 mole of NaCl from 1.0 mole of NaOH, and 1.5 moles of NaCl from 1.5 moles of HCl. However, we cannot have two different yields of NaCl from the reaction. When 1.0 mole of NaOH and 1.5 moles of HCl are mixed, there is insufficient NaOH to react with all of the HCl. Therefore, HCl is the reagent in excess and NaOH is the limiting reagent. Since the NaCl formed is dependent on the limiting reagent, the amount of NaCl formed will be 1.0 mole. Since 1.0 mole of NaOH reacts with 1.0 mole of HCl, 0.5 mole of HCl remains unreacted.

Problems giving the amounts of two reactants are generally of the limiting reagent type and may be solved in the following manner:

1. Calculate the number of moles of each substance used.
2. Determine which substance is the limiting reagent.
   (a) Compare the ratio of the moles calculated with the mole ratio of the two substances in the balanced equation. Or
   (b) Determine the number of moles of the product that can be formed from each substance to see which produces the least amount.
3. Using the moles of the limiting reagent, calculate the amount of the product in the units asked for in the problem.

*Problem 11.7*

How many grams of silver bromide, AgBr, can be formed when solutions containing 50.0 g of $MgBr_2$ and 100 g of $AgNO_3$ are mixed together?

$$MgBr_2(aq) + 2\ AgNO_3(aq) \longrightarrow 2\ AgBr\downarrow + Mg(NO_3)_2(aq)$$

*Step 1.* Moles of $MgBr_2$ and $AgNO_3$:

$$50.0\ \text{g}\ MgBr_2 \times \frac{1\ \text{mole}\ MgBr_2}{184.1\ \text{g}\ MgBr_2} = 0.272\ \text{mole}\ MgBr_2$$

$$100\ \text{g}\ AgNO_3 \times \frac{1\ \text{mole}\ AgNO_3}{169.9\ \text{g}\ AgNO_3} = 0.589\ \text{mole}\ AgNO_3$$

*Step 2.* Determine which is the limiting reagent, $MgBr_2$ or $AgNO_3$.

$$\text{Ratio of moles calculated} = \frac{0.589\ \text{mole}\ AgNO_3}{0.272\ \text{mole}\ MgBr_2} = \frac{2.16\ \text{mole}\ AgNO_3}{1\ \text{mole}\ MgBr_2}$$

$$\text{Mole ratio in the equation} = \frac{2\ \text{moles}\ AgNO_3}{1\ \text{mole}\ MgBr_2}$$

Therefore, $AgNO_3$ is in excess, and $MgBr_2$ is the limiting reagent.

*Step 3.* Calculate the grams of AgBr using the moles of the limiting reagent.

$$\text{Moles}\ MgBr_2 \longrightarrow \text{Moles AgBr} \longrightarrow \text{Grams AgBr}$$

$$0.272\ \text{mole}\ MgBr_2 \times \frac{2\ \text{moles AgBr}}{1\ \text{mole}\ MgBr_2} \times \frac{187 \cdot 8\ \text{g AgBr}}{1\ \text{mole AgBr}} = 102\ \text{g AgBr}$$

Many reactions, especially those involving organic substances, fail to give a 100% yield of product. The main reasons for this failure are the side reactions that give products other than the main product and the fact that many reactions are reversible. In addition, some product may be lost in handling and transferring from one vessel to another. The yield calculated from the chemical equation is commonly known as the **theoretical yield**; this is the maximum amount of product that can be produced according to the equation. The **actual yield** is the amount of product that we finally obtain.

theoretical yield

actual yield

The percentage yield is calculated as follows:

$$\frac{\text{Actual yield}}{\text{Theoretical yield}} \times 100\% = \text{Percentage yield}$$

*Problem 11.8*

Carbon tetrachloride was prepared by reacting 100 g of carbon disulfide and 100 g of chlorine. Calculate the percentage yield if 65.0 g of $CCl_4$ was obtained from the reaction.

$$CS_2 + 3\ Cl_2 \longrightarrow CCl_4 + S_2Cl_2$$

In this problem we need to determine the limiting reagent in order to calculate the quantity of $CCl_4$ (theoretical yield) that can be formed according to the equation for the reaction. Then we can compare this amount with the 65.0 g $CCl_4$ actual yield to calculate the percentage yield.

*Step 1.* Moles of $CS_2$ and $Cl_2$ and the limiting reagent:

$$100\ \text{g}\ CS_2 \times \frac{1\ \text{mole}\ CS_2}{76.2\ \text{g}\ CS_2} = 1.31\ \text{moles}\ CS_2$$

$$100\ \text{g}\ Cl_2 \times \frac{1\ \text{mole}\ Cl_2}{71.0\ \text{g}\ Cl_2} = 1.41\ \text{moles}\ Cl_2$$

From the equation we see that 3 moles of $Cl_2$ are needed to react with 1 mole of $CS_2$. But we have only

$$\frac{1.41\ \text{moles}\ Cl_2}{1.31\ \text{moles}\ CS_2} = 1.08\ \text{moles}\ Cl_2\ \text{per mole}\ CS_2.$$

Therefore $Cl_2$ is the limiting reagent and is used to calculate the theoretical yield of $CCl_4$.

*Step 2.* Grams of $CCl_4$:

$$\text{Moles}\ Cl_2 \longrightarrow \text{Moles}\ CCl_4 \longrightarrow \text{Grams}\ CCl_4$$

$$1.41\ \text{moles}\ Cl_2 \times \frac{1\ \text{mole}\ CCl_4}{3\ \text{moles}\ Cl_2} \times \frac{154\ \text{g}\ CCl_4}{1\ \text{mole}\ CCl_4} = 72.4\ \text{g}\ CCl_4$$

*Step 3.* Percentage yield. According to the equation, 72.4 g $CCl_4$ is the maximum amount or theoretical yield of $CCl_4$ possible from 100 g $Cl_2$. Actual yield is 65.0 g $CCl_4$.

$$\text{Percentage yield} = \frac{65.0\ \text{g}}{72.4\ \text{g}} \times 100\% = 89.8\%$$

---

When solving problems, you will achieve better results if at first you do not try to take shortcuts. Write the data and numbers in a logical, orderly manner. Make certain that the equations are balanced and that the computations are accurate and expressed to the correct number of significant figures. Remember that units are very important; a number without units has little meaning. Finally, an electronic calculator can save you many hours of tedious computations.

## Questions

*A. Review the meaning of the new terms introduced in this chapter.*

1. Stoichiometry
2. Mole ratio
3. Limiting reagent
4. Theoretical yield
5. Actual yield

*B. Review problems.*

(In some of the following problems the equations shown are not balanced.)

1. Calculate the number of moles in each of the following quantities:
   (a) 25.0 g $MnO_2$
   (b) 350 g $H_2SO_4$
   (c) 45.0 g $Br_2$
   (d) 1.00 g $CCl_4$
   (e) 1.00 kg NaCl
   (f) 11.5 g $C_2H_6O$
   (g) 410.0 g $CO_2$
   (h) 60.0 g $O_2$
   (i) 100 millimoles $HNO_3$
   (j) 250 ml concentrated HCl ($d = 1.19$ g/ml, 37% HCl by weight)
2. Calculate the number of grams in each of the following quantities:
   (a) 0.400 mole $C_3H_8$
   (b) 1.50 moles Al
   (c) 7.50 moles $H_2$
   (d) 0.100 mole $AgNO_3$
   (e) 1.00 mole $FeSO_4$
   (f) 0.250 mole $AlCl_3$
   (g) 0.600 mole Au
   (h) 50 ml $Ni(NO_3)_2$ solution ($d = 1.107$ g/ml, 12.0% $Ni(NO_3)_2$ by weight)
3. Set up all the possible mole ratios between the reactants and the products in the following equations:
   (a) $2\ Mg + O_2 \rightarrow 2\ MgO$
   (b) $2\ Al + 6\ HCl \rightarrow 2\ AlCl_3 + 3\ H_2$
   (c) $3\ Zn + N_2 \rightarrow Zn_3N_2$
   (d) $2\ C_2H_6 + 7\ O_2 \rightarrow 4\ CO_2 + 6\ H_2O$
4. Which contains the larger number of molecules?
   (a) 8.0 g of $CH_4$ or 30.0 g of $SO_2$ (b) 4.0 g of $CO_2$ or 4.0 g of CO
5. An early method of producing chlorine was by the reaction of pyrolusite, $MnO_2$, and hydrochloric acid. How many moles of HCl will react with 0.85 mole of $MnO_2$? (Balance the equation first.)

   $$MnO_2(s) + HCl(aq) \longrightarrow Cl_2(g) + MnCl_2(aq) + H_2O$$

6. How many moles of carbon dioxide and how many moles of sulfur dioxide can be formed by the reaction of 4.0 moles of carbon disulfide with oxygen?

   $$CS_2 + O_2 \longrightarrow CO_2 + SO_2$$

7. How many moles of oxygen are needed to react with 3.5 moles of isopropyl alcohol, $C_3H_7OH$?

   $$2\ C_3H_7OH + 9\ O_2 \longrightarrow 6\ CO_2 + 8\ H_2O$$

8. How many moles of each reactant are required to produce 0.80 mole of sodium sulfate, $Na_2SO_4$, according to the following equation?

   $$3\ Na_2S_2O_3 + 8\ KMnO_4 + H_2O \rightarrow 3\ Na_2SO_4 + 3\ K_2SO_4 + 8\ MnO_2 + 2\ KOH$$

9. What weight of nitric acid is needed to convert 0.125 mole of copper to copper(II) nitrate, $Cu(NO_3)_2$?

   $$3\ Cu(s) + 8\ HNO_3(aq) \longrightarrow 3\ Cu(NO_3)_2(aq) + 2\ NO(g) + 4\ H_2O$$

10. How many moles of oxygen are needed to completely burn 500 g of pentane, $C_5H_{12}$?

    $$C_5H_{12}(l) + O_2(g) \longrightarrow CO_2(g) + H_2O(g)$$

11. Given the equation

    $$6\ FeCl_2 + 14\ HCl + K_2Cr_2O_7 \longrightarrow 6\ FeCl_3 + 2\ KCl + 2\ CrCl_3 + 7\ H_2O$$

    (a) How many moles of $FeCl_3$ will be formed from 1.0 mole of $FeCl_2$?
    (b) How many moles of HCl will react with 2.5 moles of $FeCl_2$?
    (c) How many grams of $K_2Cr_2O_7$ will react with 0.50 mole of $FeCl_2$?
    (d) How many grams of $CrCl_3$ will be formed from 1.00 g of $K_2Cr_2O_7$?
    (e) How many moles of $H_2O$ are formed when 2.0 moles of $FeCl_3$ are formed?
    (f) How many grams of $FeCl_3$ are formed from 6.00 g of $FeCl_2$?

12. Oyster shells are essentially pure limestone, $CaCO_3$. When heated, they form quicklime, CaO, and carbon dioxide. What weight of quicklime can be produced from 1.00 kg of oyster shells?

    $$CaCO_3 \xrightarrow{\Delta} CaO + CO_2$$

13. The essential reactions in the production of steel (Fe) from iron ore in a blast furnace are

    $$\underset{\text{Coke}}{2\ C} + O_2 \xrightarrow{\Delta} 2\ CO$$

    $$3\ CO + Fe_2O_3 \longrightarrow 2\ Fe + 3\ CO_2$$

    How many kilograms of coke are needed to produce a metric ton (1000 kg) of steel?

14. What weight of steam and iron must react in order to produce 250 g of magnetic iron oxide, $Fe_3O_4$?

    $$Fe(s) + H_2O(g) \longrightarrow Fe_3O_4(s) + H_2\uparrow$$

15. Oxygen gas is obtained when potassium nitrate, $KNO_3$, is heated.

    $$KNO_3(s) \xrightarrow{\Delta} KNO_2(s) + O_2(g)$$

    What will be the weight loss when 22.4 g of $KNO_3$ are heated?

16. Calculate the weight of the *first product* listed in each equation that could be obtained by starting with 15.0 g of the *first reactant* listed in each equation.
    (a) $2\ Na_3PO_4 + 3\ H_2SO_4 \rightarrow 2\ H_3PO_4 + 3\ Na_2SO_4$
    (b) $4\ FeS_2 + 11\ O_2 \rightarrow 2\ Fe_2O_3 + 8\ SO_2$
    (c) $11\ Si + 4\ NpF_3 \rightarrow 3\ SiF_4 + 4\ NpSi_2$
    (d) $6\ B_2Cl_4 + 3\ O_2 \rightarrow 2\ B_2O_3 + 8\ BCl_3$

17. Both $CaCl_2$ and $MgCl_2$ react with $AgNO_3$ to precipitate AgCl. When solutions containing equal weights of $CaCl_2$ and $MgCl_2$ are reacted, which salt will produce the most AgCl? Show proof.

18. In the following equations, determine which reactant is the limiting reagent and which reactant is in excess. The amounts mixed together are shown below each reactant. Show evidence for your answers.
    (a) $\underset{3.00\ g}{KOH} + \underset{2.20\ g}{HCl} \rightarrow KCl + H_2O$
    (b) $\underset{25.0\ g}{2\ Bi(NO_3)_3} + \underset{8.0\ g}{3\ H_2S} \rightarrow Bi_2S_3 + 6\ HNO_3$

(c) $3\,Fe + 4\,H_2O \rightarrow Fe_3O_4 + 4\,H_2$
55.8 g 18.0 g

(d) $2\,C_2H_6 + 7\,O_2 \rightarrow 4\,CO_2 + 6\,H_2O$
100 g 400 g

19. Methyl alcohol, $CH_3OH$, is made by reacting carbon monoxide and hydrogen in the presence of certain metal oxide catalysts. How much alcohol can be obtained by reacting 500 g CO and 100 g $H_2$?

 $CO(g) + 2\,H_2(g) \longrightarrow CH_3OH(l)$

20. Iron was reacted with a solution containing 100 g of copper(II) sulfate. The reaction was stopped after 1 hour, and 37.4 g of copper were obtained. Calculate the percentage yield of copper obtained.

 $Fe(s) + CuSO_4(aq) \longrightarrow Cu(s) + FeSO_4(aq)$

21. Calcium carbide, $CaC_2$, is used for generating acetylene. It is made by reacting lime, CaO, and coke, C, in an electric furnace at 3000°C.

 $CaO + 3\,C \longrightarrow CaC_2 + CO$

 How many grams of $CaC_2$ can be prepared by reacting 1000 g of CaO and 500 g of C?

22. An astronaut excretes about 2500 g of water a day. If lithium oxide, $Li_2O$, is used in the spaceship to absorb this water, how many kilograms of $Li_2O$ must be carried for a 30-day space trip for three astronauts?

 $Li_2O + H_2O \longrightarrow 2\,LiOH$

23. The equation representing the reaction used for the commercial preparation of hydrogen cyanide is

 $2\,CH_4 + 3\,O_2 + 2\,NH_3 \longrightarrow 2\,HCN + 6\,H_2O$

 Which of the following statements are correct?

 (a) Three moles of $O_2$ are required for 2 moles of $NH_3$.
 (b) Twelve moles of HCN are produced for every 16 moles of $O_2$ that react.
 (c) The mole ratio between $H_2O$ and $CH_4$ is

 $$\frac{6 \text{ moles } H_2O}{2 \text{ moles } CH_4}$$

 (d) When 12 moles of HCN are produced, 4 moles of $H_2O$ will also be formed.
 (e) When 10 moles $CH_4$, 10 moles $O_2$, and 10 moles $NH_3$ are mixed and reacted, $O_2$ is the limiting reagent.
 (f) When 3 moles each of $CH_4$, $O_2$, and $NH_3$ are mixed and reacted, 3 moles of HCN will be produced.

# 12 The Gaseous State of Matter

*After studying Chapter 12 you should be able to:*

1. Understand the terms listed in Question A at the end of the chapter.
2. State the principal assumptions of the Kinetic-Molecular Theory (KMT).
3. Estimate the relative rates of diffusion of two gases of known molecular weights.
4. Sketch and explain the operation of a mercury barometer.
5. Tell what two factors determine gas pressure in a vessel of fixed volume.
6. Work problems involving (a) Boyle's and (b) Charles' gas laws.
7. State what is meant by standard temperature and pressure (STP).
8. Give the equation for the combined gas law that deals with the pressure, volume, and temperature relationships expressed in Boyle's, Charles', and Gay-Lussac's gas laws.
9. Use Dalton's Law of Partial Pressures and the general gas laws to calculate the dry STP volume of a gas collected over water.
10. State Avogadro's hypothesis.
11. Determine the density of any gas at STP.
12. Determine the molecular weight of a gas from its density at a known temperature and pressure.
13. Make mole-to-volume, weight-to-volume, and volume-to-volume stoichiometric calculations from a balanced chemical equation.
14. Define an ideal gas.
15. State two valid reasons why real gases may deviate from the behavior predicted for an ideal gas.
16. Solve problems involving the ideal gas equation.

## 12.1 General Properties of Gases

In Chapter 3, solids, liquids, and gases are described in a brief outline. In this chapter we will consider the behavior of gases in greater detail.

Of the three states of matter, gases are the least compact and most mobile. A solid has a rigid structure and its particles remain in essentially fixed positions. When a solid absorbs sufficient heat, it melts and changes into a liquid. Melting occurs because the molecules (or ions) have absorbed enough energy to break out of the rigid crystal lattice structure of the solid. The molecules or ions in the liquid are more energetic than they were in the solid, as shown by their increased mobility. Molecules in the liquid state are coherent—that is, they cling to one another. When the liquid absorbs additional heat, the more energetic molecules break away from the liquid surface and go into the gaseous state. Gases represent the most mobile state of matter. Gas molecules move with very high velocities and have high kinetic energy (KE). The average velocity of hydrogen molecules at 0°C is over 1600 metres (1 mile) per second. Because of the high velocities of their molecules, mixtures of gases are uniformly distributed within the container in which they are confined.

A quantity of a substance occupies a much greater volume as a gas than does a like quantity of the substance as a liquid or a solid. For example, 1 mole of water (18 g) has a volume of 18 ml at 4°C. This same amount of water would occupy about 22,400 ml in the gaseous state—more than a 1200-fold increase in volume. We may assume from this difference in volume that (1) gas molecules are relatively far apart, (2) gases are capable of being greatly compressed, and (3) the volume occupied by a gas is mostly empty space.

## 12.2 The Kinetic-Molecular Theory

Careful scientific studies of the behavior and properties of gases were begun in the 17th century by Robert Boyle (1627–1691). This work was carried forward by many investigators after Boyle. The accumulated data were used in the second half of the 19th century to formulate a general theory to explain the behavior and properties of gases, called the **Kinetic-Molecular Theory (KMT)**. The KMT has since been extended to cover, in part, the behavior of liquids and solids. It ranks today with the atomic theory as one of the greatest generalizations of modern science.

Kinetic-Molecular Theory (KMT)

The KMT is based on the motion of particles, particularly gas molecules. A gas that behaves exactly as outlined by the theory is known as an **ideal**, or **perfect**, **gas**. Actually, there are no ideal gases, but under certain conditions of temperature and pressure, gases approach ideal behavior, or at least show only small deviations from it. Under extreme conditions, such as very high pressure and low temperature, real gases may deviate greatly from ideal behavior. For example, at low temperature and high pressures many gases become liquids.

ideal, or perfect, gas

The principal assumptions of the Kinetic-Molecular Theory are

1. Gases consist of tiny (submicroscopic) molecules.
2. The distance between molecules is large compared to the size of the molecules themselves. The volume occupied by a gas consists mostly of empty space.
3. Gas molecules have no attraction for each other.
4. Gas molecules move in straight lines in all directions, colliding frequently with each other and with the walls of the container.
5. No energy is lost by the collision of a gas molecule with another gas molecule or with the walls of the container. All collisions are perfectly elastic.
6. The average kinetic energy for molecules is the same for all gases at the same temperature, and its value is directly proportional to the Kelvin temperature.

Let us consider the facts supporting the theory. Assumption 1 above is based on the size of atoms and molecules, already established in previous chapters. Assumptions 2 and 3 are based on the comparison of volumes occupied by equal masses of the solid, liquid, and gaseous states of a substance and the fact that gases continue to expand and completely fill any size container. Assumption 4, that gases are in constant motion, is shown by the fact that gases exert pressure, expand into larger containers, and diffuse.

diffusion

The property of **diffusion**, the ability of two or more gases to spontaneously mix, also supports the assumption that gas molecules have very little attraction for each other. The diffusion of gases may be illustrated by use of the apparatus shown in Figure 12.1. Two large flasks, one containing reddish-brown bromine vapors and the other dry air, are connected by a side tube. When the stopcock between the flasks is opened, the bromine and air will diffuse into each other. After standing awhile, both flasks will contain bromine and air.

Figure 12.2 shows that a gas exerts the same pressure at all parts of a container. The three gauges, located at different parts of the cylinder, show the same pressure.

With billions of molecules present in even a very small mass of gas, it is safe to assume that there will be collisions between these molecules as well as collisions with the walls of the container. Assumption 5 above is borne out by the fact that gases do not change temperature upon standing (external causes excepted). This shows that the molecules do not suffer loss of energy by collisions. Although one molecule may transfer energy to another molecule in a collision, the average or total energy of the system remains the same. The kinetic energy (KE) of a molecule is one-half of its mass times its velocity

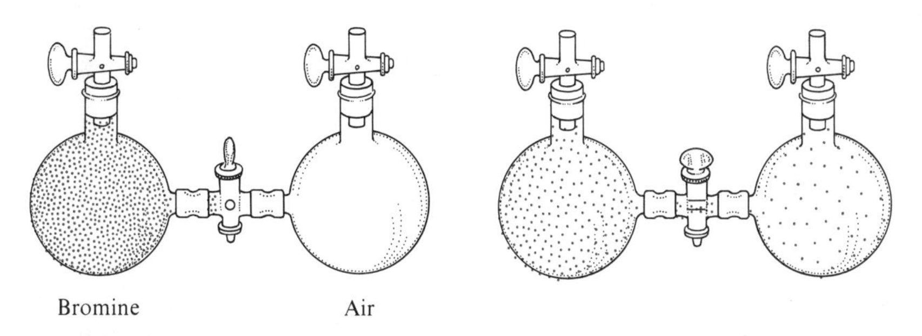


*Figure 12.1.* Diffusion of gases. When the stopcock between the two flasks is opened, colored bromine molecules can be seen diffusing into the flask containing air.

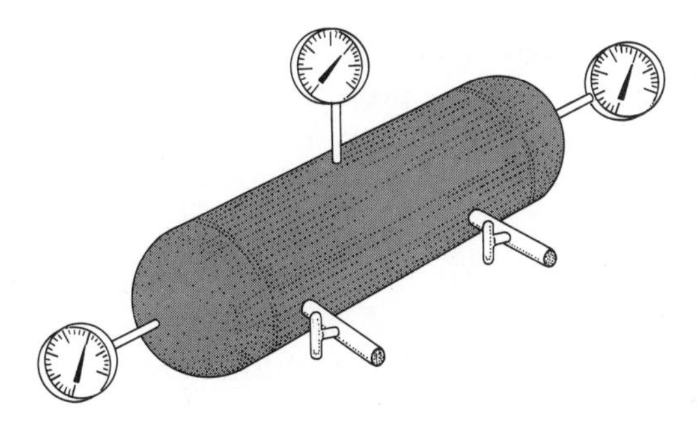

*Figure 12.2.* A gas moves in all directions and exerts the same pressure in all directions.

squared. It is expressed by the equation

$$KE = \frac{1}{2} mv^2$$

where $m$ is the mass and $v$ is the velocity of the molecule.

Experimental evidence shows that 2.0 g $H_2$ (1 mole) and 32.0 g $O_2$ (1 mole) in containers of equal volume at the same temperature exert the same pressure. This evidence supports assumption 6—that the kinetic energy for all gases is the same at the same temperature—and leads us to reason that the molecules of different gases, because of differing masses, will have different average velocities. The *relative* molecular velocities of different gases can be calculated from their kinetic energies. For example, the mass of any oxygen molecule is 32 amu and that of hydrogen is 2 amu. From the Kinetic-Molecular Theory, we have

$$\text{KE of } H_2 = \text{KE of } O_2$$

$$\frac{1}{2} m_{H_2} v^2_{H_2} = \frac{1}{2} m_{O_2} v^2_{O_2}$$

$$\frac{1}{2} \times 2 \times v^2_{H_2} = \frac{1}{2} \times 32 \times v^2_{O_2}$$

$$\frac{v^2_{H_2}}{v^2_{O_2}} = \frac{16}{1}$$

Taking the square root of both sides of the equation, we have

$$\frac{v_{H_2}}{v_{O_2}} = \frac{4}{1}$$

These calculations show that the average velocity of a hydrogen molecule is four times greater than that of an oxygen molecule.

The rates of diffusion of different gases are directly proportional to their molecular velocities. Inspection of the foregoing equations shows that molecular velocities—and therefore the rates of diffusion—of different gases are inversely proportional to the square roots of their molecular weights. This principle was first introduced by the Scottish chemist Thomas Graham (1805–1869) and is known as **Graham's law of diffusion**: The rates of diffusion of different gases are inversely proportional to the square roots of their molecular weights (or densities).

Graham's law of diffusion

These properties of an ideal gas are independent of the molecular constitution of the gas. Mixtures of gases also obey the Kinetic-Molecular Theory if the gases in the mixture do not enter into a chemical reaction with each other.

## 12.3 Measurement of Pressure of Gases

pressure

**Pressure** is defined as force per unit area. Do gases exert pressure? Yes. When a rubber balloon is inflated with air, it stretches and maintains an abnormally large size because the pressure on the inside is greater than that on the

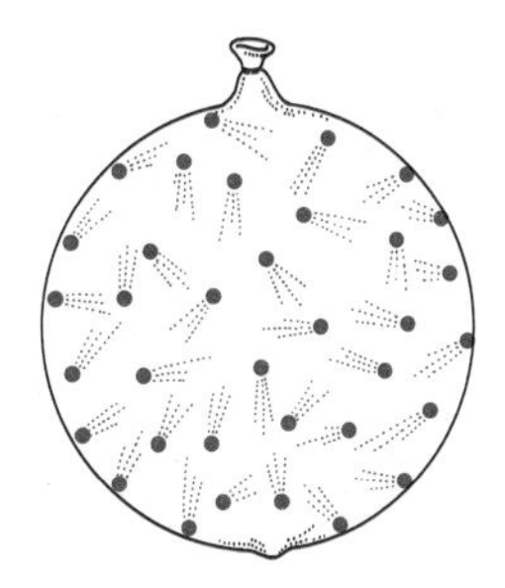

*Figure 12.3.* Pressure resulting from the collisions of gas molecules with the walls of the balloon keep the balloon inflated.

outside. Pressure results from the collisions of gas molecules with the walls of the balloon (see Figure 12.3). When the gas is released, the force or pressure of the air escaping from the small neck propels the balloon in a rapid, irregular path. If the balloon is inflated until it bursts, the gas escaping all at once causes a small explosive noise. This pressure that gases display can be measured; it can also be transformed into useful work. Steam under pressure, as used in the locomotive, played an important role in the early development of the United States. Compressed steam is used today to generate at least part of the electricity for many cities. Compressed air is used to operate many different kinds of mechanical equipment.

The mass of air surrounding the earth is called the *atmosphere*. It is composed of about 78% nitrogen, 21% oxygen, and 1% argon and other minor constituents (see Table 12.1). The outer boundary of the atmosphere is not known precisely, but more than 99% of the atmosphere is below an altitude of 20 miles (32 km). Thus, the concentration of gas molecules in the atmosphere decreases with altitude, and at about 4 miles, there is insufficient oxygen to sustain human life. The gases in the atmosphere exert a pressure known as **atmospheric pressure**. The pressure exerted by a gas depends on the number of molecules of gas present, the temperature, and the volume in which the gas is confined. Gravitational forces confine the atmosphere relatively close to the earth and act to prevent air molecules from flying off into outer space. Thus, the atmospheric pressure at any point is due to the weight of the atmosphere pressing downward at that point.

atmospheric pressure

The pressure of a gas can be measured with a pressure gauge, a manometer, or a **barometer**. A mercury barometer is commonly used in the laboratory

barometer

*Table 12.1.* Average composition of normal dry air.

| Gas | Percent by volume | Gas | Percent by volume |
|---|---|---|---|
| $N_2$ | 78.08 | He | 0.0005 |
| $O_2$ | 20.95 | $CH_4$ | 0.0002 |
| Ar | 0.93 | Kr | 0.0001 |
| $CO_2$ | 0.033 | Xe, $H_2$, and $N_2O$ | Trace |
| Ne | 0.0018 | | |

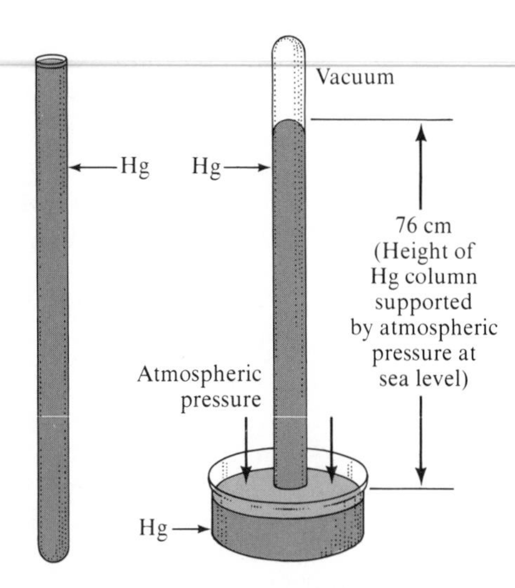


*Figure 12.4.* Preparation of a mercury barometer. The full tube of mercury at the left is inverted in a dish of mercury.

to measure atmospheric pressure. A simple barometer of this type may be prepared by filling a long tube with pure, dry mercury and inverting the open end into an open dish of mercury. If the tube is longer than 76 cm, the mercury level will drop to a point at which the column of mercury in the tube is just supported by the pressure of the atmosphere. If the tube is properly prepared, a vacuum will exist above the mercury column. The weight of mercury, per unit area, is equal to the pressure of the atmosphere. The column of mercury is supported by the pressure of the atmosphere, and the height of the column is a measure of this pressure (see Figure 12.4). The mercury barometer was invented in 1643 by the Italian physicist E. Torricelli (1608–1647), for whom the unit of pressure *torr* was named.

1 atmosphere

The average pressure of the atmosphere at sea level is **1 atmosphere** (abbreviated as atm). This pressure is equivalent to that of a 76 cm column of mercury. Other units for expressing pressure are inches of mercury, millimeters of mercury, the torr, the millibar, and pounds per square inch (lb/in.$^2$). The meteorologist uses inches of mercury in reporting atmospheric pressure. The values of these units equivalent to 1 atm are summarized in Table 12.2. 1 atm $\equiv$ 76 cm Hg $\equiv$ 760 mm Hg $\equiv$ 29.9 in. Hg $\equiv$ 760 torr $\equiv$ 1013 mbar $\equiv$ 14.7 lb/in.$^2$

*Table 12.2.* Pressure units equivalent to 1 atmosphere.

| 1 atm |
|---|
| 76 cm Hg |
| 760 mm Hg |
| 760 torr |
| 1013 mbar |
| 29.9 in. Hg |
| 14.7 lb/in.$^2$ |

Atmospheric pressure varies with altitude. The average pressure at Denver, Colorado, 1 mile above sea level, is 63 cm Hg (0.83 atm). Other liquids besides mercury may be employed for barometers, but they are not as useful as mercury because of the difficulty of maintaining a vacuum above the liquid and because of impractical heights of the liquid column. For example, a pressure of 1 atm will support a column of water about 10,336 mm (33.9 ft) high.

## 12.4 Dependence of Pressure on Number of Molecules

Pressure is produced by gas molecules colliding with the walls of a container. At a specific temperature and volume, the number of collisions depends on the number of gas molecules present. The number of collisions may be increased by increasing the number of gas molecules present. If we double the number of molecules, the frequency of collisions and the pressure should double. We find, for an ideal gas, that this doubling is actually what happens. The pressure, therefore, when the temperature and volume are kept constant, is directly proportional to the number of moles or molecules of gas present. Figure 12.5 illustrates this.

A good example of this molecule–pressure relationship may be observed on an ordinary cylinder of compressed gas that is equipped with a pressure

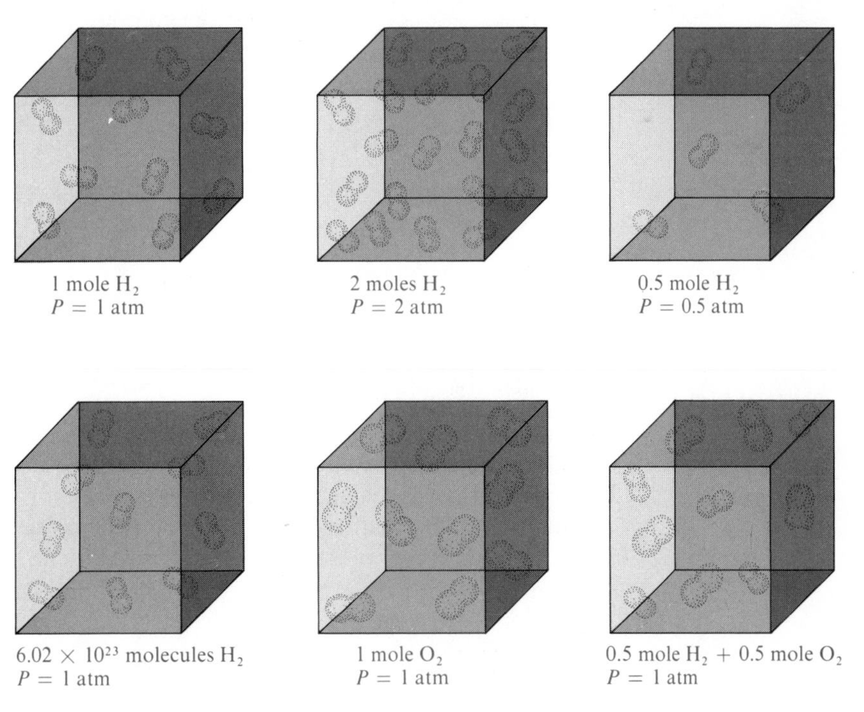


*Figure 12.5.* The pressure exerted by a gas is directly proportional to the number of molecules present. In each case, the volume is 22.4 litres and the temperature is 0°C.

gauge. When the valve is opened, gas escapes from the cylinder. The volume of the cylinder is constant and the decrease in quantity of gas is registered by a drop in pressure indicated on the gauge.

## 12.5 Boyle's Law—The Relationship of the Volume and Pressure of a Gas

Boyle's law

Robert Boyle demonstrated experimentally that, at constant temperature, $T$, the volume, $V$, of a fixed mass of a gas is inversely proportional to the pressure, $P$. This relationship of $P$ and $V$ is known as **Boyle's law**. Mathematically, Boyle's law may be expressed as

$$V \propto \frac{1}{P} \quad \text{(Mass and Temperature are constant)}$$

This equation says that the volume varies inversely as the pressure, at constant mass and temperature. When the pressure on a gas is increased, its volume will decrease, and vice versa. The inverse relationship of pressure and volume is shown graphically in Figure 12.6.

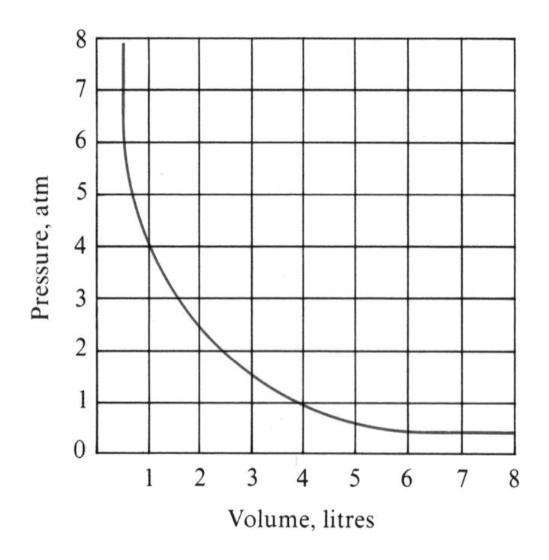


*Figure 12.6.* Graph of pressure versus volume showing inverse $PV$ relationship of an ideal gas.

Boyle demonstrated that when he doubled the pressure on a specific quantity of a gas, keeping the temperature constant, the volume was reduced to one-half the original volume; when he tripled the pressure on the system, the new volume was one-third the original volume; and so on. His demonstration shows that the product of volume and pressure is constant if the temperature is not changed:

$$PV = \text{Constant} \quad \text{or} \quad PV = k \quad \text{(Mass and } T \text{ are constant)}$$

Let us demonstrate this law by taking a cylinder of gas with a movable piston, so that the volume may be varied by changing the external pressure (see Figure 12.7). We assume that there is no change in temperature or the

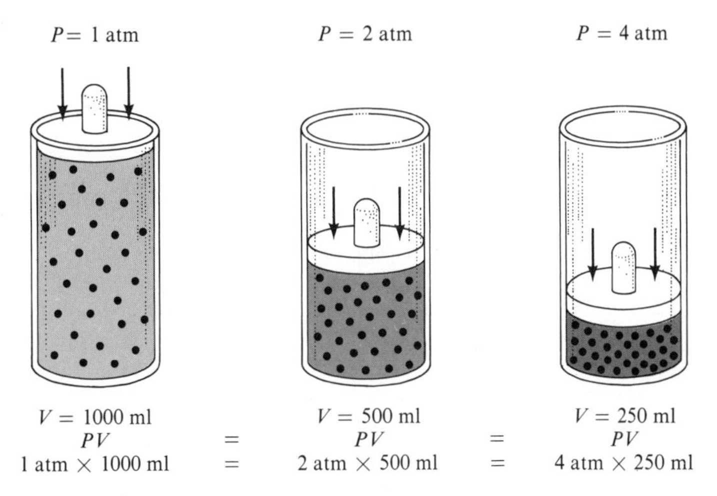


*Figure 12.7.* The effect of pressure on the volume of a gas.

number of molecules. Let us start with a volume of 1000 ml and a pressure of 1 atm. When we change the pressure to 2 atm, the gas molecules are crowded closer together and the volume is reduced to 500 ml. If we increase the pressure to 4 atm, the volume becomes 250 ml.

Note that the product of the pressure times the volume in each case is the same number, substantiating Boyle's law. We may then say that

$$P_1V_1 = P_2V_2$$

where $P_1V_1$ is the pressure–volume product at one set of conditions and $P_2V_2$, at another set of conditions. In each case, the new volume may be calculated by multiplying the starting volume by a ratio of the two pressures involved. Of course, the ratio of pressures used must reflect the direction in which the volume should change. When the pressure is changed from 1 atm to 2 atm, the ratio to be used is 1 atm/2 atm. Now we can verify the results given in Figure 12.7.

(a) Starting volume: 1000 ml; pressure change (1 atm → 2 atm)

$$1000\,\text{ml} \times \frac{1\,\text{atm}}{2\,\text{atm}} = 500\,\text{ml}$$

(b) Starting volume: 1000 ml; pressure change (1 atm → 4 atm)

$$1000\,\text{ml} \times \frac{1\,\text{atm}}{4\,\text{atm}} = 250\,\text{ml}$$

(c) Starting volume: 500 ml; pressure change (2 atm → 4 atm)

$$500\,\text{ml} \times \frac{2\,\text{atm}}{4\,\text{atm}} = 250\,\text{ml}$$

In summary, a change in the volume of a gas due to a change in pressure may be calculated by multiplying the original volume by a ratio of the two pressures. If the pressure is increased, the ratio should have the smaller pressure in the numerator and the larger pressure in the denominator. If the pressure is decreased, the larger pressure should be in the numerator and the smaller pressure in the denominator.

New volume = Original volume × Ratio of pressures

Examples of problems based on Boyle's law follow. If no mention is made of temperature, assume that it remains constant.

*Problem 12.1*

What volume will 2.50 litres of a gas occupy if the pressure is changed from 760 mm Hg to 630 mm Hg?
First we must determine whether the pressure is being increased or decreased. In this case it is being decreased. This decrease in pressure should result in an increase in the volume. Therefore, we need to multiply 2.50 litres by a ratio of the pressures, which will give us an increase in volume. This ratio is 760 mm Hg/630 mm Hg. The calculation is

$$V = 2.50 \text{ litres} \times \frac{760 \text{ mm Hg}}{630 \text{ mm Hg}} = 3.02 \text{ litres} \quad \text{(New volume)}$$

Alternatively, an algebraic approach may be used, solving $P_1V_1 = P_2V_2$ for $V_2$:

$$V_2 = V_1 \times \frac{P_1}{P_2} = 2.50 \text{ litres} \times \frac{760 \text{ mm Hg}}{630 \text{ mm Hg}} = 3.02 \text{ litres}$$

where $V_1 = 2.50$ litres, $P_1 = 760$ mm Hg, and $P_2 = 630$ mm Hg.

*Problem 12.2*

A given mass of hydrogen occupies 40.0 litres at 760 mm Hg pressure. What volume will it occupy at 5 atm pressure?
Since the units of the two pressures are not the same they must be made the same; otherwise, the units will not cancel in the final calculation. Since the pressure is increased, the volume should decrease. Therefore, we need to multiply 40.0 litres by a ratio of the pressures that will give us a decrease in volume.

First, convert 760 mm Hg to atmospheres by multiplying by the conversion factor 1 atm/760 mm Hg:

$$760 \text{ mm Hg} \times \frac{1 \text{ atm}}{760 \text{ mm Hg}} = 1 \text{ atm}$$

Second, set up a ratio of the pressures that will give a volume decrease:

$$\frac{1 \text{ atm}}{5 \text{ atm}}$$

Third, multiply the volume (40.0 litres) by this pressure ratio:

$$V = 40.0 \text{ litres} \times \frac{1 \text{ atm}}{5 \text{ atm}} = 8.00 \text{ litres} \quad \text{(Answer)}$$

*Problem 12.3*

A gas occupies a volume of 200 ml at 400 mm Hg pressure. To what pressure must the gas be subjected in order to change the volume to 75.0 ml?
In order to reduce the volume from 200 ml to 75.0 ml, it will be necessary to increase the pressure. In the same way we calculated volume change affected by a change in pressure, we must multiply the original pressure by a ratio of the two volumes. The volume ratio in this case should be 200 ml/75.0 ml. The calculation is

$$P = 400 \text{ mm Hg} \times \frac{200 \text{ ml}}{75.0 \text{ ml}} = 1067, \quad (1.07 \times 10^3) \text{ mm Hg} \quad \text{(New pressure)}$$

Algebraically, $P_1V_1 = P_2V_2$ may be solved for $P_2$:

$$P_2 = P_1 \times \frac{V_1}{V_2} = 400 \text{ mm Hg} \times \frac{200 \text{ ml}}{75.0 \text{ ml}} = 1.07 \times 10^3 \text{ mm Hg}$$

where $P_1 = 400$ mm Hg, $V_1 = 200$ ml, and $V_2 = 75.0$ ml.
In problems of this type, it is good practice to check the answers to see if they are consistent with the given facts. For example, if the data indicate that the pressure is increased, the final volume should be smaller than the initial volume.

## 12.6 Charles' Law—The Effect of Temperature on the Volume of a Gas

The effect of temperature on the volume of a gas was observed in about 1787 by the French physicist J. A. C. Charles (1746–1823). Charles found that various gases expanded by the same fractional amount when heated through the same temperature interval. Later it was found that if a given volume of any gas initially at 0°C was cooled by 1°C, the volume decreased by $\frac{1}{273}$; if cooled by 2°C, by $\frac{2}{273}$; if cooled by 20°C, by $\frac{20}{273}$; and so on. Since each degree of cooling reduced the volume by $\frac{1}{273}$, it was apparent that any quantity of any gas would have zero volume, if it could only be cooled to −273°C. Of course, no real gas can be cooled to −273°C for the simple reason that it liquefies before that temperature is reached. However, −273°C (more precisely −273.16°C) is referred to as *absolute zero*; this temperature is the zero point on the Kelvin (Absolute) temperature scale. It is the temperature at which the volume of an ideal or perfect gas would become zero.

The volume–temperature relationship for gases is shown graphically in Figure 12.8. Experimental data show the graph to be a straight line which when extrapolated crosses the temperature axis at −273.16°C, or absolute zero.

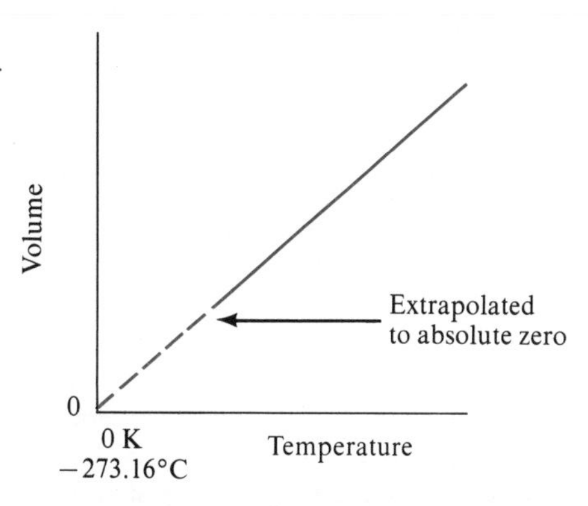


*Figure 12.8.* Volume–temperature relationship of gases. Extrapolated portion of the graph is shown by the broken line.

Charles' law

In modern form, **Charles' law** states that at constant pressure the volume of a fixed weight of any gas is directly proportional to the absolute temperature. Mathematically, Charles' law may be expressed as

$$V \propto T \quad (P \text{ is constant})$$

which means that the volume of a gas varies directly with the absolute temperature when the pressure remains constant. In equation form Charles' law may also be written as

$$V = kT \quad (\text{At constant pressure})$$

where $k$ is a constant for a fixed weight of the gas. If the absolute temperature of a gas is doubled, the volume will double. (A capital $T$ is usually used for absolute temperature, K, and a small $t$ for °C.)

To illustrate, let us return to the gas cylinder with the movable or free-floating piston (see Figure 12.9). Assume that the cylinder labeled (a) contains a quantity of gas and the pressure on it is 1 atm. When the gas is heated, the molecules move faster and their kinetic energy increases. This action should increase the number of collisions per unit of time and thereby the pressure. However, the increased internal pressure will cause the piston to rise to a level at which the internal and external pressures again equal 1 atm (cylinder b). The net result is an increase in volume due to an increase in temperature. Another equation relating the volume of a gas at two different temperatures is

$$\frac{V_1}{T_1} = \frac{V_2}{T_2} \quad (\text{Constant } P)$$

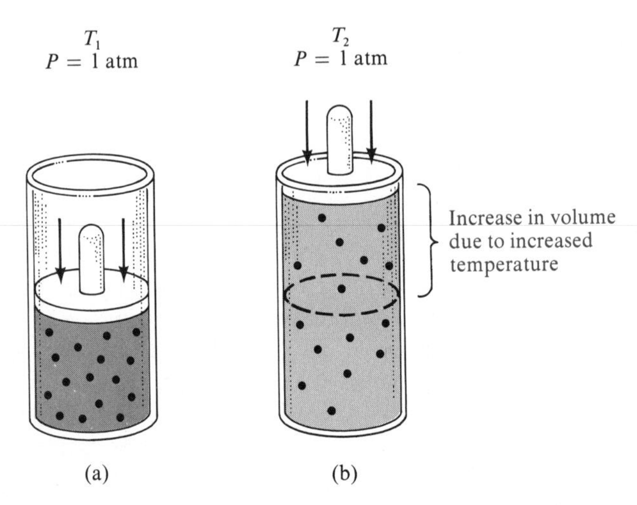


*Figure 12.9.* The effect of temperature on the volume of a gas. The gas in cylinder (a) is heated from $T_1$ to $T_2$. With the external pressure constant at 1 atm, the free-floating piston rises, resulting in an increased volume, as shown in cylinder (b).

where $V_1$ and $T_1$ are one set of conditions and $V_2$ and $T_2$ are another set of conditions.

A simple experiment showing the variation of the volume of a gas with temperature is illustrated in Figure 12.10. A small balloon attached to a bottle is immersed in either ice water or hot water. In ice water the volume is reduced, as shown by the collapse of the balloon; in hot water the gas expands and the balloon increases in size.

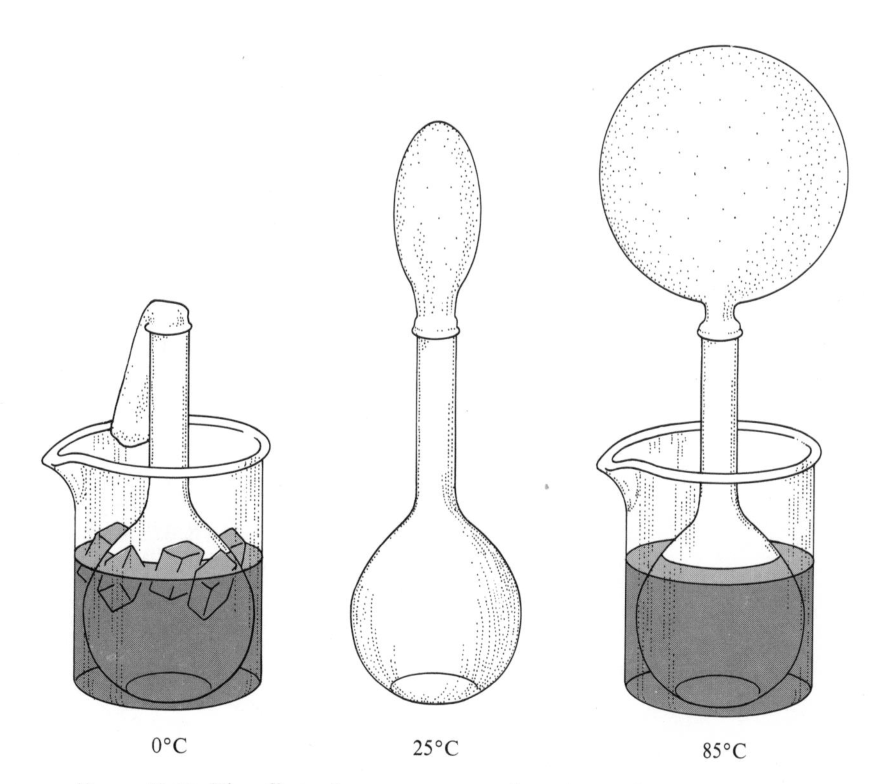


*Figure 12.10.* The effect of temperature on the volume of a gas. A volume decrease occurs when a balloon attached to a flask is immersed in ice water; the volume increases when the flask is immersed in hot water.

The calculation of changes in volume due to changes in temperature involves two basic steps: (1) changing the temperatures to K and (2) multiplying the original volume by a ratio of the initial and final temperatures. If the temperature is increased, the higher temperature is placed in the numerator of the ratio and the lower temperature in the denominator. If the temperature is decreased, the lower temperature is placed in the numerator of the ratio and the higher temperature in the denominator.

New volume = Original volume × Ratio of temperatures (K)

Problems based on Charles' law follow.

*Problem 12.4*

Three litres of hydrogen at $-20°C$ are allowed to warm to a room temperature of 27°C. What is the volume at room temperature if the pressure remains constant?
First change °C to K.

$$°C + 273 = K$$
$$-20°C + 273 = 253 \text{ K}$$
$$27°C + 273 = 300 \text{ K}$$

Since the temperature is increased, the volume should increase. The original volume should be multiplied by the temperature ratio of 300 K/253 K. The calculation is

$$V = 3.00 \text{ litres} \times \frac{300 \text{ K}}{253 \text{ K}} = 3.56 \text{ litres} \quad \text{(New volume)}$$

To obtain the answer by algebra, solve $V_1/T_1 = V_2/T_2$ for $V_2$:

$$V_2 = V_1 \times \frac{T_2}{T_1} = 3.00 \text{ litres} \times \frac{300 \text{ K}}{253 \text{ K}} = 3.56 \text{ litres}$$

where $V_1 = 3.00$ litres, $T_1 = 253$ K, and $T_2 = 300$ K.

---

*Problem 12.5*

If 20.0 litres of nitrogen are cooled from 100°C to 0°C, what is the new volume? Since no mention is made of pressure, assume that there is no pressure change.
First change °C to K.

$$100°C + 273 = 373 \text{ K}$$
$$0°C + 273 = 273 \text{ K}$$

The ratio of temperature to be used is 273 K/373 K, since the final volume should be smaller than the original volume. The calculation is

$$V = 20.0 \text{ litres} \times \frac{273 \text{ K}}{373 \text{ K}} = 14.6 \text{ litres} \quad \text{(New volume)}$$

---

Three variables—pressure, $P$; volume, $V$; and temperature, $T$—are needed to describe a fixed amount of a gas. Boyle's law, $PV = k$, relates pressure and volume at constant temperature; Charles' law, $V = kT$, relates volume and temperature at constant pressure. A third relationship involving pressure and temperature at constant volume is also known and is stated: The pressure of a fixed weight of a gas, at constant volume, is directly proportional to the Kelvin temperature. In equation form, the relationship is

$$P = kT \quad \text{(At constant volume)}$$

This relationship is a modification of Charles' law and is sometimes called Gay-Lussac's law.

We may summarize the effect of changes in pressure, temperature, and quantity of a gas as follows:

1. In the case of a fixed or constant volume,
   (a) when the temperature is increased, the pressure increases.
   (b) when the quantity of a gas is increased, the pressure increases ($T$ remaining constant).

2. In the case of a variable volume,
   (a) when the external pressure is increased, the volume decreases ($T$ remaining constant).
   (b) when the temperature of a gas is increased, the volume increases ($P$ remaining constant).
   (c) when the quantity of a gas is increased, the volume increases ($P$ and $T$ remaining constant).

## 12.7 Standard Temperature and Pressure

standard conditions

standard temperature and pressure (STP)

In order to compare volumes of gases, common reference points of temperature and pressure were selected and called **standard conditions** or **standard temperature and pressure** (abbreviated **STP**). Standard temperature is 273 K (0°C) and standard pressure is 1 atm, or 760 mm Hg. For purposes of comparison, volumes of gases are usually reduced or changed to STP conditions.

STP = *273* K (0°C) *and 1 atm or 760 mm* Hg

## 12.8 Simultaneous Changes in Pressure, Volume, and Temperature (Combined Gas Laws)

When both temperature and pressure change at the same time, the new volume may be calculated by multiplying the initial volume by the correct ratios of both pressure and temperature:

$$\text{Final volume} = \text{Initial volume} \times \begin{pmatrix}\text{Ratio of}\\ \text{pressures}\end{pmatrix} \times \begin{pmatrix}\text{Ratio of}\\ \text{temperatures}\end{pmatrix}$$

This equation combines both Boyle's and Charles' laws, and the same considerations for the pressure and the temperature ratios should be used in the calculation. There are four possible variations:

1. both $T$ and $P$ cause an increase in volume,
2. both $T$ and $P$ cause a decrease in volume,
3. $T$ causes an increase and $P$ a decrease in volume, and
4. $T$ causes a decrease and $P$ an increase in volume.

The $P$, $V$, and $T$ relationships for a given weight of any gas, in fact, may be expressed as a single equation, $PV/T = k$. For problem solving, this equation is usually written

$$\frac{P_1V_1}{T_1} = \frac{P_2V_2}{T_2}$$

where $P_1V_1/T_1$ are the initial conditions and $P_2V_2/T_2$ are the final conditions.

This equation may be solved for any one of the six variables represented and is very generally useful in dealing with the pressure–volume–temperature relationships of gases. Note that when $T$ is constant ($T_1 = T_2$), Boyle's law is represented; when $P$ is constant ($P_1 = P_2$), Charles' law is represented; and when $V$ is constant $V_1 = V_2$, the modified Charles' or Gay-Lussac's law is represented.

*Problem 12.6*

Given 20.0 litres of ammonia gas at 5°C and 730 mm Hg pressure, calculate the volume at 50°C and 800 mm Hg.
In order to get a better look at the data, tabulate the initial and final conditions:

| | *Initial* | *Final* |
|---|---|---|
| $V$ | 20.0 litres | $V_2$ |
| $T$ | 5°C | 50°C |
| $P$ | 730 mm Hg | 800 mm Hg |

Change °C to K:

$$5°\text{C} + 273 = 278 \text{ K}$$

$$50°\text{C} + 273 = 323 \text{ K}$$

Set up ratios of $T$ and $P$:

$$T \text{ ratio} = \frac{323 \text{ K}}{278 \text{ K}} \qquad \text{(Increase in } T \text{ should increase } V\text{)}$$

$$P \text{ ratio} = \frac{730 \text{ mm Hg}}{800 \text{ mm Hg}} \qquad \text{(Increase in } P \text{ should decrease } V\text{)}$$

The calculation is

$$V_2 = 20.0 \text{ litres} \times \frac{323 \text{ K}}{278 \text{ K}} \times \frac{730 \text{ mm Hg}}{800 \text{ mm Hg}} = 21.2 \text{ litres}$$

The algebraic solution is:

solve $\dfrac{P_1V_1}{T_1} = \dfrac{P_2V_2}{T_2}$ for $V_2$ by multiplying both sides of the equation by $T_2/P_2$ and rearranging to obtain

$$V_2 = V_1 \times \frac{P_1}{P_2} \times \frac{T_2}{T_1}$$

Tabulate the known values:

$V_1 = 20.0$ litres $\qquad V_2 = ?$

$T_1 = 5°\text{C} + 273 = 278\text{ K} \qquad T_2 = 50°\text{C} + 273 = 323\text{ K}$

$P_1 = 730$ mm Hg $\qquad P_2 = 800$ mm Hg

Substitute these values in the equation and calculate the value of $V_2$:

$$V_2 = 20.0 \text{ litres} \times \frac{730 \text{ mm Hg}}{800 \text{ mm Hg}} \times \frac{323 \text{ K}}{278 \text{ K}} = 21.2 \text{ litres}$$

*Problem 12.7*

To what temperature (°C) must 10.0 litres of nitrogen at 25°C and 700 mm Hg be heated in order to have a volume of 15.0 litres and a pressure of 760 mm Hg?

This problem is conveniently handled by an algebraic solution.

Solve $\frac{P_1V_1}{T_1} = \frac{P_2V_2}{T_2}$ for $T_2$ to obtain $T_2 = T_1 \times \frac{P_2}{P_1} \times \frac{V_2}{V_1}$

Tabulate the known values:

$P_1 = 700$ mm Hg  $P_2 = 760$ mm Hg

$V_1 = 10.0$ litres  $V_2 = 15.0$ litres

$T_1 = 25°C + 273 = 298$ K  $T_2 = ?$

Substitute these known values in the equation and evaluate $T_2$:

$$T_2 = 298 \text{ K} \times \frac{760 \text{ mm Hg}}{700 \text{ mm Hg}} \times \frac{15.0 \text{ litres}}{10.0 \text{ litres}} = 485 \text{ K}$$

485 K − 273 = 212°C (Answer)

*Problem 12.8*

The volume of a gas is 50.0 litres at 20°C and 742 mm Hg. What volume will it occupy at standard temperature and pressure (STP)? Tabulate the data.

| | *Initial* | *Final* |
|---|---|---|
| $V$ | 50.0 litres | $V_2$ |
| $T$ | 20°C | 0°C |
| $P$ | 742 mm Hg | 760 mm Hg |

STP conditions are 0°C and 760 mm Hg. First change °C to K.

20°C + 273 = 293 K

0°C + 273 = 273 K

Then set up ratios of $T$ and $P$.

$T$ ratio $= \frac{273 \text{ K}}{293 \text{ K}}$ (Decrease in $T$ should decrease $V$)

$P$ ratio $= \frac{742 \text{ mm Hg}}{760 \text{ mm Hg}}$ (Increase in $P$ should decrease $V$)

The calculation is

$$V_2 = 50.0 \text{ litres} \times \frac{273 \text{ K}}{293 \text{ K}} \times \frac{742 \text{ mm Hg}}{760 \text{ mm Hg}}$$

$V_2 = 45.5$ litres

## 12.9 Dalton's Law of Partial Pressures

If gases behave according to the Kinetic-Molecular Theory, there should be no difference in their pressure–volume–temperature relationships, whether the gas molecules are all the same or different from each other. This similarity

Dalton's Law of Partial Pressures

is the basis for an understanding of **Dalton's Law of Partial Pressures**, which states that in a mixture of gases, each gas exerts a pressure independent of the other gases present, and the total pressure is the sum of the partial pressures exerted by each gas in the mixture. Thus, if we have a mixture of three gases, *A*, *B*, and *C*, exerting 50 mm, 150 mm, and 400 mm Hg pressure, respectively, the total pressure will be 600 mm Hg.

$$P_{\text{Total}} = P_A + P_B + P_C$$
$$P_{\text{Total}} = 50 \text{ mm Hg} + 150 \text{ mm Hg} + 400 \text{ mm Hg} = 600 \text{ mm Hg}$$

We can see an application of Dalton's law in the collection of gases over water. When oxygen is prepared in the laboratory, it is commonly collected over water (see Figure 12.11). The $O_2$, collected by the downward displacement of water, is not pure but contains water vapor mixed with it. When the water level is adjusted to be the same inside and outside the bottle, the pressure of the oxygen plus water vapor inside the bottle is equal to the atmospheric pressure:

$$P_{\text{atm}} = P_{O_2} + P_{H_2O}$$
$$P_{O_2} = P_{\text{atm}} - P_{H_2O}$$

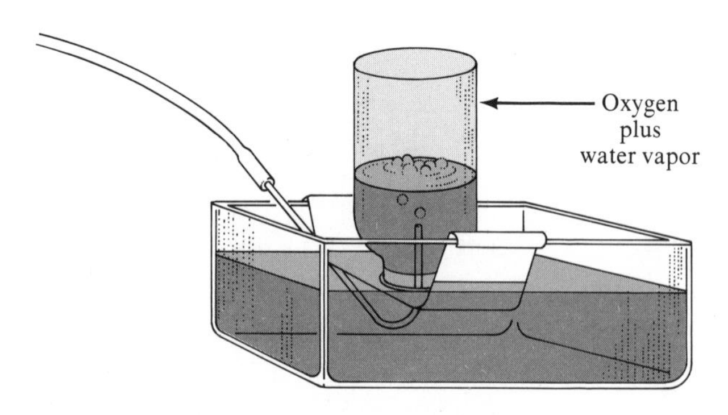


*Figure 12.11.* Oxygen collected over water.

To determine the amount of $O_2$ or any other gas collected over water, we must subtract the pressure of the water vapor from the total pressure of the gas. The vapor pressure of water at various temperatures is tabulated in Appendix II.

An illustrative problem follows.

*Problem 12.9*

A 500 ml sample of oxygen, $O_2$, was collected over water at 23°C and 760 mm Hg pressure. What volume will the dry $O_2$ occupy at 23°C and 760 mm Hg? The vapor pressure of water at 23°C is 21.0 mm Hg.

To solve this problem, we must first find the pressure of the $O_2$ alone by subtracting the pressure of the water vapor present.

$$P_{total} = 760 \text{ mm Hg} = P_{O_2} + P_{H_2O}$$
$$P_{O_2} = 760 \text{ mm Hg} - 21.0 \text{ mm Hg} = 739 \text{ mm Hg}$$

Thus, the pressure of dry $O_2$ is 739 mm Hg.

The problem is now of the Boyle's law type. It is treated as if we had 500 ml of dry $O_2$ at 739 mm Hg pressure, which is then changed to 760 mm Hg pressure, with the temperature remaining constant. The calculation is

$$V = 500 \text{ ml} \times \frac{739 \text{ mm Hg}}{760 \text{ mm Hg}} = 486 \text{ ml dry } O_2$$

This means that 486 ml of the 500 ml mixture of $O_2$ and water vapor is pure $O_2$. Figure 12.12 depicts the pressure and volume changes involved in this problem.

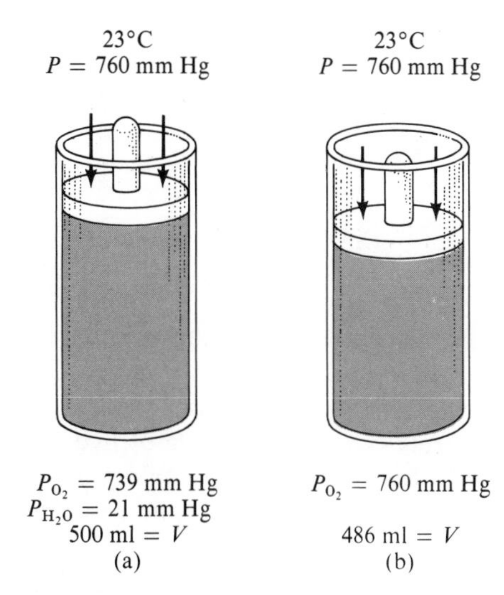


*Figure 12.12.* A 500 ml sample of oxygen was collected over water at 23°C and 760 mm Hg pressure. The original gas collected is shown in cylinder (a). When the water vapor is removed (cylinder b), the volume is reduced. The external pressure, being greater than the pressure of the oxygen, forces the cylinder lid downward until the pressure of the oxygen is 760 mm Hg. The volume of dry oxygen is 486 ml.

## 12.10 Avogadro's Hypothesis

Gay-Lussac's Law of Combining Volumes of Gases

Early in the 19th century, J. L. Gay-Lussac (1778–1850) of France studied the volume relationships of reacting gases. His results, published in 1809, were summarized in a statement known as **Gay-Lussac's Law of Combining Volumes of Gases**: *When measured at constant temperature and pressure, the ratios of the volumes of reacting gases are small whole numbers.* Thus, $H_2$ and $O_2$ combine in a volume ratio of 2:1; $H_2$ and $Cl_2$ react in a volume ratio of 1:1; $H_2$ and $N_2$ react in a volume ratio of 3:1; and so on.

Avogadro's hypothesis

Two years later, in 1811, Amadeo Avogadro of Italy used the Law of Combining Volumes of Gases to make a simple but very significant and far-reaching generalization concerning gases. **Avogadro's hypothesis** states:

*Equal volumes of different gases at the same temperature and pressure contain the same number of molecules.*

This hypothesis was a real breakthrough in understanding the nature of gases. (1) It offered a rational explanation of Gay-Lussac's Law of Combining Volumes of Gases and indicated the diatomic nature of such elemental gases as hydrogen, chlorine, and oxygen; (2) it provided a method for determining the molecular weights of gases and for comparing the densities of gases of known molecular weight (see Section 12.11); and (3) it afforded a firm foundation for the development of the Kinetic-Molecular Theory.

On a volume basis, hydrogen and chlorine react thus:

| Hydrogen + | Chlorine → | Hydrogen chloride |
|---|---|---|
| 1 volume | 1 volume | 2 volumes |

By Avogadro's hypothesis, equal volumes of hydrogen and chlorine must contain the same number of molecules. Therefore, hydrogen molecules react with chlorine molecules in a 1:1 ratio. Since two volumes of hydrogen chloride are produced, one molecule of hydrogen and one of chlorine must produce two molecules of hydrogen chloride. Therefore, each hydrogen molecule and each chlorine molecule is made up of two atoms. The coefficients of the balanced equation for the reaction give the correct ratios for volumes, molecules, and moles of reactants and products:

$$H_2 + Cl_2 \longrightarrow 2\,HCl$$

| $H_2$ | $Cl_2$ | $2\,HCl$ |
|---|---|---|
| 1 volume | 1 volume | 2 volumes |
| 1 molecule | 1 molecule | 2 molecules |
| 1 mole | 1 mole | 2 moles |

By like reasoning, oxygen molecules must contain at least two atoms because one volume of oxygen reacts with two volumes of hydrogen to produce two volumes of steam ($H_2O$).

## 12.11 Weight–Volume Relationship of Gases

A mole of any gas contains $6.02 \times 10^{23}$ molecules (Avogadro's number). Therefore, a gram-molecular weight (1 mole) of any gas, at STP conditions, occupies about the same volume—namely, 22.4 litres. This volume, 22.4 litres, occupies a cube about 28.2 cm (11.1 in.) on a side (see Figure 12.13) and is called the molar volume or **gram-molecular volume**. The gram-molecular weights of several gases, each occupying 22.4 litres at STP, are also shown in Figure 12.13.

gram molecular volume

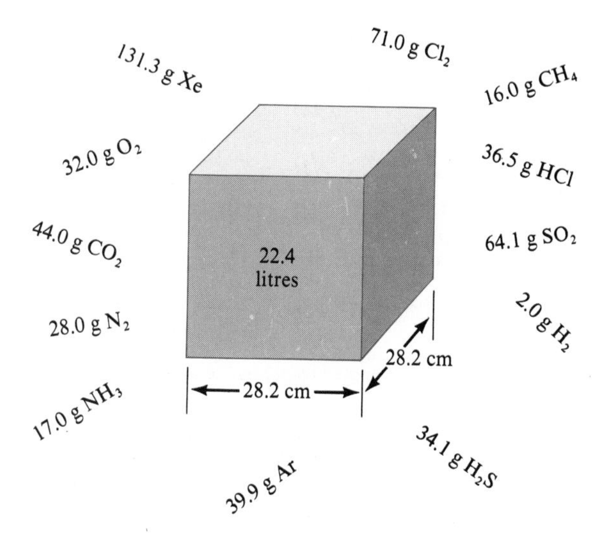


*Figure 12.13.* The gram-molecular weight of a gas at STP occupies 22.4 litres. The weight given for each gas is its gram-molecular weight.

*One gram-molecular weight (one mole) of a gas occupies 22.4 litres at STP.*

This relationship is useful for determining the molecular weight of a gas or of substances that can be easily vaporized into gases. If the weight and the volume of a gas at STP are known, we can calculate its molecular weight. One litre of pure oxygen at STP weighs 1.429 g. The molecular weight of oxygen may be calculated by multiplying the weight of 1 litre by 22.4 litres/mole.

$$\frac{1.429\text{ g}}{\cancel{\text{litre}}} \times \frac{22.4\ \cancel{\text{litres}}}{\text{mole}} = 32.0\text{ g/mole} \quad \text{(g-mol. wt)}$$

If the weight and volume are at other than standard conditions, we first change the volume to STP and then calculate the molecular weight. Note that we do not correct the weight to standard conditions—only the volume.

The gram-molecular volume, 22.4 litres/mole, is used as a conversion factor to convert g/litre to g/mole and also to convert litres to moles. The two conversion factors are

$$\frac{22.4\text{ litres}}{\text{mole}} \quad \text{and} \quad \frac{1\text{ mole}}{22.4\text{ litres}}$$

These conversions must be done at STP conditions except under certain special circumstances. Examples of problems follow.

*Problem 12.10*

If 2.00 litres of a gas measured at STP weigh 3.23 g, what is the molecular weight of the gas?

g-mol. wt at STP = 22.4 litres/mole

If 2.00 litres weigh 3.23 g then 22.4 litres will weigh 22.4/2.00 times as much as 3.23 g, or

$$\text{g-mol. wt} = \frac{3.23\ \text{g}}{2.00\ \text{litres}} \times \frac{22.4\ \text{litres}}{\text{mole}} = 36.2\ \text{g/mole}$$

---

*Problem 12.11*

Measured at 40°C and 630 mm Hg, 691 ml of ethyl ether weigh 1.65 g. Calculate the gram-molecular weight of ethyl ether.

In order to use 22.4 litres = g-mol. wt, we must first correct the volume to standard conditions. Thus,

$$V = 691\ \text{ml} \times \frac{273\ \text{K}}{313\ \text{K}} \times \frac{630\ \text{mm Hg}}{760\ \text{mm Hg}}$$

$$V\ \text{at (STP)} = 500\ \text{ml} = 0.500\ \text{litre}$$

The weight of the gas has not been altered by correcting the volume to STP, so that 500 ml at STP weigh 1.65 g. Then

$$\text{g-mol. wt} = \frac{1.65\ \text{g}}{0.500\ \text{litre}} \times \frac{22.4\ \text{litres}}{\text{mole}} = 73.9\ \text{g/mole}$$

---

## 12.12 Density of Specific Gravity of Gases

The density, $d$, of a gas is its mass per unit volume, which is generally expressed in grams per litre (g/litre):

$$d = \frac{\text{Mass}}{\text{Volume}} = \frac{\text{g}}{\text{litre}}$$

Because the volume of a gas depends on temperature and pressure, both of these should be given when stating the density of a gas. The volume of solids and liquids is hardly affected by changes in pressure and is changed only by a small degree when the temperature is varied. Increasing the temperature from 0°C to 50°C will reduce the density of a gas by about 18% if the gas is allowed to expand. In comparison, a 50°C rise in the temperature of water (0°C → 50°C) will change its density by less than 0.2%.

The density of a gas at any temperature and pressure may be determined by calculating the weight of gas present in 1 litre. At STP, in particular, the density may be calculated by dividing the gram-molecular weight of the gas by 22.4 litres/mole.

$$\text{Density (at STP)} = \frac{\text{g-mol. wt}}{22.4\ \text{litres/mole}} = \frac{\text{g/mole}}{\text{litre/mole}} = \frac{\text{g}}{\text{litre}}$$

Or

$$d(\text{at STP}) \times 22.4\ \text{litres/mole} = \text{g-mol. wt}$$

The density of $Cl_2$ at STP is calculated as follows:

$$\text{g-mol. wt of } Cl_2 = 71.0\ \text{g/mole}$$

$$d = \frac{\text{g-mol. wt}}{22.4\ \text{litres/mole}} = \frac{71.0\ \text{g/mole}}{22.4\ \text{litres/mole}} = 3.17\ \text{g/litre}$$

The specific gravity (sp gr) of a gas is the ratio of the mass of any volume of the gas to the mass of an equal volume of some reference gas. Specific gravities of gases are commonly quoted in reference to air = 1.00. The actual mass of air at STP is 1.29 g/litre, which is the density of air.

The specific gravity can be calculated by dividing the density of gas by the density of air. Both gases must be the same temperature and pressure.

$$\text{sp gr} = \frac{\text{Density of a gas}}{\text{Density of air}}$$

The specific gravity of $Cl_2$, for example, is

$$\text{sp gr of } Cl_2 = \frac{\text{Density of } Cl_2}{\text{Density of air}} = \frac{3.17 \text{ g/litre}}{1.29 \text{ g/litre}} = 2.46$$

This indicates that $Cl_2$ is 2.46 times as heavy as air. Table 12.3 lists the density and specific gravity of some common gases.

*Table 12.3.* Density and specific gravity of common gases at STP.

| Gas | Molecular weight | Density (g/litre at STP) | Specific gravity (air = 1.00) |
|---|---|---|---|
| $H_2$ | 2.0 | 0.090 | 0.070 |
| $CH_4$ | 16.0 | 0.714 | 0.553 |
| $NH_3$ | 17.0 | 0.760 | 0.589 |
| $C_2H_2$ | 26.0 | 1.16 | 0.899 |
| HCN | 27.0 | 1.21 | 0.938 |
| CO | 28.0 | 1.25 | 0.969 |
| $N_2$ | 28.0 | 1.25 | 0.969 |
| $O_2$ | 32.0 | 1.43 | 1.11 |
| $H_2S$ | 34.1 | 1.52 | 1.18 |
| HCl | 36.5 | 1.63 | 1.26 |
| $F_2$ | 38.0 | 1.70 | 1.32 |
| $CO_2$ | 44.0 | 1.96 | 1.52 |
| $C_3H_8$ | 44.1 | 1.96 | 1.52 |
| $O_3$ | 48.0 | 2.14 | 1.66 |
| $SO_2$ | 64.1 | 2.86 | 2.22 |
| $Cl_2$ | 71.0 | 3.17 | 2.46 |

The volume of a gas depends on the temperature, the pressure, and the number of gas molecules. Two or more gases at the same temperature have the same average kinetic energy. If these gases occupy the same volume, they will exhibit the same pressure. Such a system of identical *PVT* properties can only be produced by the same number of molecules having the same average kinetic energy.

## 12.13 Calculations from Chemical Equations Involving Gases

(a) Mole–volume (gas) and weight–volume (gas) calculations. Stoichiometric problems involving gas volumes can be solved by the general mole-ratio method outlined in Chapter 11. The factors 1 mole/22.4 litres and 22.4 litres/1 mole are used as needed for converting volume to moles and moles to volume, respectively. These conversion factors are used under the assumptions that the gases are at STP and behave as ideal gases. In actual practice, gases are measured at other than STP conditions, and the volumes are converted to STP for stoichiometric calculations.

In a balanced equation, the number in front of the formula of a gaseous substance represents the number of moles or molar volumes (22.4 litres at STP) of that substance.

The following are examples of typical problems involving gases and chemical equations:

*Problem 12.12*

What volume of oxygen (at STP) can be formed from 0.500 mole of potassium chlorate?

*Step 1.* Write the balanced equation:

$$\underset{2\text{ moles}}{2\,KClO_3} \rightarrow 2\,KCl + \underset{3\text{ moles}}{3\,O_2\uparrow}$$

*Step 2.* The moles of starting substance is 0.500 mole $KClO_3$.

*Step 3.* Calculate the moles of $O_2$, using the mole-ratio method:

$$0.500\text{ mole }KClO_3 \times \frac{3\text{ moles }O_2}{2\text{ moles }KClO_3} = 0.750\text{ mole }O_2$$

*Step 4.* Convert moles of $O_2$ to litres of $O_2$. The moles of a gas at STP are converted to litres by multiplying by the molar volume, 22.4 litres per mole:

$$0.750\text{ mole }O_2 \times \frac{22.4\text{ litres}}{\text{mole}} = 16.8\text{ litres }O_2 \quad \text{(Answer)}$$

Setting this up in a continuous calculation, we obtain

$$0.500\,\cancel{\text{mole }KClO_3} \times \frac{3\,\cancel{\text{moles }O_2}}{2\,\cancel{\text{moles }KClO_3}} \times \frac{22.4\text{ litres }O_2}{\cancel{\text{mole }O_2}} = 16.8\text{ litres }O_2$$

---

*Problem 12.13*

How many grams of zinc must react with sulfuric acid to product 1000 ml of hydrogen gas at STP?

*Step 1.* The equation is

$$\underset{1\text{ mole}}{Zn} + H_2SO_4 \longrightarrow ZnSO_4 + \underset{1\text{ mole}}{H_2\uparrow}$$

*Step 2.* Moles of $H_2$. The equation states that 1 mole of $H_2$ is produced from 1 mole of Zn; 1000 ml of $H_2$ equals 1 litre of $H_2$ and represents a fraction of a mole.

$$1000\text{ ml }H_2 \times \frac{1\text{ litre }H_2}{1000\text{ ml }H_2} \times \frac{1\text{ mole}}{22.4\text{ litres}} = 0.0446\text{ mole }H_2$$

*Step 3.* Convert to moles of Zn:

$$0.0446 \text{ mole } H_2 \times \frac{1 \text{ mole Zn}}{1 \text{ mole } H_2} = 0.0446 \text{ mole Zn}$$

*Step 4.* Convert to grams of Zn:

$$0.0446 \text{ mole Zn} \times \frac{65.4 \text{ g Zn}}{\text{mole Zn}} = 2.92 \text{ g Zn} \quad \text{(Answer)}$$

The continous calculation setup is

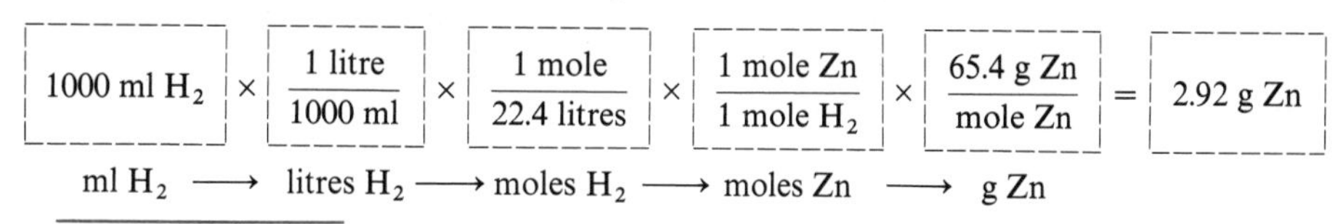


---

*Problem 12.14*

What volume of hydrogen, collected at 30°C and 700 mm Hg pressure, will be formed by reacting 50.0 g of aluminum with hydrochloric acid?

$$\underset{2 \text{ moles}}{2\,Al} + 6\,HCl \longrightarrow 2\,AlCl_3 + \underset{3 \text{ moles}}{3\,H_2}$$

In this problem, the volume of $H_2$ is first calculated from the equation, as we have done before. But, because the volume calculated by use of the equation is at STP, it must be changed to the conditions at which the gas is collected.

*Step 1.* Change grams of Al to moles of Al:

$$50.0 \text{ g Al} \times \frac{1 \text{ mole Al}}{27.0 \text{ g Al}} = 1.85 \text{ moles Al}$$

*Step 2.* Calculate litres of $H_2$ using the mole-ratio method:

$$1.85 \text{ moles Al} \times \frac{3 \text{ moles } H_2}{2 \text{ moles Al}} \times \frac{22.4 \text{ litres } H_2}{1 \text{ mole } H_2} = 62.2 \text{ litres } H_2 \quad \text{(at STP)}$$

*Step 3.* Calculate the volume of $H_2$ at 30°C and 700 mm Hg pressure:

$$\text{Volume} = 62.2 \text{ litres} \times \frac{303 \text{ K}}{273 \text{ K}} \times \frac{760 \text{ mm Hg}}{700 \text{ mm Hg}} = 75.0 \text{ litres } H_2 \quad \text{(Answer)}$$

---

(b) Volume–volume calculations. When all substances in a reaction are in the gaseous state, simplifications in the calculation can be made based on Avogadro's hypothesis that gases under identical conditions of temperature and pressure contain the same number of molecules and occupy the same volume. Using this same hypothesis, we can state that, under the same conditions of temperature and pressure, the volumes of gases reacting are proportional to the number of moles of the gases in the balanced equation. Consider the reaction

| $H_2(g)$ | + $Cl_2(g)$ | ⟶ $2\,HCl(g)$ |
|---|---|---|
| 1 mole | 1 mole | 2 moles |
| 22.4 litres | 22.4 litres | 2 × 22.4 litres |
| 1 volume | 1 volume | 2 volumes |
| *Y* volume | *Y* volume | 2*Y* volumes |

In this reaction, 22.4 litres of hydrogen will react with 22.4 litres of chlorine to give 2 × 22.4, or 44.8, litres of hydrogen chloride gas. This is true because these volumes are equivalent to the number of reacting moles in the equation. Therefore, $Y$ volume of $H_2$ will combine with $Y$ volume of $Cl_2$ to give $2Y$ volume of HCl. For example, 100 litres of $H_2$ reacts with 100 litres of $Cl_2$ to give 200 litres of HCl; if the 100 litres of $H_2$ and $Cl_2$ are at 50°C, they will give 200 litres of HCl at 50°C. When the temperature and pressure before and after a reaction are the same, calculation of volumes can be done without correcting the volumes to STP.

*For reacting gases: Volume-volume relationships are the same as the mole-mole relationships.*

*Problem 12.15*

What volume of oxygen will react with 150 litres of hydrogen to form water vapor? What volume of water vapor will be formed? Assume that both reactants and products are measured at the same conditions. Let us compare the two methods for solving this problem, using the mole method first and then the principle of reacting volumes.

$$2\,H_2(g) + O_2(g) \longrightarrow 2\,H_2O(g)$$

2 moles 1 mole 2 moles

Mole method:

*Step 1.* Moles of $H_2 = 150 \text{ litres } H_2 \times \frac{1 \text{ mole}}{22.4 \text{ litres}} = 6.70 \text{ moles } H_2$

*Step 2.* Moles of $O_2 = 6.70 \text{ moles } H_2 \times \frac{1 \text{ mole } O_2}{2 \text{ moles } H_2} = 3.35 \text{ moles } O_2$

*Step 3.* Volume of $O_2 = 3.35 \text{ moles } O_2 \times \frac{22.4 \text{ litres}}{\text{mole}} = 75.0 \text{ litres } O_2$

*Step 4.* Volume of $H_2O(g) = 150 \text{ litres } H_2 \times \frac{1 \text{ mole } H_2}{22.4 \text{ litres } H_2} \times \frac{2 \text{ moles } H_2O}{2 \text{ moles } H_2} \times \frac{22.4 \text{ litres}}{\text{mole}}$

$= 150 \text{ litres } H_2O(g)$ (Answer)

Calculation by reacting volumes:

| $2\,H_2(g)$ + | $O_2(g)$ | $\longrightarrow$ $2\,H_2O(g)$ |
|---|---|---|
| 2 moles | 1 mole | 2 moles |
| 2 × 22.4 litres | 22.4 litres | 2 × 22.4 litres |
| 2 volumes | 1 volume | 2 volumes |
| 150 litres | 1/2 × 150 = 75 litres | 2/2 × 150 = 150 litres |

Thus, 150 litres of $H_2$ will react with 75 litres of $O_2$ to produce 150 litres of $H_2O(g)$. The calculation by reacting volumes, which may be done by inspection, is certainly simpler and more direct.

*Problem 12.16*

The equation for the preparation of ammonia is

$$3\,H_2 + N_2 \xrightarrow{400^\circ C} 2\,NH_3$$

Assuming that the reaction goes to completion,

(a) What volume of $H_2$ will react with 50 litres of $N_2$?
(b) What volume of $NH_3$ will be formed from 50 litres of $N_2$?
(c) What volume of $N_2$ will react with 100 ml of $H_2$?
(d) What volume of $NH_3$ will be produced from 100 ml of $H_2$?
(e) If 600 ml of $H_2$ and 400 ml of $N_2$ are sealed in a flask and allowed to react, what amounts of $H_2$, $N_2$, and $NH_3$ are in the flask at the end of the reaction?

The answers to parts (a)–(d), shown in the boxes below, can be determined from the equation by inspection, using the principle of reacting volumes.

| | $3H_2$ + | $N_2$ ⟶ | $2NH_3$ |
|---|---|---|---|
| | 3 volumes | 1 volume | 2 volumes |
| (a) | 150 litres | 50 litres | |
| (b) | | 50 litres | 100 litres |
| (c) | 100 ml | 33.3 ml | |
| (d) | 100 ml | | 66.7 ml |

(e) Volume ratio from the equation $= \dfrac{3 \text{ volumes } H_2}{1 \text{ volumes } N_2}$

$$\text{Volume ratio used} = \frac{600 \text{ ml } H_2}{400 \text{ ml } N_2} = \frac{3 \text{ volumes } H_2}{2 \text{ volumes } N_2}$$

Comparing these two ratios, we see that an excess of $N_2$ is present in the gas mixture. Therefore, the reagent limiting the amount of $NH_3$ that can be formed is $H_2$:

$$3H_2 + N_2 \longrightarrow 2NH_3$$

600 ml  200 ml  400 ml

In order to have a 3:1 ratio of volumes reacting, 600 ml of $H_2$ will react with 200 ml of $N_2$ to produce 400 ml of $NH_3$, leaving 200 ml of $N_2$ unreacted. At the end of the reaction, the flask will contain 400 ml of $NH_3$ and 200 ml of $N_2$.

## 12.13 Ideal Gas Equation

We have used four variables in calculations involving gases: the volume, $V$; the pressure, $P$; the absolute temperature, $T$; and the number of molecules or moles, $n$. Combining these variables into a single expression, we obtain

$$V \propto \frac{nT}{P} \qquad \text{or} \qquad V = \frac{nRT}{P}$$

where $R$ is a proportionality constant known as the *ideal gas constant*. The equation is commonly written as

$$PV = nRT$$

ideal gas equation

and is known as the **ideal gas equation**. This equation states in a single expression what we have considered earlier in our discussions—that the volume of a gas varies directly with the number of gas molecules and the absolute temperature and varies inversely with the pressure. The value and units of $R$ depend on the units of $P$, $V$, and $T$. We can calculate one value of $R$ by taking 1 mole of a gas at STP conditions. Solve the equation for $R$:

$$R = \frac{PV}{nT} = \frac{1 \text{ atm} \times 22.4 \text{ litres}}{1 \text{ mole} \times 273 \text{ K}} = 0.0821 \frac{\text{litre-atm}}{\text{mole-K}}$$

The units of $R$ in this case are litre-atmospheres per mole-K.

The ideal gas equation can be used to calculate any one of the four variables if the other three are known. When the value of $R$ is 0.0821 litre-atm/mole-K, the other units must be as follows: $P$ in atm, $V$ in litres, $n$ in moles, and $T$ in K. Any problem that can be solved by the ideal gas equation can also be solved by direct application of the gas laws.

*Problem 12.17*

What pressure will be exerted by 0.400 mole of a gas in a 5.00-litre container at 17°C? First solve the ideal gas equation for $P$.

$$PV = nRT \quad \text{or} \quad P = \frac{nRT}{V}$$

Then substitute the data in the problem into the equation and solve.

$$P = \frac{0.400 \text{ mole} \times 0.0821 \text{ litre-atm} \times 290 \text{ K}}{5.00 \text{ litre} \times \text{mole-K}} = 1.90 \text{ atm} \quad \text{(Answer)}$$

*Problem 12.18*

How many moles of oxygen gas are in a 50.0 litre tank at 22°C if the pressure gauge reads 2000 lb/in$^2$?

First change to pressure in atmospheres. Then solve the ideal gas equation for $n$ (moles), and then substitute the data in the equation to complete the calculation.

*Step 1.* Pressure in atmospheres:

$$\frac{2000 \text{ lb}}{\text{in.}^2} \times \frac{1 \text{ atm}}{14.7 \text{ lb/in.}^2} = 136 \text{ atm}$$

*Step 2.* Solve for moles using the ideal gas equation:

$$PV = nRT \quad \text{or} \quad n = \frac{PV}{RT}$$

$$n = \frac{136 \text{ atm} \times 50.0 \text{ litres}}{(0.0821 \text{ litre-atm/mole-K}) \times 295 \text{ K}} = 281 \text{ moles } O_2 \quad \text{(Answer)}$$

The ideal gas equation can also be used for problems involving a specific mass of gas by substituting the mass–mole relationship, $n$ = g/g-mol. wt, into the equation.

All the gas laws are based on the behavior of an ideal gas—that is, a gas with a behavior that is described exactly by the gas laws for all possible values of $P$, $V$, and $T$. Most real gases actually do behave as predicted by the gas laws over a fairly wide range of temperatures and pressures. However, when conditions are such that the gas molecules are crowded closely together (high pressure and/or low temperature), they show marked deviations from ideal behavior. Deviations occur because molecules have finite volumes and also exhibit intermolecular attractions. This results in less compressibility at high pressures and greater compressibility at low temperatures than predicted by the gas laws.

## Questions

A. *Review the meanings of the new terms introduced in this chapter.*

1. Kinetic-Molecular Theory (KMT)
2. Ideal, or perfect gas
3. Diffusion
4. Graham's law of diffusion
5. Pressure
6. Atmospheric pressure
7. Barometer
8. 1 atmosphere
9. Boyle's law
10. Charles' law
11. Standard conditions
12. Standard temperature and pressure (STP)
13. Dalton's Law of Partial Pressures
14. Gay-Lussac's Law of Combining Volumes of Gases
15. Avogadro's hypothesis
16. Gram-molecular volume
17. Ideal gas equation

B. *Answers to the following questions will be found in tables and figures.*

1. What evidence is used to show diffusion in Figure 12.1? If methane and oxygen were substituted for bromine and air, how could we prove that diffusion had taken place?
2. Given the situation represented in Figure 12.3, are all the gas molecules traveling at the same speed? Explain.
3. In the preparation of a barometer at sea level, as in Figure 12.4, the height of the mercury column was found to be less than 76 cm. Suggest possible reasons for this discrepancy.
4. What is the weight of a column of mercury 76 cm long with a cross-sectional area of one square inch?
5. List in order of decreasing abundance the five most common gases found in normal dry air.
6. What would happen to the balloon in Figure 12.10 if the flask were immersed in a mixture of dry ice and acetone at $-78°C$?
7. List the gases in Table 12.3 that have densities at least 1.5 times greater than the density of air.
8. What volume would the box of Figure 12.13 have if it contained 14.0 g $N_2$, 8.0 g $CH_4$, and 66.0 g $CO_2$ all together at STP?
9. Explain why the pressure of $O_2$, $P_{O_2}$, has changed from 739 mm Hg to 760 mm Hg in Figure 12.12.

C. *Review questions.*

1. Outline the basic assumptions of the Kinetic-Molecular Theory.
2. What two factors determine the gas pressure in a container of fixed volume?
3. What determines the rate at which a gas diffuses?
4. List the following gases in order of increasing relative molecular velocities: $CH_4$, $SO_2$, Ne, $H_2$, $CO_2$. What is your basis for determining this order? (Assume that all the gases are at the same temperature and pressure.)
5. At 100°C, which, if any, of the gases in Question 4 (above) has the highest kinetic energy? Explain.
6. What is an ideal gas?

7. Under what kind of conditions are real gases likely to deviate widely from the behavior of an ideal gas?
8. Some aerosol cans that are pressurized with a noncombustible gas bear a warning: "Caution: keep away from fire; do not incinerate empty container." Explain the logic of this warning in terms of the KMT.
9. What is the reason for referring gases to STP conditions?
10. Explain Dalton's Law of Partial Pressures in terms of the KMT.
11. What major exception can you visualize where mixtures of gases will not obey Dalton's Law of Partial Pressures?
12. Which of the following statements are correct? (Try to answer this question without the use of your text.)
    (a) When the pressure on a sample of gas is increased with the temperature kept constant, the gas will be compressed.
    (b) $V_1/V_2 = P_1/P_2$ is a statement of Boyle's law.
    (c) To calculate the volume of a gas resulting from a change in pressure, the original volume is multiplied by the ratio of the final pressure over the initial pressure.
    (d) If the pressure of a gas is kept constant, the volume can be changed by changing the number of molecules of the gas or by changing the temperature.
    (e) $PV = k$ is a statement of Charles' law.
    (f) If the temperature on a sample of gas is increased from 25°C to 50°C, the volume of the gas will increase by 100%.
    (g) According to Charles' law, the volume of a gas would be zero at −273°C.
    (h) Increasing the temperature of a fixed volume of a gas causes the pressure of the gas to decrease.
    (i) One mole of chlorine, $Cl_2$, at 20°C and 600 mm Hg pressure contains $6.02 \times 10^{23}$ molecules.
    (j) At a given $P$ and $T$, the volume of an ideal gas will be determined by the number of molecules in the sample.
    (k) A mixture of 0.5 mole $CH_4$ and 0.5 mole $CO_2$ will occupy 11.2 litres at STP.
    (l) Although a nitrogen molecule is 14 times as heavy as a hydrogen molecule, they both have the same kinetic energy at the same temperature and pressure.
    (m) The expression $n = PV/RT$ can be derived from the ideal gas equation.
    (n) When the pressure on a sample of gas is halved with the temperature kept constant, the density of the gas is also halved.
    (o) When the temperature of a sample of gas is increased at constant pressure, the density of the gas will decrease.
    (p) In a mixture containing an equal number of ammonia and oxygen molecules, the oxygen molecules, on the average, are moving faster than the ammonia molecules.

*D. Review problems.*

1. The barometer reads 630 mm Hg. Calculate the corresponding atmospheric pressure in:
    (a) Atmospheres
    (b) Inches of Hg
    (c) Pounds per square inch
    (d) Torrs
    (e) Millibars

2. The barometric pressure at the top of Pike's Peak in Colorado was recorded at 525 mm Hg. What is this pressure in atmospheres?
3. A gas occupies a volume of 200 ml at 600 mm Hg pressure. What will be its volume if the pressure is changed to the following, with the temperature remaining constant? (a) 800 mm Hg (b) 200 mm Hg
4. A 500 ml sample of gas is at a pressure of 720 mm Hg. What must be the pressure, with the temperature remaining constant, if the volume is changed to (a) 700 ml; (b) 350 ml?
5. What pressure would be required to compress 2000 litres of hydrogen at 1 atm into a 25 litre tank? (Assume constant temperature.)
6. Given 3.00 litres of nitrogen gas at $-20°C$, what volume will the nitrogen occupy at the following temperatures? (Assume constant pressure.)
   (a) 0°C (b) $-80°C$ (c) 200 K (d) 300 K
7. Given a 250 ml sample of oxygen at 22°C, at what temperature (°C) would the volume of oxygen be doubled if the pressure remains constant?
8. Early in the morning, the pressure in a tire is 25 pounds per square inch (psi) and its temperature is 18°C. At noon, after hard driving, the temperature of the tire is 46°C. What is the pressure in the tire? (Assume constant volume.)
9. The volume of a gas at STP is 650 ml. What volume will the gas occupy at 50°C and 380 mm Hg pressure?
10. Given 500 ml of a gas at STP. At what pressure will the gas volume be 200 ml at 30°C?
11. An expandable balloon contains 1000 litres of a gas at 1 atm pressure and 25°C. What will be the volume of the balloon when it rises to 22 miles altitude where the pressure is 4 mm Hg and the temperature is 2°C?
12. What volume would 1 mole of a gas occupy at 100°C and 630 mm Hg pressure?
13. What volume will a mixture of 4.00 moles of $H_2$ and 1.00 mole of $CO_2$ occupy at STP?
14. Four moles of $O_2$ in a small tank exert a pressure of 1520 lb/in.$^2$. How many moles are in the tank if the pressure reads 900 lb/in.$^2$? (Assume constant $T$.)
15. How many moles of hydrogen are present in 2500 ml of pure $H_2$ at STP?
16. If 350 ml of a gas at STP weigh 0.726 g, what is its gram-molecular weight?
17. Calculate the gram-molecular weight of a gas if 1.52 g occupy 425 ml at 10°C and 720 mm Hg pressure.
18. Calculate the volume of 0.500 g of $CO_2$ at STP.
19. Calculate the density of the following gases at STP:
    (a) $C_2H_6$ (b) $SO_3$ (c) $NF_3$ (d) He
20. (a) Calculate the density of chlorine, $Cl_2$, at 100°C and 760 mm Hg pressure. [*Hint*: First change the volume of $Cl_2$ from STP to 100°C.]
    (b) At what temperature (°C) will the density of methane be 1.0 g/litre?
21. The density of a certain gas is 3.55 g/litre at STP. Calculate its gram-molecular weight.
22. Given the weight of the following gases, what volume will each occupy at STP?
    (a) 34.1 g $H_2S$ (b) 0.525 g $CO_2$ (c) 8.50 g $NH_3$ (d) 6.00 g HCl
23. What volume will a mixture of 4.60 g of $CH_4$ and 7.00 g of $N_2$ occupy at 200°C and 3.00 atm pressure?
24. An equilibrium mixture is composed of hydrogen, nitrogen, and ammonia, where the pressure of each gas is as follows: $H_2$, 650 mm Hg; $N_2$, 250 mm Hg; $NH_3$, 450 mm Hg. What is the total pressure of the system?

25. How many total gas molecules are present in Problem 24 (above) if the volume of the mixture is 22.4 litres at 0°C?
26. Suppose that a sample of $H_2$ was collected over water at 23°C and 740 mm Hg pressure. What is the partial pressure of $H_2$ in this system? (Check Appendix II for the vapor pressure of water.)
27. A sample of $O_2$ collected over water at 30°C and 742 mm Hg pressure occupied a volume of 455 ml. Calculate the volume the dry $O_2$ would occupy at STP.
28. A 600 ml sample of methane, $CH_4$, was collected over water at 30°C and 750 mm Hg. What was the weight of the methane?
29. A mixture of noble gases at 760 mm Hg pressure consists of 50% He, 30% Ne, and 20% Ar. What is the partial pressure of each gas?
30. What volume of hydrogen at STP can be produced by reacting 3.60 moles of aluminum with sulfuric acid according to the following equation?

$$2\ Al(s) + 3\ H_2SO_4(aq) \rightarrow Al_2(SO_4)_3(aq) + 3\ H_2(g)$$

31. Given the equation: $4\ NH_3(g) + 5\ O_2(g) \rightarrow 4\ NO(g) + 6\ H_2O(g)$
    (a) How many moles of $NH_3$ must react to produce 5.0 moles of NO?
    (b) How many moles of $O_2$ must react to produce 5.0 moles of NO?
    (c) How many litres of $NH_3$ and $O_2$ must react to produce 100 litres of NO?
    (d) How many litres of $O_2$ will react with 100 grams of $NH_3$?
    (e) How many litres of NO are formed by reacting 10 moles of $NH_3$ with 10 moles of $O_2$?
32. Given the equation: $4\ FeS_2(s) + 11\ O_2(g) \xrightarrow{\Delta} 2\ Fe_2O_3(s) + 8\ SO_2(g)$:
    (a) How many litres of $O_2$ at STP will react with 1.00 kg of $FeS_2$?
    (b) How many litres of $SO_2$ will be produced from 1.00 kg of $FeS_2$?
33. Assume that the reaction $2\,CO(g) + O_2(g) \rightarrow 2\,CO_2(g)$ goes to completion. When 15 moles of carbon monoxide and 10 moles of oxygen are mixed and reacted in a closed flask, how many moles of CO, $O_2$, and $CO_2$ are present in flask at the end of the reaction?
34. In the preparation of hydrogen from methane, $CH_4$, and steam, what volume of $H_2$ is produced per cubic foot of $CH_4$ reacted? What volume of carbon monoxide is also produced as a by-product?

$$2\,CH_4(g) + 2\,H_2O(g) \xrightarrow{\Delta} 2\,CO(g) + 6\,H_2(g)$$

35. How many litres of air (21% oxygen) are needed to burn 5.00 litres of methane gas, $CH_4$, to carbon dioxide and water? Assume STP conditions.
36. A 25.00 g mixture of KCl and $KClO_3$ was heated, driving off all the oxygen. The volume of oxygen collected was 4.50 litres at STP. What is the percentage of $KClO_3$ in the mixture?
37. Calculate, using the ideal gas equation:
    (a) Volume of 0.820 mole of oxygen at 27°C and 720 mm Hg pressure
    (b) Weight of 16.0 litres of ethane, $C_2H_6$, at 20°C and 40.0 atm pressure
    (c) Weight of 10.2 litres of $N_2$ at 427°C and 2 atm pressure
    (d) Volume of 88.0 g of $CO_2$ at 700 mm Hg pressure and 27°C

# 13 Water and the Properties of Liquids

*After studying Chapter 13 you should be able to:*

1. Understand the terms listed in Question A at the end of the chapter.
2. Describe a water molecule with respect to electron-dot structure, bond angle, and polarity.
3. Make sketches showing hydrogen bonding (a) between water molecules and (b) between hydrogen fluoride molecules.
4. Explain the effect of hydrogen bonding on the physical properties of water.
5. Determine whether a compound will or will not form hydrogen bonds.
6. Complete and balance equations showing the formation of water (a) from hydrogen and oxygen, (b) by neutralization, and (c) by combustion of hydrogen-containing compounds.
7. Complete and balance equations for (a) the electrolysis of water, (b) the reactions of water with Groups IA and IIA metals, (c) the reactions of steam with other metals, (d) the reaction of steam with carbon, and (e) the reactions of water with halogens.
8. Identify metal oxides as basic anhydrides and write balanced equations for their reactions with water.
9. Identify nonmetal oxides as acid anhydrides and write balanced equations for their reactions with water.
10. Deduce the formula of the acid anhydride or of the basic anhydride when given the formula of the corresponding acid or base.
11. Identify, name, and write equations for the complete dehydration of hydrates.
12. Distinguish clearly between peroxides and ordinary oxides.
13. Discuss the occurrence of ozone and its effects on humans.
14. Outline the processes needed to prepare a potable water supply from a contaminated river source.
15. Describe how water may be softened by distillation, chemical precipitation, ion exchange, and demineralization—including chemical equations where appropriate.
16. Explain the process of evaporation from the standpoint of kinetic energy.
17. Relate vapor pressure data or vapor pressure curves of different substances to their relative rates of evaporation and to their relative boiling points.

18. Explain what is happening in the different segments of the time-temperature phase diagram of water.

## 13.1 Occurrence of Water

Water is our most common natural resource; it covers about three-fourths of the earth's surface. Not only is it found in the oceans and seas, in lakes, rivers, streams, and in glacial ice deposits; it is also always present in the atmosphere and in cloud formations. Moreover, water is an essential constituent of all living matter.

About 97% of the earth's water is in the oceans. This is saline water that contains vast amounts of dissolved minerals. The world's fresh water comprises the other 3%, of which about two-thirds is locked up in polar ice caps and glaciers. The remaining fresh water is found in ground water, lakes, and the atmosphere. More than 70 elements have been detected in the mineral content of seawater. Only four of these—chlorine, sodium, magnesium, and bromine—are now commercially obtained from the sea.

## 13.2 Physical Properties of Water

heat of fusion

heat of vaporization

Water is a colorless, odorless, tasteless liquid with a melting point of 0°C and a boiling point of 100°C at 1 atm pressure. Two additional physical properties of matter are introduced with the study of water: heat of fusion and heat of vaporization. **Heat of fusion** is the amount of heat required to change one gram of a solid into a liquid at its melting point. The heat of fusion of water is 80 calories per gram. The temperature of the solid–liquid system does not change during the absorption of this heat. The heat of fusion is the energy used in breaking down the crystalline lattice of ice from a solid to a liquid. **Heat of vaporization** is the amount of heat required to change one gram of liquid to a vapor at its normal boiling point. The value for water is 540 calories per gram. Once again, there is no change in temperature during the absorption of this heat. The heat of vaporization is the energy needed to overcome the attractive forces between molecules in changing them from the liquid to the gaseous state. The values for water for both the heat of fusion and the heat of vaporization are relatively high compared to those for other substances; these high values indicate that strong attractive forces are acting between the molecules.

Ice and water exist together in equilibrium at 0°C, as shown in Figure 13.1. When ice at 0°C melts, it absorbs 80 cal/g in changing into a liquid; the temperature remains at 0°C. In order to refreeze the water, we have to remove 80 cal/g from the liquid at 0°C.

Both boiling water and steam are shown in Figure 13.2 to have a temperature of 100°C. It takes 100 cal to heat 1 g of water from 0°C to 100°C, but water at its boiling point absorbs 540 cal/g in changing to steam. Although boiling water and steam are both at the same temperature, steam contains

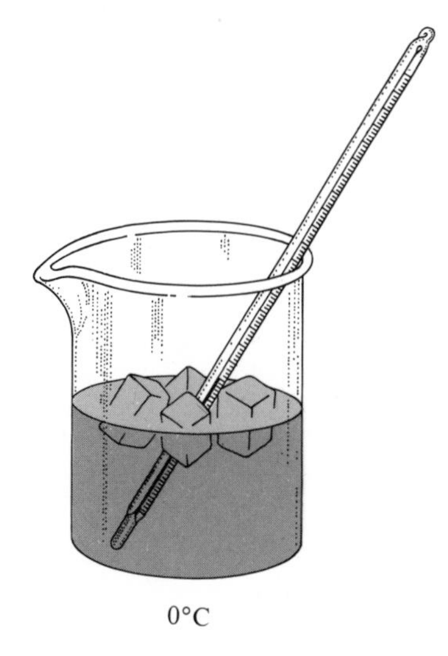


*Figure 13.1.* Ice and water in equilibrium at 0°C.

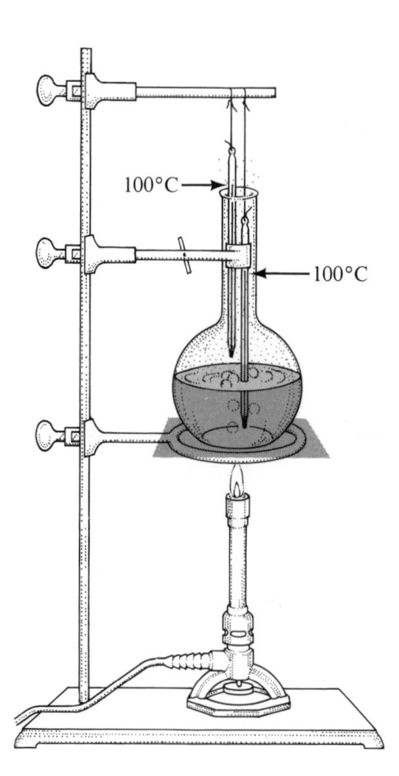


*Figure 13.2.* Boiling water and steam in equilibrium at 100°C.

considerably more heat per gram and can cause more severe burns than hot water. The physical properties of water are tabulated and compared with other hydrogen compounds of Group VIA elements in Table 13.1.

*Table 13.1.* Physical properties of water and other hydrogen compounds of Group VIA elements.

| Formula | Color | Molecular weight | Melting point (°C) | Boiling point, 760 mm Hg (°C) | Heat of fusion (cal/g) | Heat of vaporization (cal/g) |
|---|---|---|---|---|---|---|
| $H_2O$ | Colorless | 18.0 | 0.00 | 100.0 | 80.0 | 540 |
| $H_2S$ | Colorless | 34.1 | −85.5 | −60.3 | 16.7 | 131 |
| $H_2Se$ | Colorless | 81.0 | −65.7 | −41.3 | 7.4 | 57.0 |
| $H_2Te$ | Colorless | 129.6 | −51 | −2.3 | — | 42.8 |

The maximum density of water is 1.000 g/ml at 4°C. Water has the unusual property of contracting in volume as it is cooled to 4°C and then expanding when cooled from 4°C to 0°C. (Most liquids contract in volume all the way down to the point at which they solidify.) Therefore, 1 g of water occupies a volume greater than 1 ml at all temperatures above and below 4°C. Water, on the other hand, shows a large increase (about 9%) in volume when water is changed from a liquid at 0°C to a solid (ice) at 0°C. The density of ice at 0°C is 0.915 g/ml, which means that ice, being less dense than water, will float in water.

## 13.3 Structure of the Water Molecule

A single water molecule consists of two hydrogen atoms and one oxygen atom. Each H atom is attached to the O atom by a single covalent bond. This bond is formed by the overlap of the 1*s* orbital of hydrogen with an unpaired 2*p* electron orbital of oxygen. The average distance between the two nuclei is known as the *bond length*. The O–H bond length in water is 0.96 Å. The water molecule is nonlinear and has a bent structure with an angle of about 105° between the two bonds (see Figure 13.3).

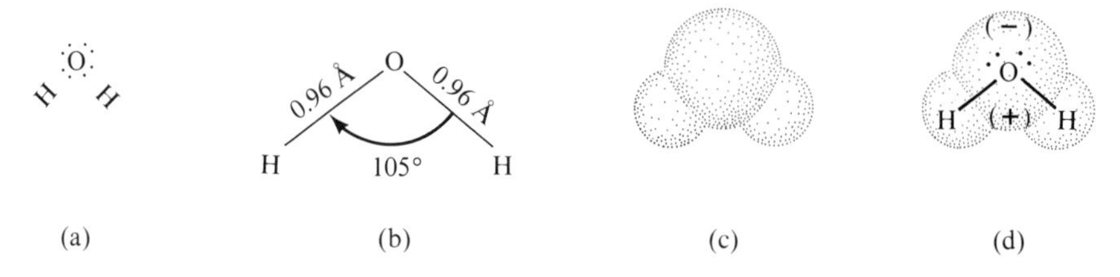


*Figure 13.3.* Diagrams of a water molecule: (a) electron distribution, (b) bond angle and O–H bond length, (c) molecular orbital structure, (d) dipole representation.

Oxygen is the second most electronegative element. As a result, the two covalent OH bonds in water are polar. If the three atoms in a water molecule were aligned in a linear structure such as H⟶O⟵H, the two polar bonds would be acting in equal and opposite directions and the molecule would be nonpolar. However, water is a highly polar molecule. Therefore, it does not have a linear structure; instead, it has a bent structure. When atoms are bonded together in a nonlinear fashion, the angle formed by the bonds is called the *bond angle*. In water the HOH bond angle is 105°. The two polar covalent bonds and the bent structure result in the oxygen atom having a partial negative charge and each hydrogen atom having a partial positive charge. The polar nature of water is responsible for many of its properties, including its behavior as a solvent.

## 13.4 The Hydrogen Bond

Table 13.1 compares the physical properties of $H_2O$, $H_2S$, $H_2Se$, and $H_2Te$. From this comparison it is apparent that four physical properties of water—melting point, boiling point, heat of fusion, and heat of vaporization—are extremely high and do not fit the trend for molecular weight. If water fitted the trend shown by the other compounds, we would expect the melting point of water to be below −85°C and the boiling point to be below −60°C.

hydrogen bond

Why does water have these anomalous physical properties? The answer is that liquid water molecules are linked together by hydrogen bonds. A **hydrogen bond** is a chemical bond that is formed between polar molecules that contain hydrogen covalently bonded to a small, highly electronegative atom such as fluorine, oxygen, or nitrogen. The bond is actually the dipole–dipole attraction of polar molecules.

What is a hydrogen bond, or H-bond? Because a hydrogen atom has only one electron, it can form only one covalent bond. When it is attached to a strong electronegative atom such as oxygen, a hydrogen atom will also be attracted to an oxygen atom of another molecule, forming a bond (or bridge) between the two molecules. Water has two types of bonds: covalent bonds that exist between hydrogen and oxygen atoms within a molecule, and hydrogen bonds that exist between hydrogen and oxygen atoms in different water molecules.

Hydrogen bonds are *intermolecular* bonds; that is, they are formed between atoms in different molecules. They are somewhat ionic in character because they are formed by electrostatic attraction. Hydrogen bonds are much weaker than the ionic or covalent bonds that unite atoms to atoms to form compounds. Despite their weakness, hydrogen bonds are of great chemical importance.

Figure 13.4, part (a), shows two water molecules linked by a hydrogen bond, and part (b) shows six water molecules linked together by hydrogen bonds. A dash (—) is used for the covalent bond and a dotted line (----) for the hydrogen bond. In water, one molecule is linked to another through hydrogen bonds, forming a three-dimensional aggregate of water molecules. This molecular

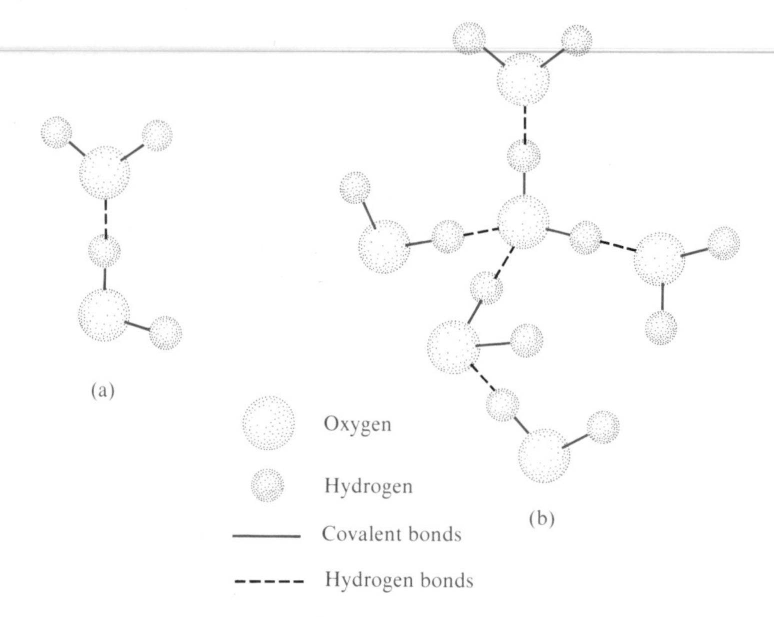


*Figure 13.4.* Hydrogen bonding: Water in the liquid and solid states exists as aggregates in which the water molecules are linked together by hydrogen bonds.

bonding effectively gives water the properties of a much larger, heavier molecule, explaining in part its relatively high melting point, boiling point, heat of fusion, and heat of vaporization. As water is heated and energy is absorbed, hydrogen bonds are continually being broken until at 100°C, with the absorption of an additional 540 cal/g, water separates into individual molecules, going into the gaseous state. Sulfur, selenium, and tellurium are not sufficiently electronegative for their hydrogen compounds to behave like water. As a result, H-bonding in $H_2S$, $H_2Se$, and $H_2Te$ is only of small consequence (if any) to their physical properties.

Fluorine, the most electronegative element, forms the strongest hydrogen bonds. This bonding is strong enough to link hydrogen fluoride molecules together as *dimers*, $H_2F_2$. The dimer structure may be represented in this way:

H—F---H—F
↑
Hydrogen bond

The existence of salts, such as $KHF_2$ and $NH_4HF_2$, verifies the hydrogen fluoride (bifluoride) structure, $HF_2^-$ $(F—H----F)^-$, where one H atom is bonded to two F atoms through one covalent bond and one hydrogen bond.

Hydrogen bonding can occur between two different atoms that are capable of forming H-bonds. Thus, we may have an O----H—N or O—H----N linkage in which the H-bond is between an oxygen and a nitrogen atom. This form of the H-bond exists in certain types of protein molecules and many biologically active substances.

## 13.5 Formation of Water and Chemical Properties of Water

Water is very stable to heat; it decomposes to the extent of only about 1% at temperatures up to 2000°C. Pure water is a nonconductor of electricity. But when a small amount of sulfuric acid or sodium hydroxide is added, the solution is readily decomposed into hydrogen and oxygen by an electric current. Two volumes of hydrogen are produced for each volume of oxygen:

$$2\,H_2O(l) \xrightarrow[H_2SO_4 \text{ or } NaOH]{\text{Electrical energy}} 2\,H_2(g) + O_2(g)$$

### Formation

Water is formed when hydrogen burns in air. Pure hydrogen burns very smoothly in air, but mixtures of hydrogen and air (or oxygen) are dangerous and explode when ignited. The reaction is strongly exothermic:

$$2\,H_2(g) + O_2(g) \longrightarrow 2\,H_2O(g) + 115.6\text{ kcal}$$

Water is produced by a variety of other reactions, especially by (1) acid–base neutralizations, (2) combustion of hydrogen-containing materials, and (3) metabolic oxidation in living cells.

1. $HCl(aq) + NaOH(aq) \longrightarrow NaCl(aq) + H_2O$
2. $\underset{\text{Acetylene}}{2\,C_2H_2(g)} + 5\,O_2(g) \longrightarrow 4\,CO_2(g) + 2\,H_2O(g) + 289.6\text{ kcal}$
3. $\underset{\text{Glucose}}{C_6H_{12}O_6} + 6\,O_2 \xrightarrow{\text{Enzymes}} 6\,CO_2 + 6\,H_2O + 673\text{ kcal}$

The reaction represented by equation 2 is strongly exothermic and is capable of producing very high temperatures. It is used in oxygen–acetylene torches to cut and weld steel and other metals. The overall reaction of glucose with oxygen represented by equation 3 is the reverse of photosynthesis. It is the overall reaction by which living cells obtain needed energy by metabolizing glucose to carbon dioxide and water.

### Reactions with Metals and Nonmetals

The reactions of metals with water at different temperatures show that these elements vary greatly in their reactivity. Metals such as sodium, potassium, and calcium react with cold water to produce hydrogen and a metal hydroxide. A small piece of sodium added to water melts from the heat produced by the reaction, forming a silvery metal ball, which rapidly flits back and forth on the surface of the water. One must use caution when experimenting with this reaction, since the hydrogen produced is frequently ignited by the sparking of the sodium, and it will explode, spattering sodium. Potassium reacts even more vigorously than sodium. Calcium sinks in water and only liberates a gentle stream of hydrogen. The equations for these reactions are

$$2\,Na(s) + 2\,H_2O(l) \longrightarrow H_2\uparrow + 2\,NaOH(aq)$$
$$2\,K(s) + 2\,H_2O(l) \longrightarrow H_2\uparrow + 2\,KOH(aq)$$
$$Ca(s) + 2\,H_2O(l) \longrightarrow H_2\uparrow + Ca(OH)_2(aq)$$

Zinc, aluminum, and iron do not react with cold water but will react with steam at high temperatures, forming hydrogen and a metallic oxide. The equations are

$$Zn(s) + H_2O(steam) \longrightarrow H_2\uparrow + ZnO(s)$$
$$2\,Al(s) + 3\,H_2O(steam) \longrightarrow 3H_2\uparrow + Al_2O_3(s)$$
$$3\,Fe(s) + 4\,H_2O(steam) \longrightarrow 4H_2\uparrow + Fe_3O_4(s)$$

Copper, silver, and mercury are examples of metals that do not react with cold water or steam to produce hydrogen. We conclude from these reactions that sodium, potassium, and calcium are chemically more reactive than zinc, aluminum, and iron, which are more reactive than copper, silver, and mercury.

Certain nonmetals react with water under various conditions. For example, fluorine reacts violently with cold water, producing hydrogen fluoride and free oxygen. The reactions of chlorine and bromine are much milder, producing what is commonly known as "chlorine water" and "bromine water," respectively. Chlorine water contains HCl, HOCl, and dissolved $Cl_2$; the free chlorine gives it a yellow-green color. Bromine water contains HBr, HOBr, and dissolved $Br_2$; the free bromine gives it a red-brown color. Steam passed over hot coke (carbon) produces a mixture of carbon monoxide and hydrogen that is known as "water gas." Since water gas is combustible, it is useful as a fuel. It is also the starting material for the commercial production of several alcohols. The equations for these reactions are

$$2\,F_2(g) + 2\,H_2O(l) \longrightarrow 4\,HF(aq) + O_2(g)$$
$$Cl_2(g) + H_2O(l) \longrightarrow HCl(aq) + HOCl(aq)$$
$$Br_2(l) + H_2O(l) \longrightarrow HBr(aq) + HOBr(aq)$$
$$C(s) + H_2O(g) \xrightarrow{1000^\circ C} CO(g) + H_2(g)$$

### *Reactions with Metal and Nonmetal Oxides*

basic anhydride

Metal oxides that react with water to form bases are known as **basic anhydrides**. If we heat the corresponding base, we can reverse the direction of the reaction and drive off water, forming the anhydride again.

$$CaO(s) + H_2O \longrightarrow Ca(OH)_2(aq)$$
$$Na_2O(s) + H_2O \longrightarrow 2\,NaOH(aq)$$
$$Ca(OH)_2(s) \xrightarrow{\Delta} CaO(s) + H_2O\uparrow$$

Certain metal oxides, such as $CuO$ and $Al_2O_3$, do not form basic solutions because the oxides are insoluble in water.

Anhydrides do not contain any hydrogen. Thus, to determine the formula of an anhydride, the elements of water, $H_2O$, are removed from the designated acid or base formulas until all the hydrogen is removed. The formula of the anhydride then consists of the remaining metal or nonmetal and oxygen atoms. In calcium hydroxide, removal of water as indicated leaves $CaO$ as the anhydride:

$$Ca(OH)_2 \xrightarrow{\Delta} CaO + H_2O$$

In sodium hydroxide, $H_2O$ cannot be removed from one formula unit, so two formula units of $NaOH$ must be used, leaving $Na_2O$ as the formula of the anhydride:

$$2\,NaOH \xrightarrow{\Delta} Na_2O + H_2O$$

acid anhydride

Nonmetal oxides that react with water to form acids are known as **acid anhydrides**. Examples are

$$SO_2(g) + H_2O \rightleftharpoons H_2SO_3(aq)$$

$$N_2O_5(s) + H_2O \longrightarrow 2\,HNO_3(aq)$$

The foregoing are examples of typical reactions of water but are by no means a complete list of the known reactions of water.

## 13.6 *Hydrates*

hydrate

Solids that contain water molecules as part of their crystalline structure are known as **hydrates**. Formulas for hydrates are expressed by first writing the usual anhydrous (without water) formula for the compound, then adding a dot, followed by the number of water molecules present. An example is $BaCl_2 \cdot 2H_2O$. Sometimes the formula for water is enclosed in parentheses and the formula written as $BaCl_2(H_2O)_2$. These formulas tell us that each formula unit of this salt contains one barium ion, two chloride ions, and two water molecules. A crystal of the salt contains many of these units in its crystalline lattice.

In naming hydrates, we first name the compound exclusive of the water and then add the term *hydrate*, with the proper prefix representing the number of water molecules in the formula. For example, $BaCl_2 \cdot 2H_2O$ is called *barium chloride dihydrate*. Hydrates are true compounds and follow the Law of Definite Composition. The gram-formula weight of $BaCl_2 \cdot 2H_2O$ is 244.3 g; it contains 56.20% barium, 29.06% chlorine, and 14.74% water.

water of hydration

Water in a hydrate is known as **water of hydration** or **water of crystallization**. Water molecules are bonded by electrostatic forces between polar water molecules and the positive or negative ions of the compound. These

water of crystallization

forces are not as strong as covalent or ionic chemical bonds. As a result, water of crystallization may be removed by moderate heating of the crystal. A partially dehydrated or completely anhydrous compound may result. When $BaCl_2 \cdot 2H_2O$ is heated, it loses its water at about 100°C:

$$BaCl_2 \cdot 2H_2O \xrightarrow{100°C} BaCl_2 + 2\,H_2O\uparrow$$

When a solution of copper(II) sulfate ($CuSO_4$) is allowed to evaporate, beautiful blue crystals containing 5 moles of water per mole of $CuSO_4$ are formed. The formula for this hydrate is $CuSO_4 \cdot 5H_2O$; it is called *cupric sulfate pentahydrate*, or *copper(II) sulfate pentahydrate*. When $CuSO_4 \cdot 5H_2O$ is heated, water is lost and a pale green-white powder, anhydrous $CuSO_4$, is formed:

$$CuSO_4 \cdot 5H_2O \xrightarrow{250°C} CuSO_4 + 5\,H_2O\uparrow$$

When water is added to anhydrous copper(II) sulfate, the above reaction is reversed and the salt turns blue again. The formation of the hydrate is noticeably exothermic. Because of this outstanding color change, anhydrous copper(II) sulfate has been used as an indicator to detect small amounts of water.

The formula for plaster of paris is $(CaSO_4)_2 \cdot H_2O$. When mixed with the proper quantity of water, plaster of paris sets to a hard mass; it is therefore useful for making patterns for the reproduction of art objects, molds, and surgical casts. The chemical reaction is

$$(CaSO_4)_2 \cdot H_2O(s) + 3\,H_2O(l) \longrightarrow 2\,CaSO_4 \cdot 2\,H_2O(s)$$

The occurrence of hydrates is very commonplace in salts. Table 13.2 lists a number of common hydrates.

*Table 13.2.* Selected hydrates.

| Hydrate | Name |
|---|---|
| $CaCl_2 \cdot 2H_2O$ | Calcium chloride dihydrate |
| $Ba(OH)_2 \cdot 8H_2O$ | Barium hydroxide octahydrate |
| $MgSO_4 \cdot 7H_2O$ | Magnesium sulfate heptahydrate |
| $SnCl_2 \cdot 2H_2O$ | Tin(II) chloride dihydrate |
| $CoCl_2 \cdot 6H_2O$ | Cobalt(II) chloride hexahydrate |
| $Na_2CO_3 \cdot 10H_2O$ | Sodium carbonate decahydrate |
| $(NH_4)_2C_2O_4 \cdot H_2O$ | Ammonium oxalate monohydrate |
| $NaC_2H_3O_2 \cdot 3H_2O$ | Sodium acetate trihydrate |
| $Na_2B_4O_7 \cdot 10H_2O$ | Sodium tetraborate decahydrate |
| $K_4Fe(CN)_6 \cdot 3H_2O$ | Potassium ferrocyanide trihydrate |

## 13.7 Hygroscopic Substances; Deliquescence; Efflorescence

hygroscopic substances

Many anhydrous salts and other substances readily absorb water from the atmosphere. Such substances are said to be **hygroscopic**. This property can be observed in the following simple experiment: Spread a weighed 10–20 g sample of anhydrous copper(II) sulfate on a watch glass and set it aside so that the salt is exposed to the air. Then weigh the sample periodically for 24 hours, noting the increase in weight, the change in color, and the formation of the blue pentahydrate.

deliquescence

Some compounds continue to absorb water beyond the hydrate stage to form solutions. A substance that absorbs water from the air until it forms a solution is said to be **deliquescent**. A few granules of anhydrous calcium chloride or pellets of sodium hydroxide exposed to the air will appear moist in a few minutes, and within an hour will absorb enough water to form a puddle of solution. Phosphorus pentoxide ($P_2O_5$) picks up water so rapidly that it cannot be weighed accurately unless it is weighed in an anhydrous atmosphere.

Compounds that absorb water are very useful as drying agents. Refrigeration systems must be kept dry with such agents or the moisture will freeze and clog the tiny orifices in the mechanism. Bags of drying agents are often enclosed in packages containing iron or steel parts to absorb moisture and prevent rusting. Anhydrous calcium chloride, magnesium sulfate, sodium sulfate, calcium sulfate, silica gel, and phosphorus pentoxide are some of the compounds commonly used for drying liquids and gases containing small amounts of moisture.

efflorescence

The process by which crystalline materials spontaneously lose water when exposed to the air is known as **efflorescence**. Glauber's salt ($Na_2SO_4 \cdot 10H_2O$), a transparent crystalline salt, loses water when exposed to the air. One can actually observe these well-defined, large crystals crumbling away as they lose water, forming a white, noncrystalline-appearing powder. From our discussion of the decomposition of hydrates, we can predict that heat will increase the rate of efflorescence. The rate also depends on the concentration of moisture in the air. A dry atmosphere will allow the process to take place more rapidly.

## 13.8 Hydrogen Peroxide

Although both water and hydrogen peroxide are compounds of hydrogen and oxygen, their properties are very different. A hydrogen peroxide molecule ($H_2O_2$) is composed of two H atoms and two O atoms. Its composition by weight is 94.1% oxygen and 5.9% hydrogen. Pure hydrogen peroxide is a pale blue liquid that freezes at −1.7°C, boils at 151°C, and has a density of 1.44 g/ml at 20°C. It is miscible with water in all proportions. Water solutions of hydrogen peroxide are slightly acid. The structure of hydrogen peroxide may be represented as

```
  H                      H
 .. ..                  /
:O:O:     or     :Ö—O:
 .. ..           /   ..
 H              H
```

Hydrogen peroxide is a common, useful source of oxygen, since it decomposes easily to give oxygen and water:

$$2\,H_2O_2(l) \longrightarrow 2\,H_2O(l) + O_2\uparrow + 46.0\text{ kcal}$$

This decomposition is accelerated by heat and light, but may be minimized by storing peroxide solutions in brown bottles, keeping them cold, and adding stabilizers. The decomposition may also be accelerated by catalysts such as manganese dioxide.

The peroxide group, like an oxygen molecule ($O_2$), contains two oxygen atoms linked by a covalent bond. It has a $-2$ oxidation number and is written $O_2^{2-}$ or $:\ddot{\underset{..}{O}}:\ddot{\underset{..}{O}}:^{2-}$; each O atom is considered to have an oxidation number of $-1$. Metal dioxides also contain two O atoms, but each is bonded individually to the metal ion. Thus, a metal dioxide contains two oxide ($:\ddot{\underset{..}{O}}:^{2-}$) ions.

Some discretion must be used when working with peroxide formulas. From their formulas, $BaO_2$ and $TiO_2$ appear to be similar compounds, but $BaO_2$ is a peroxide consisting of a $+2$ barium ion and a $-2$ peroxide ion whereas titanium dioxide is an oxide consisting of a $+4$ titanium ion and two $-2$ oxide ions. Peroxides may be distinguished from dioxides chemically because they generally yield $H_2O_2$ or $O_2$ when treated with acids or water. Two examples are the reactions of barium peroxide and sodium peroxide:

$$BaO_2(s) + H_2SO_4(aq) \longrightarrow H_2O_2(aq) + BaSO_4\downarrow$$

$$Na_2O_2(s) + 2\,H_2O(l) \longrightarrow H_2O_2(aq) + 2\,NaOH(aq)$$

A 3% solution of $H_2O_2$, which is commonly available at drug stores, is used as an antiseptic to cleanse open wounds. Somewhat stronger $H_2O_2$ solutions are widely used as bleaching agents for cotton, wood, and hair. For certain oxidation processes, the chemical industry uses a 30% solution. Concentrations of 85% and higher are used for oxidizing fuels in rocket propulsion. These highly concentrated solutions are extremely sensitive to decomposition and represent a fire hazard if allowed to come into contact with organic material.

Ozone is another compound containing multiple oxygen linkages. One molecule, $O_3$, contains three atoms of oxygen:

$$:\ddot{O}::\ddot{O}: \qquad \cdot\underset{..}{O}\cdot\cdot\overset{..}{O}\cdot\cdot\underset{..}{O}\cdot$$

Oxygen    Ozone

Ozone can be prepared by passing air or oxygen through an electrical discharge:

$$3\,O_2 \xrightarrow[\text{discharge}]{\text{Electrical}} 2\,O_3 \qquad -68.4\text{ kcal}$$

The characteristic pungent odor of ozone is noticeable in the vicinity of electrical machines and power transmission lines. Ozone is formed in the atmosphere during electrical storms and by the photochemical action of ultraviolet radiation on a mixture of nitrogen dioxide and oxygen. Ozone is not a desirable low altitude constituent of the atmosphere, since it is known to cause plant damage, cracking of rubber, and the formation of eye-irritating substances. However, in the stratosphere ozone interacts with ultraviolet radiation to form molecular and atomic ozygen and thus prevents most of this harmful radiation from reaching the earth's surface.

$$O_3 \xrightarrow[\text{radiation}]{\text{Ultraviolet}} O_2 + O$$

allotropy

Many elements exist in two or more molecular or crystalline forms. This phenomenon is known as **allotropy** (from the Greek *allotropia*, meaning "variety"). The individual forms of the element are known as allotropic forms or allotropes. Oxygen ($O_2$) and ozone ($O_3$) are allotropic forms of the element oxygen. Two other common elements that exhibit allotropy are sulfur and carbon. (Diamond and graphite are allotropic forms of carbon.)

## 13.9 Natural Waters

Natural fresh waters are not pure, but contain dissolved minerals, suspended matter, and sometimes harmful bacteria. The water supplies of large cities are usually drawn from rivers or lakes. Such water is generally unsafe to drink without treatment. To make such water potable (that is, safe to drink), it is treated by some or all of the following processes:

1. *Screening.* Removal of relatively large objects, such as trash, fish, and so on.
2. *Flocculation and sedimentation.* Chemicals, usually lime and alum (aluminum sulfate), are added to form a flocculent jellylike precipitate of aluminum hydroxide. This precipitate enmeshes most of the suspended fine matter in the water and carries it to the bottom of the sedimentation basin.
3. *Sand filtration.* Water is drawn from the top of the sedimentation basin and passed downward through fine sand filters. Nearly all the remaining suspended matter and bacteria are removed by the sand filters.
4. *Aeration.* Water is drawn from the bottom of the sand filters and is aerated by spraying. The purpose of this process is to remove objectionable odors and tastes.
5. *Disinfection.* In the final stage, chlorine gas is injected into the water to kill harmful bacteria before it is distributed to the public. Ozone is also used in some countries to disinfect water. In emergencies, water may be disinfected by simply boiling for a few minutes.

If the drinking water of children contains an optimum amount of fluoride ion, their teeth will be more resistant to decay. Therefore, in many communities NaF or $Na_2SiF_6$ is added to the water supply to bring the fluoride ion concentration up to the optimum level of about 1.0 part per million (ppm). Excessively high concentrations of fluoride ion can cause mottling of the teeth.

Water that contains dissolved calcium and magnesium salts is called *hard water*. One drawback of hard water is that ordinary soap does not lather well in it; the soap reacts with the calcium and magnesium ions to form an insoluble greasy scum. However, synthetic soaps, known as detergents or syndets, are available; they have excellent cleaning qualities and do not form precipitates with hard water. Hard water is also undesirable because it causes "boiler scale" to form on the walls of water heaters and steam boilers, which greatly reduces their efficiency.

Three techniques used to "soften" hard water are distillation, chemical precipitation, and ion exchange. In distillation, the water is boiled and the steam thus formed is then condensed to a liquid again, leaving the minerals behind in the distilling vessel. Figure 13.5 illustrates a simple laboratory distillation apparatus. Commercial stills are available that are capable of producing hundreds of litres of distilled water per hour.

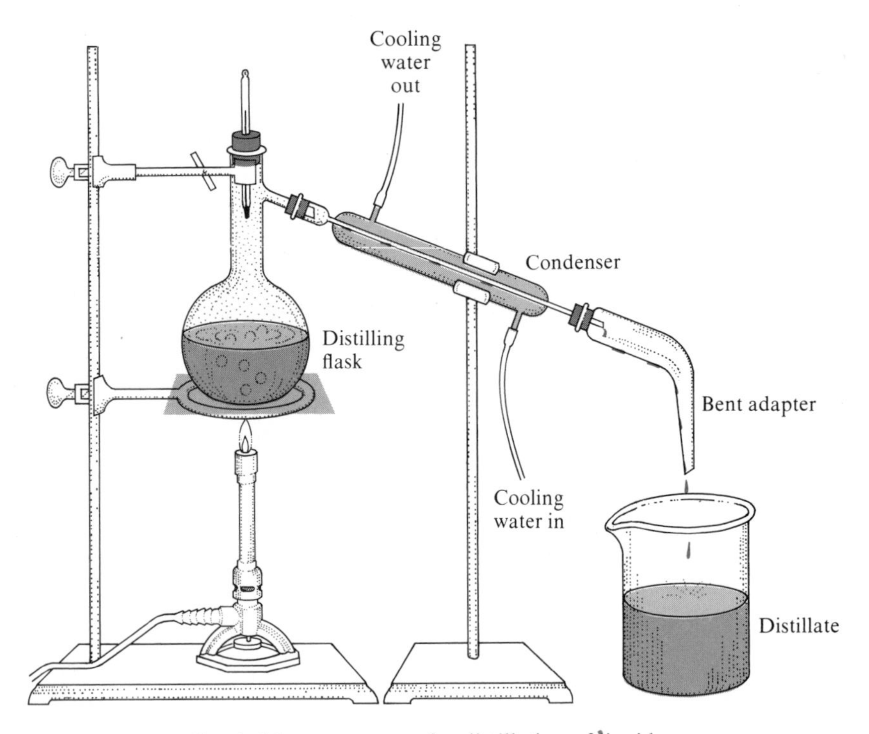


*Figure 13.5.* Simple laboratory setup for distillation of liquids.

Calcium and magnesium ions are precipitated from hard water by adding sodium carbonate and lime. Insoluble calcium carbonate and magnesium hydroxide are precipitated and are removed by filtration or sedimentation.

In the ion-exchange method, used in many households, hard water is effectively softened as it is passed through a bed or tank of zeolite. Zeolite is a complex sodium aluminum silicate. In this process, sodium ions replace objectionable calcium and magnesium ions, and the water is thereby softened:

$$Na_2Zeolite(s) + Ca^{2+}(aq) \longrightarrow CaZeolite(s) + 2\ Na^{+}(aq)$$

The zeolite is regenerated by back-flushing with concentrated sodium chloride solution, reversing the above reaction.

The sodium ions that are present in water softened either by chemical precipitation or by the zeolite process are not objectionable to most users of soft water.

In demineralization, both cations and anions are removed by a two-stage ion-exchange system. Special synthetic organic resins are used in the ion-exchange beds. In the first stage, metal cations are replaced by hydrogen ions. In the second stage, anions are replaced by hydroxide ions. The hydrogen and hydroxide ions react, and essentially pure, mineral-free water leaves the second stage (see Figure 13.6).

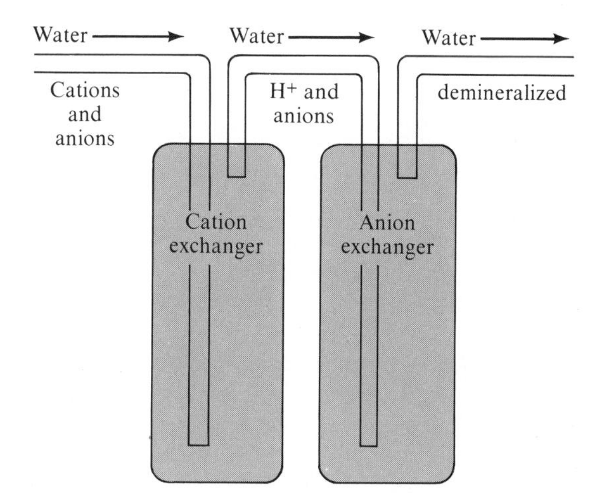


*Figure 13.6.* Demineralization of water: Water is passed through two beds of synthetic resin. In the cation exchanger, metal ions are exchanged for hydrogen ions. In the anion exchanger, anions are exchanged for hydroxide ions. The $H^+$ and $OH^-$ ions react to form water, giving essentially pure, demineralized water.

The oceans represent an inexhaustible source of water; however, seawater contains about 3.5 lb of salts per 100 lb of water. This 35,000 parts per million (ppm) of dissolved salts makes seawater unfit for agricultural and domestic uses. Water that contains less than 1000 ppm of salts is considered reasonably good for drinking, and potable (fresh) water is already being obtained from the sea in many parts of the world. Continuous research is being done in an effort to make usable water from the oceans more abundant and economical.

## 13.10 Water Pollution

Polluted water was formerly thought of as water that was unclear, had a bad odor or taste, and contained disease-causing bacteria. However, such factors as increased population, industrial requirements for water, atmospheric pollution, and use of pesticides have greatly modified the problem of water pollution.

Many of the "newer" pollutants are not removed or destroyed by the usual water-treatment processes. For example, among the 66 organic compounds found in the drinking water of a major city on the Mississippi River, 3 are labelled slightly toxic, 17 moderately toxic, 15 very toxic, 1 extremely toxic, and 1 supertoxic. Two are known carcinogens (cancer-producing agents), 11 are suspect, and 3 are metabolized to carcinogens. The United States Public Health Service classifies water pollutants under eight broad categories. These are shown in Table 13.3.

*Table 13.3.* Classification of water pollutants.

| Type of pollutant | Examples |
|---|---|
| Oxygen-demanding wastes | Decomposable organic wastes from domestic sewage and industrial wastes of plant and animal origin |
| Infectious agents | Bacteria, viruses, and other organisms from domestic sewage, animal wastes, and animal process wastes |
| Plant nutrients | Principally compounds of nitrogen and phosphorus |
| Organic chemicals | Large numbers of chemicals synthesized by industry; pesticides; chlorinated organic compounds |
| Other minerals and chemicals | Inorganic chemicals from industrial operations, mining, oil field operations, and agriculture |
| Radioactive substances | Waste products from mining and processing of radioactive materials, airborne radioactive fallout, increased use of radioactive materials in hospitals and research |
| Heat from industry | Large quantities of heated water returned to water bodies from power plants and manufacturing facilities after use for cooling purposes |
| Sediment from land erosion | Solid matter washed into streams and oceans by erosion, rain, and water runoff |

Many outbreaks of disease or poisoning such as typhoid, dysentery, and cholera have been attributed directly to drinking water. Rivers and streams are a natural means for municipalities to dispose of their domestic and industrial waste products. Much of this water is used again by people downstream, and then discharged back into the water source. Then another community still farther downstream draws the same water and discharges its own wastes. Thus, along waterways such as the Mississippi and Delaware rivers, water is withdrawn and discharged many times. If this water is not properly treated, harmful pollutants will build up, causing epidemics of various diseases.

Mercury and its compounds have long been known to be highly toxic. Mercury gets into the body primarily through the foods we eat. Although it is not an essential mineral for the body, mercury accumulates in the blood, kidneys, liver, and brain tissues. Mercury in the brain causes serious damage to the central nervous system. The sequence of events that have led to incidents of mercury poisoning is as follows: Mercury and its compounds are used in many industries and in agriculture, primarily as a fungicide in the treatment of seeds. One of the largest uses is in the electrochemical conversion of sodium chloride brines to chlorine and sodium hydroxide, as represented by this equation:

$$2\,NaCl + 2\,H_2O \xrightarrow{\text{Electrolysis}} Cl_2 + 2\,NaOH + H_2$$

Although no mercury is shown in the chemical equation, it is used in the process for electrical contact, and small amounts are discharged along with spent brine solutions. Thus, considerable quantities of mercury, in low concentrations, have been discharged into lakes and other surface waters from the effluents of these manufacturing plants. The mercury compounds discharged into the water are converted by bacterial action and other organic compounds to methyl mercury, $(CH_3)_2Hg$, which then accumulates in the bodies of fish. Several major episodes of mercury poisoning that have occurred in the past years were the result of eating mercury-contaminated fish. The best way to control this contaminant is at the source, and much has been done since 1970 to eliminate the discharge of mercury in industrial wastes. In 1976, the Environmental Protection Agency banned the use of all mercury-containing insecticides and fungicides.

Many other major water pollutants have been recognized and steps have been taken to eliminate them. Three of these that pose serious problems are lead, detergents, and chlorine-containing organic compounds. Lead poisoning, for example, has been responsible for many deaths in past years. The toxic action of lead in the body is the inhibition of the enzyme necessary for the production of hemoglobin in the blood. The normal intake of lead into the body is through food. However, extraordinary amounts of lead can be ingested from water running through lead pipes and by using lead-containing ceramic containers for storage of food and beverages.

Keeping our lakes and rivers free from pollution is a very costly and complicated process. However, it has been clearly demonstrated that waterways rendered so polluted that the water is neither fit for human use nor able to sustain marine life can be successfully restored.

## 13.11 Evaporation

When beakers of water, ethyl ether, and ethyl alcohol are allowed to stand uncovered in an open room, the volumes of these liquids gradually decrease. The process by which this takes place is called *evaporation*.

Attractive forces exist between molecules in the liquid state. All these molecules, however, do not have the same kinetic energy. Molecules that have greater than average kinetic energy may overcome the attractive forces and break away from the surface of the liquid, flying off and becoming a gas.

evaporation

**Evaporation** is the escape of molecules from the liquid state to the gas or vapor state.

In evaporation, molecules of higher than average kinetic energy escape from a liquid, leaving it cooler than it was before they escaped. For this reason, evaporation of perspiration is one way the human body cools itself and keeps its temperature constant. When volatile liquids such as ethyl chloride ($C_2H_5Cl$) are sprayed on the skin, they evaporate rapidly, cooling the area by removing heat. The numbing effect of the low temperature produced by evaporation of ethyl chloride allows it to be used as a local anesthetic for minor surgery.

Solids such as iodine, camphor, naphthalene (moth balls), and, to a small extent, even ice, will go directly from the solid to the gaseous state, bypassing the liquid state. This change is a form of evaporation and is called **sublimation**.

sublimation

$$\text{Liquid} \xrightarrow{\text{Evaporation}} \text{Vapor}$$

$$\text{Solid} \xrightarrow{\text{Sublimation}} \text{Vapor}$$

## 13.12 Vapor Pressure

When a liquid evaporates in a closed system as shown in Figure 13.7, part (b), some of the molecules in the vapor or gaseous state strike the surface and return to the liquid state by the process of *condensation*. The rate of condensation increases until it is equal to the rate of evaporation. At this point, the space above the liquid is said to be saturated with vapor, and an equilibrium, or steady state, exists between the liquid and the vapor. The equilibrium equation is:

$$\text{Liquid} \underset{\text{Condensation}}{\overset{\text{Evaporation}}{\rightleftarrows}} \text{Vapor}$$

This equilibrium is dynamic; both processes—evaporation and condensation—are taking place, even though one cannot visually observe or measure a change. The number of molecules leaving the liquid in a given time interval is equal to the number of molecules returning to the liquid.

At the point of equilibrium, the molecules in the vapor exert a pressure like any other gas. The pressure exerted by a vapor in equilibrium with its

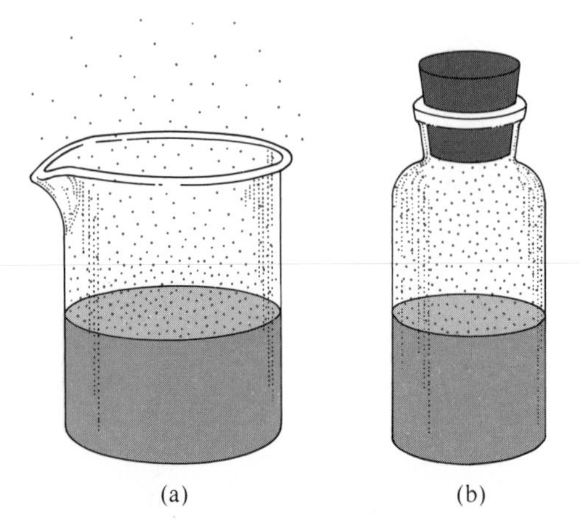


*Figure 13.7.* (a) Molecules in an open beaker may evaporate from the liquid and be dispersed into the atmosphere. Under this condition, evaporation will continue until all the liquid is gone. (b) Molecules leaving the liquid are confined to a limited space. With time, the concentration in the vapor phase will increase to a point at which an equilibrium between liquid and vapor is established.

vapor pressure

liquid is known as the **vapor pressure** of the liquid. The vapor pressure may be thought of as an internal pressure, a measure of the escaping tendency that molecules have to go from the liquid to the vapor state. The vapor pressure of a liquid is independent of the amount of liquid and vapor present, but it increases as the temperature rises (see Table 13.4).

*Table 13.4.* The vapor pressure of water, ethyl alcohol, and ethyl ether at various temperatures.

| | Vapor pressure (mm Hg) | | |
|---|---|---|---|
| Temperature (°C) | Water | Ethyl alcohol | Ethyl ether[a] |
| 0 | 4.6 | 12.2 | 185.3 |
| 10 | 9.2 | 23.6 | 291.7 |
| 20 | 17.5 | 43.9 | 442.2 |
| 30 | 31.8 | 78.8 | 647.3 |
| 40 | 55.3 | 135.3 | 921.3 |
| 50 | 92.5 | 222.2 | 1276.8 |
| 60 | 152.9 | 352.7 | 1729.0 |
| 70 | 233.7 | 542.5 | 2296.0 |
| 80 | 355.1 | 812.6 | 2993.6 |
| 90 | 525.8 | 1187.1 | 3841.0 |
| 100 | 760.0 | 1693.3 | 4859.4 |
| 110 | 1074.6 | 2361.3 | 6070.1 |

[a] Note that the vapor pressure of ethyl ether at temperatures of 40°C and higher exceeds standard pressure, 760 mm Hg. This indicates that the substance has a low boiling point and therefore should be stored in a cool place in a tightly sealed container.

When equal volumes of water, ethyl ether, and ethyl alcohol are placed in separate beakers and allowed to evaporate at the same temperature, we observe that the ether evaporates faster than the alcohol, which in turn evaporates faster than the water. This order of evaporation is consistent with the fact that ether has a higher vapor pressure at any particular temperature than ethyl alcohol or water. One reason for this higher vapor pressure is that there is less attraction between ether molecules than there is between alcohol molecules or between water molecules. The vapor pressures of these three compounds at various temperatures are compared in Table 13.4.

volatile

Substances that evaporate readily are said to be **volatile**. A volatile liquid has a relatively high vapor pressure at room temperature. Ethyl ether is a very volatile liquid; water is not too volatile; mercury, which has a vapor pressure of 0.0012 mm Hg at 20°C, is essentially a nonvolatile liquid. Most substances that are normally solids are nonvolatile (solids that sublime are exceptions).

## 13.13 Boiling Point

The boiling temperature of a liquid is associated with its vapor pressure. We have seen that the vapor pressure increases as the temperature increases. When the internal or vapor pressure of a liquid becomes equal to the external pressure, the liquid boils. (By external pressure we mean the pressure of the atmosphere above the liquid.) The boiling temperature of a pure liquid remains constant as long as the external pressure does not vary.

The boiling point of water is 100°C. Table 13.4 shows that the vapor pressure of water at 100°C is 760 mm Hg, a figure we have seen many times before. The significant fact here is that the boiling point is the temperature at which the vapor pressure of the water or other liquid is equal to standard, or atmospheric, pressure at sea level. These relationships lead to the following definition: The **boiling point** is the temperature at which the vapor pressure of a liquid is equal to the external pressure above the liquid.

boiling point

We can readily see that a liquid has an infinite number of boiling points. When we give the boiling point of a liquid, we should also state the pressure. When we express the boiling point without stating the pressure, we mean it to be the **standard** or **normal boiling point** at standard pressure (760 mm Hg). Using Table 13.4 again, we see that the normal boiling point of ethyl ether is between 30°C and 40°C, and for ethyl alcohol it is between 70°C and 80°C, because, for each compound, 760 mm Hg pressure lies within these stated temperature ranges. At the normal boiling point, one gram of a liquid changing to a vapor (gas) absorbs an amount of energy equal to its heat of vaporization (see Table 13.5).

standard or normal boiling point

*Table 13.5.* Physical properties of ethyl chloride, ethyl ether, ethyl alcohol, and water.

| | Boiling point (°C) | Melting point (°C) | Heat of vaporization (cal/g) | Heat of fusion (cal/g) |
|---|---|---|---|---|
| Ethyl chloride | 13 | −139 | 92.5 | — |
| Ethyl ether | 34.6 | −116 | 83.9 | — |
| Ethyl alcohol | 78.4 | −112 | 204.3 | 24.9 |
| Water | 100.0 | 0 | 540 | 80 |

The boiling point at various pressures may be evaluated by plotting the data of Table 13.4 on the graph in Figure 13.8, where temperature is plotted horizontally along the $x$ axis and vapor pressure vertically along the $y$ axis. The resulting curves are known as **vapor pressure curves**. Any point on these curves represents a vapor–liquid equilibrium at a particular temperature and pressure. We may find the boiling point at any pressure by tracing a horizontal line from the designated pressure to a point on the vapor pressure curve. From this point we draw a vertical line to obtain the boiling point on the temperature axis. Four such points are shown in Figure 13.8; they represent the normal boiling points of the four compounds at 760 mm Hg pressure.

vapor pressure curve

See if you can verify from the graph that the boiling points of ethyl chloride, ethyl ether, ethyl alcohol, and water at 600 mm Hg pressure are 8.5°C, 28°C, 73°C, and 93°C, respectively. By reversing this process, you can ascertain at what pressure a substance will boil at a specific temperature. The boiling point is one of the most commonly used physical properties for characterizing and identifying substances.

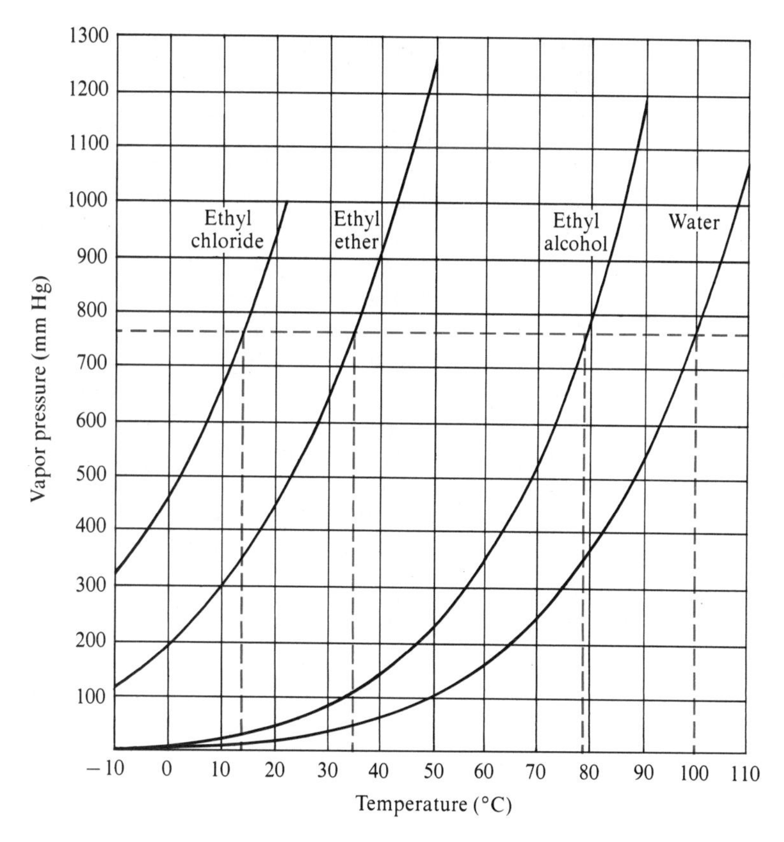


*Figure 13.8.* Vapor pressure–temperature curves for ethyl chloride, ethyl ether, ethyl alcohol, and water.

## 13.14 Freezing Point or Melting Point

As heat is removed from a liquid, the liquid becomes colder and colder, until a temperature is reached at which it begins to solidify. A liquid changing into a solid is said to be *freezing*, or *solidifying*. When a solid is heated continually, a temperature is reached at which the solid begins to liquefy. A solid changing into a liquid is said to be *melting*. The temperature at which the solid phase of a substance is in equilibrium with its liquid phase is known as the **freezing point** or **melting point** of that substance. The equilibrium equation is

freezing or melting point

$$\text{Solid} \underset{\text{Freezing}}{\overset{\text{Melting}}{\rightleftarrows}} \text{Liquid}$$

When a solid is slowly and carefully heated so that a solid–liquid equilibrium is maintained, the temperature will remain constant as long as both phases are present. One gram of a solid in changing into a liquid absorbs an amount of energy equal to its *heat of fusion* (see Table 13.5). The melting point is another physical property that is commonly used for characterizing substances.

The most common example of a solid–liquid equilibrium is ice and water (see Figure 13.1). In a well-stirred system of ice and water, the temperature remains at 0°C as long as both phases are present. The melting point is subject to changes in pressure, but is hardly affected unless the pressure change is very large.

It has been known for a long time that dissolved substances markedly decrease the freezing point of a liquid. For example, salt–water–ice equilibrium mixtures may be obtained at temperatures as low as −20°C, 20 degrees below the usual freezing point of water.

If, after all the solid has been melted, the liquid is heated continually, the temperature will rise until the liquid boils. The temperature will remain constant at the boiling point until all the liquid has boiled away. One gram of a liquid in changing into a gas at the normal boiling point absorbs an amount of energy equal to its *heat of vaporization* (see Table 13.5). The whole process of heating a substance (water) is illustrated graphically in Figure 13.9,

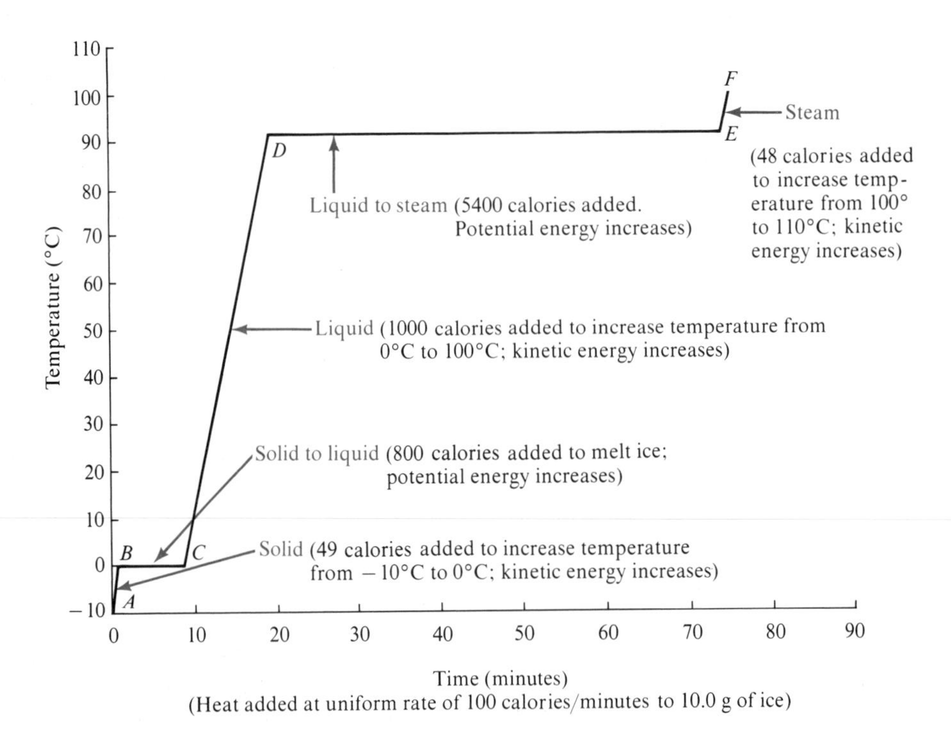


*Figure 13.9.* Time–temperature phase diagram for the absorption of heat by a substance from the solid state to the gaseous state. Using water as an example, the interval *AB* represents the ice phase; *BC* interval, the melting of ice to water; *CD* interval, the elevation of the temperature of water from 0°C to 100°C; *DE* interval, the boiling of water to steam; and *EF* interval, the heating of steam.

where line $AB$ represents the solid being heated and line $BC$ represents the time during which the solid is melting and is in equilibrium with the liquid. Along line $CD$ the liquid absorbs heat and finally, at point $D$, boils and continues to boil at a constant temperature (line $DE$). In the interval $EF$, all the water exists as steam and is being further heated or superheated.

## Questions

A. *Review the meanings of the new terms introduced in this chapter.*

1. Heat of fusion
2. Heat of vaporization
3. Hydrogen bond
4. Basic anhydride
5. Acid anhydride
6. Hydrate
7. Water of hydration
8. Water of crystallization
9. Hygroscopic substances
10. Deliquescence
11. Efflorescence
12. Allotropy
13. Evaporation
14. Sublimation
15. Vapor pressure
16. Volatile
17. Boiling point
18. Standard or normal boiling point
19. Vapor pressure curve
20. Freezing or melting point

B. *Answers to the following questions will be found in tables and figures.*

1. Compare the potential energy of the two states of water shown in Figure 13.1.
2. In what state (solid, liquid, or gas) would $H_2S$, $H_2Se$, and $H_2Te$ be at 0°C? (See Table 13.1.)
3. The two thermometers in Figure 13.2 read 100°C. What is the pressure of the atmosphere?
4. Draw a diagram of a water molecule and point out the areas that are the negative and positive ends of the dipole.
5. Would the distillation setup in Figure 13.5 be satisfactory for separating salt and water? Ethyl alcohol and water? Explain.
6. If the liquid in the flask in Figure 13.5 is ethyl alcohol and the atmospheric pressure is 543 mm Hg, what temperature would show on the thermometer?
7. If water were placed in both containers in Figure 13.7, would they both have the same vapor pressure at the same temperature? Explain.
8. In Figure 13.7, in which case, (a) or (b), will the atmosphere above the liquid reach a point of saturation?
9. Suppose that a solution of ethyl ether and ethyl alcohol were placed in the closed bottle in Figure 13.7.
   (a) Would both substances be present in the vapor?
   (b) If the answer to part (a) is yes, which would have more molecules in the vapor?
10. At approximately what temperature would each of the substances listed in Table 13.5 boil when the pressure is 30 mm Hg?
11. Use the graph in Figure 13.9 to find the following:
    (a) The boiling point of water at 1000 mm Hg pressure
    (b) The normal boiling point of ethyl chloride
    (c) The boiling point of ethyl alcohol at 0.5 atm
12. Consider Figure 13.9.
    (a) Why is line $BC$ horizontal? What is happening in this interval?
    (b) What phases are present in the interval $BC$?

(c) When heating is continued after point $C$, another horizontal line, $DE$, is reached at a higher temperature. What does this line represent?

*C. Review questions.*

1. List six physical properties of water.
2. What condition is necessary for water to have its maximum density? What is its maximum density?
3. Account for the fact that an ice–water mixture remains at 0°C until all the ice is melted, even though heat is applied to it.
4. Which contains less heat, ice at 0°C or water at 0°C? Explain.
5. Why does ice float in water? Would ice float in ethyl alcohol ($d = 0.79$ g/ml)? Explain.
6. If water molecules were linear instead of bent, would the heat of vaporization be higher or lower? Explain.
7. The heat of vaporization for ethyl ether is 83.9 cal/g and that for ethyl alcohol is 204.3 cal/g. Which of these compounds has hydrogen bonding? Explain.
8. Would there be more or less H-bonding if water molecules were linear instead of bent? Explain.
9. Which would show hydrogen bonding, ammonia ($NH_3$) or methane ($CH_4$)? Explain.
10. Hydrogen fluoride is believed to exist as $H_6F_6$ in the liquid state. Draw a possible structure for this molecule, illustrating the different bond formations. What would be its molecular weight?
11. Write equations to show how the following metals react with water: aluminum, calcium, iron, sodium, zinc. State the conditions for each reaction.
12. Is the formation of hydrogen and oxygen from water an exothermic or endothermic reaction? How do you know?
13. What is chlorine water and what chemical species are present in it?
14. (a) Write the formulas for the anhydrides of the following acids:
    $H_2SO_3$, $H_2SO_4$, $HNO_3$, $HClO_4$, $H_2CO_3$
    (b) Write the formulas for the anhydrides of the following bases:
    $NaOH$, $KOH$, $Ba(OH)_2$, $Ca(OH)_2$, $Mg(OH)_2$
15. Complete and balance the following equations:

    (a) $Mg(OH)_2 \xrightarrow{\Delta}$
    (b) $\underset{\text{Methyl alcohol}}{CH_3OH} + O_2 \xrightarrow{\Delta}$
    (c) $Li + H_2O \longrightarrow$
    (d) $MgSO_4 \cdot 7H_2O \xrightarrow{\Delta}$
    (e) $HNO_3 + NaOH \longrightarrow$
    (f) $KOH \xrightarrow{\Delta}$
    (g) $Ba + H_2O \longrightarrow$
    (h) $Cl_2 + H_2O \longrightarrow$
    (i) $SO_3 + H_2O \longrightarrow$

16. Is the conversion of oxygen to ozone an exothermic or an endothermic reaction? How do you know?
17. Distinguish among an oxygen atom, an oxygen molecule, and an ozone molecule. How many electrons are in a peroxide ion?
18. How does ozone in the stratosphere protect the earth from excessive ultraviolet radiation?
19. Name each of the following hydrates:
    (a) $BaBr_2 \cdot 2\,H_2O$
    (b) $AlCl_3 \cdot 6\,H_2O$
    (c) $FePO_4 \cdot 4\,H_2O$
    (d) $MgNH_4PO_4 \cdot 6\,H_2O$
    (e) $FeSO_4 \cdot 7\,H_2O$
    (f) $SnCl_4 \cdot 5\,H_2O$
20. Explain how anhydrous copper(II) sulfate ($CuSO_4$) can act as an indicator for moisture.

21. Compare the types of bonds in metal dioxides and metal peroxides.
22. Distinguish between deionized water and:
    (a) Hard water (b) Soft water (c) Distilled water
23. How can soap function to make soft water from hard water? What objections are there to using soap for this purpose?
24. What substance is commonly used to destroy bacteria in water?
25. What chemical, other than chlorine or chlorine compounds, can be used to disinfect water for domestic use?
26. Some organic pollutants in water can be oxidized by dissolved molecular oxygen. What harmful effect can result from this depletion of oxygen in the water?
27. Why should you not drink liquids that are stored in ceramic containers, especially unglazed ones?
28. Write the chemical equation showing how magnesium ions are removed by a zeolite water softener.
29. Write an equation to show how hard water containing calcium chloride ($CaCl_2$) is softened by using sodium carbonate ($Na_2CO_3$).
30. The vapor pressure at 20°C is given for the following compounds:

| | | | |
|---|---|---|---|
| Methyl alcohol | 96 mm Hg | Water | 17.5 mm Hg |
| Acetic acid | 11.7 mm Hg | Carbon tetrachloride | 91 mm Hg |
| Benzene | 74.7 mm Hg | Mercury | 0.0012 mm Hg |
| Bromine | 173 mm Hg | Toluene | 23 mm Hg |

    (a) Arrange these compounds in their order of increasing rate of evaporation.
    (b) Which substance listed would have the highest and which the lowest boiling point?
31. Explain why rubbing alcohol, warmed to body temperature, still feels cold when applied to your skin.
32. Suggest a method whereby water could be made to boil at 50°C.
33. If a dish of water initially at 20°C is placed in a living room maintained at 20°C, the water temperature will fall below 20°C. Explain.
34. Explain why a higher temperature is obtained in a pressure cooker than in an ordinary cooking pot.
35. What is the relationship between vapor pressure and boiling point?
36. From the point of view of the Kinetic-Molecular Theory, explain why vapor pressure increases with temperature.
37. Why does water have such a relatively high boiling point?
38. The boiling point of ammonia ($NH_3$) is −33.4°C and that of sulfur dioxide ($SO_2$) is −10.0°C. Which will have the higher vapor pressure at −40°C?
39. Explain what is occurring physically when a substance is boiling.
40. Explain why HF (bp 19.4°C) has a higher boiling point than HCl (bp −85°C), whereas $F_2$ (bp −188°C) has a lower boiling point than $Cl_2$ (bp −34°C).
41. Under which conditions are freezing point and melting point of a pure substance equal to each other?
42. Which of the following statements are correct?
    (a) The process of a substance changing directly from a solid to gas is called sublimation.
    (b) When water is decomposed, the volume ratio of $H_2$ to $O_2$ is 2:1, but the mass ratio of $H_2$ to $O_2$ is 1:8.
    (c) Hydrogen sulfide is a larger molecule than water.
    (d) The changing of ice into water is an exothermic process.
    (e) Water and hydrogen fluoride are both nonpolar molecules.
    (f) The main use of hydrogen peroxide is as an oxidizing agent.

(g) $H_2O_2 \rightarrow 2\,H_2O + O_2$ represents a balanced equation for the decomposition of hydrogen peroxide.
(h) Steam at 100°C can cause more severe burns than liquid water at 100°C.
(i) The density of water is independent of temperature.
(j) Liquid A boils at a lower temperature than liquid B. This indicates that liquid A has a lower vapor pressure than liquid B at any particular temperature.
(k) Water boils at a higher temperature in the mountains than at sea level.
(l) No matter how much heat you put under an open pot of pure water on a stove, you cannot heat the water above its boiling point.
(m) The vapor pressure of a liquid at its boiling point is equal to the prevailing atmospheric pressure.
(n) The normal boiling temperature of water is 273°C.
(o) The amount of heat needed to change 1 mole of ice at 0°C to a liquid at 0°C is 1.44 kcal.

D. *Review problems.*

1. How many moles of water can be obtained from 100 g of each of these hydrates?
   (a) $CuSO_4 \cdot 5\,H_2O$  (b) $BaCl_2 \cdot 2\,H_2O$
2. When a person purchases washing soda ($Na_2CO_3 \cdot 10\,H_2O$) to use as a water softener, what percent of water is being bought?
3. How many calories are required to change 125 g of ice at 0°C to steam at 100°C?
4. How many calories of energy must be removed to change 75.0 g of water at 25°C to ice at 0°C?
5. The *molar heat of vaporization* is the number of calories required to change 1 mole of a substance at its boiling point to a vapor at the same temperature.
   (a) What is the molar heat of vaporization of water?
   (b) How many calories would be required to change 4.00 moles of water at 25°C to steam at 100°C?
6. Suppose 100 g of ice at 0°C is added to 250 g of water at 25°C. Is there sufficient ice to lower the temperature of the system to 0°C and still have ice remaining? Show evidence.
7. What weight of water must be decomposed to produce 50.0 litres of oxygen at STP?
8. (a) How many moles of oxygen can be obtained by decomposing 5.6 moles of hydrogen peroxide?
   (b) What volume will this oxygen occupy at 22°C and 650 mm Hg pressure?
9. How many litres of $O_2$ at STP can be obtained by decomposing 1.00 kg of 3.0% hydrogen peroxide?
10. Cadmium bromide ($CdBr_2$) forms a hydrate containing 20.9% water. Calculate the formula of this hydrate.
11. How many grams of $CuSO_4 \cdot 5\,H_2O$ need to be decomposed to obtain 22.0 ml of water ($d_{H_2O} = 1.00$ g/ml)?
12. What volume of hydrogen gas (at STP) can be obtained from 1000 g of aluminum reacting with steam? (See Section 13.5.)
13. How many grams of water will react with each of the following?
    (a) 1.00 mole of Na  (c) 1.00 g of Na
    (b) 1.00 mole of K  (d) 1.00 g of K
14. What is the pressure in a 1 litre vessel containing 0.50 g of water at 100°C?
15. Suppose one mole of water evaporates in one day. How many water molecules, on the average, leave the liquid each second?

16. A quantity of sulfuric acid is added to 100 ml of water. The final volume of the solution is 139 ml and it has a density of 1.38 g/ml. What weight of acid was added? Assume that the density of water is 1.00 g/ml.

E. *Review exercises.*

1. Can ice be colder than 0°C? Explain.
2. Why does a boiling liquid maintain a constant temperature when heat is continually being added?
3. At what temperature will copper have a vapor pressure of 760 mm Hg?
4. Why does a lake freeze from the top down?
5. What water temperature would you theoretically expect to find at the bottom of a very deep lake? Explain.
6. What reasons can you give for two compounds having approximately the same molecular weight but having very different boiling points?

# 14 Solutions

*After studying Chapter 14 you should be able to:*

1. Understand the terms listed under Question A at the end of the chapter.
2. Describe the different types of solutions that are possible based on the three states of matter.
3. List the general properties of solutions.
4. Outline the solubility rules for common mineral substances.
5. Describe and illustrate the process by which an ionic substance like sodium chloride dissolves in water.
6. Tell how temperature changes affect the solubilities of solids and gases in liquids.
7. Tell how changes of pressure affect the solubility of a gas in a liquid.
8. Identify and discuss the variables that affect the rate at which a solid dissolves in a liquid.
9. Determine by using a solubility graph or table whether a given solution is unsaturated, saturated, or supersaturated at a given temperature.
10. Calculate the weight percent or volume percent composition of a solution from appropriate data.
11. Calculate the amount of solute in a given quantity of a solution when given the weight percent or volume percent composition.
12. Calculate the molarity of a solution when given the volume of solution and moles of solute.
13. Calculate the weight of a substance needed to prepare a solution of specified volume and molarity.
14. Determine the resulting molarity when a given volume of a solution of known molarity is mixed with a specified volume of water or is mixed with a solution of different molarity.
15. From the equation for a reaction, relate the given weight, moles, solution volume, or gas volume of one substance to the corresponding quantities of any other substance appearing in the equation.
16. Understand the concepts of equivalent weight and normality and do calculations involving these concepts.
17. Relate the effect of a solute on the vapor pressure of a solvent to the freezing point and the boiling point of a solution.
18. Calculate the boiling point or freezing point of a solution from appropriate concentration data.

19. Calculate molality and molecular weight of a solute from boiling point or freezing point and weight concentration data.
20. Explain the phenomenon of osmosis.
21. Predict the direction of net solvent flow between solutions of known concentration separated by a semipermeable membrane.

## 14.1 Components of a Solution

solution

solute

solvent

The term **solution** is used in chemistry to describe a system in which one or more substances are homogeneously mixed or dissolved in another substance. A simple solution has two components, a solute and a solvent. The **solute** is the substance that is dissolved. The **solvent** is the dissolving agent and usually makes up the greater proportion of the solution. For example, when salt is dissolved in water to form a solution, salt is the solute and water is the solvent. Complex solutions containing more than one solute and/or more than one solvent are common.

## 14.2 Types of Solutions

From the three states of matter—solid, liquid, and gas—it is possible to have nine different types of solutions: solid dissolved in solid, solid dissolved in liquid, solid dissolved in gas, liquid dissolved in liquid, and so on. Of these, the most common solutions are solid dissolved in a liquid, liquid dissolved in liquid, gas dissolved in a liquid, and gas dissolved in a gas.

## 14.3 General Properties of Solutions

A true solution is one in which the dissolved solute is molecular or ionic in size, generally in the range of 1 to 10 Angstrom units ($10^{-8}$ to $10^{-7}$ cm). The properties of a true solution are as follows:

1. It is a homogeneous mixture of two or more substances, solute and solvent.
2. It has a variable composition.
3. The dissolved solute is either molecular or ionic in size.
4. It may be either colored or colorless but is usually transparent.
5. The solute remains uniformly distributed throughout the solution and will not settle out with time.
6. The solute generally may be separated from the solvent by purely physical means (for example, by evaporation).

These properties are illustrated by water solutions of sugar and of potassium permanganate. Suppose that we prepare two sugar solutions, the first containing 10 g of sugar added to 100 ml of water and the second containing 20 g of sugar added to 100 ml of water. Each solution is stirred until all the solute dissolves, demonstrating that we may vary the composition of a solution. Every portion of the solution has the same sweet taste because the sugar molecules are uniformly distributed throughout. If confined so that no solvent is lost, the solution will taste and appear the same a week or a month later. The properties of the solution are unaltered after the solution is passed through filter paper. But by carefully evaporating the water, we may recover the sugar from the solution.

To observe the dissolving of potassium permanganate ($KMnO_4$) we affix a few crystals of $KMnO_4$ to paraffin wax or rubber cement at the end of a glass rod and submerge the entire rod, with the wax–permanganate end up, in a cylinder of water. Almost at once the beautiful purple color of dissolved permanganate ions ($MnO_4^-$) appears at the top of the rod and streams to the bottom of the cylinder as the crystals dissolve. The purple color at first is mostly at the bottom of the cylinder because potassium permanganate is denser than water. But after a while, the purple color disperses until it is evenly distributed throughout the solution. This demonstrates that molecules and ions move about freely and spontaneously (diffuse) in a liquid or solution. Once the solution is formed, it is permanent; the solute does not settle out.

The permanency of a solution is explained by the Kinetic-Molecular Theory (KMT). All matter, according to the KMT, is in some kind of motion at all temperatures above absolute zero, and the intensity of motion increases with increasing temperature. Once prepared, a solution remains homogeneous because of this constant random or thermal motion of the solute and solvent particles. This constant random motion is responsible for diffusion in liquids and gases.

## 14.4 Solubility

solubility

We use the term **solubility** to describe the amount of one substance that will dissolve in another. For example, 36.0 g of sodium chloride (NaCl) will dissolve in 100 g of water at 20°C. We say, then, that the solubility of NaCl in water is 36.0 g per 100 g of water at 20°C.

Solubility is often used in a relative way. We say that a substance is very soluble, moderately soluble, slightly soluble, or insoluble. Although these terms do not accurately indicate how much solute will dissolve, they are frequently used to describe the solubility of a substance qualitatively.

miscible

immiscible

Two other terms often used to describe solubility are miscible and immiscible. Liquids that are capable of mixing and forming a solution are **miscible**; those that do not form solutions or are generally insoluble in each other are **immiscible**. Methyl alcohol and water are miscible in each other in all proportions. Carbon tetrachloride and water are immiscible, forming two separate layers when they are mixed. Miscible and immiscible systems are illustrated in Figure 14.1.

The general rules for the solubility of common mineral substances are given in Table 14.1. The solubility of over 200 compounds is given in the

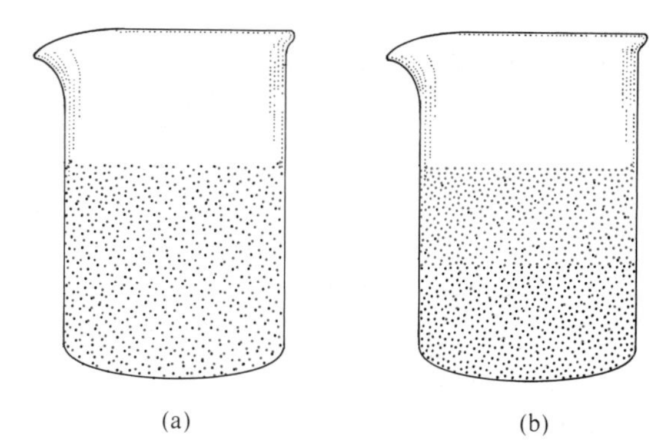


*Figure 14.1.* Miscible and immiscible systems: (a) miscible: $H_2O$ and $CH_3OH$; (b) immiscible: $H_2O$ and $CCl_4$. In a miscible system, a solution is formed, consisting of a single phase with the solute and solvent uniformly dispersed. An immiscible system is heterogeneous, and, in the case of two liquids, forms two liquid layers.

*Table 14.1.* General solubility rules for common mineral substances.[a]

| Class | Solubility in cold water |
|---|---|
| Nitrates | All nitrates are soluble. |
| Acetates | All acetates are soluble. |
| Chlorides, Bromides, Iodides | All chlorides, bromides, and iodides are soluble except those of Ag, Hg(I), and Pb(II); $PbCl_2$ and $PbBr_2$ are slightly soluble in hot water. |
| Sulfates | All sulfates are soluble except those of Ba, Sr, and Pb; Ca and Ag sulfates are slightly soluble. |
| Carbonates, Phosphates | All carbonates and phosphates are insoluble except those of Na, K, and $NH_4^+$. Many bicarbonates and acid phosphates are soluble. |
| Hydroxides | All hydroxides are insoluble except those of the alkali metals and $NH_4OH$; $Ba(OH)_2$ and $Ca(OH)_2$ are slightly soluble. |
| Sodium salts, Potassium salts, Ammonium salts | All common salts of these ions are soluble. |
| Sulfides | All sulfides are insoluble except those of the alkali metals, ammonium, and the alkaline earth metals (Ca, Mg, Ba). |

[a] When we say a substance is soluble, we mean that the substance is reasonably soluble. All substances have some solubility in water, although the amount of solubility may be very small; the solubility of silver iodide, for example, is about $1 \times 10^{-8}$ mole AgI/litre $H_2O$.

Solubility Table in Appendix IV. Solubility data for thousands of compounds may be found by consulting standard reference sources.[1]

concentration of a solution

The quantitative expression of the amount of dissolved solute in a particular quantity of solvent is known as the **concentration of a solution**. Several methods of expressing concentration will be described in Section 14.7.

## 14.5 Factors Related to Solubility

The entire concept of predicting solubilities is, at best, very complex and difficult. There are many variables, such as size of ions, charge on ions, interaction between ions, interaction between solute and solvent, and temperature, all of which bear upon the problem. Because of the factors involved, there are many exceptions to the general rules of solubility given in Table 14.1. However, the rules are very useful, because they do apply to a good many of the more common compounds that we encounter in the study of chemistry. Keep in mind that these are rules, not laws, and are therefore subject to exceptions. Fortunately, the solubility of a solute is relatively easy to determine experimentally. Four factors related to solubility are discussed below.

1. The nature of the solute and solvent. The old adage that "like dissolves like" has merit, in a general way. Polar substances tend to be more miscible, or soluble, with other polar substances. Nonpolar substances tend to be miscible with other nonpolar substances and less miscible with polar substances. Thus, mineral acids, bases, and salts, which are polar, tend to be much more soluble in water, which is polar, than in solvents such as ether, carbon tetrachloride, or benzene, which are essentially nonpolar. Sodium chloride, a very polar substance, is soluble in water, slightly soluble in ethyl alcohol (less polar than water), and insoluble in ether and benzene. Pentane ($C_5H_{12}$), a nonpolar substance, is only slightly soluble in water but is very soluble in benzene and ether.

At the molecular level, the formation of a solution from two nonpolar substances, such as carbon tetrachloride and benzene, can be visualized as a process of simple mixing. The nonpolar molecules, having little tendency to either attract or repel one another, easily intermingle to form a homogeneous mixture.

Solution formation between polar substances is much more complex. For example, the process by which sodium chloride dissolves in water is illustrated in Figure 14.2. Water molecules are very polar and are attracted to other polar molecules or ions. When salt crystals (NaCl) are put into water, polar water molecules become attracted to the sodium and chloride ions on the crystal surfaces and weaken the attraction between $Na^+$ and $Cl^-$ ions. The positive end of

[1] Two commonly used handbooks are *Lange's Handbook of Chemistry*, 11th ed. (New York: McGraw-Hill, 1973), and *Handbook of Chemistry and Physics*, 61st ed. (Cleveland; Chemical Rubber Co., 1980).

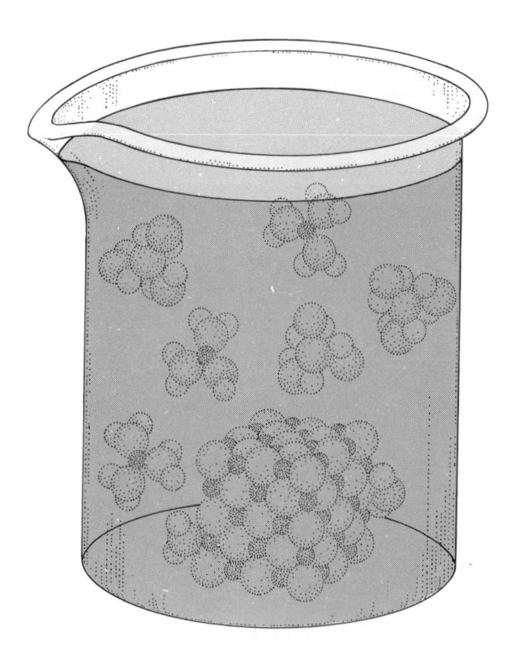

*Figure 14.2.* Dissolution of sodium chloride in water. Polar water molecules are attracted to $Na^+$ and $Cl^-$ ions in the salt crystal, weakening the attraction between the ions. As the attraction between the ions weakens, the ions move apart and become surrounded by water dipoles. The hydrated ions slowly diffuse away from the crystal to become dissolved in solution.

the water dipole is attracted to the $Cl^-$ ions, and the negative end of the water dipole to the $Na^+$ ions. The weaker attraction permits the ions to move apart, making room for more water dipoles. Thus, the surface ions are surrounded by water molecules, becoming hydrated ions, $Na^+(aq)$ and $Cl^-(aq)$, and slowly diffuse away from the crystals and dissolve in solution.

$$\text{NaCl (crystal)} \xrightarrow{H_2O} Na^+(aq) + Cl^-(aq)$$

Examination of the data in Table 14.2 reveals some of the complex questions relating to solubility. For example, some questions that may arise on examining the table are: Why are lithium halides, except for lithium fluoride (LiF), more soluble than sodium and potassium halides? Why, indeed, are the solubilities of LiF and sodium fluoride (NaF) so low in comparison to those of the other salts? Why does not the solubility of LiF, NaF, and NaCl increase proportionately with temperature, as the solubilities of the other salts do? Sodium chloride is appreciably soluble in water, but is insoluble in concentrated hydrochloric acid (HCl) solution. On the other hand, LiF and NaF are not very soluble in water but are quite soluble in hydrofluoric acid (HF) solution—why? These questions will not be answered directly here, but it is hoped that your curiosity will be aroused to the point that you will do some reading and research on the properties of solutions.

*Table 14.2.* Solubility of alkali metal halides in water.

| Salt | Solubility (g salt/100 g $H_2O$) 0°C | 100°C |
|---|---|---|
| LiF | 0.12 | 0.14 (at 35°C) |
| LiCl | 67 | 127.5 |
| LiBr | 143 | 266 |
| LiI | 151 | 481 |
| NaF | 4 | 5 |
| NaCl | 35.7 | 39.8 |
| NaBr | 79.5 | 121 |
| NaI | 158.7 | 302 |
| KF | 92.3 (at 18°C) | Very soluble |
| KCl | 27.6 | 57.6 |
| KBr | 53.5 | 104 |
| KI | 127.5 | 208 |

2. The effect of temperature on solubility. Most solutes have a limited solubility in a specific amount of solvent at a fixed temperature. The temperature of the solvent has a marked effect on the amount of solute that will dissolve. For *most solids* dissolved in a liquid, an increase in temperature results in an increase in solubility (see Figure 14.3). However, the solubility of a *gas* in a liquid always

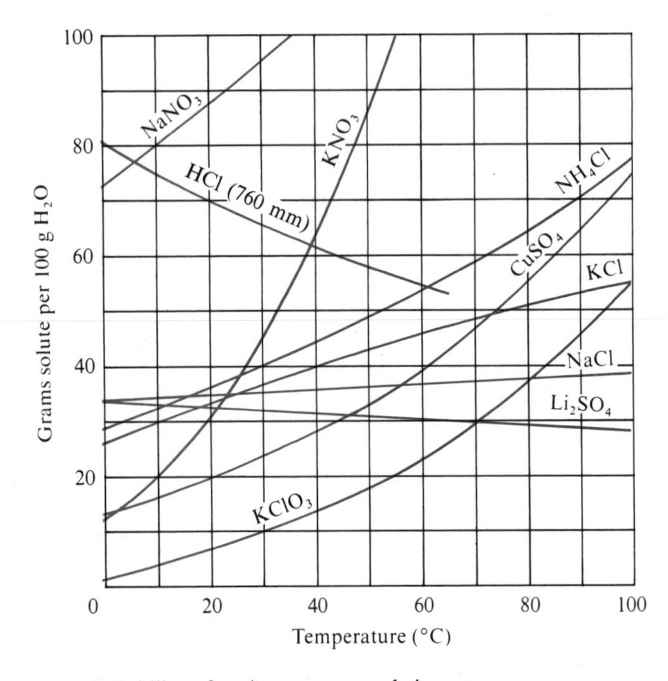


*Figure 14.3.* Solubility of various compounds in water.

*decreases* as the temperature *increases*. This is explained in terms of the KMT by assuming that in order to dissolve, the gas molecules must form "bonds" of some sort with the molecules of the liquid. An increase in temperature decreases the solubility of the gas because it increases the kinetic energy (speed) of the gas molecules and thereby decreases their ability to form "bonds" with the liquid molecules.

3. The effect of pressure on solubility. Small changes in pressure have little effect on the solubility of solids in liquids but have a marked effect on the solubility of gases in liquids. The solubility of a gas in a liquid is directly proportional to the pressure of that gas above the solution. Thus, the amount of a gas that is dissolved in solution will double if the pressure of that gas over the solution is doubled. For example, carbonated beverages contain dissolved carbon dioxide at pressures greater than atmospheric pressure. When a bottle of carbonated soda is opened, the pressure is immediately reduced to the atmospheric pressure, and the excess dissolved carbon dioxide bubbles out of the solution.

4. Rate of dissolving. The rate at which a solid solute dissolves is affected by (a) particle size of the solute, (b) temperature, (c) agitation or stirring, and (d) concentration of the solution.

(a) *Particle size.* A solid can dissolve only at the surface that is in contact with the solvent. Because the surface:volume ratio increases as size decreases, smaller crystals dissolve faster than large ones. For example, if a salt crystal 1 cm on a side (6 $cm^2$ surface area) is divided into 1000 cubes, each 0.1 cm on a side, the total surface of the smaller cubes is 60 $cm^2$—a tenfold increase in surface area (see Figure 14.4).

(b) *Temperature.* In most cases, the rate of dissolving of a solid increases with temperature. This is because of kinetic effects. The solvent molecules, moving more rapidly at higher temperatures, strike the solid surfaces more often and harder causing the rate of dissolving to increase.

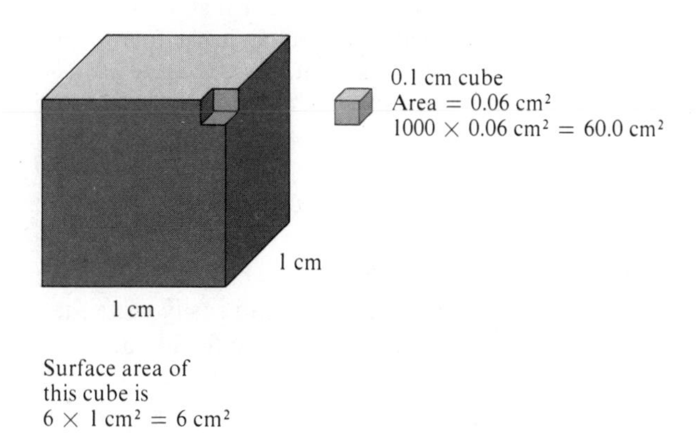


*Figure 14.4.* Surface area of crystals: A crystal 1 cm on a side has a surface area of 6 $cm^2$. Subdivided into 1000 smaller crystals, each with 0.1 cm on a side, the total surface area is increased to 60.0 $cm^2$.

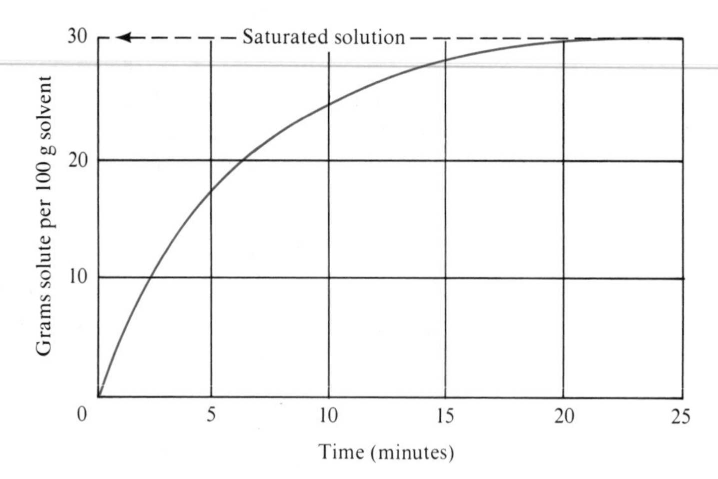


*Figure 14.5.* Rate of dissolution of a solid solute in a solvent. The rate is maximum at the beginning and decreases as the concentration approaches the saturation point.

(c) *Agitation or stirring.* The effect of agitation or stirring is kinetic. When a solid is first put into water, the only solvent with which it comes in contact is in the immediate vicinity. As the solid dissolves, the amount of dissolved solute around the solid becomes more and more concentrated and the rate of dissolving slows down. If the mixture is not stirred, the dissolved solute diffuses very slowly throughout the entire solution; weeks may pass before the solid is entirely dissolved. Through stirring, the dissolved solute is distributed rapidly throughout the solution and more solvent is brought into contact with the solute, causing it to dissolve more rapidly.

(d) *Concentration of the solution.* When the solute and solvent are first mixed, the rate of dissolving is at its maximum. As the concentration of the solution increases and the solvent becomes more nearly saturated with the solute, the rate of dissolving decreases greatly. The rate of dissolving is pictured graphically in Figure 14.5. Note that about 17 g dissolve in the first 5 minute interval but only about 1 g dissolves in the fourth 5 minute interval. Although different solutes show different rates, the rate of dissolving always becomes very slow as the concentration approaches the saturation point.

## 14.6 Solutions: A Reaction Zone

Many solids must be put in solution in order to undergo appreciable chemical reactions. We can easily write the equation for the double-replacement reaction between sodium chloride and silver nitrate:

$$NaCl + AgNO_3 \longrightarrow AgCl + NaNO_3$$

But suppose we mix solid NaCl and solid $AgNO_3$ and look for a chemical change. Some reaction may occur, but if it does, it is quite slow and essentially undetectable. In fact, the crystalline structures of NaCl and $AgNO_3$ are so different that we can separate them by tediously picking out the two different kinds of crystals from the mixture. But if we dissolve the sodium chloride and silver nitrate separately in water and mix the two solutions, we observe the immediate formation of a white, curdy precipitate of silver chloride.

Molecules or ions must come into intimate contact or collide with one another in order to react. In the foregoing example, the two solids did not react because the ions were securely locked within the crystal structures. But when the sodium chloride and silver nitrate are dissolved, their crystal lattices are broken down and the ions become mobile. When the two solutions are mixed, the mobile $Ag^+$ and $Cl^-$ ions come into contact and react to form insoluble AgCl, which precipitates out of solution. The soluble $Na^+$ and $NO_3^-$ ions remain in solution but form the crystalline salt $NaNO_3$ when the water is evaporated:

$$NaCl(aq) + AgNO_3(aq) \longrightarrow AgCl\downarrow + NaNO_3(aq)$$

$$\underset{\text{Sodium chloride solution}}{(Na^+ + Cl^-)} + \underset{\text{Silver nitrate solution}}{(Ag^+ + NO_3^-)} \xrightarrow{H_2O} \underset{\text{Silver chloride}}{AgCl\downarrow} + \underset{\text{Sodium nitrate in solution}}{Na^+ + NO_3^-}$$

The mixture of the two solutions provided a zone or space in which the $Ag^+$ and $Cl^-$ ions could react. (See Chapter 15 for further discussion of ionic reactions.)

Solutions also function as diluents in reactions in which the undiluted reactants would combine with each other too violently. Moreover, a solution of known concentration provides a convenient method for delivering specific amounts of reagents.

## 14.7 Concentration of Solutions

The concentration of a solution gives us information concerning the amount of solute dissolved in a unit volume of solution. Because reactions are often conducted in solution, it is important to understand the methods of expressing concentration and to know how to prepare solutions of particular concentrations.

1. Dilute and concentrated solutions. When we say that a solution is *dilute* or *concentrated*, we are expressing, in a relative way, the amount of solute present. One gram of salt and 2 g of salt in solution are both dilute solutions when compared to the same volume of a solution containing 20 g of salt. Ordinary concentrated hydrochloric acid (HCl) contains 12 moles of HCl per litre of solution. In some laboratories, the dilute acid is made by mixing equal volumes of water and the concentrated acid. In other laboratories, the concentrated acid is diluted with two or three volumes of water, depending on its use. The term **dilute solution**, then, describes a solution that contains a relatively small amount of dissolved solute. Conversely, a **concentrated solution** contains a relatively large amount of dissolved solute.

dilute solution

concentrated solution

2. Saturated, unsaturated, and supersaturated solutions. At a specific temperature there is a limit to the amount of solute that will dissolve in a given amount of solvent. When this limit is reached, the resulting solution is said to be *saturated.* For example, when we put 40.0 g KCl into 100 g $H_2O$ at 20°C, we find that 34.0 g KCl dissolve and 6.0 g KCl remain undissolved. The solution formed is a saturated solution of KCl.

Two processes are occurring simultaneously in a saturated solution. The solid is dissolving into solution and, at the same time, the dissolved solute is crystallizing out of solution. This may be expressed as

$$\text{Solute (undissolved)} \rightleftharpoons \text{Solute (dissolved)}$$

saturated solution

When these two opposing processes are occurring at the same rate, the amount of solute in solution is constant and a condition of equilibrium is established between dissolved and undissolved solute. A **saturated solution** contains dissolved solute in equilibrium with undissolved solute.

It is especially important to state the temperature of a saturated solution. A solution that is saturated at one temperature may not be saturated at another. If the temperature of a saturated solution is changed, the equilibrium is disturbed, and the amount of dissolved solute will change to reestablish the equilibrium.

A saturated solution may be either dilute or concentrated, depending on the solubility of the solute. A saturated solution can be conveniently prepared by dissolving the solute at a temperature somewhat higher than room temperature. The amount of solute in solution should be in excess of its solubility at room temperature. When the solution cools, the excess solute crystallizes, leaving the solution saturated. In this case, the solute must be more soluble at higher temperatures and must not form a supersaturated solution. Examples expressing the solubility of saturated solutions at two different temperatures are given in Table 14.3.

unsaturated solution

A solution containing less solute per unit of volume than does its corresponding saturated solution is said to be **unsaturated**. In other words, more solute can be dissolved into an unsaturated solution without altering other conditions. Consider a solution made from 40 g of KCl and 100 g of water

*Table 14.3.* Saturated solutions at 20°C and 50°C.

| | Solubility (g solute/100 g $H_2O$) | |
|---|---|---|
| | 20°C | 50°C |
| NaCl | 36.0 | 37.0 |
| KCl | 34.0 | 42.6 |
| $NaNO_3$ | 88.0 | 114.0 |
| $KClO_3$ | 7.4 | 19.3 |
| $AgNO_3$ | 222.0 | 455.0 |
| $C_{12}H_{22}O_{11}$ | 203.9 | 260.4 |

at 20°C (see Table 14.3). The solution formed will certainly be saturated and will contain about 6 g of undissolved salt, because the maximum amount of KCl that can dissolve in 100 g of water at 20°C is 34 g. If the solution is now heated and maintained at 50°C, all the salt will dissolve and, in fact, more can be dissolved. Thus, the solution at 50°C is unsaturated.

In some instances, solutions can be prepared that contain more solute than that of the saturated solution at a particular temperature. Such solutions are said to be **supersaturated**. However, we must qualify this definition by noting that a supersaturated solution is unstable. Disturbances such as jarring, stirring, scratching the walls of the container, or dropping in a "seed" crystal cause the supersaturation to break. When a supersaturated solution is disturbed, the excess solute crystallizes out rapidly, returning the solution to a saturated state.

supersaturated solution

Supersaturated solutions, while not easy to prepare, may be formed from selected substances by dissolving, in warm solvent, an amount of solute greater than that needed for a saturated solution at room temperature. The warm solution is then allowed to cool very slowly. With the proper solute and careful work, a supersaturated solution will result. Two substances commonly used to demonstrate this property are sodium thiosulfate pentahydrate, $Na_2S_2O_3 \cdot 5H_2O$, and sodium sulfate, $Na_2SO_4$ (from a saturated solution at 30°C).

3. Weight percent solution. This expression of concentration gives the percentage of solute by weight in a solution. It says that for a given weight of solution, a certain percentage of that weight is solute. Suppose that we take a bottle from the reagent shelf that reads "Sodium hydroxide, NaOH, 10%" (see Figure 14.6). This means that for every 100 g of this solution we use, 10 g will be NaOH and 90 g will be water. (Note that this is 100 g and not 100 ml

*Figure 14.6.* Weight percent concentration of solutions. The bottle contains 10 g of NaOH per 90 g of $H_2O$.

of solution.) We could also make this same concentration of solution by dissolving 2 g of NaOH in 18 grams of water. Weight percent concentrations are most generally used for solids dissolved in liquids.

$$\text{Weight percent} = \frac{\text{g solute}}{\text{g solute} + \text{g solvent}} \times 100\%$$

Illustrative problems follow.

*Problem 14.1*

What is the weight percent of sodium hydroxide in a solution that is made by dissolving 8.00 g of NaOH in 50.0 g of $H_2O$?

$$\text{Grams of solute (NaOH)} = 8.00 \text{ g}$$

$$\text{Grams of solvent } (H_2O) = 50.0 \text{ g}$$

$$\frac{8.00 \text{ g NaOH}}{8.00 \text{ g NaOH} + 50.0 \text{ g } H_2O} \times 100\% = 13.8\% \text{ NaOH solution}$$

*Problem 14.2*

What weights of potassium chloride (KCl) and water are needed to make 250 g of 5.00% solution?

The percentage expresses the weight of the solute.

$$250 \text{ g} = \text{Total weight of solution}$$

$$5.00\% \text{ of } 250 \text{ g} = 0.0500 \times 250 \text{ g} = 12.5 \text{ g KCl (solute)}$$

$$250 \text{ g} - 12.5 \text{ g} = 237.5 \text{ g } H_2O$$

Dissolving 12.5 g KCl in 237.5 g of $H_2O$ gives a 5.00% KCl solution.

*Problem 14.3*

Suppose that 2.50 ml of 20.0% silver nitrate solution ($d = 1.19$ g/ml) was used to precipitate the chloride in a sample of salt water. What weight of $AgNO_3$ was used?

In this problem, a volume of solution is used. First, the weight of the solution is calculated from the volume and the density. Then, the weight of $AgNO_3$ can be determined from the weight percent.

$$d = \frac{\text{Mass}}{\text{Volume}} = \frac{\text{g}}{\text{ml}}$$

So,

$$\text{g} = \text{ml} \times d = 2.50 \text{ ml} \times \frac{1.19 \text{ g}}{\text{ml}} = 2.98 \text{ g}$$

$$\text{Weight of solution} = 2.98 \text{ g of } 20.0\% \; AgNO_3$$

Taking 20.0% of 2.98 g gives us the weight of $AgNO_3$ used:

$$2.98 \text{ g } AgNO_3 \times 0.200 = 0.596 \text{ g } AgNO_3 \quad \text{(Answer)}$$

When the concentration is given in percentage, it is assumed to be weight percent unless otherwise stated. The student should also be aware that the concentration expressed as weight percent is independent of the formula of the solute.

4. Volume percent. Solutions that are formulated from two liquids are often expressed as *volume percent* with respect to the solute. The label on a bottle of ordinary rubbing alcohol reads "Isopropyl alcohol, 70% by volume." Such a solution could be made by mixing 70 ml of alcohol and 30 ml of water. If we assume that these volumes are additive (which they are not, exactly), 1 litre of 70% isopropyl alcohol by volume will contain 700 ml of the alcohol.

$$\text{Volume percent} = \frac{\text{Volume of liquid in question}}{\text{Total volume of solution}} \times 100\%$$

5. Molarity. Weight percent solutions do not equate or express the number of formula or molecular weights of the solute in solution. For example, 1000 g of 10% NaOH solution contain 100 g of NaOH; 1000 g of 10% KOH solution contain 100 g of KOH. In terms of moles of NaOH and KOH, these solutions contain

$$\text{moles NaOH} = 100 \text{ g NaOH} \times \frac{1 \text{ mole NaOH}}{40.0 \text{ g NaOH}} = 2.50 \text{ moles NaOH}$$

$$\text{moles KOH} = 100 \text{ g KOH} \times \frac{1 \text{ mole KOH}}{56.1 \text{ g KOH}} = 1.78 \text{ moles KOH}$$

From the above figures, we see that the two 10% solutions do not contain the same number of moles of NaOH and KOH. Yet one mole of each of these two bases will neutralize the same amount of acid. As a result, we find that a 10% NaOH solution has more reactive alkali than a 10% KOH solution.

We need a method expressing concentration that will easily identify how many moles or formula weights of solute are present per unit of volume of solution. For this purpose, the molar method of expressing concentration is used.

1 molar solution

A **1 molar solution** contains 1 mole, or 1 gram-molecular weight, or 1 gram-formula weight of solute per litre of solution. For example, to make a 1 molar solution of sodium hydroxide (NaOH), we dissolve 40.0 g of NaOH (1 mole) in water and dilute the solution with more water to a volume of 1 litre. The solution contains 1 mole of the solute in 1 litre of solution and is said to be 1 molar (1 $M$) in concentration. Figure 14.7 illustrates the preparation of a 1 molar solution. Note that the volume of the solute and the solvent together is 1 litre.

molarity

The concentration of a solution may, of course, be varied by using more or less solute or solvent; but in any case, the **molarity** of a solution is the number of moles of solute per litre of solution. A capital $M$ is the abbreviation for molarity. The units of molarity are moles per litre. The expression "2.0 $M$ NaOH" means a 2.0 molar solution of NaOH (2.0 moles, or 80 g, of NaOH dissolved in 1 litre of solution).

$$\text{Molarity} = M = \frac{\text{number of moles of solute}}{\text{litre of solution}} = \frac{\text{moles}}{\text{litre}}$$

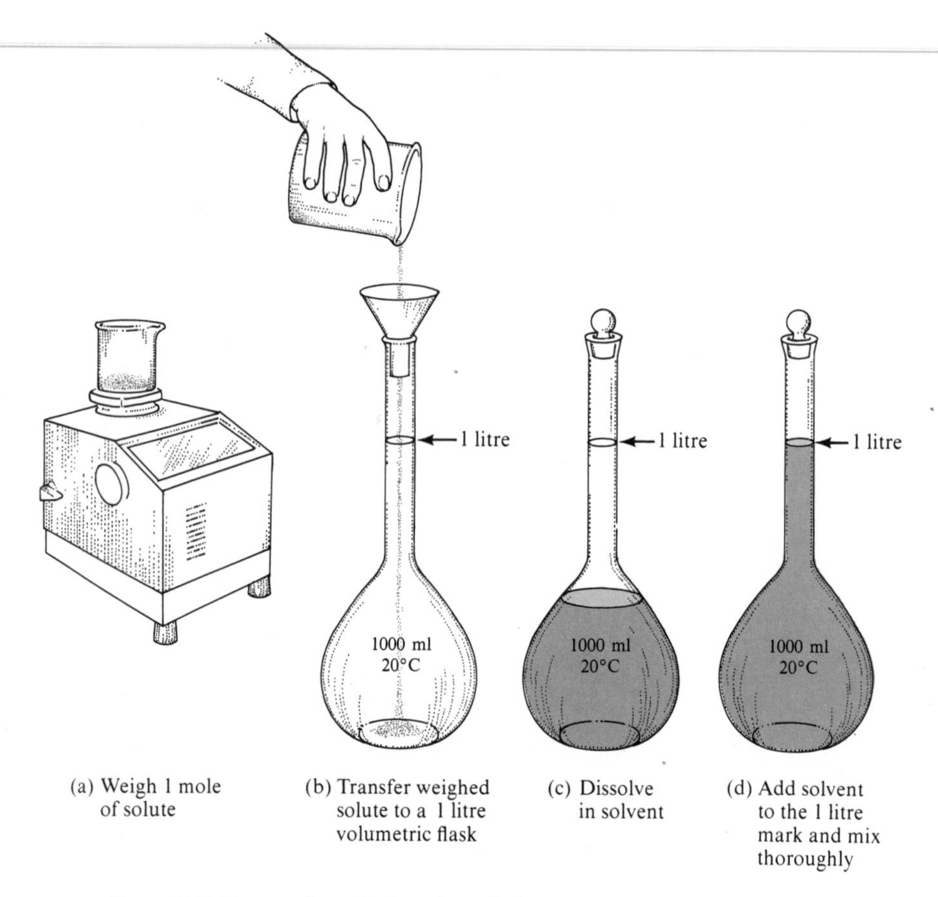


*Figure 14.7.* Preparation of a 1 molar solution.

Flasks that are calibrated to contain specific volumes at a particular temperature are used to prepare solutions of a desired concentration. These *volumetric flasks* have a calibration mark on the neck to indicate accurately the measured volume.

Suppose we want to make 500 ml of 1 $M$ solution. This solution can be prepared by weighing 0.5 mole of the solute and diluting with water in a 500 ml volumetric flask. The molarity will be

$$M = \frac{0.5 \text{ mole solute}}{0.5 \text{ litre}} = 1 \text{ molar}$$

Thus, you can see that it is not necessary to have a litre of solution to express molarity. All we need to know is the number of moles of dissolved solute and the volume of solution. Thus, 0.001 mole NaOH in 10 ml of solution is 0.1 $M$:

$$\frac{0.001 \text{ mole}}{10 \text{ ml}} \times \frac{1000 \text{ ml}}{\text{litre}} = 0.1\ M$$

When we stop to think that a balance is not calibrated in moles but in grams, we see that we really need to incorporate grams into the molarity formula. This is done by using the relationship

$$\text{Number of moles} = \frac{\text{g}}{\text{g-mol. wt}}$$

Substituting this relationship into our expression for molarity, we get

$$M = \frac{\text{moles}}{\text{litre}} = \frac{\text{g solute}}{\text{g-mol. wt solute} \times \text{litre solution}} = \frac{\text{g}}{\text{g-mol. wt} \times \text{litre}}$$

We can now weigh any amount of a solute that has a known formula, dilute it to any volume, and calculate the molarity of the solution using this formula.

The molarities of the concentrated acids commonly used in the laboratory are

| | | | |
|---|---|---|---|
| $HCl$ | 12 $M$ | $HC_2H_3O_2$ | 17 $M$ |
| $HNO_3$ | 16 $M$ | $H_2SO_4$ | 18 $M$ |

Illustrative problems follow.

*Problem 14.4*

What weight of potassium hydroxide (KOH) is needed to prepare 1.00 litre of 1.00 $M$ solution?

From the definition for molarity, 1 mole of KOH will be needed. This is 56.1 g, the gram-formula weight.

*Data*: Volume = 1.00 litre $\quad M = \frac{1.00\text{ mole}}{\text{litre}} \quad \text{g-mol. wt} = \frac{56.1\text{ g}}{\text{mole}}$

litres $\longrightarrow$ moles $\longrightarrow$ grams

$$1.00\text{ litres} \times \frac{1.00\text{ mole}}{\text{litre}} \times \frac{56.1\text{ g}}{\text{mole}} = 56.1\text{ g KOH} \quad \text{(Answer)}$$

---

*Problem 14.5*

What is the molarity of a solution containing 1.4 moles of acetic acid ($HC_2H_3O_2$) in 250 ml of solution?

We may reason that if there are 1.4 moles in 250 ml, there will be four times that amount, or 5.6 moles, in 1 litre, because 250 ml is contained four times in 1 litre:

$$1\text{ litre} = 1000\text{ ml}; \qquad \frac{1000\text{ ml}}{250\text{ ml}} = 4$$

The concentration of the solution would be 5.6 $M$ (4 × 1.4).

By the unit conversion method, we note that the concentration given in the problem statement is 1.4 moles per 250 ml (moles/ml). Since molarity = moles/litre, the needed conversion is:

$$\frac{\text{moles}}{\text{ml}} \longrightarrow \frac{\text{moles}}{\text{litre}}$$

$$\frac{1.40\text{ moles}}{250\text{ ml}} \times \frac{1000\text{ ml}}{\text{litre}} = \frac{5.6\text{ moles}}{\text{litre}} = 5.6\ M \quad \text{(Answer)}$$

---

*Problem 14.6*

What is the molarity of a solution made by dissolving 2.00 g of potassium chlorate ($KClO_3$) in enough water to make 150 ml of solution?

Use the formula

$$M = \frac{\text{g}}{\text{g-mol. wt} \times \text{litre}}$$

and substitute the given data:

$$\text{g} = 2.00$$
$$\text{g-mol. wt } KClO_3 = 122.6 \text{ g/mole}$$
$$150 \text{ ml} = 0.150 \text{ litre}$$

$$M = 2.00 \text{ g } KClO_3 \times \frac{1 \text{ mole } KClO_3}{122.6 \text{ g } KClO_3} \times \frac{1}{0.150 \text{ litre}}$$

$$= \frac{0.109 \text{ mole } KClO_3}{\text{litre}} = 0.109\ M \quad \text{(Answer)}$$

This problem may also be solved using the unit conversion method. The steps in the conversions must lead to units of moles/litres.

$$\frac{\text{g } KClO_3}{\text{ml}} \longrightarrow \frac{\text{g } KClO_3}{\text{litre}} \longrightarrow \frac{\text{moles } KClO_3}{\text{litre}} = M$$

$$\frac{2.00 \text{ g } \cancel{KClO_3}}{150 \cancel{\text{ml}}} \times \frac{1000 \cancel{\text{ml}}}{\text{litre}} \times \frac{1 \text{ mole}}{122.6 \text{ g } \cancel{KClO_3}} = \frac{0.109 \text{ mole}}{\text{litre}} = 0.109\ M$$

---

*Problem 14.7*

How many grams of potassium hydroxide are required to prepare 600 ml of 0.450 *M* KOH solution?

*Data*:

$$\text{Volume} = 0.600 \text{ litre} \qquad M = \frac{0.450 \text{ mole}}{\text{litre}} \qquad \text{g-mol. wt KOH} = \frac{56.1 \text{ g KOH}}{\text{mole}}$$

The calculation is

$$0.600 \text{ litre} \times \frac{0.450 \text{ mole}}{\text{litre}} \times \frac{56.1 \text{ g KOH}}{\text{mole}} = 15.1 \text{ g KOH} \quad \text{(Answer)}$$

---

*Problem 14.8*

How many millilitres of 2.0 *M* HCl will react with 28.0 g of NaOH?

*Step 1.* Write and balance the equation for the reaction:

$$HCl(aq) + NaOH(aq) \longrightarrow NaCl(aq) + H_2O(aq)$$

The equation states that 1 mole of HCl reacts with 1 mole of NaOH.

*Step 2.* Find the number of moles of NaOH in 28.0 g of NaOH.

$$\text{moles} = \frac{\text{g}}{\text{g-mol. wt}} = \frac{28.0 \text{ g NaOH}}{40.0 \text{ g/mole}} = 0.70 \text{ mole NaOH}$$

$$28.0 \text{ g NaOH} = 0.70 \text{ mole NaOH}$$

*Step 3.* Solve for moles and volume of HCl needed. From Steps 1 and 2 we see that 0.70 mole of HCl will react with 0.70 mole of NaOH, because the ratio of moles reacting is 1:1. We know that 2.0 *M* HCl contain 2.0 moles of HCl per litre; therefore, the volume that contains 0.70 mole of HCl will be less than 1 litre.

$$\text{moles NaOH} \longrightarrow \text{moles HCl} \longrightarrow \text{litres HCl} \longrightarrow \text{ml HCl}$$

$$0.70 \text{ mole NaOH} \times \frac{1 \text{ mole HCl}}{1 \text{ mole NaOH}} \times \frac{1 \text{ litre HCl}}{2 \text{ moles HCl}} = 0.350 \text{ litre HCl}$$

$$0.350 \text{ litre HCl} \times 1000 \text{ ml/litre} = 350 \text{ ml HCl}$$

Therefore, 350 ml of 2.0 *M* HCl contain 0.70 mole of HCl and will react with 0.70 mole, or 28.0 g, of NaOH.

*Problem 14.9* What volume of 0.250 *M* solution can be prepared from 16.0 g of potassium carbonate ($K_2CO_3$)?

*Step 1.* Solving the equation

$$M = \frac{\text{g}}{\text{g-mol. wt} \times \text{litre}}$$

for litres, we obtain

$$\text{litres} = \frac{\text{g}}{\text{g-mol. wt} \times M}$$

*Step 2.* Substitute the data given into the equation and solve.

*Data*: g $K_2CO_3$ = 16.0 g-mol. wt $K_2CO_3$ = 138.2 g/mole

$M$ = 0.250 mole/litre

$$\text{litres} = \frac{16.0 \text{ g}}{138.2 \text{ g/mole} \times 0.250 \text{ mole/litre}} = 0.463 \text{ litre, or } 463 \text{ ml}$$

Or, by the unit conversion method,

$$16.0 \text{ g } K_2CO_3 \times \frac{1 \text{ mole } K_2CO_3}{138.1 \text{ g } K_2CO_3} \times \frac{1 \text{ litre}}{0.250 \text{ mole } K_2CO_3} = 0.463 \text{ litre}$$

Thus, a 0.250 *M* solution can be made by dissolving 16.0 g of $K_2CO_3$ in water and diluting to 463 ml.

Chemists often find it necessary to dilute solutions from one concentration to another by adding more solvent to the solution. If a solution is diluted by adding pure solvent, the volume of the solution increases but the number of moles of solute in the solution remains the same. Thus, the moles/litre (molarity) of the solution decreases. It is important to read a problem carefully to distinguish between (1) how much solvent must be added to dilute a solution to a particular concentration and (2) to what volume a solution must be diluted to prepare a solution of a particular concentration.

*Problem 14.10* Calculate the molarity of a sodium hydroxide solution that is prepared by mixing 100 ml of 0.20 *M* NaOH with 150 ml of water.

This is a dilution problem. If we double the volume of a solution by adding water, we cut the concentration in half. Therefore, the concentration of the above solution should be less than 0.10 *M*. In the dilution, the moles of NaOH remain constant; the molarity and volume change.

*Step 1.* Calculate the moles of NaOH in the original solution.

$$\text{moles} = \text{litres} \times M = 0.100 \text{ litre} \times \frac{0.20 \text{ mole NaOH}}{\text{litre}} = 0.020 \text{ mole NaOH}$$

*Step 2.* Solve for the new molarity, taking into account that the total volume of the solution is 250 ml (0.250 litre).

$$M = \frac{0.020 \text{ mole NaOH}}{0.250 \text{ litre}} = 0.080\ M \text{ NaOH} \quad \text{(Answer)}$$

*Alternate Solution*: When the moles of solute in a solution before and after dilution are the same, then the moles before and after dilution may be set equal to each other:

$$\text{moles}_1 = \text{moles}_2$$

where $\text{moles}_1$ = moles before dilution and $\text{moles}_2$ = moles after dilution. Then

$$\text{moles}_1 = \text{litres}_1 \times M_1 \qquad \text{moles}_2 = \text{litres}_2 \times M_2$$
$$\text{litres}_1 \times M_1 = \text{litres}_2 \times M_2$$

or $V_1 \times M_1 = V_2 \times M_2$, where both volumes are in the same units. For this problem,

$$V_1 = 0.100 \text{ litre} \qquad M_1 = 0.20\ M$$
$$V_2 = 0.250 \text{ litre} \qquad M_2 = M_2 \text{ (unknown)}$$

Then:

$$0.100 \text{ litre} \times 0.20\ M = 0.250 \text{ litre} \times M_2$$

Solving for $M_2$, we get

$$M_2 = \frac{0.100 \text{ litre} \times 0.20\ M}{0.250 \text{ litre}} = 0.080\ M \text{ NaOH}$$

---

*Problem 14.11* What weight of silver chloride (AgCl) will be precipitated by adding sufficient silver nitrate ($AgNO_3$) to react with 1500 ml of 0.400 *M* $BaCl_2$ (barium chloride) solution?

$$\underset{}{2\ AgNO_3(aq)} + \underset{\text{1 mole}}{BaCl_2(aq)} \longrightarrow \underset{\text{2 moles}}{2\ AgCl\downarrow} + Ba(NO_3)_2(aq)$$

The fact that $BaCl_2$ is in solution means that we need to consider the volume and concentration of the solution in order to know the number of moles of $BaCl_2$ reacting.

*Step 1.* Determine the number of moles of $BaCl_2$ in 1500 ml of 0.400 *M* solution:

$$M = \frac{\text{moles}}{\text{litre}}$$

$$\text{moles} = \text{litre} \times M$$

$$1.500 \text{ litres} \times \frac{0.400 \text{ mole } BaCl_2}{\text{litre}} = 0.600 \text{ mole } BaCl_2$$

*Step 2.* Calculate the moles of AgCl formed by using the mole-ratio method:

$$0.600 \text{ mole } BaCl_2 \times \frac{2 \text{ moles AgCl}}{1 \text{ mole } BaCl_2} = 1.20 \text{ moles AgCl}$$

*Step 3.* Convert the moles of AgCl to grams:

$$1.20 \text{ moles AgCl} \times \frac{143.3 \text{ g AgCl}}{\text{mole AgCl}} = 172 \text{ g AgCl} \quad \text{(Answer)}$$

---

normality

1 normal solution

6. Normality. Normality is another way of expressing the concentration of a solution. It is based on an alternate chemical unit of mass called the *equivalent weight.* The **normality** of a solution is the concentration expressed as the number of equivalent weights (equivalents) of solute per litre of solution. A **1 normal (1 *N*) solution** contains 1 equivalent weight of solute per litre of solution. Normality is widely used in analytical chemistry because it simplifies many of the calculations involving solution concentration.

Every substance may be assigned an equivalent weight; it may be equal either to its molecular weight or to some small integral fractional part of its molecular weight (that is, the molecular weight divided by 2, 3, 4, and so on). To gain an understanding of the meaning of equivalent weight, let us start by considering these two reactions:

$$\underset{\substack{1 \text{ mole} \\ (36.5 \text{ g})}}{HCl(aq)} + \underset{\substack{1 \text{ mole} \\ (40.0 \text{ g})}}{NaOH(aq)} \longrightarrow NaCl(aq) + H_2O$$

$$\underset{\substack{1 \text{ mole} \\ (98.1 \text{ g})}}{H_2SO_4(aq)} + \underset{\substack{2 \text{ moles} \\ (80.0 \text{ g})}}{2\,NaOH(aq)} \longrightarrow Na_2SO_4(aq) + 2\,H_2O$$

We note first that 1 mole of hydrochloric acid (HCl) reacts with 1 mole of sodium hydroxide (NaOH) and 1 mole of sulfuric acid ($H_2SO_4$) reacts with 2 moles of NaOH. If we make 1 molar solutions of these acids, 1 litre of 1 *M* HCl will react with 1 litre of 1 *M* NaOH. From this, we can see that $H_2SO_4$ has twice the chemical capacity of HCl when reacting with NaOH. We can, however, adjust these acid solutions to be equal in reactivity by dissolving only 0.5 mole of $H_2SO_4$ per litre of solution. By doing this, we find that we are required to use 49.0 g of $H_2SO_4$ per litre (instead of 98.1 g of $H_2SO_4$ per litre) to make a solution that is equivalent to one made from 36.5 g of HCl per litre. These weights, 49.0 g of $H_2SO_4$ and 36.5 g of HCl, are chemically equivalent and are known as the equivalent weights of these substances, because they react with the same amount of NaOH (40.0 g). The equivalent weight of HCl is equal to its molecular weight, but that of $H_2SO_4$ is one-half its molecular weight. Table 14.4 summarizes these relationships.

*Table 14.4.* Comparison of molar and normal solutions of HCl and $H_2SO_4$ reacting with NaOH.

| | Molecular weight | Concentration | Volumes that react | Equivalent weight | Concentration | Volumes that react |
|---|---|---|---|---|---|---|
| HCl | 36.5 | 1 $M$ | 1 litre | 36.5 | 1 $N$ | 1 litre |
| NaOH | 40.0 | 1 $M$ | 1 litre | 40.0 | 1 $N$ | 1 litre |
| $H_2SO_4$ | 98.1 | 1 $M$ | 1 litre | 49.0 | 1 $N$ | 1 litre |
| NaOH | 40.0 | 1 $M$ | 2 litres | 40.0 | 1 $N$ | 1 litre |

Expressions for normality follow. Notice the similarity to the molar solution definition.

$$\text{Normality} = N = \frac{\text{Number of equivalents of solute}}{\text{1 litre of solution}} = \frac{\text{equivalents}}{\text{litre}}$$

where

$$\text{Number of equivalents of solute} = \frac{\text{grams of solute}}{\text{equivalent weight of solute}}$$

Then

$$N = \frac{\text{g solute}}{\text{eq wt solute} \times \text{litre solution}} = \frac{\text{g}}{\text{eq wt} \times \text{litre}}$$

Thus, 1 litre of solution containing 36.5 g of HCl would be 1 $N$, and 1 litre of solution containing 49.0 g of $H_2SO_4$ would also be 1 $N$. A solution containing 98.1 g of $H_2SO_4$ (1 mole) per litre would be 2 $N$ when reacting with NaOH in the above equation.

Consider the following reactions, in which an excess of HCl is present. Hydrogen actually exists as $H_2$ molecules, but for convenience of considering the data, the hydrogen produced is shown as the number of atomic weights of hydrogen released per atomic weight of metal reacting. Table 14.5 summarizes the pertinent data for these reactions.

$$\text{Na}(s) + \text{HCl}(aq) \longrightarrow \text{NaCl}(aq) + \text{H}^\circ(g)$$
$$\text{Ca}(s) + 2\,\text{HCl}(aq) \longrightarrow \text{CaCl}_2(aq) + 2\,\text{H}^\circ(g)$$
$$\text{Al}(s) + 3\,\text{HCl}(aq) \longrightarrow \text{AlCl}_3(aq) + 3\,\text{H}^\circ(g)$$

In each of the above reactions, the equivalent weight of the reacting metals is the weight that reacts with 1 equivalent weight of the acid, liberates 1 atomic weight of H atoms, or involves the transfer of 1 mole of electrons in the reaction. One atomic weight of Na metal lost 1 electron per atom in going to NaCl; 1 at. wt of Ca metal lost 2 electrons per atom in going to $CaCl_2$; 1 at. wt of Al metal lost 3 electrons per atom in going to $AlCl_3$. In each reaction, 1 at. wt of $H^+$ gained 1 electron per atom in going to free hydrogen.

$$\text{eq wt} = \frac{\text{at. wt Na}}{1} = \frac{\text{at. wt Ca}}{2} = \frac{\text{at. wt Al}}{3} = \frac{\text{at. wt H}}{1}$$

*Table 14.5.* Equivalent weight of sodium, calcium, and aluminum in reaction with hydrochloric acid.

| Metal | Atomic weight (amu) | Number of atomic weights of hydrogen liberated per atomic weight of metal | Equivalent weight of metal (amu) |
|---|---|---|---|
| Na | 23.0 | 1 | $\frac{23.0}{1} = 23.0$ |
| Ca | 40.1 | 2 | $\frac{40.1}{2} = 20.0$ |
| Al | 27.0 | 3 | $\frac{27.0}{3} = 9.0$ |

equivalent weight

Two definitions of **equivalent weight** can now be stated:

1. The equivalent weight is that weight of a substance which will react with, combine with, contain, replace, or in any other way be equivalent to 1 gram-atomic weight of hydrogen.
2. In oxidation–reduction reactions, the gram-equivalent weight is that weight of a substance which loses or gains Avogadro's number of electrons.

The equivalent weight of a substance may be variable; its value is dependent on the reaction that the substance is undergoing. Consider the reactions represented by these equations:

$$NaOH + H_2SO_4 \longrightarrow NaHSO_4 + H_2O \quad (1)$$

$$2\,NaOH + H_2SO_4 \longrightarrow Na_2SO_4 + 2\,H_2O \quad (2)$$

In reaction (1), 1 mole of sulfuric acid furnishes 1 g-at. wt of hydrogen. Therefore, the equivalent weight of sulfuric acid is the formula weight, namely 98.1 g. But in reaction (2), 1 mole of $H_2SO_4$ furnishes 2 g-at. wt of hydrogen. Therefore, the equivalent weight of the sulfuric acid is one-half the formula weight, or 49.0 g.

## 14.8 Colligative Properties of Solutions

When two solutions are prepared, one containing 1 mole (60.0 g) of urea ($NH_2CONH_2$) and the other containing one mole (342 g) of sucrose ($C_{12}H_{22}O_{11}$) in 1 kg of water, the freezing point of each solution is $-1.86°C$, not 0°C as in pure water. Urea and sucrose are very different substances, yet each lowers the freezing point of the water by the same amount. The only thing apparently common to these two solutions is that each contains 1 mole ($6.02 \times 10^{23}$ molecules) of solute and 1 kg of solvent. In fact, if we dissolved one mole of any other solute (provided that it is one that does not produce ions in solution) in 1 kg of water, the freezing point of the resulting solution would be $-1.86°C$.

This leads us to conclude that the freezing point depression for a solution containing $6.02 \times 10^{23}$ solute molecules (particles) and 1 kg of water is a constant, namely 1.86°. Freezing point depression is a general property of solutions. When solutions are prepared from 1 kg of any solvent and 1 mole of any (nonionized) solute, the freezing point will be lower than that of the pure solvent. Furthermore the amount by which the freezing point is depressed is the same for all solutions made with a given solvent; that is, each solvent shows a characteristic *freezing point depression constant*. Freezing point depression constants for several solvents are given in Table 14.6.

*Table 14.6.* Freezing point depression and boiling point elevation constants of selected solvents.

| Solvent | Freezing point of pure solvent (°C) | Freezing point depression constant $\left(\frac{°C, \text{kg solvent}}{\text{mole solute}}\right)$ | Boiling point of pure solvent (°C) | Boiling point elevation constant $\left(\frac{°C, \text{kg solvent}}{\text{mole solute}}\right)$ |
|---|---|---|---|---|
| Water | 0.00 | 1.86 | 100.0 | 0.52 |
| Acetic acid | 16.6 | 3.90 | 118.5 | 3.07 |
| Benzene | 5.5 | 5.1 | 80.1 | 2.53 |
| Camphor | 178 | 40 | 208.2 | 5.95 |

The solution formed by the addition of a nonvolatile solute to a solvent has a lower freezing point, a higher boiling point, and a lower vapor pressure than that of the pure solvent. All these effects are related and are known as colligative properties. The **colligative properties** are properties that depend only upon the number of solute atoms or molecules in a solution and not on the nature of those atoms or molecules. Freezing point depression, boiling point elevation, vapor pressure lowering, and osmotic pressure are colligative properties of solutions.

colligative properties

The colligative properties of a solution can be considered in terms of vapor pressure. The vapor pressure of a pure liquid depends on the tendency of molecules to escape from its surface. Thus, if 10% of the molecules in a solution are nonvolatile solute molecules, the vapor pressure of the solution is 10% lower than that of the pure solvent. The vapor pressure is lower because the surface of the solution contains 10% nonvolatile molecules and 90% of the volatile solvent molecules. A liquid boils when its vapor pressure equals the pressure of the atmosphere. Thus, we can see that the solution just described as having a lower vapor pressure will have a higher boiling point than the pure solvent. The solution with a lowered vapor pressure does not boil until it has been heated above the boiling point of the solvent (see Figure 14.8). Each solvent has its own characteristic boiling point elevation constant (see Table 14.6). The boiling point elevation constant is based on a solution that contains 1 mole of solute particles per kilogram of solvent. For example, the boiling point elevation for a solution containing 1 mole of solute particles per kilogram of water is 0.52°C. This means that this water solution will boil at 100.52°C.

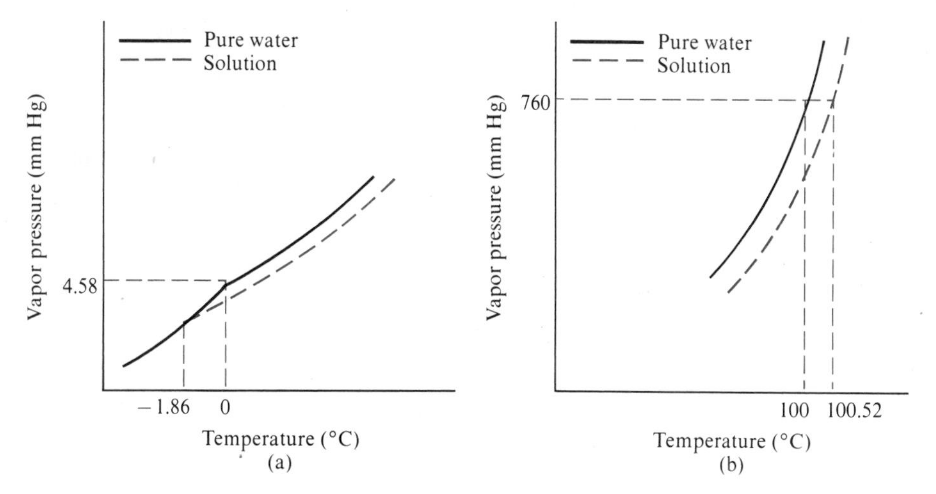


*Figure 14.8.* Vapor pressure curves of pure water and water solutions, showing (a) freezing point depression and (b) boiling point elevation effects. (Concentration: 1 mole of solute per kilogram of water.)

The freezing behavior of a solution can also be considered in terms of lowered vapor pressure. Figure 14.8 shows the vapor pressure relationships of ice, water, and a solution containing 1 mole of solute per kilogram of water. The freezing point of water is at the intersection of the water and ice vapor pressure curves; that is, at the point where water and ice have the same vapor pressure. Because the vapor pressure of water is lowered by the solute, the vapor pressure curve of the solution does not intersect the vapor pressure curve of ice until the solution has been cooled below the freezing point of pure water. Thus, it is necessary to cool the solution below 0°C in order to freeze out ice.

The foregoing discussion dealing with freezing point depressions is restricted to *un-ionized* substances. The discussion of boiling point elevations is restricted to *nonvolatile* and un-ionized substances. The colligative properties of ionized substances (Electrolytes, Chapter 15) are not explored at this point.

Some practical applications involving colligative properties are (1) use of salt–ice mixtures to provide low freezing temperatures for homemade ice cream, (2) use of salt or calcium chloride to melt ice from streets, (3) use of ethylene glycol and water mixtures as antifreeze in automobile radiators (ethylene glycol also raises the boiling point of radiator fluid and thus allows the engine to operate at a higher temperature).

molality

Both the freezing point depression and the boiling point elevation are directly proportional to the number of moles of solute per kilogram of solvent. When we deal with the colligative properties of solutions, another concentration expression, *molality*, is used. The **molality** ($m$) of a solute is the number of moles of solute per kilogram of solvent:

$$m = \frac{\text{moles solute}}{\text{kg solvent}}$$

Note that a lowercase $m$ is used for molality concentrations, while a capital $M$ is used for molarity. The difference between molality and molarity is that molality refers to moles of solute *per kilogram of solvent*, whereas molarity refers to moles of solute *per litre of solution.* For un-ionized substances, the colligative properties of a solution are directly proportional to its molality.

The following equations show the relationship of freezing point depression, molecular weight, and solution concentration. The symbol $\Delta t_f$ indicates the change in the freezing point of the solution with respect to the freezing point of the pure solvent; $K_f$ is the freezing point depression constant of the solvent.

$$\Delta t_f = K_f m \qquad \text{or} \qquad m = \frac{\Delta t_f}{K_f}$$

$$\Delta t_f = K_f \times \frac{\text{moles solute}}{\text{kg solvent}} = K_f \times \frac{\text{g solute}}{\text{Mol. wt solute}} \times \frac{1}{\text{kg solvent}}$$

This equation is commonly used to calculate the freezing points of solutions and the molecular weights of compounds. For boiling point elevation calculations substitute $\Delta t_b$ for $\Delta t_f$ and $K_b$ for $K_f$, where $\Delta t_b$ is the observed boiling point elevation and $K_b$ is the boiling point elevation constant of the solvent.

*Problem 14.12*

A solution is made by dissolving 100 g of ethylene glycol ($C_2H_6O_2$) in 200 g of water. What is the freezing point of this solution?

*Data*: $K_f$(for water) = 1.86°C/mole/kg     Mol. wt $C_2H_6O_2$ = 62.0 g/mole

$$\Delta t_f = 1.86°\text{C/mole/kg} \times \frac{100\ \text{g}}{62.0\ \text{g/mole}} \times \frac{1}{0.200\ \text{kg}} = 15.0°\text{C}$$

Since 15.0°C is the freezing point depression, it must be subtracted from 0°C, the freezing point of the pure solvent. Therefore, the freezing point of the solution is −15°C.

*Problem 14.13*

A solution made by dissolving 3.25 g of a compound of unknown molecular weight in 100.0 g of water has a freezing point of −1.46°C. What is the molecular weight of the compound?

*Data*: $\Delta t_f = 1.46°\text{C}$     $K_f = 1.86°\text{C/mole/kg}$

Solving the general formula for the molecular weight of the solute, we obtain

$$\text{Mol. wt solute} = K_f \times \frac{\text{g solute}}{\Delta t_f} \times \frac{1}{\text{kg solvent}}$$

$$= 1.86°\text{C/mole/kg} \times \frac{3.25\ \text{g}}{1.46°\text{C}} \times \frac{1}{0.100\ \text{kg}} = 41.4\ \text{g/mole} \quad \text{(Answer)}$$

## 14.9 Osmosis and Osmotic Pressure

When red blood cells are put in distilled water, they gradually swell and, in time, may burst. If red blood cells are put in a 5% urea (or a 5% salt) solution, they gradually shrink and take on a wrinkled appearance. The cells behave in

semipermeable membrane

osmosis

this fashion because they are enclosed in semipermeable membranes. A **semipermeable membrane** is one that allows the passage of water (solvent) molecules through it in either direction but prevents the passage of solute molecules or ions. When two solutions of different concentrations (or water and a water solution) are separated by a semipermeable membrane, water diffuses through the membrane from the solution of lower concentration into the solution of higher concentration. The diffusion of water, either from a dilute solution or from pure water, through a semipermeable membrane into a solution of higher concentration is called **osmosis**.

All solutions exhibit *osmotic pressure*. Osmotic pressure is another colligative property; it is dependent only on the concentration of the solute particles and is independent of their nature. The osmotic pressure of a solution can be measured by determining the amount of counterpressure needed to prevent osmosis; this pressure can be very large. The osmotic pressure of a solution containing 1 mole of solute particles in 1 kg of water is about 22.4 atm, which is about the same as the pressure exerted by 1 mole of a gas confined in a volume of 1 litre at 0°C.

Osmosis has a role in many biological processes. Semipermeable membranes occur commonly in living organisms. But artificial or synthetic membranes can also be made. Ordinary cellophane that has been treated to remove the waterproof coating is a good semipermeable membrane. The roots of plants are covered with tiny structures called root hairs; soil water enters the plant by osmosis, passing through the semipermeable membranes covering the root hairs.

Osmosis can be demonstrated with the simple laboratory setup shown in Figure 14.9. As a result of osmotic pressure, water passes through the cellophane membrane into the thistle tube, causing the solution level to rise. In osmosis, water is always transferred from a less concentrated to a more concentrated solution; that is, the effect is toward equalization of the concentration on both sides of the membrane. It should also be noted that the effective movement of water in osmosis is always from the region of *higher water concentration* to the region of *lower water concentration*.

Osmosis can be explained by assuming that a semipermeable membrane has passages that permit water molecules—but no other molecules or ions—to pass in either direction. Both sides of the membrane are constantly being struck by water molecules in random motion. The number of water molecules entering the membrane is proportional to the number of water molecule–membrane impacts per unit of time. Since solute molecules or ions reduce the concentration of water, there are more water molecules, and more water molecule impacts, on the side with the lower solute concentration (more dilute solution). The greater number of water molecule–membrane impacts on the dilute side is thus responsible for the net transfer of water to the more concentrated solution. Again note that the overall process involves the net transfer, by diffusion through the membrane, of water molecules from a region of higher water concentration (dilute solution) to one of lower water concentration (more concentrated solution).

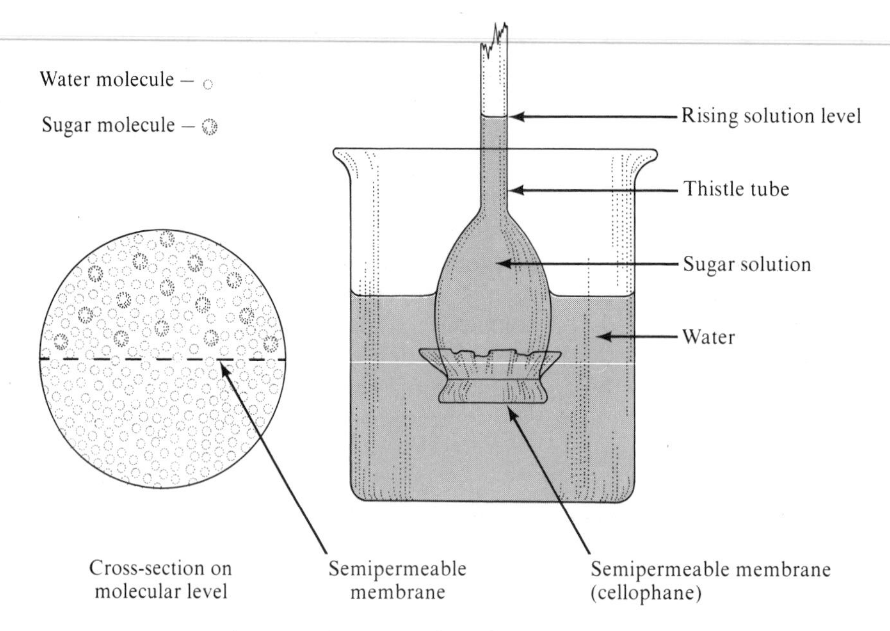


*Figure 14.9.* Laboratory demonstration of osmosis; as a result of osmosis, water passes through the membrane into the thistle tube, causing the solution level to rise.

This explanation is a simplified picture of osmosis. Remember that no one has ever seen the hypothetical passages that allow water molecules, and no other kinds of molecules or ions, to pass through them! Alternative explanations have been proposed. Our discussion has been confined to water solutions, but osmotic pressure is a general colligative property and osmosis is known to occur in nonaqueous systems.

physiological saline solution

A 0.90% (0.15 *M*) sodium chloride solution is known as a **physiological saline solution** because it is *isotonic* with blood plasma; that is, it has the same osmotic pressure as blood plasma. Since each mole of NaCl yields about 2 moles of ions when in solution, the solute particle concentration in physiological saline solution is nearly 0.30 molar. Five percent glucose solution (0.28 molar) is also approximately isotonic with blood plasma. Blood cells neither swell nor shrink in an isotonic solution. The cells described in the first paragraph of this section swelled in water because the water was *hypotonic* to cell plasma. The cells shrank in 5% urea solution because the urea solution was *hypertonic* to the cell plasma. In order to prevent possible injury to blood cells by osmosis, fluids for intravenous use are usually made up at approximately isotonic concentration.

## Questions

A. *Review the meanings of the new terms introduced in this chapter.*

1. Solution
2. Solute
3. Solvent
4. Solubility
5. Miscible
6. Immiscible

7. Concentration of a solution
8. Dilute solution
9. Concentrated solution
10. Saturated solution
11. Unsaturated solution
12. Supersaturated solution
13. 1 molar solution
14. Molarity
15. Normality
16. 1 normal solution
17. Equivalent weight
18. Colligative properties
19. Molality
20. Semipermeable membrane
21. Osmosis
22. Physiological saline solution

B. *Answers to the following questions will be found in tables and figures.*

1. Make a sketch indicating the orientation of water molecules (a) about a single sodium ion and (b) about a single chloride ion in solution.
2. Which of the substances listed below are reasonably soluble and which are insoluble?
   (a) KOH (b) $NiCl_2$ (c) ZnS (d) $AgC_2H_3O_2$ (e) $Na_2CrO_4$ (f) $PbI_2$ (g) $MgCO_3$ (h) $CaCl_2$ (i) $Fe(NO_3)_3$ (j) $BaSO_4$
3. Estimate the number of grams of sodium fluoride that would dissolve in 100 g of water at 50°C.
4. What is the solubility of each of the substances listed below at 25°C (see Figure 14.3)?
   a) Potassium chloride (b) Potassium chlorate (c) Potassium nitrate
5. What would be the total surface area if the 1 cm cube in Figure 14.4 were cut into cubes 0.01 cm on a side?
6. At which temperatures—10°C, 20°C, 30°C, 40°C, and 50°C—would you expect a solution made from 63 g of ammonium chloride and 150 g of water to be unsaturated? (See Figure 14.3.)
7. Explain why the rate of dissolving decreases as shown in Figure 14.5.
8. Does the bottle in Figure 14.6 necessarily contain 90 g of water? Explain.
9. Would the volumetric flasks in Figure 14.7 be satisfactory for preparing normal solutions? Explain.
10. Assume the thistle tube in Figure 14.9 contains 1.0 *M* sugar solution and that the water in the beaker has just been replaced by a 2.0 *M* solution of urea. Would the solution level in the thistle tube (a) continue to rise, (b) remain constant, or (c) fall? Explain.

C. *Review questions.*

1. Name and distinguish between the two components of a solution.
2. Explain why the solute does not settle out of a solution.
3. Is it possible to have one solid dissolved in another solid? Explain.
4. Explain how a colored salt such as $KMnO_4$ can be used to demonstrate the spontaneous movement (diffusion) of ions in solution.
5. Why is air considered to be a solution?
6. Explain why carbon tetrachloride will dissolve benzene but will not dissolve sodium chloride.
7. Tea and Coca Cola are popular drinks over most of the globe. Tea is drunk either hot or iced, but Coca Cola is never served hot. Why not?
8. In which will a teaspoonful of sugar dissolve more rapidly, 200 ml of iced tea or 200 ml of hot coffee? Explain your answer in terms of the KMT.
9. Which will dissolve more rapidly in 200 ml of water, 25 g of rock salt (large crystals) or 25 g of table salt? Explain.
10. What is the effect of pressure on the solubility of the following:
    (a) Solids in liquids (b) Gases in liquids
11. Is the rate of dissolving zero in a saturated solution? Explain.

12. What is the effect of an increase in temperature on the solubility of a gas in a liquid?
13. Explain why there is no apparent reaction when crystals of $AgNO_3$ and NaCl are mixed, but a reaction is apparent immediately when solutions of $AgNO_3$ and NaCl are mixed.
14. What do we mean when we say that concentrated hydrochloric acid (HCl) is 12 *M*?
15. Will 1 litre of 1 *M* NaCl contain more chloride ions than 0.5 litre of 1 *M* $MgCl_2$? Explain.
16. Will 1 litre of 1 *M* HCl neutralize 1 litre of 1 *M* KOH?
17. Will 1 litre of 1 *N* $H_2SO_4$ neutralize 1 litre of 1 *N* NaOH? Explain.
18. Explain in terms of vapor pressure why the boiling point of a solution containing a nonvolatile solute is higher than that of the pure solvent.
19. Explain in terms of vapor pressure why the freezing point of a solution containing a solute is lower than that of the pure solvent.
20. Which would be colder, a glass of water and crushed ice or a glass of Seven-Up and crushed ice? Explain.
21. Which would be the most effective in lowering the freezing point of 400 ml of water?
    (a) 100 g of ethyl alcohol ($C_2H_5OH$) or 100 g of ethylene glycol [$C_2H_4(OH)_2$]
    (b) 75.0 g of ethyl alcohol or 60.0 g of methyl alcohol ($CH_3OH$)
22. Explain in terms of the KMT how a semipermeable membrane functions when placed between pure water and a 10% sugar solution.
23. Which has the higher osmotic pressure, a solution containing 100 g of urea ($CH_4ON_2$) in 1 kg of $H_2O$ or one containing 150 g of glucose ($C_6H_{12}O_6$) in 1 kg of $H_2O$?
24. Explain why a lettuce leaf in contact with a salad dressing containing salt and vinegar soon becomes wilted and limp while another lettuce leaf in contact with plain water remains crisp.
25. Which of the following statements are correct?
    (a) A solution differs from a compound in that it can have a variable concentration.
    (b) Benzene and water do not mix; they are said to be miscible.
    (c) Most common nitrates and acetates are soluble in water.
    (d) Compared to 12 *M* HCl, 3 *M* HCl is a dilute solution.
    (e) A solution may be saturated at several different temperatures.
    (f) Hydrogen bromide, a gas, is more soluble in water at 40°C than at 20°C.
    (g) A 10% solution of NaCl contains 10 g of NaCl in 100 ml of solution.
    (h) 1 mole of solute in 1 litre of solution has the same concentration as 0.1 mole of solute in 100 ml of solution.
    (i) When 100 ml of 0.200 *M* HCl is diluted by adding 100 ml of water, the number of moles of HCl in the final solution is one-half the number of moles of HCl in the original solution.
    (j) 1 mole of HCl will react with the same amount of NaOH as 1 mole of $HNO_3$.
    (k) 1 mole of $H_2SO_4$ will react with twice as much NaOH as 1 mole of $HNO_3$.
    (l) A solution containing 0.2 mole $CH_3OH$ in 100 g of water will freeze at −1.86°C.
    (m) A solution containing a nonvolatile solute has a higher boiling point than the pure solvent as a result of having a lower vapor pressure than the pure solvent.

*D. Review problems.*

1. What is the weight percent of a saturated solution of copper(II) sulfate ($CuSO_4$) at 20°C?
2. Calculate the weight percent of the following solutions:
   (a) 10 g KCl + 100 g $H_2O$ (c) 30 g $MgCl_2$ + 150 g $H_2O$
   (b) 20 g KCl + 100 g $H_2O$ (d) 0.50 g $KMnO_4$ + 7.5 g $H_2O$
3. How much solute is present in the following solutions?
   (a) 50 g of 6.0% NaCl (b) 250 g of 15.0% KCl
4. Physiological saline solutions used in intravenous injections have a concentration of 0.90% NaCl. How many grams of NaCl are needed to prepare 400 g of this solution?
5. A solution made from 10.0 g of $NaNO_3$ and 100 g of $H_2O$ is allowed to evaporate in an open beaker at 20°C. What weight of water must evaporate for the solution to become saturated?
6. Will 100 g of 6.0% KOH solution be a sufficient amount of base to neutralize 100 ml of 1.0 *M* HCl? Show proof.
7. A sugar syrup solution contains 22.0% sugar and has a density of 1.09 g/ml. How many grams of sugar are there in 1.00 litre of this syrup?
8. The density of 24.0% $H_2SO_4$ solution is 1.17 g/ml. What weight and what volume of this solution will contain 50.0 g of $H_2SO_4$?
9. Determine the molarity of the following solutions:
   (a) 140 g $CaBr_2$ in 1.00 litre of solution
   (b) 48.0 g $NH_4Cl$ in 400 ml of solution
   (c) 12.0 g NaOH in 200 ml of solution
   (d) 25.0 g $BaCl_2 \cdot 2\,H_2O$ in 1500 ml of solution
10. Calculate the number of moles of solute in each of the following solutions:
    (a) 2.00 litres of 3.00 *M* $CaCl_2$ (d) 15.0 ml of 12.0 *M* $HNO_3$
    (b) 225 ml of 1.50 *M* $KC_2H_3O_2$ (e) 25.0 ml of 0.320 *M* NaOH
    (c) 2000 ml of 0.250 *M* $AgNO_3$ (f) 2.50 litres of 1.50 *M* KF
11. Determine the weight of solute in each of these solutions:
    (a) 2.5 litres of 0.10 *M* KCl (c) 1.0 litre of 1.0 *M* $H_3PO_4$
    (b) 125 ml of 1.50 *M* $Na_2SO_4$ (d) 4.00 ml of 0.600 *M* $Zn(NO_3)_2$
12. What is the molarity of concentrated hydrobromic acid if the solution is 48.0% HBr and has a density of 1.50 g/ml?
13. Calculate the volume of concentrated reagent required to prepare the diluted solutions indicated:
    (a) 12 *M* HCl to prepare 100 ml of 4.0 *M* HCl
    (b) 16 *M* $HNO_3$ to prepare 200 ml of 6.0 *M* $HNO_3$
    (c) 18 *M* $H_2SO_4$ to prepare 1000 ml of 3.0 *M* $H_2SO_4$
    (d) 17 *M* $HC_2H_3O_2$ to prepare 25 ml of 3.0 *M* $HC_2H_3O_2$
14. To what volume must 26.0 g of zinc nitrate [$Zn(NO_3)_2$] be diluted to make a 1.50 *M* solution?
15. What will be the molarity of each of the solutions made by mixing 200 ml of 0.50 *M* $H_2SO_4$ with the following?
    (a) 200 ml of $H_2O$
    (b) 200 ml of 0.80 *M* $H_2SO_4$
    (c) 300 ml of 0.80 *M* $H_2SO_4$
16. In the reaction
    $$6FeCl_2 + K_2Cr_2O_7 + 14HCl \longrightarrow 6FeCl_3 + 2CrCl_3 + 2KCl + 7H_2O$$
    (a) How many moles of $FeCl_3$ will be produced from 1.0 mole of $FeCl_2$?
    (b) How many moles of $CrCl_3$ will be produced from 1.0 mole of $FeCl_2$?

(c) How many moles of $K_2Cr_2O_7$ will react with 0.040 mole of $FeCl_2$?
(d) How many millilitres of 0.080 *M* $K_2Cr_2O_7$ will react with 0.040 mole of $FeCl_2$?
(e) What volume of 6.0 *M* HCl is needed to react with 0.040 mole $FeCl_2$?

17. Using the equation below, calculate
$2\,KMnO_4 + 16\,HCl \longrightarrow 2\,MnCl_2 + 5\,Cl_2 + 8\,H_2O + 2\,KCl$
(a) The moles of $MnCl_2$ produced from 1.0 mole of $KMnO_4$
(b) The moles of $Cl_2$ produced from 1.0 mole of $KMnO_4$
(c) The moles of HCl that will react with 1.0 mole of $KMnO_4$
(d) The number of millilitres of 0.100 *M* HCl that will react with 50.0 ml of 0.250 *M* $KMnO_4$
(e) The volume of $Cl_2$ (gas) at STP that will be produced when 100 ml of 3.00 *M* HCl react

18. $BaCl_2(aq) + K_2CrO_4(aq) \longrightarrow BaCrO_4\downarrow + 2\,KCl(aq)$
(a) What weight of $BaCrO_4$ can be obtained by reacting 50.0 ml of 0.250 *M* $BaCl_2$ solution?
(b) What volume of 1.0 *M* $K_2CrO_4$ solution is needed to react with the 50.0 ml of 0.250 *M* $BaCl_2$ solution?

19. $3\,Cu(s) + 8\,HNO_3(aq) \longrightarrow 3\,Cu(NO_3)_2(aq) + 2\,NO\uparrow + 4\,H_2O$
(a) How many grams of Cu will react with 100 ml of 4.00 *M* $HNO_3$ solution?
(b) What volume of NO gas (at STP) will be produced in the reaction in part (a)?

20. How many moles of hydrogen will be liberated from 100 ml of 3.00 *M* HCl reacting with an excess of magnesium?
$Mg(s) + 2\,HCl(aq) \longrightarrow MgCl_2(aq) + H_2\uparrow$

21. When 250 ml of hydrochloric acid were treated with an excess of magnesium, 6.00 litres (at STP) of hydrogen gas were liberated. Calculate the molarity of the HCl solution.

22. The compounds $Mg(OH)_2$ and $Al(OH)_3$ are used in antacid formulations to neutralize excess acid in the stomach. On an equal mass basis, show which base is more effective in neutralizing hydrochloric acid in the stomach.

23. Calculate the equivalent weight of the acid and the base in each of the reactions below:
(a) $2\,HCl + Ca(OH)_2 \longrightarrow CaCl_2 + 2\,H_2O$
(b) $3\,HCl + Al(OH)_3 \longrightarrow AlCl_3 + 3\,H_2O$
(c) $H_2SO_4 + Mg(OH)_2 \longrightarrow MgSO_4 + 2\,H_2O$
(d) $H_2SO_4 + LiOH \longrightarrow LiHSO_4 + H_2O$
(e) $H_3PO_4 + KOH \longrightarrow KH_2PO_4 + H_2O$

24. Calculate the equivalent weight of the metal in each of the following reactions:
(a) $2\,Rb + 2\,H_2O \longrightarrow 2\,RbOH + H_2\uparrow$
(b) $Mg + H_2O \xrightarrow{\Delta} MgO + H_2\uparrow$
(c) $Zn + 2\,HBr \longrightarrow ZnBr_2 + H_2\uparrow$
(d) $Pb + CuSO_4 \longrightarrow PbSO_4 + Cu$
(e) $Fe_2O_3 + 3\,H_2 \xrightarrow{\Delta} 2\,Fe + 3\,H_2O$
(f) $2\,Ga + 3\,Cl_2 \longrightarrow 2\,GaCl_3$

25. (a) What is the freezing point of a solution that contains 1.200 g of urea ($CH_4ON_2$) in 10.0 g of $H_2O$?
(b) Calculate the molality of this solution.

26. (a) What is the freezing point of a solution that contains 40.0 g of ethylene glycol ($C_2H_6O_2$) in 100.0 g of $H_2O$?
(b) What is the boiling point of this solution?

27. What is (a) the freezing point and (b) the boiling point of a solution containing 3.07 g of naphthalene ($C_{10}H_8$) in 20.0 g of benzene?
28. The freezing point of a solution of 4.50 g of an unknown compound dissolved in 20.0 g of acetic acid is 13.2°C. Calculate the molecular weight of the compound.
29. When 5.40 g of a compound are dissolved in 60.0 g of $H_2O$, the solution has a freezing point of $-0.930$°C. The empirical formula of the compound is $CH_2O$. What is its molecular formula?
30. What (a) weight and (b) volume of ethylene glycol ($C_2H_6O_2$, $d = 1.11$ g/ml) should be added to 10.0 litres of $H_2O$ in an automobile radiator to protect it from freezing at $-20.0$°C?

# 15 Ionization, Acids, Bases, Salts

*After studying Chapter 15 you should be able to:*

1. Understand the terms listed in Question A at the end of the chapter.
2. Define an acid and a base in terms of Arrhenius, Brønsted–Lowry, and Lewis theories.
3. When given the reactants, complete and balance equations for the reactions of acids with bases, metals, metal oxides, and carbonates.
4. When given the reactants, complete and balance equations for the reaction of an amphoteric hydroxide with either a strong acid or a strong base.
5. Write balanced equations for the reaction of sodium hydroxide or potassium hydroxide with zinc and with aluminum.
6. Classify common compounds as electrolytes or nonelectrolytes.
7. Write equations for the dissociation or ionization of acids, bases, and salts in water.
8. Describe and write equations for the ionization of water.
9. Given pH as an integer, give the $H^+$ molarity and vice versa.
10. Use the simplified log scale given in the chapter to estimate pH values from corresponding $H^+$ molarities.
11. Calculate the molarity or volume of an acid or base solution from appropriate titration data.
12. Write balanced molecular, total ionic, and net ionic equations for neutralization reactions.
13. List the rules for writing net ionic equations.
14. Discuss colloids and describe methods for their preparation.
15. Describe the characteristics that distinguish true solutions, colloidal dispersions, and mechanical suspensions.
16. Tell how colloidal dispersions can be (a) stabilized and (b) how they can be precipitated.

## 15.1 Acids and Bases

The word *acid* is derived from the Latin *acidus*, meaning "sour" or "tart," and is also related to the Latin word *acetum*, meaning "vinegar." Vinegar has been known since antiquity as the product of the fermentation of wine and apple cider. The sour constituent of vinegar is acetic acid ($HC_2H_3O_2$).

Some of the characteristic properties commonly associated with acids are the following: Water solutions of acids are sour to the taste and are capable of changing the color of litmus, a vegetable dye, from blue to red. Water solutions of nearly all acids are able to react with: (1) metals such as zinc and magnesium to produce hydrogen gas; (2) bases to produce water and a salt; and (3) carbonates to produce carbon dioxide. These properties are due to hydrogen ions, $H^+$, released by the acid in a water solution.

Classically, a *base* is a substance capable of liberating hydroxide ions, $OH^-$, in water solution. Hydroxides of the alkali metals (Group IA) and alkaline earth metals (Group IIA), such as LiOH, NaOH, KOH, $Ca(OH)_2$, and $Ba(OH)_2$, are the most common inorganic bases. Water solutions of bases are called *alkaline solutions* or *basic solutions*. They have the following properties: a bitter or caustic taste; a slippery, soapy feeling; the ability to change litmus from red to blue; and the ability to interact with acids to form a salt and water.

Several theories have been proposed to answer the question "What is an acid and a base?" One of the earliest, most significant of these theories was advanced in a doctoral thesis in 1884 by Svante Arrhenius (1859–1927), a Swedish scientist, who stated that an acid is a hydrogen-containing substance that dissociates to produce hydrogen ions, and that a base is a hydroxide-containing substance that dissociates to produce hydroxide ions in aqueous solutions. Arrhenius postulated that the hydrogen ions were produced by the dissociation of acids in water; and hydroxide ions were produced by the dissociation of bases in water.

$$\underset{\text{Acid}}{HA} \longrightarrow H^+ + A^-$$

$$\underset{\text{Base}}{MOH} \longrightarrow M^+ + OH^-$$

Thus, an acid solution contains an excess of hydrogen ions and a base an excess of hydroxide ions.

In 1923, the Brønsted–Lowry proton transfer theory was introduced by J. N. Brønsted, a Danish chemist (1897–1947), and T. M. Lowry, an English chemist (1874–1936). This theory states that an acid is a proton donor and a base is a proton acceptor.

Consider the reaction of hydrogen chloride gas with water to form hydrochloric acid.

$$HCl(g) + H_2O(l) \longrightarrow H_3O^+(aq) + Cl^-(aq) \quad (1)$$

In the course of the reaction, HCl donates, or gives up, a proton to form a $Cl^-$ ion and $H_2O$ accepts a proton to form the $H_3O^+$ ion. Thus, HCl is an acid and $H_2O$ is a base, according to the Brønsted–Lowry theory.

A hydrogen ion, $H^+$, is nothing more than a bare proton and does not exist by itself in an aqueous solution. In water, a proton combines with a polar water molecule to form a hydrated hydrogen ion, $H_3O^+$ $[H(H_2O)^+]$,

hydronium ion

commonly called a **hydronium ion**. The proton is attracted to a polar water molecule, forming a bond with one of the two pairs of unshared electrons:

$$H^+ + H:\ddot{\underset{\cdot\cdot}{O}}: \longrightarrow \left[ H:\ddot{\underset{\cdot\cdot}{O}}:H \right]^+$$

$$\qquad\qquad\quad H \qquad\qquad\quad H$$

Hydronium ion

Note the electron structure of the hydronium ion. For simplicity of expression in equations, we often use $H^+$ instead of $H_3O^+$, with the explicit understanding that $H^+$ is always hydrated in solution.

Whereas the Arrhenius theory is restricted to aqueous solutions, the Brønsted–Lowry approach has application in all media and has become the more important theory when the chemistry of substances in solutions other than water is studied. Ammonium chloride ($NH_4Cl$) is a salt, yet its water solution has an acidic reaction. From this test we must conclude that $NH_4Cl$ has acidic properties. The Brønsted–Lowry explanation shows that the ammonium ion, $NH_4^+$, is a proton donor, and water is the proton acceptor:

$$\underset{\text{Acid}}{NH_4^+} \rightleftharpoons \underset{\text{Base}}{NH_3} + \underset{\text{Acid}}{H^+} \tag{2}$$

$$\underset{\text{Acid}}{NH_4^+} + \underset{\text{Base}}{H_2O} \longrightarrow \underset{\text{Acid}}{H_3O^+} + \underset{\text{Base}}{NH_3} \tag{3}$$

The Brønsted–Lowry theory also applies to certain cases where no solution is involved. For example, in the reaction of hydrogen chloride and ammonia gases, HCl is the proton donor and $NH_3$ is the base. (Remember that "(*g*)" after a formula in equations stands for a gas.)

$$\underset{\text{Acid}}{HCl(g)} + \underset{\text{Base}}{NH_3(g)} \longrightarrow \underset{\text{Acid}}{NH_4^+} + \underset{\text{Base}}{Cl^-} \tag{4}$$

In equations (1), (3), and (4), a conjugate acid and base are produced as products. The formulas of a conjugate acid–base pair differ by one proton ($H^+$). In equation (1), the conjugate base of the acid HCl is $Cl^-$, and the conjugate acid of the base $H_2O$ is $H_3O^+$. In equation (3), the conjugate acid of the base $H_2O$ is $H_3O^+$, and the conjugate base of the acid $NH_4^+$ is $NH_3$. In equation (4), HCl–$Cl^-$ and $NH_4^+$–$NH_3$ are the conjugate acid–base pairs.

$$\underset{\text{Acid}}{HCl} + \underset{\text{Base}}{H_2O} \longrightarrow \underset{\substack{\text{Conjugate} \\ \text{acid}}}{H_3O^+} + \underset{\substack{\text{Conjugate} \\ \text{base}}}{Cl^-}$$

A more general concept of acids and bases was introduced by Gilbert N. Lewis (1875–1946). The Lewis theory deals with the way in which a substance with an unshared pair of electrons reacts in an acid–base type of reaction. According to this theory a base is any substance that has an unshared pair of electrons (electron-pair donor) and an acid is any substance that will attach itself to or accept a pair of electrons. In the reaction

$$\begin{array}{lclcl} & \text{H} & & \text{H} \\ \text{H}^+ + & :\ddot{\underset{\cdot\cdot}{\text{N}}}:\text{H} & \longrightarrow & \text{H}:\ddot{\underset{\cdot\cdot}{\text{N}}}:\text{H}^+ \\ & \text{H} & & \text{H} \\ \text{Acid} & \text{Base} & & \end{array}$$

$H^+$ is a Lewis acid and $:NH_3$ is a Lewis base. According to the Lewis theory, substances other than proton donors (for example, $BF_3$) behave as acids:

$$\begin{array}{lclcl} \text{F} & & \text{H} & & \text{F H} \\ \text{F}:\ddot{\underset{\cdot\cdot}{\text{B}}} & + & :\ddot{\underset{\cdot\cdot}{\text{N}}}:\text{H} & \longrightarrow & \text{F}:\ddot{\underset{\cdot\cdot}{\text{B}}}:\ddot{\underset{\cdot\cdot}{\text{N}}}:\text{H} \\ \text{F} & & \text{H} & & \text{F H} \\ \text{Acid} & & \text{Base} & & \end{array}$$

The Lewis and Brønsted–Lowry bases are identical, because to accept a proton, a base must have an unshared pair of electrons.

The three theories are summarized in Table 15.1. These theories explain how acid–base reactions occur. We will generally use the theory that best explains the reaction that is under consideration. Most of our examples will refer to aqueous solutions. It is important to realize that in an aqueous acidic solution, the $H^+$ ion concentration is always greater than the $OH^-$ ion concentration. And, vice versa, in an aqueous basic solution, the $OH^-$ ion concentration is always greater than the $H^+$ ion concentration.

*Table 15.1.* Summary of acid–base definitions according to Arrhenius, Brønsted–Lowry, and G. N. Lewis theories.

| Theory | Acid | Base |
|---|---|---|
| Arrhenius | A hydrogen-containing substance that produces hydrogen ions in aqueous solution | A hydroxide-containing substance that produces hydroxide ions in aqueous solution |
| Brønsted–Lowry | A proton ($H^+$) donor | A proton ($H^+$) acceptor |
| Lewis | Any species that will bond to an unshared pair of electrons (electron-pair acceptor) | Any species that has an unshared pair of electrons (electron-pair donor) |

## 15.2 Reactions of Acids

In aqueous solutions it is the $H^+$ or $H_3O^+$ ions that are responsible for the characteristic reactions of acids. All the following reactions are in an aqueous medium.

(a) Reaction with metals. Acids react with metals that lie above hydrogen in the activity series of elements to produce hydrogen and a salt (see Section 17.5).

$$\text{Acid} + \text{Metal} \longrightarrow \text{Hydrogen} + \text{Salt}$$
$$2\,HCl(aq) + Ca(s) \longrightarrow H_2\uparrow + CaCl_2(aq)$$
$$H_2SO_4(aq) + Mg(s) \longrightarrow H_2\uparrow + MgSO_4(aq)$$
$$6\,HC_2H_3O_2(aq) + 2\,Al(s) \longrightarrow 3\,H_2\uparrow + 2\,Al(C_2H_3O_2)_3(aq)$$

Acids such as nitric acid ($HNO_3$) are oxidizing substances (see Chapter 17) and react with metals to produce water instead of hydrogen. For example,

$$3\,Zn(s) + 8\,HNO_3(\text{dilute}) \longrightarrow 3\,Zn(NO_3)_2(aq) + 2\,NO(g) + 4\,H_2O$$

(b) Reaction with bases. The interaction of an acid and a base is called a *neutralization reaction.* In aqueous solutions, the products of this reaction are water and a salt.

$$\text{Acid} + \text{Base} \longrightarrow \text{Salt} + \text{Water}$$
$$HBr(aq) + KOH(aq) \longrightarrow KBr(aq) + H_2O$$
$$2\,HNO_3(aq) + Ca(OH)_2(aq) \longrightarrow Ca(NO_3)_2(aq) + 2\,H_2O$$
$$2\,H_3PO_4(aq) + 3\,Ba(OH)_2(aq) \longrightarrow Ba_3(PO_4)_2\downarrow + 6\,H_2O$$

(c) Reaction with metal oxides. This reaction is closely related to that of an acid with a base. With an aqueous acid, the products are water and a salt.

$$\text{Acid} + \text{Metal oxide} \longrightarrow \text{Salt} + \text{Water}$$
$$2\,HCl(aq) + Na_2O(s) \longrightarrow 2\,NaCl(aq) + H_2O$$
$$H_2SO_4(aq) + MgO(s) \longrightarrow MgSO_4(aq) + H_2O$$
$$6\,HCl(aq) + Fe_2O_3(s) \longrightarrow 2\,FeCl_3(aq) + 3\,H_2O$$

(d) Reaction with carbonates. Many acids react with carbonates to produce carbon dioxide, water, and a salt. Carbonic acid ($H_2CO_3$) is not the product because it is unstable and decomposes into water and carbon dioxide.

$$\text{Acid} + \text{Carbonate} \longrightarrow \text{Salt} + \text{Water} + \text{Carbon dioxide}$$
$$2\,HCl(aq) + Na_2CO_3(aq) \longrightarrow 2\,NaCl(aq) + H_2O + CO_2\uparrow$$
$$H_2SO_4(aq) + MgCO_3(s) \longrightarrow MgSO_4(aq) + H_2O + CO_2\uparrow$$

## 15.3 Reactions of Bases

The $OH^-$ ions are responsible for the characteristic reactions of bases. All the following reactions are in an aqueous medium.

(a) Reaction with acids. Bases react with acids to produce a salt and water. See reaction of acids with bases in Section 15.2(b).

amphoteric

(b) Amphoteric hydroxides. Hydroxides of certain metals, such as zinc, aluminum, and chromium, are **amphoteric**; that is, they are capable of reacting as either an acid or a base. When treated with a strong acid, they behave like bases; when reacted with a strong base, they behave like acids.

$$Zn(OH)_2(s) + 2\,HCl(aq) \longrightarrow ZnCl_2(aq) + 2\,H_2O$$
$$Zn(OH)_2(s) + 2\,NaOH(aq) \longrightarrow Na_2ZnO_2(aq) + 2\,H_2O$$

(c) Reaction of NaOH and KOH with certain metals. Some amphoteric metals react directly with the strong bases sodium hydroxide and potassium hydroxide to produce hydrogen and a salt.

$$\text{Base} + \text{Metal} \longrightarrow \text{Salt} + \text{Hydrogen}$$
$$2\,NaOH(aq) + Zn(s) \longrightarrow Na_2ZnO_2(aq) + H_2\uparrow$$
$$6\,KOH(aq) + 2\,Al(s) \longrightarrow 2\,K_3AlO_3(aq) + 3\,H_2\uparrow$$

(d) Reaction with salts. Bases will react with many salts in solution due to the formation of insoluble metal hydroxides.

$$\text{Base} + \text{Salt} \longrightarrow \text{Metal hydroxide}\downarrow + \text{Salt}$$
$$2\,NaOH(aq) + MnCl_2(aq) \longrightarrow Mn(OH)_2\downarrow + 2\,NaCl(aq)$$
$$3\,NH_4OH(aq) + FeCl_3(aq) \longrightarrow Fe(OH)_3\downarrow + 3\,NH_4Cl(aq)$$
$$2\,KOH(aq) + CuSO_4(aq) \longrightarrow Cu(OH)_2\downarrow + K_2SO_4(aq)$$

## 15.4 Salts

Salts are very abundant in nature. Most of the rocks and minerals of the earth's mantle are salts of one kind or another. Huge quantities of dissolved salts also exist in the oceans. Salts may be considered to be compounds that have been derived from acids and bases. They consist of positive metal or ammonium ions ($H^+$ excluded) combined with negative nonmetal ions ($OH^+$ and $O^{2-}$ excluded). The positive ion is the base counterpart and the nonmetal ion is the acid counterpart:

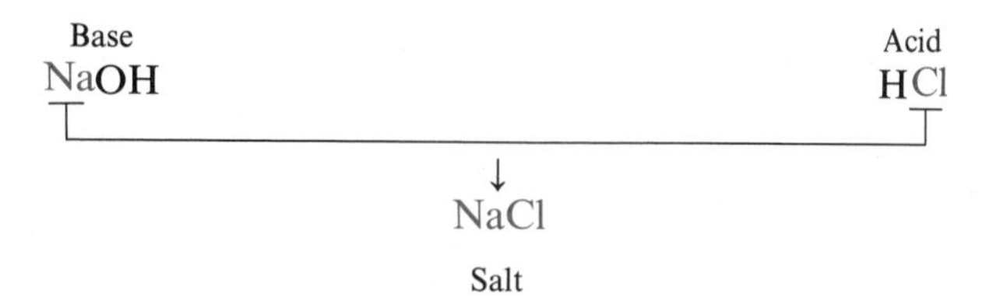


Salts are generally crystalline and have high melting and boiling points.

From a single acid such as hydrochloric acid (HCl), we may produce many chloride salts by replacing the hydrogen with a metal ion (for example, NaCl, KCl, RbCl, $CaCl_2$, $NiCl_2$). The number of known salts greatly exceeds the number of known acids and bases. Salts are ionic compounds. If the hydrogen atoms of a binary acid are replaced by a nonmetal, the resulting compound has covalent bonding and is therefore not considered to be a salt (for example, $PCl_3$, $S_2Cl_2$, $Cl_2O$, $NCl_3$, ICl).

A review of Chapter 8 on the nomenclature of acids, bases, and salts may be beneficial at this point.

## 15.5 Electrolytes and Nonelectrolytes

Some of the most convincing evidence as to the nature of chemical bonding within a substance is the ability (or lack of ability) of a water solution of the substance to conduct electricity.

It can be readily shown that solutions of certain substances are conductors of electricity. A simple apparatus to demonstrate conductivity consists of a pair of electrodes connected to a voltage source through a light bulb and a switch (see Figure 15.1). When the switch is closed and the medium between the electrodes is a conductor of electricity, the light bulb glows. When chemically pure water is placed in the beaker and the switch is closed, the light does not glow, indicating that water is a nonconductor. When we dissolve a small amount of sugar in water and test the resulting solution, the light does not glow, showing that a sugar solution is also a nonconductor. But when a small amount of salt, NaCl, is dissolved in water and this solution is tested, the light glows brilliantly.

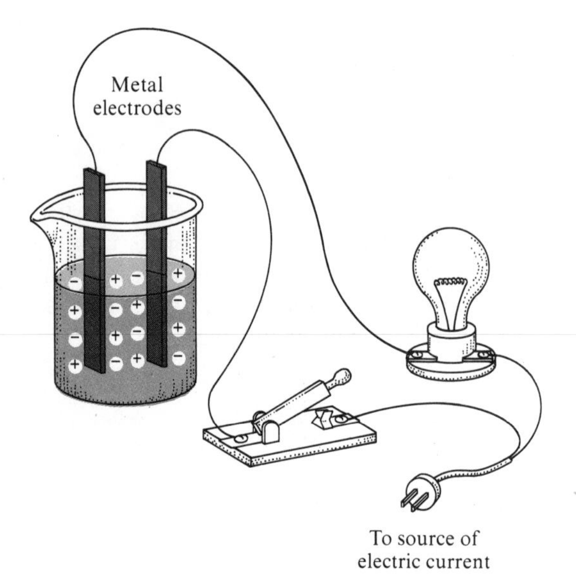


*Figure 15.1.* A simple conductivity apparatus for testing electrolytes and nonelectrolytes in solution. If the solution contains an electrolyte, the light will glow when the switch is closed.

Thus, the salt solution conducts electricity. A fundamental difference exists between the chemical bonding of sugar and that of salt. Sugar is a covalently bonded (molecular) substance; salt is an electrovalently bonded (ionic) substance.

electrolyte

nonelectrolyte

Substances whose aqueous solutions are conductors of electricity are called **electrolytes**. Substances whose solutions are nonconductors are known as **nonelectrolytes**. The classes of compounds that are electrolytes are acids, bases, and salts. Solutions of certain oxides are also conductors because they form an acid or a base when dissolved in water. One major difference between electrolytes and nonelectrolytes is that electrolytes exist as ions or are capable of producing ions in solution, whereas nonelectrolytes do not have this property. Solutions that contain a sufficient number of ions will conduct an electric current. Although pure water is a nonconductor, many city water supplies contain enough dissolved ionic matter to cause the light to glow dimly when tested in a conductivity apparatus. Table 15.2 lists some common electrolytes and nonelectrolytes.

*Table 15.2.* Representative electrolytes and nonelectrolytes.

| Electrolytes | Nonelectrolytes |
|---|---|
| $H_2SO_4$ | $C_{12}H_{22}O_{11}$ (sugar) |
| $HCl$ | $C_2H_5OH$ (ethyl alcohol) |
| $HNO_3$ | $C_2H_4(OH)_2$ (ethylene glycol) |
| $NaOH$ | $C_3H_5(OH)_3$ (glycerol) |
| $HC_2H_3O_2$ | $CH_3OH$ (methyl alcohol) |
| $NH_4OH$ | $CO(NH_2)_2$ (urea) |
| $K_2SO_4$ | $O_2$ |
| $NaNO_3$ | |

*Acids, bases, and salts are electrolytes.*

## 15.6 Dissociation and Ionization of Electrolytes

Arrhenius received the 1903 Nobel Prize in chemistry for his work on electrolytes. He stated that a solution conducts electricity because the solute dissociates immediately upon dissolving into electrically charged particles called *ions*. The movement of these ions toward oppositely charged electrodes causes the solution to be a conductor. According to his theory, solutions that are relatively poor conductors contain electrolytes that are partly dissociated. Arrhenius also believed that ions exist in solution whether or not there is an electric current. In other words, the electric current does not cause the formation of ions. Positive ions, attracted to the cathode, are cations; and negative ions, attracted to the anode, are anions.

*Positive ions are called cations.*
*Negative ions are called anions.*

We have seen that sodium chloride crystals consist of sodium and chloride ions held together by ionic bonds. When placed in water, these ions are attracted by polar water molecules, which surround each ion as it dissolves. In water, the salt dissociates, forming hydrated sodium and chloride ions (see Figure 15.2). The sodium and chloride ions in solution are bonded to a specific number of water dipoles and have less attraction for each other than they had in the crystalline state. The equation representing this dissociation is

$$NaCl(s) + (x + y)H_2O \longrightarrow Na^+(H_2O)_x + Cl^-(H_2O)_y$$

A simplified dissociation equation in which the water is omitted but understood to be present is

$$NaCl \longrightarrow Na^+ + Cl^-$$

It is important to remember that sodium chloride exists in an aqueous solution as hydrated ions and not as NaCl units, although the formula as such is very often used in equations.

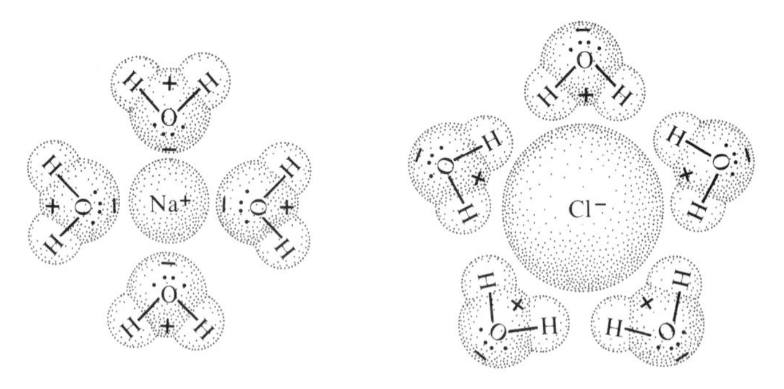


*Figure 15.2.* Hydrated sodium and chloride ions. When sodium chloride dissolves in water, each $Na^+$ and $Cl^-$ ion becomes surrounded by water molecules. The negative end of the water dipole is attracted to the $Na^+$ ion, and the positive end is attracted to the $Cl^-$ ion.

The chemical reactions of salts in solution are the reactions of their ions. For example, when sodium chloride and silver nitrate react and form a precipitate of silver chloride, only the $Ag^+$ and $Cl^-$ ions participate in the reaction. The $Na^+$ and $NO_3^-$ remain as ions in solution.

$$Ag^+ + Cl^- \longrightarrow AgCl\downarrow$$

In many cases, the number of molecules of water associated with a particular ion is known. For example, the blue color of the copper(II) ion is due to the hydrated ion $Cu(H_2O)_4^{2+}$. The hydration of ions can be demonstrated in a striking way with cobalt(II) chloride. When cobalt(II) chloride hexahydrate is dissolved in water, a pink solution forms due to the $Co(H_2O)_6^{2+}$ ions. If concentrated hydrochloric acid is added to this pink solution, the color gradually changes to blue. If water is now added to the blue solution, the color changes to pink again. These color changes are due to the exchange of water molecules and chloride ions on the cobalt ion. The complex ion $CoCl_4^{2-}$ is blue. Thus, the hydration of the cobalt ion is a reversible or equilibrium reaction (see Chapter 16). The equilibrium equation representing these changes is

$$\underset{\text{Pink}}{Co(H_2O)_6^{2+}} + 4\,Cl^- \rightleftharpoons \underset{\text{Blue}}{CoCl_4^{2-}} + 6\,H_2O$$

Ionization is the formation of ions; it may occur as a result of a chemical reaction of certain substances with water. Glacial acetic acid (100% $HC_2H_3O_2$) is a liquid that behaves as a nonelectrolyte when tested by the method described in Section 15.5. But a water solution of acetic acid conducts an electric current, as indicated by the dull-glowing light of the conductivity apparatus. The equation for the reaction with water, forming hydronium and acetate ions, is

$$\underset{\text{Acid}}{HC_2H_3O_2} + \underset{\text{Base}}{H_2O} \rightleftharpoons \underset{\text{Acid}}{H_3O^+} + \underset{\text{Base}}{C_2H_3O_2^-}$$

or, in the simplified equation,

$$HC_2H_3O_2 \leftrightarrows H^+ + C_2H_3O_2^-$$

In the above ionization reaction, water serves not only as a solvent but also as a base according to the Brønsted–Lowry theory.

The bond in hydrogen chloride is predominantly covalent, but when dissolved in water, it reacts, forming hydronium and chloride ions:

$$HCl(g) + H_2O \longrightarrow H_3O^+ + Cl^-$$

When a hydrogen chloride solution is tested for conductivity, the light glows brilliantly, indicating many ions in the solution.

In each of the above two reactions with water, ionization occurs, producing ions in solution. The necessity for water in the ionization process may be demonstrated by dissolving hydrogen chloride in a nonpolar solvent such as benzene, and testing the solution for conductivity. The solution fails to conduct electricity, indicating that no ions are produced.

The terms *dissociation* and *ionization* are often used interchangeably to describe processes taking place in water. But, strictly speaking, the two are different. In the **dissociation** of a salt, the salt already exists as ions, but when dissolved in water, the ions separate or dissociate and increase in mobility. In the **ionization** process, ions are actually produced by the reaction of a compound with water.

dissociation

ionization

Electrolytes composed of two ions per formula unit dissociate to give two ions in solution; electrolytes composed of three ions per formula unit dissociate to give three ions in solution; and so on. The dissociation equations for several electrolytes are given below. In all cases, the ions are actually hydrated.

$NaOH \xrightarrow{H_2O} Na^+ + OH^-$ 2 ions in solution per formula unit

$Ca(OH)_2 \xrightarrow{H_2O} Ca^{2+} + 2OH^-$ 3 ions in solution per formula unit

$Na_2SO_4 \xrightarrow{H_2O} 2Na^+ + SO_4^{2-}$ 3 ions in solution per formula unit

$AlCl_3 \xrightarrow{H_2O} Al^{3+} + 3Cl^-$ 4 ions in solution per formula unit

$Fe_2(SO_4)_3 \xrightarrow{H_2O} 2Fe^{3+} + 3SO_4^{2-}$ 5 ions in solution per formula unit

One mole of NaCl will give 1 mole of $Na^+$ ions and 1 mole of $Cl^-$ ions in solution, assuming complete dissociation of the salt. One mole of $CaCl_2$ will give 1 mole of $Ca^{2+}$ ions and 2 moles of $Cl^-$ ions in solution.

$$\underset{1\ \text{mole}}{NaCl} \xrightarrow{H_2O} \underset{1\ \text{mole}}{Na^+} + \underset{1\ \text{mole}}{Cl^-}$$

$$\underset{1\ \text{mole}}{CaCl_2} \xrightarrow{H_2O} \underset{1\ \text{mole}}{Ca^{2+}} + \underset{2\ \text{moles}}{2Cl^-}$$

We have learned that when 1 mole of sucrose, a nonelectrolyte, is dissolved in 1000 g of water, the solution freezes at $-1.86°C$. When 1 mole of NaCl is dissolved in 1000 g of water, the freezing point of the solution is not $-1.86°C$, as might be expected, but is closer to $-3.72°C$ ($-1.86 \times 2$). The reason for the lower freezing point is that 1 mole of NaCl in solution produces 2 moles of particles ($2 \times 6.02 \times 10^{23}$ ions) in solution. Thus, the freezing point lowering by 1 mole of NaCl is essentially equivalent to that produced by 2 moles of a nonelectrolyte. An electrolyte such as $CaCl_2$, which yields three ions in water, gives a freezing point depression about three times that of a nonelectrolyte. These freezing point data provide additional evidence that electrolytes dissociate when dissolved in water.

## 15.7 Strong and Weak Electrolytes

strong electrolyte

weak electrolyte

Electrolytes are classified as either strong or weak, depending on the degree or extent of dissociation or ionization. **Strong electrolytes** are essentially 100% ionized in solution; **weak electrolytes** are considerably less ionized (assuming 0.1 $M$ solutions). Most electrolytes are either strong or weak, with a small number being classified as moderately strong or weak. Most salts are strong electrolytes. Acids and bases that are strong electrolytes (highly ionized) are called *strong acids* and *strong bases*. Acids and bases that are weak electrolytes (slightly ionized) are called *weak acids* and *weak bases*.

For equivalent concentrations, solutions of strong electrolytes contain many more ions than solutions of weak electrolytes. As a result, solutions of strong electrolytes are better conductors of electricity. Consider the two solutions, 1 $M$ HCl and 1 $M$ $HC_2H_3O_2$. Hydrochloric acid is almost 100% ionized;

acetic acid is about 1% ionized. Thus, HCl is a strong acid and $HC_2H_3O_2$ is a weak acid. Hydrochloric acid has about 100 times as many hydronium ions in solution as acetic acid, making the HCl solution much more acidic.

One can distinguish between strong and weak electrolytes experimentally by using the apparatus described in Section 15.5. A 1 *M* HCl solution causes the light to glow brilliantly, but a 1 *M* $HC_2H_3O_2$ solution causes only a dull glow. In a similar fashion, the strong base sodium hydroxide (NaOH) may be distinguished from the weak base ammonium hydroxide ($NH_4OH$). The ionization of a weak electrolyte in water is represented by an equilibrium equation showing that both the un-ionized and ionized forms are present in solution. In the equilibrium equation of $HC_2H_3O_2$ and its ions, the equilibrium is far to the left, since relatively few hydrogen and acetate ions are present in solution:

$$HC_2H_3O_2(aq) \rightleftarrows H^+ + C_2H_3O_2^-$$

We have previously used a double arrow in an equation to represent reversible processes in the equilibrium between dissolved and undissolved solute in a saturated solution. A double arrow ($\rightleftarrows$) is also used in the ionization equation of soluble weak electrolytes to indicate that the solution contains a considerable amount of the un-ionized compound in equilibrium with its ions in solution. (See Section 16.1 for a discussion of reversible reactions.) A single arrow is used to indicate that the electrolyte is essentially all in the ionic form in the solution. For example, nitric acid is a strong acid; nitrous acid is a weak acid. Their ionization equations in water may be indicated as

$$HNO_3(aq) \xrightarrow{H_2O} H^+ + NO_3^-$$

$$HNO_2(aq) \underset{}{\overset{H_2O}{\rightleftarrows}} H^+ + NO_2^-$$

Practically all soluble salts; acids such as sulfuric, nitric, and hydrochloric acids; and bases such as sodium, potassium, calcium, and barium hydroxides are strong electrolytes. Weak electrolytes include numerous other acids and bases such as acetic acid, nitrous acid, carbonic acid, and ammonium hydroxide. The terms *strong acid*, *strong base*, *weak acid*, and *weak base* refer to whether an acid or base is a strong or weak electrolyte. A list of strong and weak electrolytes is given in Table 15.3.

*Table 15.3.* Strong and weak electrolytes.

| Strong electrolytes | Weak electrolytes |
|---|---|
| Most soluble salts | $HC_2H_3O_2$ |
| $H_2SO_4$ | $H_2CO_3$ |
| $HNO_3$ | $HNO_2$ |
| HCl | $H_2SO_3$ |
| HBr | $H_2S$ |
| $HClO_4$ | $H_2C_2O_4$ |
| NaOH | $H_3BO_3$ |
| KOH | HClO |
| $Ca(OH)_2$ | $NH_4OH$ |
| $Ba(OH)_2$ | HF |

## 15.8 Ionization of Water

The more we study chemistry, the more intriguing the little molecule of water becomes. Two equations commonly used to show how water ionizes are

$$\underset{\text{Acid}}{H_2O} + \underset{\text{Base}}{H_2O} \rightleftharpoons \underset{\text{Acid}}{H_3O^+} + \underset{\text{Base}}{OH^-}$$

and

$$H_2O \rightleftharpoons H^+ + OH^-$$

The first equation represents the Brønsted–Lowry concept, with water reacting as both an acid and a base, forming a hydronium ion and a hydroxide ion. The second equation is a simplified version, indicating that water ionizes to give a hydrogen and a hydroxide ion. Actually, the proton, $H^+$, is hydrated and exists as a hydronium ion. In either case, equal molar amounts of acid and base are produced so that water is neutral, having neither $H^+$ nor $OH^-$ ions in excess. The ionization of water at 25°C produces an $H^+$ ion concentration of $1.0 \times 10^{-7}$ mole per litre and an $OH^-$ ion concentration of $1.0 \times 10^{-7}$ mole per litre. These concentrations are usually expressed as

$$[H^+] \text{ or } [H_3O^+] = 1.0 \times 10^{-7} \text{ mole/litre}$$
$$[OH^-] = 1.0 \times 10^{-7} \text{ mole/litre}$$

These figures mean that about two out of every billion water molecules are ionized. This amount of ionization, small as it is, is a significant factor in the behavior of water in many chemical reactions.

## 15.9 Introduction to pH

The acidity of an aqueous solution depends on the concentration of hydrogen or hydronium ions. The acidity of solutions involved in a chemical reaction is often critically important, especially for biochemical reactions. The pH scale of acidity was devised to fill the need for a simple, convenient numerical way to state the acidity of a solution. Values on the pH scale are obtained by mathematical conversion of $H^+$ ion concentrations to pH by this expression:

$$\text{pH} = \log \frac{1}{[H^+]} \quad \text{or} \quad -\log [H^+]$$

pH

where $[H^+] = H^+$ or $H_3O^+$ ion concentration in moles per litre. The **pH** is defined as the logarithm (log) of the reciprocal of the $H^+$ or $H_3O^+$ ion concentration in moles per litre. The scale itself is based on the $H^+$ concentration in water at 25°C. At this temperature, water has an $H^+$ concentration of $1 \times 10^{-7}$ mole/litre and is calculated to have a pH of 7.

$$pH = \log \frac{1}{[H^+]} = \log \frac{1}{[1 \times 10^{-7}]} = \log 1 \times 10^7 = 7$$

The pH of pure water at 25°C is 7 and is said to be neutral; that is, it is neither acidic nor basic, because the concentrations of $H^+$ and $OH^-$ are equal. Solutions that contain more $H^+$ ions than $OH^-$ ions have pH values less than 7, and solutions that contain less $H^+$ ions than $OH^-$ ions have values greater than 7.

When $[H^+] = 1 \times 10^{-5}$ mole/litre, pH = 5
When $[H^+] = 1 \times 10^{-9}$ mole/litre, pH = 9

Instead of saying that the hydrogen ion concentration in the solution is $1 \times 10^{-5}$ mole/litre, it is customary to say that the pH of the solution is 5. The smaller the pH value, the more acidic the solution.

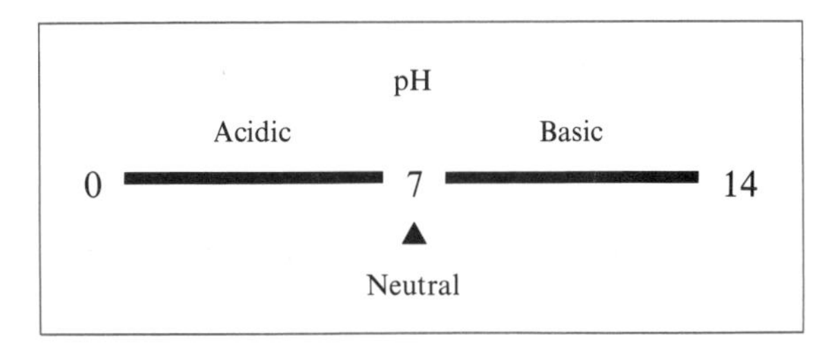


A solution of a strong acid is more acidic (has more $H^+$) than a weak acid at the same molarity. The pH of 0.1 *M* HCl is 1.00 and that of 0.1 *M* $HC_2H_3O_2$ is 2.87, indicating that hydrochloric acid is a stronger acid than acetic acid. The $[H^+]$ and thus the pH varies with the degree of dilution of a solution. The following comparative data show that although acetic acid is a weak acid, its pH approaches that of hydrochloric acid (100% ionized) as the solution becomes more dilute; this indicates that a higher percentage of acetic acid molecules ionize as the solution becomes more dilute:

| HCl solution | pH | $HC_2H_3O_2$ solution | pH |
|---|---|---|---|
| 0.100 *M* | 1.00 | 0.100 *M* | 2.87 |
| 0.0100 *M* | 2.00 | 0.0100 *M* | 3.37 |
| 0.00100 *M* | 3.00 | 0.00100 *M* | 3.90 |

The pH scale, along with its interpretation, is given in Table 15.4. Note that a change of only 1 pH unit means a tenfold increase or decrease in $H^+$ ion concentration.

$[H^+] = 1 \times 10^{-5}$

When this number is exactly 1 — pH = This number (5)
pH = 5

$$[H^+] = 2 \times 10^{-5}$$

When this number is between 1 and 10 (the 2)

pH is between this number and next lower number (4 and 5) (the −5)
pH = 4.7

*Table 15.4.* The pH scale for expressing acidity.

| $[H^+]$ (mole/litre) | pH | | |
|---|---|---|---|
| $1 \times 10^{-14}$ | 14 | ↑ | |
| $0.0000000000001 = 1 \times 10^{-13}$ | 13 | | |
| $1 \times 10^{-12}$ | 12 | | |
| $0.00000000001 = 1 \times 10^{-11}$ | 11 | Basic | |
| $1 \times 10^{-10}$ | 10 | | |
| $0.000000001 = 1 \times 10^{-9}$ | 9 | | |
| $1 \times 10^{-8}$ | 8 | ↓ | |
| $0.0000001 = 1 \times 10^{-7}$ | 7 | Neutral | Increasing acidity |
| $1 \times 10^{-6}$ | 6 | ↑ | |
| $0.00001 = 1 \times 10^{-5}$ | 5 | | |
| $1 \times 10^{-4}$ | 4 | | |
| $0.001 = 1 \times 10^{-3}$ | 3 | Acid | |
| $1 \times 10^{-2}$ | 2 | | |
| $0.1 = 1 \times 10^{-1}$ | 1 | | |
| $1.0 = 1 \times 10^{0}$ | 0 | ↓ | ↓ |

Table 15.5 lists the pH of some common solutions.

Calculation of the pH value corresponding to any $H^+$ ion concentration requires the use of logarithms. However, if you are not familiar with the use of logarithms (logs), the following simplified log scale can be used to estimate the logarithms of various numbers:

| Number | 1 | 1.5 | 2 | 2.5 | 3 | 3.5 | 4 | 4.5 | 5 | 5.5 | 6 | 6.5 | 7 | 7.5 | 8 | 8.5 | 9 | 9.5 | 1 |
|---|---|---|---|---|---|---|---|---|---|---|---|---|---|---|---|---|---|---|---|
| Log | 0 | 0.18 | 0.30 | 0.40 | 0.48 | 0.54 | 0.60 | 0.65 | 0.70 | 0.74 | 0.78 | 0.81 | 0.85 | 0.88 | 0.90 | 0.93 | 0.95 | 0.98 | 0 |

*Table 15.5* The pH of some common solutions.

| Solution | pH |
|---|---|
| 0.1 *M* HCl | 1.0 |
| 0.1 *M* $HC_2H_3O_2$ | 2.87 |
| Blood | 7.4 |
| Urine | 5.5–8.0 |
| Lemon juice | 2.3 |
| Vinegar | 2.8 |
| Milk | 6.6 |
| Carbonated water | 3.0 |
| Tomatoes | 4.0–4.5 |

Let us see how to use this log scale in calculating the pH of a solution with $[H^+] = 2 \times 10^{-5}$:

$$[H^+] = (2) \times 10^{-(5)}$$

pH = This number minus the log of this number (which must be between 1 and 10)

pH = 5 − log 2

From the log scale, log 2 = 0.30. Thus, pH = 5 − 0.30 = 4.7 (Answer)

The measurement and control of pH is extremely important in many fields of science and technology. The proper soil pH is necessary to grow certain types of plants successfully. The pH of certain foods is too acid for some diets. Many biological processes are delicately controlled pH systems. The pH of human blood is regulated to very close tolerances by the uptake or release of $H^+$ by mineral ions such as $HCO_3^-$ and $CO_3^{2-}$. Changes in the pH of the blood by as little as 0.4 pH unit result in death.

Compounds with colors that change at particular pH values are used as indicators in acid–base reactions. For example, phenolphthalein, an organic compound, is colorless in acid solution and changes to pink at a pH of 8.3. When a solution of sodium hydroxide is added to a hydrochloric acid solution containing phenolphthalein, the change in color (from colorless to pink) indicates that all the acid is neutralized. Commercially available pH test paper, such as shown in Figure 15.3, contains chemical indicators. The indicator in the paper takes on different colors when wetted with solutions of different pH. Thus, the pH of a solution can be estimated by placing a drop on the test paper and comparing the color of the test paper with a color chart calibrated at different pH values. Electronic pH meters of the type shown in Figure 15.4 are used for making rapid and highly precise pH determinations.

*Figure 15.3.* pH test paper for determining the approximate acidity of solutions. (Courtesy Micro Essential Laboratory, Inc.)

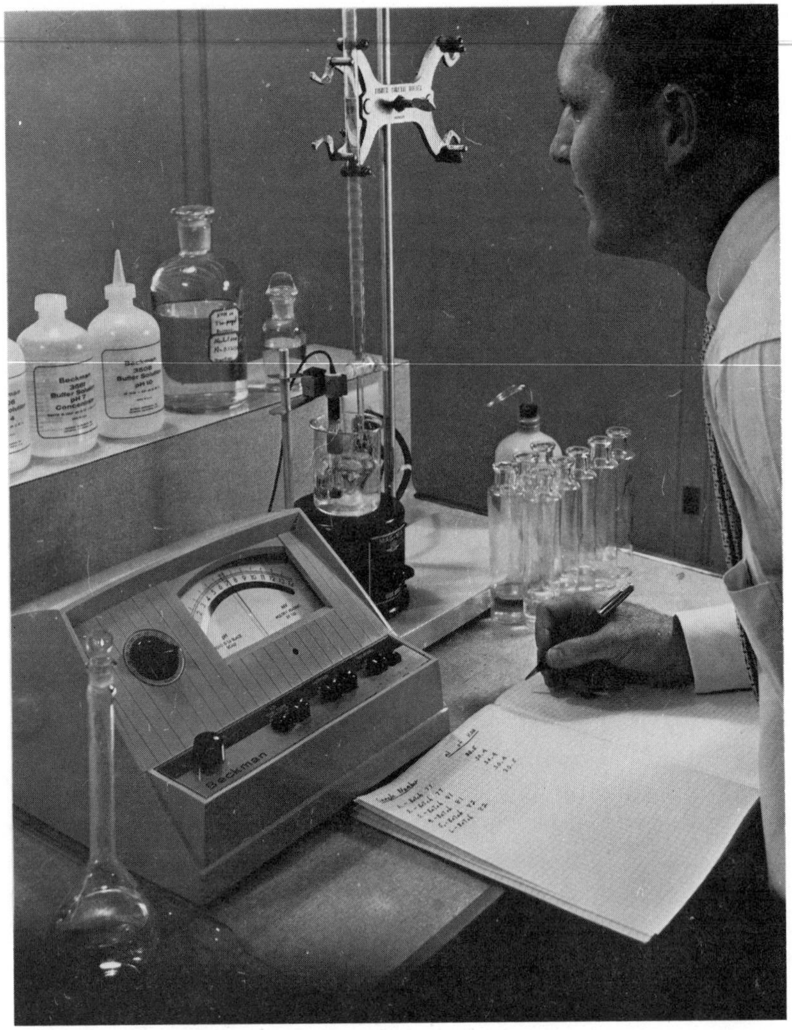

*Figure 15.4.* An electronic pH meter: Accurate measurements may be made by meters of this type. The scale is calibrated to read in both pH units and millivolts. (Courtesy Beckman Instruments, Inc. Zeromatic is a registered trademark.)

## 15.10 Neutralization

neutralization

The reaction of an acid and a base to form a salt and water is known as **neutralization**. We have seen this reaction before, but now, in the light of what we have learned about ions and ionization, let us reexamine what occurs during neutralization.

Consider the reaction that occurs when solutions of sodium hydroxide and hydrochloric acid are mixed. The ions present initially are $Na^+$ and $OH^-$

from the base and $H^+$ and $Cl^-$ from the acid. The products, sodium chloride and water, exist as $Na^+$ and $Cl^-$ ions and $H_2O$ molecules. A chemical equation representing this reaction is:

$$HCl(aq) + NaOH(aq) \longrightarrow NaCl(aq) + H_2O \qquad (1)$$

This equation is a formula equation and does not show that ions are present. The following total ionic equation gives a much better representation of the reaction:

$$(H^+ + Cl^-) + (Na^+ + OH^-) \longleftarrow Na^+ + Cl^- + H_2O \qquad (2)$$

spectator ions

Equation (2) shows that the $Na^+$ and $Cl^-$ ions did not react. These ions are **spectator ions** because they were present but did not take part in the reaction. The only reaction that occurred was that between the $H^+$ and $OH^-$ ions. Therefore, the equation for the neutralization can be written as this net ionic equation:

$$\underset{\text{Acid}}{H^+} + \underset{\text{Base}}{OH^-} \longrightarrow \underset{\text{Water}}{H_2O} \qquad (3)$$

This simple net ionic equation (3) represents not only the reaction of sodium hydroxide and hydrochloric acid but also the reaction of any acid with any base in an aqueous solution. The driving force of a neutralization reaction is the ability of an $H^+$ ion and an $OH^-$ ion to react and form a molecule of un-ionized water.

titration

The amount of acid, base, or other species in a sample may be determined by titration. **Titration** is the process of measuring the volume of one reagent required to react with a measured weight or volume of another reagent. Let us consider the titration of an acid with a base. A measured volume of acid of unknown concentration is placed into a flask and a few drops of an indicator solution are added. Base solution of known concentration is slowly added from a buret to the acid until the indicator changes color (see Figure 15.5). The indicator selected is one that changes color when the stoichiometric quantity (according to the equation) of base has been added to the acid. At this point, known as the *end point of the titration*, the titration is complete and the volume of base used to neutralize the acid is read from the buret. The concentration or amount of acid in solution can be calculated from the titration data and the chemical equation for the reaction. Illustrative problems follow.

*Problem 15.1*

Suppose that 42.00 ml of 0.15 *M* NaOH solution is required to titrate 100 ml of hydrochloric acid solution. What is the molarity of the acid solution?

The equation for the reaction is

$$NaOH(aq) + HCl(aq) \longrightarrow NaCl(aq) + H_2O$$

In this neutralization, NaOH and HCl react in a 1:1 mole ratio. Therefore, the moles of HCl in solution are equal to the moles of NaOH required to react with it. First we calculate the moles of NaOH used, and from this value, the moles of HCl.

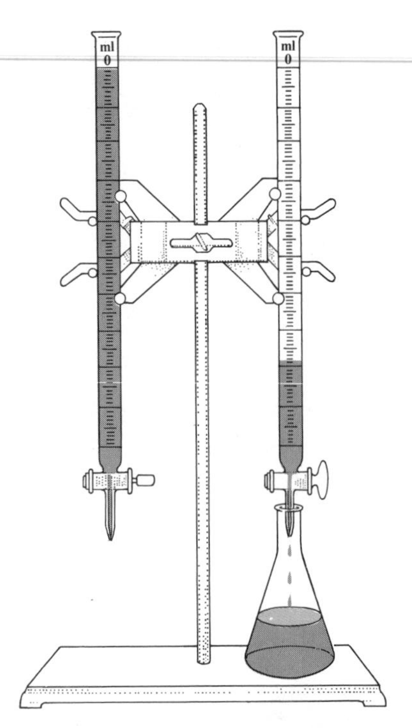


*Figure 15.5.* Graduated burets are used in titrations for neutralization of acids and bases as well as for many other volumetric determinations. Figure 15.4 illustrates a titration using a pH meter as indicator.

*Data:* 42.00 ml of 0.15 $M$ NaOH    100 ml HCl    Molarity of acid = $M$ (unknown)

Moles of NaOH:

$$M = \text{moles/litre} \qquad 42.00 \text{ ml} = 0.04200 \text{ litre}$$

$$0.04200 \text{ litre} \times \frac{0.15 \text{ mole NaOH}}{\text{litre}} = 0.0063 \text{ mole NaOH}$$

Since NaOH and HCl react in a 1:1 mole ratio, 0.0063 mole of HCl was present in the 100 ml of HCl solution. Therefore, the molarity of the HCl is

$$M = \frac{\text{moles}}{\text{litre}} = \frac{0.0063 \text{ mole HCl}}{0.100 \text{ litre}} = 0.063\ M \text{ HCl} \quad \text{(Answer)}$$

---

*Problem 15.2*

Suppose that 42.00 ml of 0.15 $M$ NaOH solution is required to titrate 100 ml of sulfuric acid ($H_2SO_4$) solution. What is the molarity of the acid solution?

The equation for the reaction is

$$2\,\text{NaOH}(aq) + \text{H}_2\text{SO}_4(aq) \longrightarrow \text{Na}_2\text{SO}_4(aq) + 2\,\text{H}_2\text{O}$$

The same amount of base (0.0063 mole of NaOH) is used in this titration as in Problem 15.1. However, the mole ratio of acid to base in the reaction is 1:2. The moles of $H_2SO_4$ reacted can be calculated by using the mole-ratio method.

*Data:* 42.00 ml of 0.15 $M$ NaOH = 0.0063 mole NaOH

$$0.0063 \text{ mole NaOH} \times \frac{1 \text{ mole } H_2SO_4}{2 \text{ moles NaOH}} = 0.00315 \text{ mole } H_2SO_4$$

Therefore, 0.00315 mole of $H_2SO_4$ was present in 100 ml of $H_2SO_4$ solution. The molarity of the $H_2SO_4$ is

$$M = \frac{\text{moles}}{\text{litre}} = \frac{0.00315 \text{ mole } H_2SO_4}{0.100 \text{ litre}} = 0.0315\ M\ H_2SO_4 \quad \text{(Answer)}$$

## 15.11 Writing Ionic Equations

molecular equation

total ionic equation

net ionic equation

In Section 15.10, we wrote the reaction of hydrochloric acid and sodium hydroxide in three different equations: (1) the molecular equation, (2) the total ionic equation, and (3) the net ionic equation. In the **molecular equation**, compounds are written in their molecular or normal formula expressions. In the **total ionic equation**, compounds are written in the form in which they are predominantly present: strong electrolytes as ions in solution; and nonelectrolytes, weak electrolytes, precipitates, and gases, in their molecular forms. In the **net ionic equation**, only those molecules or ions that have changed are included in the equation; ions or molecules that do not change are omitted. Up to now, we have been concerned only with balancing the individual elements when we balanced equations. Because we are using ions, which are electrically charged, net ionic equations are often not neutral in charge and end up with a net electrical charge. The net electrical charge of an ionic equation, as well as its atoms, should be in balance. Therefore, a balanced ionic equation will have the same net electrical charge on both sides of the equation, whether it is zero, positive, or negative.

Study the examples below. Note that all reactions are in solution.

(a) $HNO_3(aq) + KOH(aq) \longrightarrow KNO_3(aq) + H_2O$ Molecular equation

$(H^+ + NO_3^-) + (K^+ + OH^-) \longrightarrow (K^+ + NO_3^-) + H_2O$ Total ionic equation

$H^+ + OH^- \longrightarrow H_2O$ Net ionic equation

In example (a), $HNO_3$, KOH, and $KNO_3$ are soluble, strong electrolytes.

(b) $2\,AgNO_3(aq) + BaCl_2(aq) \longrightarrow 2\,AgCl\downarrow + Ba(NO_3)_2(aq)$

$(2\,Ag^+ + 2\,NO_3^-) + (Ba^{2+} + 2\,Cl^-) \longrightarrow 2\,AgCl\downarrow + (Ba^{2+} + 2\,NO_3^-)$

$Ag^+ + Cl^- \longrightarrow AgCl\downarrow$ Net ionic equation

Although silver chloride (AgCl) is an ionic salt, it is written in the molecular form because in example (b), most of the $Ag^+$ and $Cl^-$ ions are no longer in solution, but have formed a precipitate of AgCl.

(c) $Na_2CO_3(aq) + H_2SO_4(aq) \longrightarrow Na_2SO_4(aq) + H_2O + CO_2\uparrow$

$(2\,Na^+ + CO_3^{2-}) + (2\,H^+ + SO_4^{2-}) \longrightarrow (2\,Na^+ + SO_4^{2-}) + H_2O + CO_2\uparrow$

$CO_3^{2-} + 2\,H^+ \longrightarrow H_2O + CO_2\uparrow$ Net ionic equation

In example (c), carbon dioxide ($CO_2$) is a gas and evolves from solution.

(d) $HC_2H_3O_2(aq) + NaOH(aq) \longrightarrow NaC_2H_3O_2(aq) + H_2O$

$HC_2H_3O_2 + (Na^+ + OH^-) \longrightarrow (Na^+ + C_2H_3O_2^-) + H_2O$

$HC_2H_3O_2 + OH^- \longrightarrow C_2H_3O_2^- + H_2O$ Net ionic equation

In example (d), acetic acid ($HC_2H_3O_2$), a weak acid, is written in the molecular form, but sodium acetate ($NaC_2H_3O_2$), a soluble salt, is written in the ionic form. The $Na^+$ ion is the only spectator ion in this reaction. Both sides of the net ionic equation have a $-1$ electrical charge.

(e) $Mg(s) + 2\,HCl(aq) \longrightarrow MgCl_2(aq) + H_2\uparrow$

$Mg + (2\,H^+ + 2\,Cl^-) \longrightarrow (Mg^{2+} + 2\,Cl^-) + H_2\uparrow$

$Mg + 2\,H^+ \longrightarrow Mg^{2+} + H_2\uparrow$ Net ionic equation

In example (e), the net electrical charge on both sides of the equation is $+2$.

(f) $H_2SO_4(aq) + Ba(OH)_2(aq) \longrightarrow BaSO_4\downarrow + 2\,H_2O$

$(2\,H^+ + SO_4^{2-}) + (Ba^{2+} + 2\,OH^-) \longrightarrow BaSO_4\downarrow + 2\,H_2O$

$2\,H^+ + SO_4^{2-} + Ba^{2+} + 2\,OH^- \longrightarrow BaSO_4\downarrow + 2\,H_2O$ Net ionic equation

In example (f), barium sulfate ($BaSO_4$) is a highly insoluble salt. If we conduct this reaction using the apparatus described in Section 15.5, the light, which glows brightly at first, will be extinguished when the reaction is complete, because essentially no ions are left in solution. The $BaSO_4$ precipitates out of solution, and water is a nonconductor.

Here is a list of rules to observe when writing ionic equations.

1. Strong electrolytes are written in their ionic form.
2. Weak electrolytes are written in their molecular form.
3. Nonelectrolytes are written in their molecular form.
4. Insoluble substances, precipitates, and gases are written in their molecular forms.
5. The net ionic equation should include only those substances that have undergone a chemical change.
6. Equations must be balanced, both in atoms and in electrical charge.

## 15.12 Colloids: Introduction

When we add sugar to a flask of water and shake, the sugar dissolves and forms a clear homogeneous *solution*. When we do the same experiment with very fine sand and water, the sand particles form a *suspension*, which settles when the shaking stops. For a third trial, let us use ordinary corn starch. Starch does not dissolve in cold water; but when the mixture is heated and stirred, the starch forms a cloudy, opalescent *dispersion*. This dispersion does not appear to be clear and homogeneous like the sugar solution. Yet it is not obviously heterogeneous and does not settle like the sand dispersion. In short, its properties are intermediate between those of the sugar solution and those of the sand suspension. The starch dispersion is actually a *colloid.* The name "colloid" is derived from the Greek *kolla*, meaning "glue," and was coined by the English scientist Thomas Graham in 1861. Graham classified solutes as crystalloids if they readily diffused through a parchment membrane and as colloids if they did not diffuse through the membrane.

colloid

As it is now used, the word **colloid** means a dispersion in which the dispersed particles are larger than the solute ions or molecules of a true solution and smaller than the particles of a mechanical suspension. The term does not imply a gluelike quality, although most glues actually are colloidal materials. The size of colloidal particles ranges from a lower limit of about 10 Å (1 nm, or $10^{-7}$ cm) to an upper limit of about 10,000 Å (1000 nm, or $10^{-4}$ cm).

The fundamental difference between a colloidal dispersion and a true solution is the size, not the nature, of the particles. The solute particles in a solution are ordinarily single ions or molecules, which may be hydrated to varying degrees. Colloidal particles are usually aggregations of ions or molecules. However, the molecules of some polymers such as proteins are large enough to be classified as colloidal particles when in solution. To appreciate fully the differences in relative size, the volumes, not just the linear dimensions, of colloidal particles and solute particles must be compared. The difference in volumes can be approximated by assuming that the particles are spheres. A large colloidal particle has a diameter of about 5000 Å, while a fair-sized ion or molecule has a diameter of about 5 Å. Thus, the diameter of the colloidal particle is 1000 times that of the solute "particle." Because the volumes of spheres are proportional to the cubes of their diameters, we can readily calculate that the volume of a colloidal particle can be up to a billion ($10^3 \times 10^3 \times 10^3 = 10^9$) times greater than that of a solution "particle"!

Colloids are mixtures in which one component, the *dispersed phase*, exists as discrete particles in the other component, the *dispersing phase.* The components of a colloidal dispersion are also sometimes called the *discontinuous phase* and the *continuous phase*.

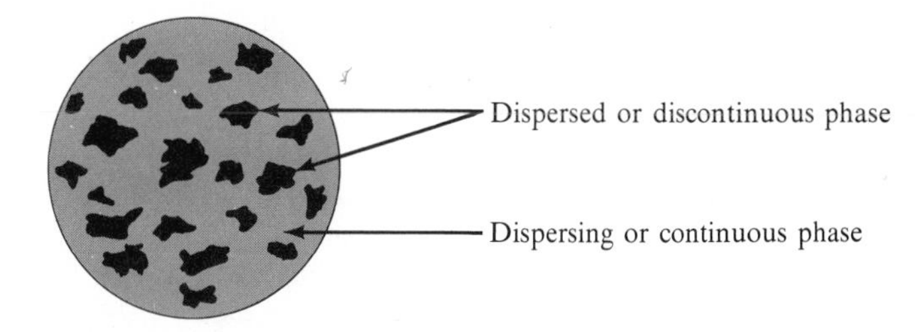


Each component, or phase, may exist as a solid, a liquid, or a gas. The components cannot be mutually soluble and both phases of the dispersion cannot be gases because such conditions would describe an ordinary solution. Hence, only eight types of colloidal dispersions, based on the physical states of the phases, are known. The eight types, together with specific examples, are listed in Table 15.6.

*Table 15.6.* Types of colloidal dispersions.

| Type | Name | Examples |
|---|---|---|
| Gas in liquid | Foam | Whipped cream, soap suds |
| Gas in solid | Solid foam | Styrofoam, foam rubber, pumice |
| Liquid in gas | Liquid aerosol | Fog, clouds |
| Liquid in liquid | Emulsion | Milk, vinegar in oil salad dressing, mayonnaise |
| Liquid in solid | Solid emulsion | Cheese, opals, jellies |
| Solid in gas | Solid aerosol | Smoke, dust in air |
| Solid in liquid | Sol | India ink, gold sol |
| Solid in solid | Solid sol | Tire rubber, certain gems (for example, rubies) |

## 15.13 Preparation of Colloids

Colloidal dispersions can be prepared by two general methods: (1) *dispersion*, the breaking down of larger particles to colloidal size; (2) *condensation*, the formation of colloidal particles from solutions.

Homogenized milk is a good example of a colloid prepared by dispersion. Milk, as drawn from the cow, is a poor emulsion of fat in water. The fat globules are so large that they rise and form a cream layer in a few hours. To avoid this separation of the cream, milk is homogenized. This is usually accomplished by pumping the milk through very small holes, or orifices, at high pressure. The violent shearing action of this treatment breaks the fat globules into particles well within the colloidal size range. The butterfat in homogenized milk remains dispersed indefinitely. Colloid mills, which reduce particles to colloidal size by a grinding or shearing action, are used in preparing many commercial products such as paints, cosmetics, and salad dressings.

The preparation of colloids by condensation frequently involves a precipitation reaction in a dilute solution. As an example, a good colloidal sulfur sol can be made by simply bubbling hydrogen sulfide into a solution of sulfur dioxide. Solid sulfur is formed and dispersed as a colloid.

$$SO_2 + 2\,H_2S \longrightarrow \underset{\substack{\text{Colloidal} \\ \text{sulfur}}}{3\,S} + 2\,H_2O$$

A colloidal dispersion is also easily made by adding iron(III) chloride solution to boiling water. The reddish-brown colloidal dispersion that is formed probably consists of iron(III) hydroxide and hydrated iron(III) oxide.

$$FeCl_3 + 3\,HOH \longrightarrow Fe(OH)_3 + 3\,HCl$$

$$Fe(OH)_3 \xrightarrow{H_2O} Fe_2O_3 \cdot xH_2O$$

A great many products for home use (insecticides, insect repellents, and deodorants, to name a few) are packaged as aerosols. The active ingredient, either a liquid or a solid, is dissolved in a liquified gas and sealed under pressure in a container fitted with a release valve. When this valve is opened, the pressurized solution is ejected. The liquified gas vaporizes, and the active ingredient is converted to a colloidal aerosol almost instantaneously.

## 15.14 Properties of Colloids

In 1827, Robert Brown (1773–1858), while observing strongly illuminated pollen under a high-powered microscope, noted that the pollen grains appeared to have a trembling, erratic motion. He determined later that this erratic motion was not confined to pollen but was characteristic of colloidal particles in general. The random motion of colloidal particles, first reported by Brown, is called **Brownian movement**. It is readily observed in cigarette smoke. The smoke is confined in a small transparent chamber and is illuminated with a strong beam of light at right angles to the optical axis of the microscope. The smoke particles appear as tiny randomly moving lights because the light is reflected from their surfaces. This motion is due to the continual bombardment of the smoke particles by air molecules. Since Brownian movement can be seen when colloidal particles are dispersed in either a gaseous or a liquid medium, it affords nearly direct visual proof that matter at the molecular level actually is moving randomly, as postulated by the Kinetic-Molecular Theory.

Brownian movement

When an intense beam of light is passed through an ordinary solution and is viewed at an angle, the beam passing through the solution is hardly visible. A beam of light, however, is clearly visible and sharply outlined when it is passed through a colloidal dispersion (see Figure 15.6). This phenomenon is known as the **Tyndall effect**. It was first described by Michael Faraday in 1857 and later amplified by John Tyndall. The Tyndall effect, like Brownian movement, can be observed in nearly all colloidal dispersions. It occurs because the colloidal particles are large enough to scatter the rays of visible light. The ions or molecules of true solutions are too small to scatter light and, therefore, do not exhibit a noticeable Tyndall effect.

Tyndall effect

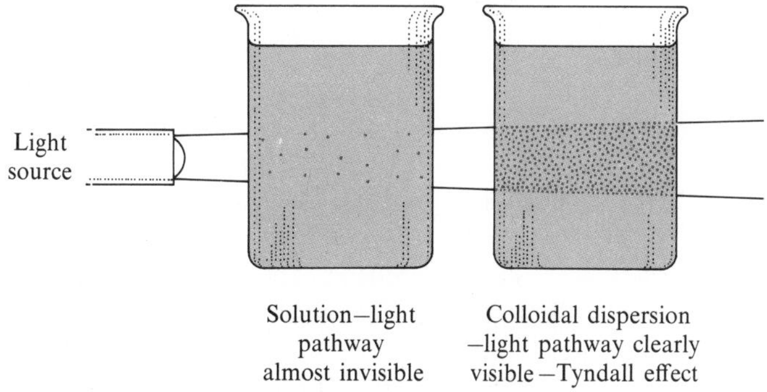

Light source

Solution—light pathway almost invisible

Colloidal dispersion —light pathway clearly visible—Tyndall effect

*Figure 15.6.* The Tyndall effect.

Another important characteristic of colloids is that the particles have relatively huge surface areas. We have seen in Section 14.5 that the surface area is increased tenfold when a 1 cm cube is divided into 1000 cubes with sides of 0.1 cm. When a 1 cm cube is divided into colloidal-size cubes measuring $10^{-6}$ cm, the surface area of the particles becomes a million times greater than that of the original cube.

Colloidal particles become electrically charged when they adsorb ions on their surfaces. This property is, of course, directly related to the large surface area of the particles. Adsorption occurs because the atoms or ions at the surface of a particle are not surrounded by other atoms or ions like those in the interior. Consequently, these surface ions attract and adsorb ions or polar molecules from the dispersion medium onto the surfaces of the colloidal particles. The particles in a given dispersion tend to adsorb ions having only one kind of charge. For example, cations are primarily adsorbed on ferric hydroxide sol, resulting in positively charged colloidal particles. On the other hand, the particles of an arsenic(III) sulfide ($As_2S_3$) sol primarily adsorb anions, resulting in negatively charged colloidal particles. The properties of true solutions, colloidal dispersions, and mechanical suspensions are summarized and compared in Table 15.7.

*Adsorption* should not be confused with *absorption*. Adsorption refers to the adhesion of molecules or ions to a surface, while absorption refers to the taking in of one material by another material.

*Table 15.7.* Comparison of the properties of true solutions, colloidal dispersions, and suspensions.

| | Particle size (Å) | Ability to pass through filter paper | Ability to pass through parchment | Exhibits Tyndall effect | Exhibits Brownian movement | Settles out on standing | Appearance |
|---|---|---|---|---|---|---|---|
| True solution | <10 | Yes | Yes | No | No | No | Transparent, homogeneous |
| Colloidal dispersion | 10–10,000 | Yes | No | Yes | Yes | Generally does not | Usually not transparent, but may appear to be homogeneous |
| Suspension | >10,000 | No | No | — | No | Yes | Not transparent, heterogeneous |

## 15.15 Stability of Colloids

The properties of the dispersed and dispersing phases are key factors in making a colloidal dispersion. The stability of different dispersions varies a great deal. We have noted that nonhomogenized milk is a colloid, yet it will separate on standing for a few hours. However, the particles of a good colloidal dispersion will remain in suspension indefinitely. As a case in point, a ruby-red gold sol has been kept in the British Museum for more than a century without noticeable settling. (This specimen is kept for historical interest because it was prepared by Michael Faraday.) The particles of a specific colloid remain dispersed because (1) they are bombarded by the molecules of the dispersing phase and (2) they are electrically charged. Molecular bombardment keeps the particles in motion (Brownian movement) so that gravity does not cause them to settle out. Since the colloidal particles have the same kind of electrical charge, they repel each other. This mutual repulsion prevents the dispersed particles from coalescing to larger particles, which would settle out of suspension.

In certain types of colloids, the presence of a material known as a protective colloid is necessary for stability. Egg yolk, for example, acts as a stabilizer or protective colloid in mayonnaise. The yolk adsorbs on the surfaces of the oil particles and prevents them from coalescing.

## 15.16 Applications of Colloidal Properties

Activated charcoal is frequently used as an adsorbent in gas masks. This material has an enormous surface area, approximately one million square centimetres per gram in some samples of charcoal. Hence, charcoal is very effective in selectively adsorbing the polar molecules of some poisonous gases. Charcoal can be used to adsorb impurities from liquids as well as from gases, and large amounts are used to remove substances that have objectionable tastes and odors from water supplies. In sugar refineries, activated charcoal is used to adsorb colored impurities from the raw sugar solutions.

The Cottrell process is widely used for dust and smoke control in many urban and industrial areas. This process, devised by an American, Frederick Cottrell (1877–1948), takes advantage of the fact that the particulate matter in dust and smoke is electrically charged. Air to be cleaned of dust or smoke is passed between electrode plates charged with high voltage. Positively charged particles are attracted to, neutralized, and thereby precipitated at the negative electrodes. Negatively charged particles are removed in the same fashion at the positive electrodes. Large Cottrell units are fitted with devices for automatic removal of precipitated material. In some installations, particularly at cement mills and smelters, the value of the dust collected may be sufficient to pay for the precipitation equipment. Small units, designed for removing dust and pollen from air in the home, are now on the market. Unfortunately, Cottrell units remove only particulate matter; they cannot remove gaseous pollutants such as carbon monoxide, sulfur dioxide, and nitrogen oxides.

dialysis

Thomas Graham found that a parchment membrane would allow the passage of true solutions but would prevent the passage of colloidal dispersions. Dissolved solutes can be removed from colloidal dispersions through the use of such a membrane by a process called **dialysis**. The membrane itself is called a *dialyzing membrane*. Many animal membranes can act as dialyzing membranes. Artificial membranes are made from such materials as parchment paper, collodion, or certain kinds of cellophane. Dialysis can be demonstrated by putting a colloidal starch dispersion and some copper(II) sulfate solution in a parchment paper bag and suspending it in running water. In a few hours, the blue color of the copper(II) sulfate has disappeared and only the starch dispersion remains in the bag.

An interesting application of dialysis has been the development of artificial kidneys. These devices are dialyzing units that are able to act as kidneys by removing soluble waste products from the blood. The blood of a patient suffering from partial kidney failure is bypassed through the artificial kidney machine for several hours. During passage through the machine, the soluble waste products are removed by dialysis.

## Questions

A. *Review the meanings of the new terms introduced in this chapter.*

1. Hydronium ion
2. Amphoteric
3. Electrolyte
4. Nonelectrolyte
5. Dissociation
6. Ionization
7. Strong electrolyte
8. Weak electrolyte
9. pH
10. Neutralization
11. Spectator ions
12. Titration
13. Molecular equation
14. Total ionic equation
15. Net ionic equation
16. Colloid
17. Brownian movement
18. Tyndall effect
19. Dialysis

B. *Answers to the following questions will be found in tables and figures.*

1. In which situations will the light glow in the apparatus shown in Figure 15.1? The switch is closed and the aqueous solution contains:
   (a) $HNO_3$
   (b) $C_2H_5OH$
   (c) $C_2H_5OH + NH_4Cl$
   (d) $HC_2H_3O_2$
   (e) $KOH + K_2SO_4$
2. List the solutions in Table 15.5 in order of decreasing acidity.
3. Is it possible for the pH scale shown in Table 15.4 to be extended on both ends? Explain.
4. Between what whole numbers is the pH of a solution that has an $H^+$ ion concentration of $3.6 \times 10^{-3}$ mole/litre?
5. Explain the orientation of the water molecules on the hydrated sodium and chloride ions shown in Figure 15.2.
6. Between what levels of $H^+$ ion concentration does the pH of blood lie?

C. *Review questions.*

1. Explain why hydrogen chloride is classified as an acid under the (a) Arrhenius, (b) Brønsted–Lowry, and (c) Lewis acid–base theories.
2. Why is a species that is a base by Brønsted–Lowry theory also a base by Lewis theory?
3. How does the Brønsted–Lowry theory define an acid? A base?
4. Can ammonia ($NH_3$) act as a base in the absence of water (a) by the Arrhenius definition, (b) by the Brønsted–Lowry definition, or (c) by the Lewis definition? Explain.
5. According to the Brønsted–Lowry theory, by what do the formulas of a conjugate acid–base pair differ?
6. Identify the conjugate acid–base pairs in the following equations:
   (a) $HNO_3 + H_2O \rightarrow H_3O^+ + NO_3^-$
   (b) $HC_2H_3O_2 + OH^- \rightarrow H_2O + C_2H_3O_2^-$

(c) $NH_3 + H_3O^+ \rightarrow NH_4^+ + H_2O$
(d) $NH_3 + HBr \rightarrow NH_4^+ + Br^-$
(e) $HC_2H_3O_2 + H_2O \rightarrow H_3O^+ + C_2H_3O_2^-$

7. Write the electron-dot structure for:
   (a) Chloride ion (b) Hydroxide ion (c) Sulfate ion
   Why are these ions considered to be bases according to the Brønsted–Lowry and Lewis acid–base theories?
8. Complete and balance the following equations:
   (a) $CaO(s) + H_2SO_4(aq) \rightarrow$
   (b) $Zn(s) + HCl(aq) \rightarrow$
   (c) $NH_3(g) + HC_2H_3O_2(g) \rightarrow$
   (d) $Ba(OH)_2(aq) + HClO_4(aq) \rightarrow$
   (e) $K_2CO_3(aq) + HI(aq)$
   (f) $Fe_2O_3(s) + HNO_3(aq) \rightarrow$
9. Complete and balance the following equations:
   (a) $NaOH(aq) + HCl(aq) \rightarrow$
   (b) $Ca(OH)_2(aq) + HBr(aq) \rightarrow$
   (c) $Ba(OH)_2(aq) + H_2SO_4(aq) \rightarrow$
   (d) $Ca(OH)_2(aq) + Na_2CO_3(aq) \rightarrow$
   (e) $NH_4OH(aq) + FeCl_3(aq) \rightarrow$
10. Which of the following are electrolytes and which are nonelectrolytes? Consider each substance to be mixed with water.
    (a) $Mg(NO_3)_2$
    (b) $CH_4$ (insoluble)
    (c) $CO_2$
    (d) $Na_2O$
    (e) $C_6H_6$ (insoluble)
    (f) $H_2$ (insoluble)
    (g) HBr
    (h) $K_2CrO_4$
    (i) HCOOH (formic acid)
    (j) $C_2H_5OH$ (ethyl alcohol)
    (k) $H_3PO_4$
11. Name all the compounds shown in Table 15.3.
12. Explain the following statement in terms of ionization and chemical bonding: When solutions of hydrogen chloride in water and in benzene are prepared, the water solution conducts an electric current but the benzene solution does not.
13. An aqueous solution of ethyl alcohol ($C_2H_5OH$) does not conduct an electric current but an aqueous solution of potassium hydroxide (KOH) does. What does this information tell us about the OH group in the alcohol?
14. Why does molten sodium chloride conduct an electric current?
15. How does a hydronium ion differ from a hydrogen ion?
16. How does hydrochloric acid differ from hydrogen chloride? Are both electrolytes?
17. Water solutions of HCl and of NaCl are strong electrolytes, but the processes by which HCl and NaCl dissolve in water are quite different. Explain.
18. Distinguish between the dissociation of ionic compounds and the ionization of molecular compounds.
19. Distinguish between weak and strong electrolytes.
20. Explain why ions are hydrated in aqueous solution.
21. Write simplified equations to show how the following compounds ionize or dissociate in water:
    (a) $NaNO_3$ (c) $HC_2H_3O_2$ (e) $HClO_4$ (g) $H_2C_2O_4$
    (b) $NH_4Br$ (d) $K_2SO_4$ (f) HClO (h) $Na_3PO_4$
22. Write the net ionic equation for the reaction of an acid with a base in an aqueous solution.

23. Rewrite the following unbalanced equations, converting them into balanced net ionic equations (all reactions are in water solution):
 (a) $BaCl_2(aq) + AgNO_3(aq) \rightarrow Ba(NO_3)_2(aq) + AgCl\downarrow$
 (b) $MgCO_3(s) + HCl(aq) \rightarrow MgCl_2(aq) + H_2O + CO_2\uparrow$
 (c) $Zn(s) + HC_2H_3O_2(aq) \rightarrow Zn(C_2H_3O_2)_2 + H_2\uparrow$
 (d) $H_2S(g) + Hg(C_2H_3O_2)_2(aq) \rightarrow HgS\downarrow + HC_2H_3O_2$
 (e) $CaCl_2(aq) + (NH_4)_2C_2O_4(aq) \rightarrow CaC_2O_4\downarrow + NH_4Cl(aq)$
24. Water contains both acid and base ions. Why is it neutral?
25. At 100°C water ionizes about ten times as much as at 25°C and contains about $1 \times 10^{-6}$ mole of $H^+$ ions/litre. Is water acidic at this temperature? Explain.
26. Which is more acidic, 1 $M$ $HNO_3$ or 1 $M$ $H_2SO_4$?
27. Two drops (0.10 ml) of 1.0 $M$ HCl are added to pure water to make 1 litre of solution. What is the pH of this solution if the HCl is 100% ionized?
28. Samples of lemon juice and tomato juice have pH values of 2.3 and 4.3, respectively. How many times greater is the $H^+$ ion concentration in the lemon juice than in the tomato juice?
29. Explain why a solution containing 1 mole of acetic acid in 1 kg of water freezes at a lower temperature than one containing 1 mole of ethyl alcohol in 1 kg of water.
30. Three solutions are made by dissolving 1 mole of each of the following substances in 1 kg of water:
 (a) HCl (b) $C_6H_{12}O_6$ (glucose) (c) $CaCl_2$
 Arrange these solutions in decreasing order of their freezing points, placing the one with the highest freezing point first.
31. Based on the physical states of the dispersed phases, what type of colloidal dispersion cannot exist? Explain.
32. Give brief descriptions of the two general methods for preparing colloidal dispersions.
33. What are the principal reasons that colloidal particles, once dispersed, do not settle out?
34. How does homogenized milk differ physically from nonhomogenized milk?
35. Ozone is a serious air pollutant in some areas. Would a small Cottrell precipitation unit be of value in reducing the ozone concentration in a home in such an area? Explain.
36. Would activated charcoal be more efficient at 0°C or at 100°C in absorbing gaseous hydrogen sulfide? Explain.
37. Which of the following statements are correct?
 (a) Seawater will boil at a higher temperature than pure water.
 (b) The concentration of $Cl^-$ in a 0.50 $M$ $AlCl_3$ solution is 1.5 $M$.
 (c) $H^+(aq) + OH^-(aq) \rightarrow H_2O$ represents a neutralization reaction.
 (d) Water can act as both a Brønsted–Lowry acid and a Brønsted–Lowry base.
 (e) The conjugate base of $NH_4^+$ is $NH_3$.
 (f) The conjugate acid of $Cl^-$ is HCl.
 (g) Sodium acetate ($NaC_2H_3O_2$) is a weak electrolyte.
 (h) $Na^+$, $Ca^{2+}$, $Al^{3+}$, and $NH_3$ are all considered to be cations.
 (i) One mole of $CaCl_2$ contains $3 \times 6.02 \times 10^{23}$ ions.
 (j) A solution with $[H^+] = 1 \times 10^{-8}$ mole/litre is acidic.
 (k) A 0.001 $M$ HCl solution has a pH of 3.
 (l) Equal volumes of 0.10 $M$ HCl and 0.10 $M$ $HC_2H_3O_2$ will react with the same volume of 0.20 $M$ NaOH solution.

(m) A 1.0 molal solution of HCl will freeze at a lower temperature than a 1.0 molal solution of $HC_2H_3O_2$.
(n) The size of colloidal particles ranges from $10^{-4}$ to $10^{-1}$ cm.
(o) Adsorption refers to the adhesion of molecules or ions to a surface.
(p) The Tyndall effect can be observed because colloidal particles are large enough to scatter the rays of visible light.

D. *Review problems.*

1. Calculate the molarity of the ions in each of the salt solutions listed below. Consider each salt to be 100% dissociated. For example, calculate both the $Na^+$ and $Cl^-$ ion molarity in a NaCl solution.
   (a) 0.10 *M* NaCl
   (b) 0.32 *M* $KNO_3$
   (c) 1.25 *M* $Na_2SO_4$
   (d) 0.68 *M* $CaCl_2$
   (e) 0.22 *M* $FeCl_3$
   (f) 0.75 *M* $MgSO_4$
   (g) 0.050 *M* $(NH_4)_3PO_4$
   (h) 0.050 *M* $Al_2(SO_4)_3$
2. What is the molar concentration of a magnesium bromide ($MgBr_2$) solution that has a bromide ion molarity of 0.526 *M*?
3. How many $Al^{3+}$ and $Cl^-$ ions are there in 10.0 ml of 0.001 *M* $AlCl_3$ solution? Assume 100% dissociation.
4. Given the data for the following five titrations, calculate the molarity of HCl in titrations (a), (b), and (c), and the molarity of NaOH in titrations (d) and (e):

| Volume of HCl (ml) | Molarity of HCl | Volume of NaOH (ml) | Molarity of NaOH |
|---|---|---|---|
| (a) 25.00 | — | 35.00 | 0.150 |
| (b) 32.40 | — | 18.50 | 0.425 |
| (c) 9.50 | — | 48.00 | 0.235 |
| (d) 24.00 | 0.260 | 24.30 | — |
| (e) 12.20 | 0.260 | 49.80 | — |

5. What would be the concentration of each ion in a solution made by dissolving 40.0 g of magnesium chloride ($MgCl_2$) in sufficient water to make 250 ml of solution? How many moles of silver chloride (AgCl) could be precipitated from this solution?
6. The concentration of a barium hydroxide solution is determined by titration with hydrochloric acid:

   $$Ba(OH)_2(aq) + 2HCl(aq) \longrightarrow BaCl_2(aq) + 2H_2O$$

   It was found that 32.8 ml of 0.200 *M* HCl were required to titrate 25.0 ml of $Ba(OH)_2$ solution. What is the molarity of the $Ba(OH)_2$ solution?
7. How many moles of solute and solvent are used in preparing a solution containing 210 g of benzene ($C_6H_6$) and 1000 g of carbon tetrachloride ($CCl_4$)?
8. What volume of each component should be used to prepare the solution in Problem 7? (Densities at 20°C: $C_6H_6 = 0.879$ g/ml; $CCl_4 = 1.595$ g/ml.)
9. A sample of pure sodium bicarbonate weighing 0.420 g was dissolved in water and neutralized with hydrochloric acid; 42.5 ml of the acid were required. Compute the molarity of the acid.

   $$NaHCO_3(aq) + HCl(aq) \longrightarrow NaCl(aq) + CO_2\uparrow + H_2O$$

10. What volume of hydrogen gas can be obtained at 27°C and 600 mm Hg pressure by reacting 3.00 g Zn metal with 100 ml of 0.300 *M* HCl?

11. Calculate the pH of the following solutions:
    (a) $[H^+] = 0.001\ M$
    (b) $[H^+] = 0.10\ M$
    (c) $[H^+] = 1 \times 10^{-7}\ M$
    (d) $[H^+] = 8.5 \times 10^{-10}\ M$
    (e) $[H^+] = 0.50\ M$
12. Calculate the pH of the following:
    (a) Orange juice, $[H^+] = 2 \times 10^{-4}\ M$
    (b) Vinegar, $[H^+] = 1.6 \times 10^{-3}\ M$
    (c) Black coffee, $[H^+] = 6.3 \times 10^{-6}\ M$
    (d) Limewater, $[H^+] = 3.2 \times 10^{-11}\ M$
13. Suppose that 30 g of acetic acid, ($HC_2H_3O_2$) are dissolved in 100 g of water, forming a solution that has a density of 1.03 g/ml. What is the volume of the solution ?
14. A solution is made by dissolving 12.0 g of barium chloride ($BaCl_2$) in sufficient water to make 50.0 ml of solution. The density of the solution is 1.203 g/ml. Calculate the weight percent and the molarity of the solution.
15. What volume of concentrated (18.0 *M*) sulfuric acid must be used to prepare 20.0 litres of 3.00 *M* solution?
16. How many grams of silver iodide (AgI) will be precipitated when 10.0 ml of 1.00 *M* KI and 19.0 ml of 0.500 *M* Ag $NO_3$ are mixed together?

# 16 Chemical Equilibrium

*After studying Chapter 16 you should be able to:*

1. Understand the terms listed in Question A at the end of the chapter.
2. Describe a reversible reaction.
3. Explain why the rate of the forward reaction decreases and the rate of the reverse reaction increases as a chemical reaction approaches equilibrium.
4. State the principle of Le Chatelier.
5. Tell how the speed of a chemical reaction is affected by the following: (a) changes in concentration of reactants, (b) changes of pressure on gaseous reactants, (c) changes of temperature, and (d) presence of a catalyst.
6. Write the equilibrium constant expression for a chemical reaction from the balanced equation.
7. Calculate the concentration of one substance in an equilibrium when given the equilibrium constant and the concentrations of all the other substances.
8. Calculate the concentrations of all the chemical species in a solution of a weak acid when given the percent ionization or the ionization constant.
9. Compare the relative strengths of acids from their ionization constants.
10. Using the ion product constant for water, $K_w$, calculate $[H^+]$, $[OH^-]$, pH, or pOH when given any one of these quantities.
11. Calculate the solubility product constant, $K_{sp}$, of a slightly soluble salt when given its solubility, or vice versa.
12. Calculate the solubility of a salt when given the $K_{sp}$ value.
13. Compare the relative solubilities of salts when given their solubility products.
14. Explain how a buffer solution is able to counteract the addition of small amounts of either $H^+$ or $OH^-$ ions.
15. Explain the relative energy diagram of a reaction in terms of activation energy, exothermic or endothermic reaction, and the effect of a catalyst.

## 16.1 Reversible Reactions

In the preceding chapters we have treated chemical reactions mainly as reactants going to products. However, many reactions do not go to completion. One reason why reactions do not go to completion is that many of them are reversible; that is, when the products are formed, they react to produce the starting reactants.

We have encountered several reversible systems. One is the vaporization of a liquid by heating and subsequent condensation by cooling:

$$\text{Liquid} + \text{Heat} \longrightarrow \text{Vapor}$$
$$\text{Vapor} + \text{Cooling} \longrightarrow \text{Liquid}$$

Another is the crystallization of an aqueous salt solution, which may be considered the reverse of the dissolving and dissociation of the salt:

$$\text{NaCl}(s) \xrightarrow{H_2O} \text{Na}^+(aq) + \text{Cl}^-(aq) \quad \text{(Dissociation)}$$
$$\text{Na}^+(aq) + \text{Cl}^-(aq) \longrightarrow \text{NaCl}(s) \quad \text{(Crystallization)}$$

Weak electrolytes are ionized to a small degree because of the reversible reaction of their ions to form the un-ionized compound. A 1 *M* solution of acetic acid illustrates this behavior:

$$HC_2H_3O_2 + H_2O \longrightarrow H_3O^+ + C_2H_3O_2^- \quad \text{(Forward reaction 1\%)}$$
$$H_3O^+ + C_2H_3O_2^- \longrightarrow HC_2H_3O_2 + H_2O \quad \text{(Reverse reaction 99\%)}$$

These two reactions may be represented by a single equation with a double arrow, $\rightleftarrows$, to indicate that they are taking place at the same time:

$$HC_2H_3O_2 + H_2O \rightleftarrows H_3O^+ + C_2H_3O_2^- \quad \text{(This single equation represents both the forward and reverse reactions)}$$

The interconversion of nitrogen dioxide ($NO_2$) and dinitrogen tetroxide ($N_2O_4$) offers visible evidence of the reversibility of a reaction. $NO_2$ is a reddish-brown gas that changes, with cooling, to $N_2O_4$, a yellow liquid boiling at 21.2°C and to a colorless solid melting at $-11.2$°C. The reaction is reversible by heating the $N_2O_4$.

$$2\,NO_2(g) \xrightarrow{\text{Cooling}} N_2O_4(s)$$
$$N_2O_4(s) \xrightarrow{\text{Heating}} 2\,NO_2(g)$$

This reversible reaction can readily be demonstrated by sealing samples of $NO_2$ in two tubes and placing one tube in warm water and the other in ice water (see Figure 16.1). Heating promotes disorder or randomness in a system, so we would expect more $NO_2$, a gas, to be present at higher temperatures.

reversible reaction

A **reversible reaction** is one in which the products formed in a chemical reaction also react to produce the original reactants. Both the forward reaction and the reverse reaction occur simultaneously. The forward reaction is also called *the reaction to the right*, and the reverse reaction is called *the reaction to the left*. A double arrow is used in the equation to indicate that the reaction is reversible.

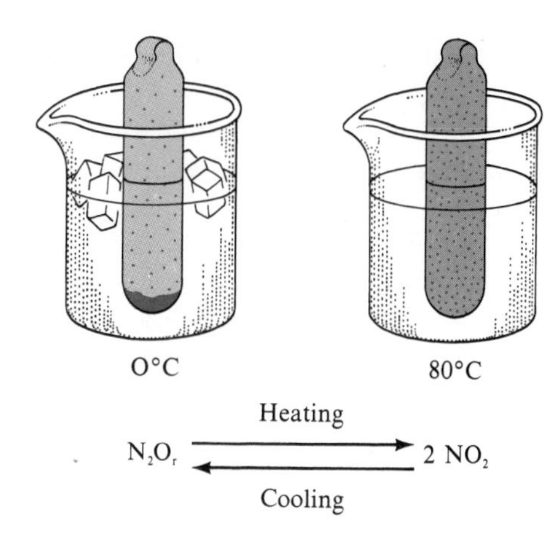


*Figure 16.1.* Reversible reaction of nitrogen dioxide ($NO_2$) and dinitrogen tetroxide ($N_2O_4$). More of the reddish-brown $NO_2$ molecules are visible in the tube that is heated than in the tube that is cooled.

## 16.2 Rates of Reaction

chemical kinetics

Every reaction has a rate, or speed, at which it proceeds. Some are fast and some are extremely slow. The study of reaction rates and reaction mechanisms is known as **chemical kinetics**.

The speed of a reaction is variable and depends on the concentration of the reacting species, the temperature, the presence of catalytic agents, and the nature of the reactants. Consider the hypothetical reaction

$$A + B \longrightarrow C + D \quad \text{(Forward reaction)}$$
$$C + D \longrightarrow A + B \quad \text{(Reverse reaction)}$$

where a collision between A and B is necessary for a reaction to occur. The rate at which A and B react depends on the concentration or the number of A and B molecules present; it will be fastest, for a fixed set of conditions, when they are first mixed. As the reaction proceeds, the number of A and B molecules available for reaction decreases, and the rate of reaction slows down. If the reaction is reversible, the speed of the reverse reaction is zero at first, and gradually increases as the concentrations of C and D increase. As the number of A and B molecules decreases, the forward rate slows down, because A and B cannot find one another as often in order to accomplish a reaction. To counteract this diminishing rate of reaction, an excess of one reagent is often used to keep the reaction from becoming unreasonably slow. Collisions between molecules may be likened to the scooters or "dodge'ems" found at amusement parks. When many cars are on the floor, collisions occur frequently, but if only a few cars are present, collisions can usually be avoided.

## 16.3 Chemical Equilibrium

equilibrium

Any system at **equilibrium** represents a dynamic state in which two or more opposing processes are taking place at the same time and at the same rate. A chemical equilibrium is a dynamic system in which two or more chemical reactions are going on at the same time and at the same rate. When the rate of the forward reaction is exactly equal to the rate of the reverse reaction, a condition of **chemical equilibrium** exists (see Figure 16.2). The concentration of the products is not changing and the system appears to be at a standstill because the products are reacting at the same rate at which they are being formed.

chemical equilibrium

*Chemical Equilibrium:*
*Rate of forward reaction = Rate of reverse reaction*

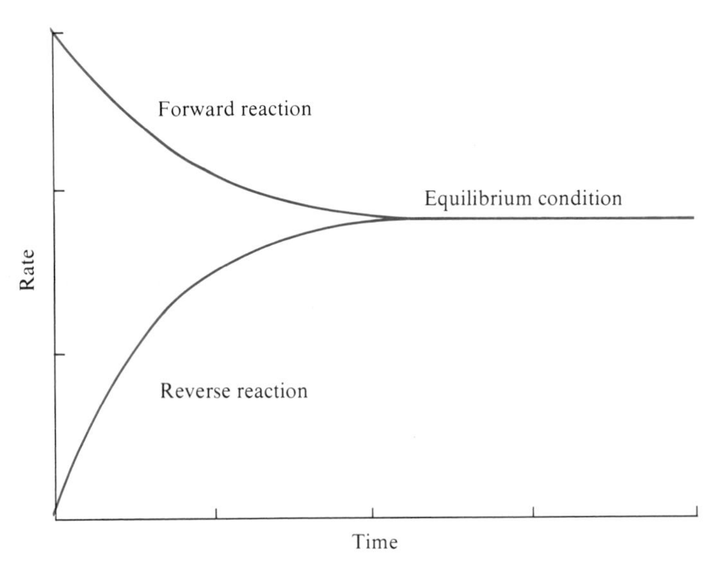


*Figure 16.2.* The graph illustrates that the rates of the forward and reverse reactions become equal at some point in time. The forward reaction rate decreases as a result of decreasing amounts of reactants. The reverse reaction rate starts at zero and increases as the amount of product increases. When the two rates become equal, a state of chemical equilibrium has been reached.

A saturated salt solution is in a condition of equilibrium:

$$NaCl(s) \rightleftharpoons Na^+(aq) + Cl^-(aq)$$

At equilibrium, salt crystals are continuously dissolving, and $Na^+$ and $Cl^-$ ions are continuously crystallizing. Both processes are occurring at the same rate.

The ionization of weak electrolytes represents another common chemical equilibrium system:

$$HC_2H_3O_2 + H_2O \rightleftharpoons H_3O^+ + C_2H_3O_2^-$$

In this reaction, the equilibrium is established in a 1 *M* solution when the forward reaction has gone about 1%—that is, when only 1% of the acetic acid molecules in solution have ionized. Therefore, only a relatively few ions are present, and the acid behaves as a weak electrolyte. In any acid–base equilibrium system, the position of equilibrium favored is toward the weaker conjugate acid and base. In the ionization of acetic acid, the $HC_2H_3O_2$ is a weaker acid than the $H_3O^+$, and $H_2O$ is a weaker base than $C_2H_3O_2^-$.

The reactions represented by

$$H_2 + I_2 \rightleftharpoons 2\,HI$$

provide another good example of a chemical equilibrium. Theoretically, 1.00 mole of hydrogen should react with 1.00 mole of iodine to yield 2.00 moles of hydrogen iodide. Actually, when 1.00 mole of $H_2$ and 1.00 mole of $I_2$ are reacted at 700 K, only 1.58 moles of HI are present when equilibrium is attained. Since 1.58 is 79% of the theoretical yield of 2.00 moles of HI, the forward reaction is only 79% complete at equilibrium. The equilibrium mixture will also contain 0.21 mole each of unreacted $H_2$ and $I_2$ because only 79% reacted (1.00 mole − 0.79 mole = 0.21 mole).

$$\underset{1.00\text{ mole}}{H_2} + \underset{1.00\text{ mole}}{I_2} \xrightarrow{700\text{ K}} \underset{2.00\text{ moles}}{2\,HI}$$

(This would represent the condition if the reaction were 100% complete; 2.00 moles of HI would be formed and no $H_2$ and $I_2$ would be left unreacted.)

$$\underset{0.21\text{ mole}}{H_2} + \underset{0.21\text{ mole}}{I_2} \underset{}{\overset{700\text{ K}}{\rightleftharpoons}} \underset{1.58\text{ moles}}{2\,HI}$$

(This represents the actual equilibrium attained starting with 1.00 mole each of $H_2$ and $I_2$. It shows that the forward reaction is only 79% complete.)

## 16.4 Principle of Le Chatelier

principle of Le Chatelier

In 1888, the French chemist Henri Le Chatelier (1850–1936) set forth a simple far-reaching generalization on the behavior of equilibrium systems. This generalization is known as the **principle of Le Chatelier** and states: If the conditions of a system in equilibrium are altered, then processes occur in the system that tend to counteract the change. In other words, if a stress is applied to a system in equilibrium, the system will behave in such a way as to relieve that stress and restore equilibrium, but under a new set of conditions.

The application of Le Chatelier's principle helps us to predict the effect of changing conditions in chemical reactions. We will examine the effect of changes in concentration, temperature, and pressure.

## 16.5 Effect of Concentration on Reaction Rate and Equilibrium

The way in which the rate of a chemical reaction depends on the concentration of the reactants must be determined experimentally. Many simple, one-step reactions occur as the result of a collision between two molecules or ions. The rate of such one-step reactions can be altered by changing the concentration of the reactants or products. An increase in concentration of the reactants provides more individual reacting species for collisions and results in an increase in the rate of reaction.

An equilibrium is disturbed when the concentration of one or more of its components is changed. As a result, the concentration of all the species will change and a new equilibrium mixture will be established. Consider the hypothetical equilibrium represented by the equation

$$A + B \rightleftharpoons C + D$$

where A and B react in one step to form C and D. When the concentration of B is increased, the following occurs:

1. The rate of the reaction to the right (forward) increases. This rate is proportional to the concentration of A times the concentration of B.
2. Therefore, the rate to the right is greater than the rate to the left.
3. Reactants A and B are used faster than they are produced; C and D are produced faster than they are used.
4. After a period of time, rates to the right and left become equal, and the system is again in equilibrium.
5. In the new equilibrium, the concentration of A is less and the concentrations of C and D are greater than in the original equilibrium.

*Conclusion:* The equilibrium has shifted to the right.

Applying this change in concentration to the equilibrium mixture of 1.00 mole of hydrogen and 1.00 mole of iodine from Section 16.3, we find that when an additional 0.20 mole of $I_2$ is added, the yield of HI is 85% (1.70 moles) instead of 79%. A comparison of the two systems after the new equilibrium mixture is reached follows.

| Original equilibrium | New equilibrium |
|---|---|
| 1.00 mole $H_2$ + 1.00 mole $I_2$ | 1.00 mole $H_2$ + 1.20 moles $I_2$ |
| Yield: 79% HI | Yield: 85% HI (based on $H_2$) |
| Equilibrium mixture contains: | Equilibrium mixture contains: |
| 1.58 moles HI | 1.70 moles HI |
| 0.21 mole $H_2$ | 0.15 mole $H_2$ |
| 0.21 mole $I_2$ | 0.35 mole $I_2$ |

Analyzing this new system, we see that when the 0.20 mole $I_2$ was added, the equilibrium shifted to the right in order to counteract the change of $I_2$

concentration. Some of the $H_2$ reacted with added $I_2$ and produced more HI, until an equilibrium mixture was established again. When $I_2$ was added, the concentration of $I_2$ increased, the concentration of $H_2$ decreased, and the concentration of HI increased. What do you think would be the effects of adding (a) more $H_2$ and (b) more HI?

The equation

$$Fe^{3+}(aq) + SCN^-(aq) \rightleftharpoons Fe(SCN)^{2+}(aq)$$

Pale yellow    Colorless    Red

represents an equilibrium that is used in certain analytical procedures as an indicator because of the readily visible, intense red color of the complex $Fe(SCN)^{2+}$ ion. A very dilute solution of iron(III), $Fe^{3+}$, and thiocyanate, $SCN^-$, is light red in color. When the concentration of $Fe^{3+}$ or $SCN^-$ is increased, the equilibrium shift to the right is observed by an increase in the intensity of the color, resulting from the formation of additional $Fe(SCN)^{2+}$.

If either $Fe^{3+}$ or $SCN^-$ is removed from solution, the equilibrium will shift to the left, and the solution will become lighter in color. When $Ag^+$ is added to the above solution, a white precipitate of silver thiocyanate (AgSCN) is formed, thus removing $SCN^-$ ion from the equilibrium:

$$Ag^+(aq) + SCN^-(aq) \longrightarrow AgSCN\downarrow$$

The system accordingly responds to counteract the change in $SCN^-$ concentration by shifting the equilibrium to the left. This shift is evident by a decrease in the intensity of the red color due to a decreased concentration of $Fe(SCN)^{2+}$.

Let us now consider the effect of changing the concentrations in the equilibrium mixture of chlorine water. The equilibrium equation is

$$Cl_2(aq) + 2\,H_2O \rightleftharpoons HOCl(aq) + H_3O^+ + Cl^-(aq)$$

The variation in concentrations and the equilibrium shifts are tabulated below. An X in the second or third column indicates the reagent that is increased or decreased. The fourth column indicates the direction of the equilibrium shift.

| | Concentration | | |
|---|---|---|---|
| Reagent | Increase | Decrease | Equilibrium shift |
| $Cl_2$ | — | X | Left |
| $H_2O$ | X | — | Right |
| HOCl | X | — | Left |
| $H_3O^+$ | — | X | Right |
| $Cl^-$ | X | — | Left |

Consider the equilibrium in a 0.1 *M* acetic acid solution:

$$HC_2H_3O_2 + H_2O \rightleftharpoons H_3O^+ + C_2H_3O_2^-$$

In this solution, the concentration of the hydronium ion ($H_3O^+$), which is a measure of the acidity, is $1.34 \times 10^{-3}$ mole/litre, corresponding to a pH of 2.87.

What will happen to the acidity when 0.1 mole of sodium acetate ($NaC_2H_3O_2$) is added to 1 litre of 0.1 *M* $HC_2H_3O_2$? When $NaC_2H_3O_2$ dissolves, it dissociates into sodium ($Na^+$) and acetate ($C_2H_3O_2^-$) ions. The acetate ion from the salt is a common ion to the acetic acid equilibrium system and increases the total acetate ion concentration in the solution. As a result, the equilibrium shifts to the left, decreasing the hydronium ion concentration and lowering the acidity of the solution. Evidence of this decrease in acidity is shown by the fact that the pH of a solution that is 0.1 *M* in $HC_2H_3O_2$ and 0.1 *M* in $NaC_2H_3O_2$ is 4.74. The pH of several different solutions of $HC_2H_3O_2$ and $NaC_2H_3O_2$ is shown below. Each time the acetate ion is increased, the pH increases, showing a further shift in the equilibrium toward un-ionized acetic acid.

| Solution | pH |
|---|---|
| 1 litre 0.1 *M* $HC_2H_3O_2$ | 2.87 |
| 1 litre 0.1 *M* $HC_2H_3O_2$ + 0.1 mole $NaC_2H_3O_2$ | 4.74 |
| 1 litre 0.1 *M* $HC_2H_3O_2$ + 0.2 mole $NaC_2H_3O_2$ | 5.05 |
| 1 litre 0.1 *M* $HC_2H_3O_2$ + 0.3 mole $NaC_2H_3O_2$ | 5.23 |

A secondary reaction of the acetate ion, which also aids in reducing the acidity, is its reaction with water, forming un-ionized $HC_2H_3O_2$ and a hydroxide ($OH^-$) ion:

$$C_2H_3O_2^- + H_2O \rightleftharpoons HC_2H_3O_2 + OH^-$$

The $OH^-$ ion produced reacts with $H_3O^+$ to decrease the acidity. Proof that sodium acetate produces $OH^-$ ions in solution is shown by the fact that a 0.1 *M* $NaC_2H_3O_2$ solution is alkaline, having a pH of 8.87.

In summary, we can say that when the concentration of a reagent on the left side of an equation is increased, the equilibrium shifts to the right. When the concentration of a reagent on the right side of an equation is increased, the equilibrium shifts to the left. In accordance with Le Chatelier's principle, the equilibrium always shifts in the direction that tends to reduce the concentration of the added reactant.

## 16.6 Effect of Pressure on Reaction Rate and Equilibrium

Changes in pressure significantly affect the reaction rate only when one or more of the reactants or products is a gas. In these cases the effect of increasing the pressure of the reacting gases is equivalent to increasing their concentrations. In the reaction

$$CaCO_3(s) \overset{\Delta}{\rightleftharpoons} CaO(s) + CO_2(g)$$

calcium carbonate decomposes into calcium oxide and carbon dioxide when heated above 825°C. Increasing the pressure of the equilibrium system by adding $CO_2$ or by decreasing the volume speeds up the reverse reaction and causes the equilibrium to shift to the left. The increased pressure gives the same effect as that caused by increasing the concentration of $CO_2$, the only gaseous substance in the reaction.

We have seen that when the pressure on a gas is increased, its volume is decreased. In a system composed entirely of gases, an increase of the pressure will cause the reaction and the equilibrium to shift to the side that contains the smaller volume or smaller number of molecules. This is because the increase in pressure is partially relieved by the system's shifting its equilibrium toward the side in which the substances occupy the smaller volume.

Prior to World War I, Fritz Haber in Germany invented the first major process for the fixation of nitrogen. In the Haber process nitrogen and hydrogen are reacted together in the presence of a catalyst at moderately high temperature and pressure to produce ammonia. The catalyst consists of iron and iron oxide with small amounts of potassium and aluminum oxides. For this process, Haber received the Nobel Prize in chemistry in 1918.

$$N_2(g) + 3H_2(g) \rightleftharpoons 2NH_3(g) + 22.1\text{ kcal}$$

| $N_2(g)$ | $3H_2(g)$ | $2NH_3(g)$ |
|---|---|---|
| 1 mole | 3 moles | 2 moles |
| 1 volume | 3 volumes | 2 volumes |

The left side of the equation in the Haber process represents four volumes of gas combining to give two volumes of gas on the right side of the equation. An increase in the total pressure on the system shifts the equilibrium to the right. This increase in pressure results in a higher concentration of both reactants and products. Since there are fewer moles of $NH_3$ than there are moles of $N_2$ and $H_2$, the equilibrium shifts to the right when the pressure is increased.

Ideal conditions for the Haber process are 200°C and 1000 atm pressure. However, at 200°C the rate of reaction is very slow, and at 1000 atm extraordinarily heavy equipment is required. As a compromise the reaction is run at 400–600°C and 200–350 atm pressure, which gives a reasonable yield at a reasonable rate. The effect of pressure on the yield of ammonia at one particular temperature is shown in Table 16.1.

When the total number of gaseous molecules on both sides of an equation is the same, a change in pressure does not cause an equilibrium shift. The

*Table 16.1.* The effect of pressure in the conversion of $H_2$ and $N_2$ to $NH_3$ at 450°C. The starting ratio of $H_2$ to $N_2$ is 3 moles to 1 mole.

| Pressure (atm) | Yield of $NH_3$ (%) |
|---|---|
| 10 | 2.04 |
| 30 | 5.80 |
| 50 | 9.17 |
| 100 | 16.4 |
| 300 | 35.5 |
| 600 | 53.4 |
| 1000 | 69.4 |

following reaction is an example:

$$N_2(g) \quad + \quad O_2(g) \rightleftharpoons \quad 2\,NO(g)$$

| $N_2(g)$ | $O_2(g)$ | $2\,NO(g)$ |
|---|---|---|
| 1 mole | 1 mole | 2 moles |
| 1 volume | 1 volume | 2 volumes |
| $6.02 \times 10^{23}$ molecules | $6.02 \times 10^{23}$ molecules | $2 \times 6.02 \times 10^{23}$ molecules |

When the pressure on this system is increased, the rate of both the forward and the reverse reactions will increase because of the higher concentrations of $N_2$, $O_2$, and NO. But the equilibrium will not shift, because the increase in concentration of molecules is the same on both sides of the equation and the decrease in volume is the same on both sides of the equation.

## 16.7 Effect of Temperature on Reaction Rate and Equilibrium

An increase in temperature is generally accompanied by an increased rate of reaction. Molecules at elevated temperatures are more energetic and have more kinetic energy; thus, their collisions are more likely to result in a reaction. However, we cannot assume that the rate of a desired reaction will keep increasing indefinitely as the temperature is raised. High temperatures may cause the destruction or decomposition of the reactants and products or may initiate reactions other than the one desired. For example, when calcium oxalate ($CaC_2O_4$) is heated to 500°C, it decomposes into calcium carbonate and carbon monoxide:

$$CaC_2O_4(s) \xrightarrow{500°C} CaCO_3(s) + CO\uparrow$$

If calcium oxalate is heated to 850°C, the products are calcium oxide, carbon monoxide, and carbon dioxide:

$$CaC_2O_4(s) \xrightarrow{850°C} CaO(s) + CO\uparrow + CO_2\uparrow$$

When heat is applied to a system in equilibrium, the reaction that absorbs heat is favored. When the process, as written, is endothermic, the forward reaction is increased. When the reaction is exothermic, the reverse reaction is favored. In this sense heat may be treated as a reactant in endothermic reactions or as a product in exothermic reactions. Therefore, temperature is analogous to concentration when applying Le Chatelier's principle to heat effects on a chemical reaction.

Hot coke (C) is a very good reducing agent. In the reaction

$$C(s) + CO_2(g) + \text{Heat} \rightleftharpoons 2\,CO\uparrow$$

very little, if any, CO is formed at room temperature. At 1000°C, the equilibrium mixture contains about an equal number of moles of CO and $CO_2$. At higher temperatures, the equilibrium shifts to the right, increasing the yield of CO. The reaction is endothermic and, as can be seen, the equilibrium is shifted to the right at higher temperatures.

Phosphorus trichloride reacts with dry chlorine gas to form phosphorus pentachloride. The reaction is exothermic:

$$PCl_3(l) + Cl_2(g) \rightleftharpoons PCl_5(s) + \text{Heat}$$

Heat must continually be removed during the reaction to obtain a good yield of the product. According to the principle of Le Chatelier, heat will cause the product, $PCl_5$, to decompose, reforming $PCl_3$ and $Cl_2$. The equilibrium mixture at 200°C contains 52% $PCl_5$ and at 300°C the mixture contains 3% $PCl_5$, verifying that heat causes the equilibrium mixture to shift to the left.

When the temperature of a system is raised, the rate of reaction increases because of increased kinetic energy and more frequent collisions of the reacting species. In a reversible reaction, the rate of both the forward and reverse reactions is increased by an increase in temperature; however, the reaction that absorbs heat increases to a greater extent, and the equilibrium shifts to favor that reaction. The following examples illustrate these effects:

$$4\,HCl + O_2 \rightleftharpoons 2\,H_2O + 2\,Cl_2 + 28.4\text{ kcal} \quad (1)$$

$$H_2 + Cl_2 \rightleftharpoons 2\,HCl + 44.2\text{ kcal} \quad (2)$$

$$CH_4 + 2\,O_2 \rightleftharpoons CO_2 + 2\,H_2O + 212.8\text{ kcal} \quad (3)$$

$$N_2O_4 + 14\text{ kcal} \rightleftharpoons 2\,NO_2 \quad (4)$$

$$2\,CO_2 + 135.2\text{ kcal} \rightleftharpoons 2\,CO + O_2 \quad (5)$$

$$H_2 + I_2 + 12.4\text{ kcal} \rightleftharpoons 2\,HI \quad (6)$$

Reactions (1), (2), and (3) are exothermic; an increase in temperature will cause the equilibrium to shift to the left. Reactions (4), (5), and (6) are endothermic; an increase in temperature will cause the equilibrium to shift to the right.

## 16.8 Effect of Catalysts on Reaction Rate and Equilibrium

catalyst

A **catalyst** is a substance that influences the speed of a reaction and that may be recovered essentially unchanged at the end of the reaction. A catalyst does not shift the equilibrium of a reaction; it affects only the speed at which the equilibrium is reached. If a catalyst does not affect the equilibrium, then it follows that it must affect the speed of both the forward and the reverse reactions equally.

The reaction between phosphorus trichloride ($PCl_3$) and sulfur is highly exothermic, but it is so slow that very little product, thiophosphoryl chloride ($PSCl_3$), is obtained, even after prolonged heating. When a catalyst, such as aluminum chloride ($AlCl_3$), is added, the reaction is complete in a few seconds:

$$PCl_3(l) + S(s) \xrightarrow{AlCl_3} PSCl_3(l)$$

We have already demonstrated that manganese dioxide used as a catalyst increases the rates of decomposition of potassium chlorate and hydrogen peroxide.

Catalysts are extremely important to industrial chemistry. Hundreds of chemical reactions that are otherwise too slow to be of practical value have been put to commercial use once a suitable catalyst was found. But it is in the area of biochemistry that catalysts are of supreme importance. Nearly all the chemical reactions associated with all forms of life are completely dependent on biochemical catalysts known as *enzymes*.

## 16.9 Equilibrium Constants

Law of Mass Action

The **Law of Mass Action** states that the rate of a chemical reaction is proportional to the concentration of the reacting species. This simply means that the higher the concentration of reactants, the more frequently they collide and form products. For the equilibrium system in which A and B react in one step to give C and D,

$$A + B \rightleftharpoons C + D$$

the rates of the forward and reverse reactions can be expressed as

$$\text{Rate}_f = k_f \times \text{Concentration of A} \times \text{Concentration of B}$$

$$\text{Rate}_r = k_r \times \text{Concentration of C} \times \text{Concentration of D}$$

where $\text{Rate}_f$ is the rate of the forward reaction, $\text{Rate}_r$ is the rate of the reverse reaction, and $k_f$ and $k_r$ are the proportionality rate constants. For dilute solutions, the unit of concentration is moles per litre; for gases, it is either moles per litre or pressure. To simplify the rate expressions, we place the formula of each substance in brackets to indicate the concentration of each substance. The concentrations of A, B, C, and D are then expressed as [A], [B], [C], and [D], respectively.

At equilibrium,

$$\text{Rate}_f = \text{Rate}_r$$

Then

$$\text{Rate}_f = k_f[A][B]$$

$$\text{Rate}_r = k_r[C][D]$$

$$\frac{k_f}{k_r} = \frac{[C][D]}{[A][B]}$$

equilibrium constant

Since both $k_f$ and $k_r$ are constants, $k_f/k_r$ is also a constant and is known as the **equilibrium constant**, abbreviated $K_{eq}$:

$$K_{eq} = \frac{[C][D]}{[A][B]}$$

This expression reads as follows: The equilibrium constant, $K_{eq}$, is equal to the product of the concentration of C and the concentration of D divided by the product of the concentration of A and the concentration of B. Consider the equilibrium

$$2A \rightleftarrows C + D$$

The forward reaction is dependent on the collision of two A molecules. Therefore,

$$\text{Rate}_f = k_f[A][A] = k_f[A]^2$$

$$\text{Rate}_r = k_r[C][D]$$

$$K_{eq} = \frac{k_f}{k_r} = \frac{[C][D]}{[A]^2}$$

In this equilibrium, the equilibrium constant is equal to the product of the concentrations of C times D divided by the concentration of A squared.

For the general reaction

$$nA + mB \rightleftarrows pC + qD$$

where $n$, $m$, $p$, and $q$ are the small whole numbers in the balanced equation, the equilibrium constant expression is

$$K_{eq} = \frac{[C]^p[D]^q}{[A]^n[B]^m}$$

Observe that the concentration of each substance is raised to a power that is the same as its numerical coefficient in the balanced equation. It is conventional that the concentrations of the substances in the numerator are those of the products (the substances on the right side of the equation as written); the concentrations of the reactants are in the denominator.

## 16.10 *Ionization Constants*

As a first application of an equilibrium constant, let us consider the constant for acetic acid in solution. Because it is a weak acid, an equilibrium is established between molecular $HC_2H_3O_2$ and its ions in solution. The constant is called the **ionization constant**, $K_i$, a special type of equilibrium constant. The concentration of water in the solution does not change appreciably, so we may use the following simplified equation to set up the constant:

ionization constant

$$HC_2H_3O_2 \rightleftarrows H^+ + C_2H_3O_2^-$$

The ionization constant expression is

$$K_i = \frac{[H^+][C_2H_3O_2^-]}{[HC_2H_3O_2]}$$

It states that the ionization constant, $K_i$, is equal to the product of the hydrogen ion ($H^+$) concentration times the acetate ion ($C_2H_3O_2^-$) concentration divided by the concentration of the un-ionized acetic acid ($HC_2H_3O_2$).

At 25°C, a 0.1 *M* $HC_2H_3O_2$ solution is 1.34% ionized and has a hydrogen ion concentration of $1.34 \times 10^{-3}$ mole/litre. From this information, we can calculate the ionization constant for acetic acid.

A 0.1 *M* solution contains 0.1 mole of acetic acid per litre. Of this 0.10 mole, only 1.34%, or $1.34 \times 10^{-3}$ mole, is ionized. This gives an $H^+$ ion concentration of $1.34 \times 10^{-3}$ mole/litre. Since each molecule of acid that ionizes yields one $H^+$ and one $C_2H_3O_2^-$, the concentration of $C_2H_3O_2^-$ ions is also $1.34 \times 10^{-3}$ mole/litre. This ionization leaves 0.10 – 0.00134 = 0.09866 mole/litre of un-ionized acetic acid.

| | Initial concentration | Equilibrium concentration |
|---|---|---|
| $[HC_2H_3O_2]$ | 0.10 mole/litre | 0.09866 mole/litre |
| $[H^+]$ | 0 | 0.00134 mole/litre |
| $[C_2H_3O_2^-]$ | 0 | 0.00134 mole/litre |

Substituting these concentrations in the equilibrium expression, we obtain the value for $K_i$:

$$K_i = \frac{[H^+][C_2H_3O_2^-]}{[HC_2H_3O_2]}$$

$$= \frac{[1.34 \times 10^{-3}][1.34 \times 10^{-3}]}{[0.09866]} = 1.8 \times 10^{-5}$$

The low magnitude of this constant indicates that the position of the equilibrium is far toward the un-ionized acetic acid. In fact, a 0.1 *M* acetic acid solution is 98.66% un-ionized.

Once the value for this constant is established, it can be used to describe any system containing $H^+$, $C_2H_3O_2^-$, and $HC_2H_3O_2$ in equilibrium at 25°C. The ionization constants for several other weak acids are listed in Table 16.2.

*Table 16.2.* Ionization constants ($K_i$) of weak acids at 25°C.

| Acid | Formula | $K_i$ |
|---|---|---|
| Acetic | $HC_2H_3O_2$ | $1.8 \times 10^{-5}$ |
| Benzoic | $HC_7H_5O_2$ | $6.3 \times 10^{-5}$ |
| Carbolic (phenol) | $HC_6H_5O$ | $1.3 \times 10^{-10}$ |
| Cyanic | $HCNO$ | $2.0 \times 10^{-4}$ |
| Formic | $HCHO_2$ | $1.8 \times 10^{-4}$ |
| Hydrocyanic | $HCN$ | $4.0 \times 10^{-10}$ |
| Hypochlorous | $HClO$ | $3.5 \times 10^{-8}$ |
| Nitrous | $HNO_2$ | $4.5 \times 10^{-4}$ |

*Problem 16.1* What is the ionization constant expression for nitrous acid?

First, write the simplified ionization equation:

$$HNO_2(aq) \rightleftharpoons H^+ + NO_2^- \quad \text{(Simplified)}$$

The format of the ionization constant expression, $K_i$, is the product of the concentrations of the substances on the right side divided by the product of the concentrations of the substances on the left side of the equation. Thus,

$$K_i = \frac{[H^+][NO_2^-]}{[HNO_2]} \quad \text{(Answer)}$$

---

*Problem 16.2* What is the $H^+$ ion concentration in a 0.50 $M$ $HC_2H_3O_2$ solution? The ionization constant, $K_i$, for $HC_2H_3O_2$ is $1.8 \times 10^{-5}$.

To solve this problem, first write the equilibrium equation and the $K_i$ expression:

$$HC_2H_3O_2 \rightleftharpoons H^+ + C_2H_3O_2^-$$

$$K_i = \frac{[H^+][C_2H_3O_2^-]}{[HC_2H_3O_2]} = 1.8 \times 10^{-5}$$

Let $[H^+] = Y$. Then $[C_2H_3O_2^-]$ will also equal $Y$, because one acetate ion is produced for each hydrogen ion. The $[HC_2H_3O_2]$ will then be $0.50 - Y$, the starting concentration minus the amount that ionized.

$$[H^+] = [C_2H_3O_2^-] = Y \qquad [HC_2H_3O_2] = 0.50 - Y$$

Substituting these values into the $K_i$ expression, we obtain

$$K_i = \frac{(Y)(Y)}{0.50 - Y} = \frac{Y^2}{0.50 - Y} = 1.8 \times 10^{-5}$$

An exact solution of this equation for $Y$ requires the use of the quadratic formula. However, an approximate solution is readily obtained if we first assume that $Y$ is small compared to 0.50. Then $0.50 - Y$ will be equal to approximately 0.50. The equation now becomes

$$\frac{Y^2}{0.50} = 1.8 \times 10^{-5}$$

$$Y^2 = 1.8 \times 10^{-5} \times 0.50 = 0.90 \times 10^{-5} = 9.0 \times 10^{-6}$$

$$Y = \sqrt{9.0 \times 10^{-6}} = 3.0 \times 10^{-3}$$

Thus, the $[H^+]$ is approximately $3.0 \times 10^{-3}$ mole/litre in a 0.50 $M$ $HC_2H_3O_2$ solution. The exact solution to this problem, using a quadratic equation, gives a value of $2.99 \times 10^{-3}$ for $[H^+]$, showing that we were justified in neglecting $Y$ compared to 0.50.

---

## 16.11 Ion Product Constant for Water

We have seen that water ionizes to a slight degree. This ionization is represented by these equilibrium equations:

$$H_2O + H_2O \quad H_3O^+ + OH^- \tag{1}$$

$$H_2O \rightleftharpoons H^+ + OH^- \tag{2}$$

Equation (1) is the more accurate representation of the equilibrium since free protons ($H^+$) do not exist in water. Equation (2) is a simplified and widely used representation of the water equilibrium. The actual concentration of $H^+$ produced in pure water is very minute and amounts to only $1 \times 10^{-7}$ mole per litre at 25°C. In pure water,

$$[H^+] = [OH^-] = 1 \times 10^{-7} \text{ mole/litre}$$

since both ions are produced in equal molar amounts, as shown in equation (2).

$K_w$

The $H_2O \rightleftarrows H^+ + OH^-$ equilibrium exists in water and in all water solutions. A special equilibrium constant called the *ion product constant for water*, $K_w$, applies to this equilibrium. The constant $K_w$ is defined as the product of the $H^+$ ion concentration and the $OH^-$ ion concentration, each in moles per litre:

$$K_w = [H^+][OH^-]$$

The numerical value of $K_w$ is $1 \times 10^{-14}$, since for pure water at 25°C

$$K_w = [H^+][OH^-] = [1 \times 10^{-7}][1 \times 10^{-7}] = 1 \times 10^{-14}$$

The value of $K_w$ for all water solutions at 25°C is the constant $1 \times 10^{-14}$. It is important to realize that as the concentration of one of these ions, $H^+$ or $OH^-$, increases, the other decreases. However, the product of the $H^+$ and $OH^-$ concentrations always equals the constant $1 \times 10^{-14}$. This relationship can be seen in the examples shown in Table 16.3. If the concentration of one ion is known, the concentration of the other can be calculated by use of the $K_w$ expression.

*Table 16.3.* Relationship of $H^+$ and $OH^-$ concentrations in water solutions.

| $[H^+]$ | $[OH^-]$ | $K_w$ | pH | pOH |
|---|---|---|---|---|
| $1 \times 10^{-2}$ | $1 \times 10^{-12}$ | $1 \times 10^{-14}$ | 2 | 12 |
| $1 \times 10^{-4}$ | $1 \times 10^{-10}$ | $1 \times 10^{-14}$ | 4 | 10 |
| $2 \times 10^{-6}$ | $5 \times 10^{-9}$ | $1 \times 10^{-14}$ | 5.7 | 8.3 |
| $1 \times 10^{-7}$ | $1 \times 10^{-7}$ | $1 \times 10^{-14}$ | 7 | 7 |
| $1 \times 10^{-9}$ | $1 \times 10^{-5}$ | $1 \times 10^{-14}$ | 9 | 5 |

*Problem 16.3*

What is the concentration of (a) $H^+$ and (b) $OH^-$ in a 0.001 *M* HCl solution? Assume that the HCl is 100% ionized.

(a) Since all the HCl is ionized, $H^+$ = 0.001 mole/litre.

$$\underset{0.001\,M}{HCl} \longrightarrow \underset{0.001\,M}{H^+} + \underset{0.001\,M}{Cl^-}$$

$$[H^+] = 1 \times 10^{-3} \quad \text{(Answer)}$$

(b) To calculate the $[OH^-]$ in this solution, solve the $K_w$ expression for $OH^-$ and substitute in the values for $K_w$ and $[H^+]$.

$$K_w = [H^+][OH^-]$$

$$[OH^-] = \frac{K_w}{[H^+]} = \frac{1 \times 10^{-14}}{1 \times 10^{-3}} = 1 \times 10^{-11} \text{ mole/litre} \quad \text{(Answer)}$$

*Problem 16.4*

What is the pH of a 0.01 $M$ NaOH solution? Assume that the NaOH is 100% ionized.

Since all the NaOH is ionized, $OH^- = 0.01$ mole/litre.

$$\underset{0.01\,M}{NaOH} \longrightarrow \underset{0.01\,M}{Na^+} + \underset{0.01\,M}{OH^-}$$

In order to find the pH of the solution, we must first calculate the $H^+$ concentration. This is done by using the $K_w$ expression. Solve for $[H^+]$ and substitute the values for $K_w$ and $[OH^-]$.

$$K_w = [H^+][OH^-]$$

$$[H^+] = \frac{K_w}{[OH^-]} = \frac{1 \times 10^{-14}}{1 \times 10^{-2}} = 1 \times 10^{-12} \text{ mole/litre}$$

$$pH = \log \frac{1}{[H^+]} = \log \frac{1}{10^{-12}} = \log 10^{12} = 12 \quad \text{(Answer)}$$

The pH can also be calculated by the method shown in Section 15.9:

$$[H^+] = 1 \times 10^{-12}$$

$$pH = 12 - \log 1 = 12 - 0 = 12 \quad \text{(Answer)}$$

---

Just as pH is used to express the acidity of a solution, pOH is used to express the basicity of an aqueous solution. The pOH is related to the $OH^-$ ion concentration in the same way that the pH is related to the $H^+$ ion concentration:

$$pOH = \log \frac{1}{[OH^-]} \quad \text{or} \quad -\log[OH^-]$$

Thus, a solution in which $[OH^-] = 1 \times 10^{-2}$, as in Problem 16.4, will have $pOH = 2$.

In pure water, where $[H^+] = 1 \times 10^{-7}$ and $[OH^-] = 1 \times 10^{-7}$, the pH is 7 and the pOH is 7. The sum of the pH and pOH is 14.

$$pH + pOH = 14$$

This relationship holds in all aqueous solutions and is illustrated in the examples in Table 16.3.

## 16.12 Solubility Product Constant

solubility product constant

The **solubility product constant**, abbreviated $K_{sp}$, is another application of the equilibrium constant. It is derived from the equilibrium between a slightly soluble substance and its ions in solution. The following example illustrates how this constant is evaluated.

The solubility of silver chloride (AgCl) in water is $1.3 \times 10^{-5}$ mole per litre at 25°C. The equation for the equilibrium between AgCl and its ions in solution is

$$AgCl(s) \rightleftharpoons Ag^+ + Cl^-$$

The equilibrium constant expression is

$$K_{eq} = \frac{[Ag^+][Cl^-]}{[AgCl(s)]}$$

The amount of solid AgCl does not affect the equilibrium system provided that some is present. In other words, the concentration of solid silver chloride is constant, whether 1 mg or 10 g of the salt is present. Therefore, the product obtained by multiplying the two constants $K_{eq}$ and $[AgCl(s)]$ is also a constant. This constant is called the solubility product constant, $K_{sp}$.

$$K_{eq} \times [AgCl(s)] = [Ag^+][Cl^-] = K_{sp}$$
$$K_{sp} = [Ag^+][Cl^-]$$

The $K_{sp}$ is equal to the product of the $Ag^+$ ion times the $Cl^-$ ion concentrations, each in moles/litre. When $1.3 \times 10^{-5}$ mole/litre of AgCl dissolves, it produces $1.3 \times 10^{-5}$ mole/litre each of $Ag^+$ and $Cl^-$. From these concentrations, the $K_{sp}$ can be evaluated.

$$[Ag^+] = 1.3 \times 10^{-5} \text{ mole/litre} \qquad [Cl^-] = 1.3 \times 10^{-5} \text{ mole/litre}$$
$$K_{sp} = [Ag^+][Cl^-] = [1.3 \times 10^{-5}][1.3 \times 10^{-5}] = 1.7 \times 10^{-10}$$

Once this $K_{sp}$ value is established, it can be used to describe other systems containing $Ag^+$ and $Cl^-$. For example, if silver nitrate, $AgNO_3$, is added to a saturated AgCl solution until the $Ag^+$ concentration is 0.1 *M*, what will be the $Cl^-$ ion concentration remaining in solution? The addition of $AgNO_3$ puts $Ag^+$ ions into solution and causes the AgCl equilibrium to shift to the left, reducing the $Cl^-$ ion concentration. This process of increasing the concentration of one of the ions in an equilibrium, thereby causing the other ion to decrease in concentration, is known as the **common ion effect**.

common ion effect

We use the $K_{sp}$ to calculate the $Cl^-$ ion concentration remaining in solution. The $K_{sp}$ is constant at a particular temperature and remains the same no matter how we change the concentration of the species involved.

$$K_{sp} = [Ag^+][Cl^-] = 1.7 \times 10^{-10} \qquad [Ag^+] = 0.1 \text{ mole/litre}$$

We then substitute the concentration of $Ag^+$ ion into the $K_{sp}$ expression and calculate:

$$[0.1][Cl^-] = 1.7 \times 10^{-10}$$
$$[Cl^-] = \frac{1.7 \times 10^{-10}}{0.1}$$
$$[Cl^-] = 1.7 \times 10^{-9} \text{ mole/litre}$$

This calculation shows a 10,000-fold reduction of $Cl^-$ ions in solution. It illustrates that $Cl^-$ ions may be quantitatively removed from solution with an excess of $Ag^+$ ions. The equilibrium equations and the $K_{sp}$ expressions for several

*Table 16.4.* Solubility product constants ($K_{sp}$) at 25°C.

| Compound | $K_{sp}$ |
|---|---|
| AgCl | $1.7 \times 10^{-10}$ |
| AgBr | $5 \times 10^{-13}$ |
| AgI | $8.5 \times 10^{-17}$ |
| $AgC_2H_3O_2$ | $2 \times 10^{-3}$ |
| $Ag_2CrO_4$ | $1.9 \times 10^{-12}$ |
| $BaCrO_4$ | $8.5 \times 10^{-11}$ |
| $BaSO_4$ | $1.5 \times 10^{-9}$ |
| $CaF_2$ | $3.9 \times 10^{-11}$ |
| CuS | $9 \times 10^{-45}$ |
| $Fe(OH)_3$ | $6 \times 10^{-38}$ |
| PbS | $7 \times 10^{-29}$ |
| $PbSO_4$ | $1.3 \times 10^{-8}$ |
| $Mn(OH)_2$ | $2.0 \times 10^{-13}$ |

other substances are given below. Table 16.4 lists $K_{sp}$ values for these and several other substances.

$$AgBr(s) \rightleftharpoons Ag^+ + Br^- \qquad K_{sp} = [Ag^+][Br^-]$$
$$BaSO_4(s) \rightleftharpoons Ba^{2+} + SO_4^{2-} \qquad K_{sp} = [Ba^{2+}][SO_4^{2-}]$$
$$Ag_2CrO_4(s) \rightleftharpoons 2\,Ag^+ + CrO_4^{2-} \qquad K_{sp} = [Ag^+]^2[CrO_4^{2-}]$$
$$CuS(s) \rightleftharpoons Cu^{2+} + S^{2-} \qquad K_{sp} = [Cu^{2+}][S^{2-}]$$
$$Mn(OH)_2(s) \rightleftharpoons Mn^{2+} + 2\,OH^- \qquad K_{sp} = [Mn^{2+}][OH^-]^2$$
$$Fe(OH)_3(s) \rightleftharpoons Fe^{3+} + 3\,OH^- \qquad K_{sp} = [Fe^{3+}][OH^-]^3$$

Note that the concentration of each substance in the $K_{sp}$ expressions is raised to a power that is the same number as its numerical coefficient in the balanced equilibrium equation, for example, $[Ag^+]^2$ in $Ag_2CrO_4$. When the product of the molar concentration of the ions in solution, each raised to its proper power, is greater than the $K_{sp}$ for that substance, precipitation should occur. If the ion product is less than the $K_{sp}$ value, no precipitation will occur.

*Problem 16.5*

Write $K_{sp}$ expressions for AgI and $PbI_2$, both of which are slightly soluble salts.

First write the equilibrium equations:

$$AgI(s) \rightleftharpoons Ag^+ + I^-$$
$$PbI_2(s) \rightleftharpoons Pb^{2+} + 2\,I^-$$

Since the concentration of the solid crystals is constant, the $K_{sp}$ equals the product of the molar concentrations of the ions in solution. In the case of $PbI_2$, the $[I^-]$ must be squared.

$$K_{sp} = [Ag^+][I^-]$$
$$K_{sp} = [Pb^{2+}][I^-]^2$$

*Problem 16.6*

The $K_{sp}$ value for lead sulfate ($PbSO_4$, g-mol. wt = 305.9 g/mole) is $1.3 \times 10^{-8}$. Calculate the solubility of $PbSO_4$ in grams per litre.

First find the solubility in moles per litre from the $K_{sp}$ value and then convert moles per litre to grams per litre. Let $S$ equal moles per litre of $PbSO_4$ in a saturated solution. Since the salt is completely dissociated,

$$PbSO_4(s) \rightleftharpoons Pb^{2+} + SO_4^{2-}$$
$$[Pb^{2+}] = S \qquad [SO_4^{2-}] = S$$
$$K_{sp} = [Pb^{2+}][SO_4^{2-}] = 1.3 \times 10^{-8}$$
$$[S][S] = 1.3 \times 10^{-8}$$
$$[S]^2 = 1.3 \times 10^{-8}$$
$$S = \sqrt{1.3 \times 10^{-8}} = 1.14 \times 10^{-4} \text{ mole/litre}$$

Therefore, the solubility of $PbSO_4$ is $1.14 \times 10^{-4}$ mole/litre. The weight in grams that dissolves is

$$g = \text{moles} \times \text{g-mol. wt}$$

$$g = \frac{1.14 \times 10^{-4} \text{ mole}}{\text{litre}} \times \frac{305.9 \text{ g}}{\text{mole}} = 3.5 \times 10^{-2} \text{ g/litre}$$

The solubility of $PbSO_4$ is 0.035 g/litre.

## 16.13 Buffer Solutions

The control of pH within narrow limits is critically important in many chemical applications and vitally important in many biological systems. For example, human blood must be maintained between a pH of 7.35 and 7.45 for the efficient transport of oxygen from the lungs to the cells. This narrow pH range is maintained by buffer systems within the blood.

buffer solution

A solution that resists changes of pH when diluted or when an acid or base is added is called a **buffer solution**. Two common types of buffer solutions are (1) a weak acid together with a salt of that weak acid and (2) a weak base together with a salt of that weak base.

The action of a buffer system can be understood by considering a solution of acetic acid and sodium acetate. The weak acid, $HC_2H_3O_2$, is mostly un-ionized and is in equilibrium with its ions in solution. The salt, $NaC_2H_3O_2$, is completely ionized.

$$HC_2H_3O_2(aq) \rightleftharpoons H^+(aq) + C_2H_3O_2^-(aq)$$
$$NaC_2H_3O_2 \longrightarrow Na^+(aq) + C_2H_3O_2^-(aq)$$

Since the salt is completely ionized, the solution contains a much higher concentration of acetate ions than would be present if only acetic acid were in solution. The acetate ion represses the ionization of acetic acid and also reacts with water, causing the solution to have a higher pH (be more basic) than an acetic acid solution (see Section 16.5). Thus, a 0.1 $M$ acetic acid solution has a pH of 2.87, but a solution that is 0.1 $M$ in acetic acid and 0.1 $M$ in sodium acetate has a pH of 4.74.

A buffer solution has a built-in mechanism that counteracts the effect of adding acid or base. Consider the effect of adding HCl or NaOH to an acetic

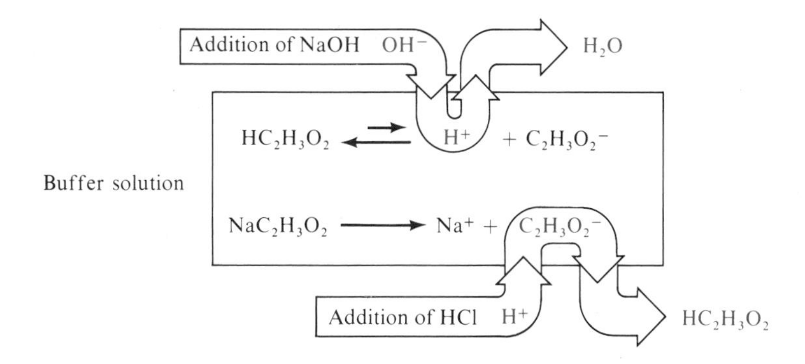


*Figure 16.3.* The effect of adding HCl and NaOH to an acetic acid–sodium acetate buffer solution. The added $H^+$ from HCl is removed from solution by forming un-ionized acetic acid. The added $OH^-$ from NaOH is removed by reacting with $H^+$ to form water.

acid–sodium acetate buffer. When HCl is added, the acetate ions of the buffer combine with the $H^+$ ions of HCl to form un-ionized acetic acid, thus neutralizing the added acid and maintaining the approximate pH of the solution. When NaOH is added, the $H^+$ ions in the buffer combine with the $OH^-$ ions to form water. Additional acetic acid then ionizes (Le Chatelier's principle) to restore the $H^+$ ions and maintain the approximate pH of the solution. The action of this buffer system in counteracting added acid or added base is illustrated in Figure 16.3.

Data comparing the changes in pH caused by adding HCl and NaOH to pure water and to an acetic acid–sodium acetate buffer solution are shown in Table 16.5.

*Table 16.5.* Changes in pH caused by the addition of HCl and NaOH to pure water and to an acetic acid–sodium acetate buffer solution.

| Solution | pH | Change in pH |
|---|---|---|
| $H_2O$ (1000 ml) | 7 | — |
| $H_2O$ + 0.01 mole HCl | 2 | 5 |
| $H_2O$ + 0.01 mole NaOH | 12 | 5 |
| Buffer solution (1000 ml), | | |
| 0.10 $M$ $HC_2H_3O_2$ + 0.10 $M$ $NaC_2H_3O_2$ | 4.74 | — |
| Buffer + 0.01 mole HCl | 4.66 | 0.08 |
| Buffer + 0.01 mole NaOH | 4.83 | 0.09 |

## 16.14 Mechanism of Reactions

mechanism of a reaction

How a reaction occurs—that is, the manner or route by which it proceeds—is known as the **mechanism of the reaction**. The mechanism shows us the path, or course, the atoms and molecules take to arrive at the products.

Our aim here is not to study the mechanisms themselves but to show that chemical reactions occur by specific routes.

When hydrogen and iodine are mixed at room temperature, we observe no appreciable reaction. In this case, the reaction takes place as a result of a collision between an $H_2$ and an $I_2$ molecule, but at room temperature the collisions do not result in reaction because the molecules do not have sufficient energy to react with each other. We might say that an energy barrier to reaction exists. As heat is added, the kinetic energy of the molecules increases. When molecules of $H_2$ and $I_2$ having sufficient energy collide, an intermediate, known as the **activated complex**, is formed. The amount of energy needed to form the activated complex is known as the **activation energy**. The complex, $H_2I_2$, is in a metastable form and has an energy level higher than that of the reactants or that of the product. It can decompose to form either the reactants or the product. Three steps constitute the mechanism of the reaction: (1) collision of an $H_2$ and an $I_2$ molecule; (2) formation of the activated complex, $H_2I_2$; and (3) decomposition to the product, HI. The various steps in the formation of HI are shown in Figure 16.4. Figure 16.5 illustrates the energy relationships in this reaction.

activated complex

activation energy

The reaction of hydrogen and chlorine proceeds by a different mechanism. When $H_2$ and $Cl_2$ are mixed and kept in the dark, essentially no product is formed. But if the mixture is exposed to sunlight or ultraviolet radiation, it reacts very rapidly. The overall reaction is

$$H_2(g) + Cl_2(g) \longrightarrow 2\ HCl(g)$$

free radical

This reaction proceeds by what is known as a *free radical mechanism.* A **free radical** is a neutral atom or group containing one or more unpaired electrons. Both atomic chlorine ($:\ddot{\underset{\cdot\cdot}{Cl}}\cdot$) and atomic hydrogen ($H\cdot$) have an unpaired electron and are free radicals. The reaction occurs in three steps.

*Step 1. Initiation:*

$$:\ddot{\underset{\cdot\cdot}{Cl}}:\ddot{\underset{\cdot\cdot}{Cl}}: + h\nu \longrightarrow :\ddot{\underset{\cdot\cdot}{Cl}}\cdot + :\ddot{\underset{\cdot\cdot}{Cl}}\cdot$$

Chlorine free radicals

In this step a chlorine molecule absorbs energy in the form of a photon, $h\nu$, of light or ultra violet radiation. The energized chlorine molecule then splits into two chlorine free radicals.

*Step 2. Propagation:*

$$:\ddot{\underset{\cdot\cdot}{Cl}}\cdot + H:H \longrightarrow HCl + H\cdot$$

Hydrogen free radical

$$H\cdot + :\ddot{\underset{\cdot\cdot}{Cl}}:\ddot{\underset{\cdot\cdot}{Cl}}: \longrightarrow HCl + :\ddot{\underset{\cdot\cdot}{Cl}}\cdot$$

This step begins when a chlorine free radical reacts with a hydrogen molecule to produce a molecule of hydrogen chloride and a hydrogen free radical. The hydrogen radical next reacts with another chlorine molecule to form hydrogen chloride and another chlorine free radical. This chlorine radical can repeat the process by reacting with another hydrogen molecule, and the reaction continues to propagate itself in this manner until one or both of the reactants is used up.

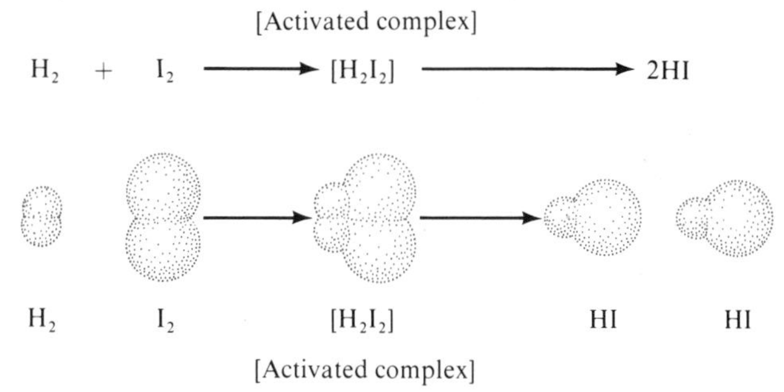

[Activated complex]

$$H_2 + I_2 \longrightarrow [H_2I_2] \longrightarrow 2HI$$

$H_2$ $I_2$ $[H_2I_2]$ HI HI

[Activated complex]

*Figure 16.4.* Mechanism of the reaction between hydrogen and iodine: $H_2$ and $I_2$ molecules of sufficient energy unite, forming the intermediate activated complex, which decomposes to the product, hydrogen iodide.

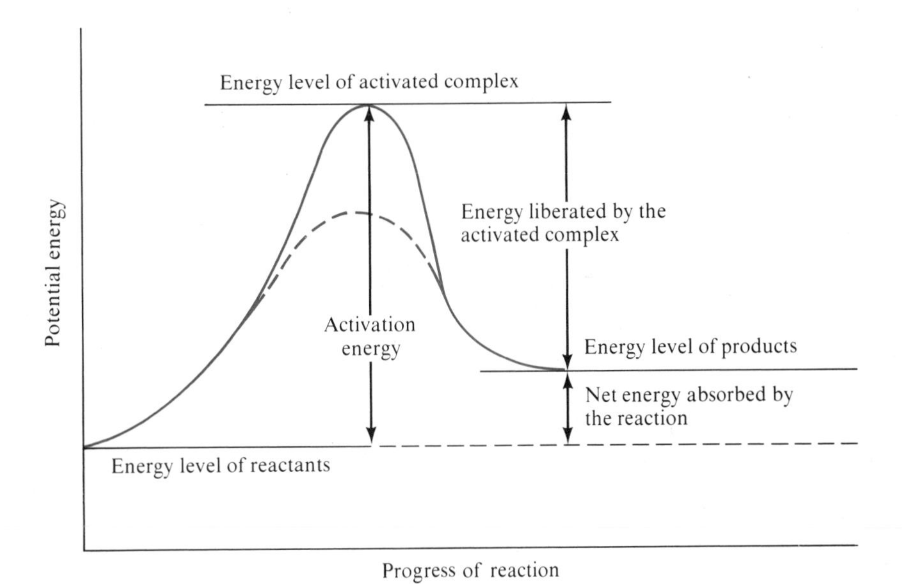


*Figure 16.5.* Relative energy diagram for the reaction between hydrogen and iodine.

$$H_2 + I_2 \longrightarrow [H_2I_2] \rightarrow 2\,HI$$

Energy equal to the activation energy is put into the system in forming the activated complex, $H_2I_2$. When this complex decomposes, it liberates energy, forming the product. In this case, the product is at a higher energy level than the reactants, indicating that the reaction is endothermic and that energy is absorbed during the reaction. The dotted line represents the effect that a catalyst would have on the reaction. The catalyst lowers the activation energy, thereby increasing the rate of the reaction.

*Step 3. Termination:*

$$:\ddot{\underset{\cdot\cdot}{Cl}}\cdot + :\ddot{\underset{\cdot\cdot}{Cl}}\cdot \longrightarrow Cl_2$$

$$H\cdot + H\cdot \longrightarrow H_2$$

$$H\cdot + :\ddot{\underset{\cdot\cdot}{Cl}}\cdot \longrightarrow HCl$$

Hydrogen and chlorine free radicals can react in any of the three ways shown. Unless further activation occurs, the formation of hydrogen chloride will terminate when the radicals form molecules. In an exothermic reaction such as that between hydrogen and chlorine, there is usually enough heat and light energy available to maintain the supply of free radicals, and the reaction will continue until at least one reactant is exhausted.

## Questions

A. *Review the meanings of the new terms introduced in this chapter.*

1. Reversible reaction
2. Chemical kinetics
3. Equilibrium
4. Chemical equilibrium
5. Principle of Le Chatelier
6. Catalyst
7. Law of Mass Action
8. Equilibrium constant, $K_{eq}$
9. Ionization constant, $K_i$
10. $K_w$
11. Solubility product constant, $K_{sp}$
12. Common ion effect
13. Buffer solution
14. Mechanism of a reaction
15. Activated complex
16. Activation energy
17. Free radical

B. *Answers to the following questions will be found in tables and figures.*

1. Explain how each tube in Figure 16.1 illustrates the principle of Le Chatelier.
2. How would the contents of the two tubes of Figure 16.1 differ in appearance if the one at the left was at $-25°C$ and the one at the right was at 25°C?
3. Using the same rate and time scale, resketch Figure 16.2 as it would appear if a catalyst were used to double the forward reaction rate.
4. Which is the strongest and which is the weakest acid listed in Table 16.2? (The strength of an acid is indicated by its degree of ionization.)
5. Classify each of the solutions in Table 16.3 as being acid, basic, or neutral.
6. How many times greater is the $H^+$ ion concentration in the first solution of Table 16.3 than in pure water?
7. Using Table 16.4, tabulate the relative order of molar solubilities of AgCl, AgBr, AgI, $AgC_2H_3O_2$, $PbSO_4$, $BaSO_4$, $BaCrO_4$, CuS, and PbS. List the most soluble first.
8. Using Figure 16.3 and Table 16.5, explain how the acetic acid–sodium acetate buffer system maintains a nearly constant pH when 0.01 mole of HCl is added to 1 litre of the buffer solution.
9. How would Figure 16.4 be modified if the reaction were exothermic?

C. *Review questions.*

1. Express the following reversible systems in equation form:
   (a) Water at 0°C
   (b) A saturated solution of potassium bromide (KBr)
   (c) A closed system containing liquid and gaseous methanol ($CH_3OH$)
2. What constitutes equilibrium in a chemical system?

3. Sodium chloride precipitates when gaseous hydrogen chloride is passed into a saturated NaCl solution. Would a precipitate form if gaseous HCl were passed into a saturated solution of potassium chloride (KCl)? Explain.
4. Will a precipitate form when gaseous hydrogen chloride is passed into a saturated ammonium chloride ($NH_4Cl$) solution? Explain.
5. Explain why the rate of a reaction increases when the concentration of one of the reactants is increased.
6. Explain why a decrease in temperature causes the rate of a chemical reaction to decrease.
7. If pure hydrogen iodide (HI) is placed in a sealed vessel at 700 K, will it decompose? Explain.
8. What changes take place in the concentration of each of the substances when some ammonia is removed from the equilibrium system shown below?

   $4\,NH_3(g) + 5\,O_2(g) \rightleftarrows 4\,NO(g) + 6\,H_2O(g)$

9. Consider the equilibrium represented by this equation:

   $N_2(g) + 3\,H_2(g) \rightleftarrows 2\,NH_3(g) + \text{Heat}$

   What will be the effect on the equilibrium when each of the following changes is made?
   (a) More $N_2$ is added. (c) The pressure is increased.
   (b) Some $NH_3$ is removed. (d) The temperature is decreased.
10. In the equations below, in which direction (left or right) will the equilibrium shift when the following changes are made? The temperature is increased; the pressure is increased; a catalyst is added.
    (a) $3\,O_2(g) + 64.8\text{ kcal} \rightleftarrows 2\,O_3(g)$
    (b) $CH_4(g) + Cl_2(g) \rightleftarrows CH_3Cl(g) + HCl(g) + 26.4\text{ kcal}$
    (c) $2\,NO(g) + 2\,H_2(g) \rightleftarrows N_2(g) + 2\,H_2O(g) + 159\text{ kcal}$
    (d) $2\,SO_3(g) + 47\text{ kcal} \rightleftarrows 2\,SO_2(g) + O_2(g)$
    (e) $4\,NH_3(g) + 3\,O_2(g) \rightleftarrows 2\,N_2(g) + 6\,H_2O(g) + 366\text{ kcal}$
11. Explain what occurs when the pure substances A and B are mixed and react to establish the equilibrium

    $A + B \leftrightarrows C + D$

12. A 1.0 $M$ $HC_2H_3O_2$ solution is 0.42% ionized. A 0.10 $M$ $HC_2H_3O_2$ solution is 1.34% ionized. Explain these figures, using the ionization equation and equilibrium principles.
13. A more concentrated acetic acid ($HC_2H_3O_2$) solution ionizes to a smaller degree than a less concentrated $HC_2H_3O_2$ solution. Does this fact contradict Le Chatelier's principle? Explain. (See Question 12.)
14. Write the equilibrium constant expression for each of the following reactions:
    (a) $H_2(g) + I_2(g) \rightleftarrows 2\,HI(g)$
    (b) $N_2(g) + 3\,H_2(g) \rightleftarrows 2\,NH_3(g)$
    (c) $PCl_5(g) \rightleftarrows PCl_3(g) + Cl_2(g)$
    (d) $HClO_2(aq) \rightleftarrows H^+ + ClO_2^-$
    (e) $NH_4OH(aq) \rightleftarrows NH_4^+ + OH^-$
    (f) $4\,NH_3(g) + 5\,O_2(g) \rightleftarrows 4\,NO(g) + 6\,H_2O(g)$
15. What effect, if any, will increasing the $OH^-$ ion concentration have upon the following?
    (a) pH (b) pOH (c) $[H^+]$ (d) $K_w$

16. Write the solubility product expression ($K_{sp}$) for each of the following substances:
    (a) $AgBr$ (c) $PbCl_2$ (e) $Al(OH)_3$ (g) $FePO_4$
    (b) $CaSO_4$ (d) $Ag_2CO_3$ (f) $As_2S_3$ (h) $Ca_3(PO_4)_2$
17. Explain why silver acetate ($AgC_2H_3O_2$) is more soluble in nitric acid than in water. (*Hint:* Write the equilibrium equation first and then consider the effect of the acid on the acetate ion.)
18. Explain why the solution becomes basic when the salt, sodium acetate, is dissolved in pure water. (*Hint:* A small amount of $HC_2H_3O_2$ is formed.)
19. One of the important pH-regulating systems in the blood consists of a carbonic acid–sodium bicarbonate buffer:

$$H_2CO_3(aq) \rightleftarrows H^+(aq) + HCO_3^-(aq)$$

$$NaHCO_3(aq) \rightarrow Na^+(aq) + HCO_3^-(aq)$$

    Explain how this buffer resists changes in pH when (a) excess acid and (b) excess base get into the blood stream.
20. The action of a catalyst does not shift the position of an equilibrium (change the value of the equilibrium constant). Why not?
21. Explain and illustrate the formation and decomposition of an activated complex in the reaction of $H_2$ and $I_2$ molecules to form HI molecules.
22. Describe and illustrate the role of free radicals in the reaction of hydrogen and chlorine to form hydrogen chloride.
23. Which of the following statements are correct?
    (a) In an equilibrium reaction, equilibrium is established when the concentrations of the reactants and products are equal.
    (b) The effect of a catalyst in an equilibrium system is to lower the activation energy of the reaction.
    (c) Increasing the pressure on a reaction in the gaseous state increases the rate of reaction.
    (d) A large value for an equilibrium constant means a high concentration of the products and a low concentration of the reactants at equilibrium.

    *Statements (e)–(k) pertain to the equilibrium system*

$$2\,NO(g) + O_2(g) \rightleftarrows 2\,NO_2(g) + 27\ \text{kcal}$$

    (e) As the temperature on the system is increased, the amount of $O_2$ increases.
    (f) As the pressure on the system is decreased, the amount of $NO_2$ decreases.
    (g) As the pressure on the system is increased, the amount of NO decreases.
    (h) If a catalyst is added to the system, the heat of reaction decreases.
    (i) When some $NO_2$ is removed from the system, the concentration of all three substances will be less when equilibrium is reestablished.
    (j) The equilibrium constant expression for the reaction is

$$K_{eq} = \frac{[NO_2]^2}{[NO]^2[O_2]^2}$$

    (k) The reaction as shown is endothermic.
    (l) A solution with an $H^+$ ion concentration of $1 \times 10^{-4}$ mole/litre has a pOH of 10.
    (m) An aqueous solution that has an $OH^-$ ion concentration of $1 \times 10^{-3}$ mole/litre has an $H^+$ ion concentration of $1 \times 10^{-11}$ mole/litre.
    (n) $K_w = [H^+][OH^-] = 1 \times 10^{-14}$
    (o) As solid $BaSO_4$ is added to a saturated solution of $BaSO_4$, the magnitude of its $K_{sp}$ increases.

*D. Review problems.*

1. What is the maximum amount of hydrogen iodide (HI) that can be produced when 1.40 moles of $H_2$ and 1.20 moles of $I_2$ are reacted?
2. (a) How many moles of hydrogen iodide (HI) will be produced when 1.50 moles of $H_2$ and 1.50 moles of $I_2$ are reacted at 700 K? (The reaction is 79% complete.)
   (b) If 0.20 mole of $I_2$ is added to this system, the yield of HI is 85%. What will be the new concentration (in moles) of $H_2$, $I_2$, and HI?
3. Three (3.00) g of hydrogen and 100 g of iodine are reacted at 500 K. After equilibrium is reached, analysis shows that there are 32.0 g of HI in the flask. How many moles of $H_2$, $I_2$, and HI are present in this equilibrium mixture? (See Section 16.3.)
4. Five (5.00) moles of pure hydrogen iodide are placed in a reaction vessel at 700 K. What concentration of $H_2$, $I_2$, and HI will be present when equilibrium is reached? (See Section 16.3.)
5. If the velocity of a reaction doubles for every 10 degrees that the temperature rises, how much faster will it proceed at 100°C than at 20°C?
6. One hundred grams of phosphorus pentachloride ($PCl_5$) are sealed in a flask and heated. At equilibrium, 12.0 g of $PCl_5$ remain undecomposed.

   $PCl_5(s) \rightleftarrows PCl_3(l) + Cl_2(g)$

   (a) How many moles of chlorine ($Cl_2$) are present in the equilibrium mixture?
   (b) What volume will this amount of $Cl_2$ occupy at 200°C and 1 atm pressure?
7. When 1.00 mole each of $H_2$ and $I_2$ are reacted in a 1.00 litre flask at 700 K to produce HI, the reaction is 79% complete at equilibrium. Calculate the $K_{eq}$ for this reaction.
8. Calculate the ionization constant for the acids listed in the table. Each is a monoprotic acid and ionizes as follows: $HA \rightleftarrows H^+ + A^-$.

| Acid | Acid concentration | $[H^+]$ |
|---|---|---|
| Propanoic, $HC_3H_5O_2$ | 0.10 *M* | $1.16 \times 10^{-3}$ mole/litre |
| Hydrofluoric, HF | 0.10 *M* | $8.5 \times 10^{-3}$ mole/litre |
| Hydrocyanic, HCN | 0.20 *M* | $8.94 \times 10^{-6}$ mole/litre |

9. Calculate the hydrogen ion concentration and the pH in 0.20 *M* $HC_2H_3O_2$ solution. The ionization constant, $K_i$, for $HC_2H_3O_2$ is $1.8 \times 10^{-5}$.
10. Calculate the nitrite ion concentration in a 0.25 *M* solution of nitrous acid ($HNO_2$). The ionization constant, $K_i$, for $HNO_2$ is $4.5 \times 10^{-4}$.
11. What is the hydrogen ion concentration in 100 ml of a solution that contains 0.50 g of acetic acid? (The ionization constant, $K_i$, for $HC_2H_3O_2$ is $1.8 \times 10^{-5}$.)
12. What is the percent ionization of acetic acid in a 0.010 *M* $HC_2H_3O_2$ solution?
13. What is the pH of a 0.010 *M* $HC_2H_3O_2$ solution?
14. Given the following solubility data, calculate the solubility product constant for each substance:
    (a) $CuCO_3$, $1.58 \times 10^{-5}$ mole/litre
    (b) ZnS, $3.5 \times 10^{-12}$ mole/litre
    (c) $MgF_2$, $2.7 \times 10^{-9}$ mole/litre
    (d) $Ag_2CO_3$, $1.27 \times 10^{-4}$ mole/litre
    (e) $MgCO_3$, $2.7 \times 10^{-6}$ g/litre
    (f) $CaSO_4$, 0.67 g/litre
    (g) $Zn(OH)_2$, $2.33 \times 10^{-4}$ g/litre
    (h) $Ag_3PO_4$, $6.73 \times 10^{-3}$ g/litre
15. Calculate the solubility of $AgC_2H_3O_2$ in moles/litre and grams/100 ml. (See Table 16.4 for $K_{sp}$.)

16. Given the following solubility products, calculate the molar solubility for each substance:
    (a) $PbCO_3$, $K_{sp} = 1.5 \times 10^{-15}$ (c) $PbCl_2$, $K_{sp} = 1.6 \times 10^{-5}$
    (b) $SrSO_4$, $K_{sp} = 7.6 \times 10^{-7}$ (d) $Mg(OH)_2$, $K_{sp} = 8.9 \times 10^{-12}$
17. Which has the greater molar solubility?
    (a) $CaCO_3$ or $MgCO_3$ (b) $BaCO_3$ or $Ag_2CO_3$
    $K_{sp}$ values: $CaCO_3$, $4.7 \times 10^{-9}$; $MgCO_3$, $1.0 \times 10^{-15}$; $BaCO_3$, $1.6 \times 10^{-9}$; $Ag_2CO_3$, $8.2 \times 10^{-12}$
18. Calculate the $H^+$ ion concentration and the pH of buffer solutions which are 0.10 $M$ in $HC_2H_3O_2$ and contain sufficient sodium acetate to make the $C_2H_3O_2^-$ ion concentration equal to:
    (a) 0.050 $M$ (b) 0.10 $M$
    Use the $K_i$ expression of $HC_2H_3O_2$ in your calculations.

# 17 Oxidation–Reduction

*After studying Chapter 17 you should be able to:*

1. Understand the terms listed in Question A at the end of the chapter.
2. Assign oxidation numbers to all the atoms in a given compound or ion.
3. Determine what is being oxidized and what is being reduced in an oxidation–reduction reaction.
4. Identify the oxidizing agent and the reducing agent in an oxidation–reduction reaction.
5. Balance oxidation–reduction equations in molecular and in ionic form.
6. List the general principles concerning the activity series of the metals.
7. Use the activity series to determine whether a proposed single-replacement reaction will occur.
8. Distinguish between an electrolytic and a voltaic cell.
9. Draw a voltaic cell that will produce electric current from an oxidation–reduction reaction involving two metals and their salts.
10. Identify the anode reaction and the cathode reaction in a given electrolytic or voltaic cell.
11. Write equations for the overall chemical reaction and for the oxidation and reduction reactions involved in the discharging or charging of a lead storage battery.
12. Explain how the charge condition of a lead storage battery can be estimated with the aid of a hydrometer.

## 17.1 Oxidation Number

The oxidation number of an atom (sometimes called its oxidation state) can be considered to represent the number of electrons lost, gained, or unequally shared by the atom. Oxidation numbers can be zero, positive, or negative. When the oxidation number of an atom is zero, the atom has the same number of electrons assigned to it as there are in the free neutral atom. When the oxidation number is positive, the atom has fewer electrons assigned to it than there are in the neutral atom. When the oxidation number is negative, the atom has more electrons assigned to it than there are in the neutral atom.

The oxidation number of an atom that has lost or gained electrons to form an ion is the same as the plus or minus charge of the ion. In covalent compounds, where electrons are shared between two atoms, the atoms are assigned oxidation numbers by a somewhat arbitrary system based on the relative electronegativities of the atoms. When two atoms share a pair of electrons, the atom with the higher electronegativity has a greater attraction for the electrons. Therefore, when a pair of electrons is shared unequally between two atoms, both electrons are assigned to the more electronegative element. Each element is then assigned an oxidation number based on the number of electrons gained or lost compared to the neutral atom.

In the ionic compound sodium chloride (NaCl), where one electron has completely transferred from a Na atom to a Cl atom, the oxidation number of Na is clearly established to be +1, and for Cl it is −1. In magnesium chloride ($MgCl_2$), two electrons have completely transferred from the Mg atom to the Cl atoms; thus, the oxidation number of Mg is +2.

In symmetrical covalent molecules such as $H_2$ and $Cl_2$,

H:H    :C̈l:C̈l:

electrons are shared equally between the two atoms. Neither atom is more positive or negative than the other, therefore, each is assigned an oxidation number of zero.

In compounds with covalent bonds, such as $NH_3$ and $H_2O$,

H
:N̈:H    Shared pairs of electrons    H:Ö:
H                                              H

the pairs of electrons are unequally shared between the atoms and are attracted toward the more electronegative elements, N and O. This unequal sharing causes the N and O atoms to be relatively negative with respect to the H atoms. At the same time, it causes the H atoms to be relatively positive with respect

*Table 17.1.* Arbitrary rules for assigning oxidation numbers.

1. All elements in their free state (uncombined with other elements) have an oxidation number of zero (for example, Na, Cu, Mg, $H_2$, $O_2$, $Cl_2$, $N_2$).
2. H is +1, except in metal hydrides, where it is −1 (for example, NaH, $CaH_2$).
3. O is −2, except in peroxides, where it is −1, and in $OF_2$, where it is +2.
4. The metallic element in an ionic compound has a positive oxidation number.
5. In covalent compounds, the negative oxidation number is assigned to the most electronegative atom.
6. The algebraic sum of the oxidation numbers of the elements in a compound is zero.
7. The algebraic sum of the oxidation numbers of the elements in a polyatomic ion is equal to the charge of the ion.

*Table 17.2.* Oxidation numbers of atoms in selected compounds.

| Ion or compound | Oxidation number |
|---|---|
| $H_2O$ | H, +1; O, −2 |
| $SO_2$ | S, +4; O, −2 |
| $CH_4$ | C, −4; H, +1 |
| $CO_2$ | C, +4; O, −2 |
| $KMnO_4$ | K, +1; Mn, +7; O, −2 |
| $Na_3PO_4$ | Na, +1; P, +5; O, −2 |
| $Al_2(SO_4)_3$ | Al, +3; S, +6; O, −2 |
| NO | N, +2; O, −2 |
| $BCl_3$ | B, +3; Cl, −1 |
| $SO_4^{2-}$ | S, +6; O, −2 |
| $NO_3^-$ | N, +5; O, −2 |
| $CO_3^{2-}$ | C, +4; O, −2 |

to the N and O atoms. In $H_2O$, both pairs of shared electrons are assigned to the O atom, giving it two electrons more than the neutral O atom. At the same time, each H atom is assigned one electron less than the neutral H atom. Therefore, the O atom is assigned an oxidation number of −2, and each H atom is assigned an oxidation number of +1. In $NH_3$, the three pairs of shared electrons are assigned to the N atom, giving it three electrons more than the neutral N atoms. At the same time, each H atom has one electron less than the neutral atom. Therefore, the N atom is assigned an oxidation number of −3, and each H atom is assigned an oxidation number of +1.

The assignment of correct oxidation numbers to elements is essential for balancing oxidation–reduction equations. Restudy Sections 7.11–7.14, regarding oxidation numbers, oxidation number tables, and the determination of oxidation numbers from formulas. Table 7.4 (page 119) lists relative electronegativities of the elements. Rules for assigning oxidation numbers are given in Section 7.11 (page 127) and are summarized in Table 17.1, given here. Examples showing oxidation numbers in compounds and ions are given in Table 17.2.

## 17.2 Oxidation–Reduction

oxidation–reduction

redox

oxidation

reduction

**Oxidation–reduction**, also known as **redox**, is a chemical process in which the oxidation number of an element is changed. The process may involve the complete transfer of electrons to form ionic bonds or only a partial transfer or shift of electrons to form covalent bonds.

**Oxidation** occurs whenever the oxidation number of an element increases as a result of losing electrons. Conversely, **reduction** occurs whenever the oxidation number of an element decreases as a result of gaining electrons. For example, a change in oxidation number from +2 to +3 or from −1 to 0 is oxidation; a change from +5 to +2 or from −2 to −4 is reduction (see Figure 17.1). Oxidation and reduction occur simultaneously in a chemical reaction; one cannot take place without the other.

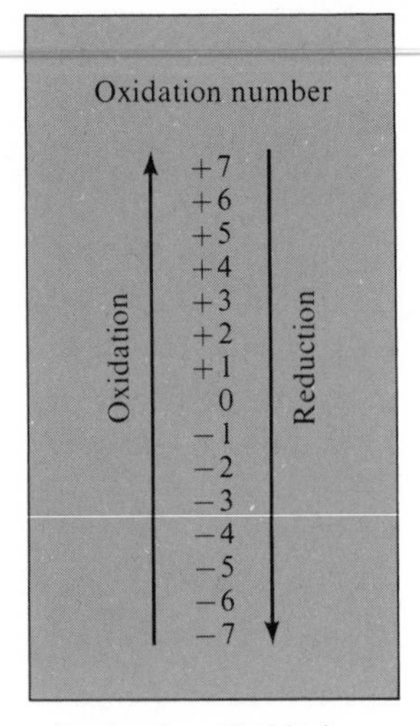


*Figure 17.1.* Oxidation and reduction: Oxidation results in an increase in the oxidation number, and reduction results in a decrease in the oxidation number.

Many combination, decomposition, and single-replacement reactions involve oxidation–reduction. Let us examine the combustion of hydrogen and oxygen from this point of view:

$$2\,H_2 + O_2 \longrightarrow 2\,H_2O$$

Both reactants, hydrogen and oxygen, are elements in the free state and have an oxidation number of zero. In the product water, hydrogen has been oxidized to +1 and oxygen reduced to −2. The substance that does the oxidizing is known as the **oxidizing agent**. The substance that does the reducing is the **reducing agent**. In this reaction, the oxidizing agent is free oxygen and the reducing agent is free hydrogen. In the reaction

oxidizing agent

reducing agent

$$Zn(s) + H_2SO_4(aq) \longrightarrow ZnSO_4(aq) + H_2\uparrow$$

metallic zinc is oxidized and hydrogen ions are reduced. Zinc is the reducing agent and hydrogen ions the oxidizing agent. Electrons are transferred from the zinc metal to the hydrogen ions. The reaction is better expressed as

$$Zn^0 + 2\,H^+ + SO_4^{2-} \longrightarrow Zn^{2+} + SO_4^{2-} + H_2^0\uparrow$$

*Oxidation:* *Increase in oxidation number*
*Loss of electrons*

*Reduction:* *Decrease in oxidation number*
*Gain of electrons*

The oxidizing agent is reduced and gains electrons. The reducing agent is oxidized and loses electrons. The loss and gain of electrons is a characteristic feature of all redox reactions.

## 17.3 Balancing Oxidation–Reduction Equations

Many simple redox equations may be easily balanced by inspection, or trial and error.

$$Na + Cl_2 \longrightarrow NaCl \quad \text{(Unbalanced)}$$
$$2\ Na + Cl_2 \longrightarrow 2\ NaCl \quad \text{(Balanced)}$$

Balancing this equation is certainly not complicated. But as we study more complex reactions and equations, such as

$$P + HNO_3 + H_2O \longrightarrow NO + H_3PO_4 \quad \text{(Unbalanced)}$$
$$3\ P + 5\ HNO_3 + 2\ H_2O \longrightarrow 5\ NO + 3\ H_3PO_4 \quad \text{(Balanced)}$$

the trial-and-error method of finding the proper numbers to balance the equation would take an unnecessarily long time.

The systematic method of balancing oxidation–reduction equations is based on the transfer of electrons between the oxidizing and reducing agents. Consider the first equation again.

$$Na^0 + Cl_2^0 \longrightarrow Na^+Cl^- \quad \text{(Unbalanced)}$$

In this reaction, sodium metal loses one electron per atom when it changes to a sodium ion. At the same time, chlorine gains one electron per atom. Because chlorine is diatomic, two electrons per molecule are needed to form a chloride ion from each atom. These electrons are furnished by two sodium atoms. Stepwise, the reaction may be written as two half-reactions, the oxidation half-reaction and the reduction half-reaction:

Oxidation half-reaction $\quad 2\ Na^0 \longrightarrow 2\ Na^+ + 2\ e^-$

Reduction half-reaction $\quad Cl_2^0 + 2\ e^- \longrightarrow 2\ Cl^-$

$$2\ Na^0 + Cl_2^0 \longrightarrow 2\ Na^+Cl^-$$

When the two half-reactions, each containing the same number of electrons, are added together algebraically, the electrons cancel out. In this reaction there are no excess electrons; the two electrons lost by the two sodium atoms are utilized by chlorine. In all redox reactions, the loss of electrons by the reducing agent must equal the gain of electrons by the oxidizing agent. Sodium is oxidized; chlorine is reduced. Chlorine is the oxidizing agent; sodium is the reducing agent.

The following examples illustrate a systematic method of balancing redox equations.

*Example 1* Balance the equation.

$$Sn + HNO_3 \longrightarrow SnO_2 + NO_2 + H_2O \quad \text{(Unbalanced)}$$

(1) The first step is to assign oxidation numbers to each element in order to identify the elements that are being oxidized and those that are being reduced. Write the oxidation

numbers below each element in order to avoid confusing them with the charge on an ion or radical.

$$\underset{0}{Sn} + \underset{+1}{H}\,\underset{+5}{N}\,\underset{-2}{O_3} \longrightarrow \underset{+4}{Sn}\,\underset{-2}{O_2} + \underset{+4}{N}\,\underset{-2}{O_2} + \underset{+1}{H_2}\,\underset{-2}{O}$$

Note that the oxidation numbers of Sn and N have changed.

(2) Now write two new equations, using only the elements that change in oxidation number. Then add electrons to bring the equations into electrical balance. One equation represents the oxidation step; the other represents the reduction step. The oxidation step produces electrons; the reduction step uses electrons.

Oxidation $Sn^0 \longrightarrow Sn^{4+} + 4e^-$ ($Sn^0$ loses 4 electrons)
Reduction $N^{5+} + 1e^- \longrightarrow N^{4+}$ ($N^{5+}$ gains 1 electron)

(3) Now multiply the two equations by the smallest integral numbers that will make the loss of electrons by the oxidation step equal to the number of electrons gained in the reduction step. In this reaction, the oxidation step is multiplied by 1 and the reduction step by 4. The equations become

Oxidation $Sn^0 \longrightarrow Sn^{4+} + 4e^-$ ($Sn^0$ loses 4 electrons)
Reduction $4N^{5+} + 4e^- \longrightarrow 4N^{4+}$ ($4N^{5+}$ gain 4 electrons)

We have now established the ratio of the oxidizing to the reducing agent as being four atoms of N to one atom of Sn.

(4) Now transfer the coefficient that appears in front of each substance in the balanced oxidation–reduction equations to the corresponding substance in the original equation. We need to use 1 Sn, 1 $SnO_2$, 4 $HNO_3$, and 4 $NO_2$:

$$Sn + 4HNO_3 \longrightarrow SnO_2 + 4NO_2 + H_2O \quad \text{(Unbalanced)}$$

(5) In the usual manner, balance the remaining elements that are not oxidized or reduced to give the final balanced equation:

$$Sn + 4HNO_3 \longrightarrow SnO_2 + 4NO_2 + 2H_2O \quad \text{(Balanced)}$$

In balancing the final elements, we must not change the ratio of the elements that were oxidized and reduced. We should make a final check to ensure that both sides of the equation have the same number of atoms of each element. The final balanced equation contains 1 atom of Sn, 4 atoms of N, 4 atoms of H, and 12 atoms of O on each side.

Since each new equation may present a slightly different problem and since proficiency in balancing equations requires practice, we will work through a few more examples.

*Example 2* Balance the equation.

$$I_2 + Cl_2 + H_2O \longrightarrow HIO_3 + HCl \quad \text{(Unbalanced)}$$

(1) Assign oxidation numbers:

$$\underset{0}{I_2} + \underset{0}{Cl_2} + \underset{+1}{H_2}\underset{-2}{O} \longrightarrow \underset{+1}{H}\,\underset{+5}{I}\,\underset{-2}{O_3} + \underset{+1}{H}\,\underset{-1}{Cl}$$

The oxidation numbers of $I_2$ and $Cl_2$ have changed.

(2) Write oxidation and reduction steps, balancing with electrons:

Oxidation $I_2 \longrightarrow 2\,I^{5+} + 10\,e^-$ ($I_2$ loses 10 electrons)
Reduction $Cl_2 + 2\,e^- \longrightarrow 2\,Cl^-$ ($Cl_2$ gains 2 electrons)

(3) Adjust loss and gain of electrons so that they are equal. Multiply oxidation step by 1 and reduction step by 5.

Oxidation $I_2 \longrightarrow 2\,I^{5+} + 10\,e^-$ ($I_2$ loses 10 electrons)
Reduction $5\,Cl_2 + 10\,e^- \longrightarrow 10\,Cl^-$ ($5\,Cl_2$ gain 10 electrons)

(4) Transfer the coefficients from the balanced redox equations into the original equation. We need to use 1 $I_2$, 2 $HIO_3$, 5 $Cl_2$, and 10 HCl.

$$I_2 + 5\,Cl_2 + H_2O \longrightarrow 2\,HIO_3 + 10\,HCl \quad \text{(Unbalanced)}$$

(5) Balance the remaining elements, H and O:

$$I_2 + 5\,Cl_2 + 6\,H_2O \longrightarrow 2\,HIO_3 + 10\,HCl \quad \text{(Balanced)}$$

*Check*: The final balanced equation contains 2 atoms of I, 10 atoms of Cl, 12 atoms of H, and 6 atoms of O on each side.

---

*Example 3* Balance the equation.

$$K_2Cr_2O_7 + FeCl_2 + HCl \longrightarrow CrCl_3 + KCl + FeCl_3 + H_2O \quad \text{(Unbalanced)}$$

(1) Assign oxidation numbers (Cr and Fe have changed):

$$\underset{+1\,+6\,-2}{K_2Cr_2O_7} + \underset{+2\,-1}{FeCl_2} + \underset{+1\,-1}{H\,Cl} \longrightarrow \underset{+3\,-1}{CrCl_3} + \underset{+1\,-1}{K\,Cl} + \underset{+3\,-1}{FeCl_3} + \underset{+1\,-2}{H_2O}$$

(2) Write the oxidation–reduction steps:

Oxidation $Fe^{2+} \longrightarrow Fe^{3+} + 1\,e^-$ ($Fe^{2+}$ loses 1 electron)
Reduction $Cr^{6+} + 3\,e^- \longrightarrow Cr^{3+}$
or $2\,Cr^{6+} + 6\,e^- \longrightarrow 2\,Cr^{3+}$ ($2\,Cr^{6+}$ gain 6 electrons)

(3) Balance the loss and gain of electrons. Multiply the oxidation step by 6 and the reduction step by 1 to equalize the transfer of electrons.

Oxidation $6\,Fe^{2+} \longrightarrow 6\,Fe^{3+} + 6\,e^-$ ($6\,Fe^{2+}$ lose 6 electrons)
Reduction $2\,Cr^{6+} + 6\,e^- \longrightarrow 2\,Cr^{3+}$ ($2\,Cr^{6+}$ gain 6 electrons)

(4) Transfer the coefficients from the balanced redox equations into the original equation. (Note that one formula unit of $K_2Cr_2O_7$ contains two Cr atoms.) We need to use 1 $K_2Cr_2O_7$, 2 $CrCl_3$, 6 $FeCl_2$, and 6 $FeCl_3$.

$$K_2Cr_2O_7 + 6\,FeCl_2 + HCl \longrightarrow 2\,CrCl_3 + KCl + 6\,FeCl_3 + H_2O \quad \text{(Unbalanced)}$$

(5) Balance the remaining elements in this order: K, Cl, H, O.

$$K_2Cr_2O_7 + 6\,FeCl_2 + 14\,HCl \longrightarrow 2\,CrCl_3 + 2\,KCl + 6\,FeCl_3 + 7\,H_2O \quad \text{(Balanced)}$$

*Check*: The final balanced equation contains 2 K atoms, 2 Cr atoms, 7 O atoms, 6 Fe atoms, 26 Cl atoms, and 14 H atoms on each side.

---

*Example 4* Try the following equation, which has a little different twist to it.

$$Cu + HNO_3 \longrightarrow Cu(NO_3)_2 + NO + H_2O \quad \text{(Unbalanced)}$$

(1) Assign oxidation numbers [Cu and N (in NO) have changed]:

$$\underset{0}{Cu} + H\underset{+5}{N}O_3 \longrightarrow \underset{+2}{Cu}(\underset{+5}{N}O_3)_2 + \underset{+2}{N}O + H_2O$$

(2) Write the oxidation–reduction steps:

Oxidation $Cu^0 \longrightarrow Cu^{2+} + 2\,e^-$ ($Cu^0$ loses 2 electrons)
Reduction $N^{5+} + 3\,e^- \longrightarrow N^{2+}$ ($N^{5+}$ gains 3 electrons)

(3) Balance the loss and gain of electrons. Multiply the oxidation step by 3 and the reduction step by 2 to equalize the loss and gain of electrons.

Oxidation $3\,Cu^0 \longrightarrow 3\,Cu^{2+} + 6\,e^-$ (3 $Cu^0$ lose 6 electrons)
Reduction $2\,N^{5+} + 6\,e^- \longrightarrow 2\,N^{2+}$ (2 $N^{5+}$ gain 6 electrons)

(4) Transfer the coefficients from the balanced redox equations into the original equation.

$$3\,Cu + 2\,HNO_3 \longrightarrow 3\,Cu(NO_3)_2 + 2\,NO + H_2O \quad \text{(Unbalanced)}$$

(5) Balance the remaining elements. In doing this, we notice that there are 8 N atoms on the right side of the equation and 2 N atoms on the left. This imbalance indicates that 6 more $HNO_3$ molecules are needed on the left and also that 6 $NO_3^-$ ions did not enter into the redox reaction. The use of 8 $HNO_3$ in the balanced equation does not destroy the ratio of 3 Cu/2 $HNO_3$ needed for oxidation–reduction. The balanced equation is

$$3\,Cu + 8\,HNO_3 \longrightarrow 3\,Cu(NO_3)_2 + 2\,NO + 4\,H_2O \quad \text{(Balanced)}$$

*Check:* The final balanced equation contains 3 Cu atoms, 8 H atoms, 8 N atoms, and 24 O atoms on each side.

## 17.4 Balancing Ionic Redox Equations

The main difference in balancing ionic versus molecular redox equations is the handling of ions. In addition to having the same number of each kind of element, the net charges on both sides of the final equation must be equal to each other. In assigning oxidation numbers, we must be careful to consider the charge on the ions. In many respects, balancing ionic equations is much simpler than balancing molecular equations.

*Example 5* Balance the equation.

$$Fe^{2+} + Br_2 \longrightarrow Fe^{3+} + Br^- \quad \text{(Unbalanced)}$$

You might try to balance this equation simply by placing a 2 in front of the $Br^-$:

$$Fe^{2+} + Br_2 \longrightarrow Fe^{3+} + 2\,Br^- \quad \text{(Unbalanced)}$$

However, the equation is not balanced, because the electrical charges on the left and the right sides of the equation are not equal. The left side has a charge of $+2$ and the right side has a charge of $(+3) + (-2) = +1$. The net charge on each side is determined by adding the charges of all the ions. The equation is correctly balanced by the use of electrons, as follows:

| | | |
|---|---|---|
| Oxidation | $Fe^{2+} \longrightarrow Fe^{3+} + 1\,e^-$ | ($Fe^{2+}$ loses 1 electron) |
| Reduction | $Br_2 + 2\,e^- \longrightarrow 2\,Br^-$ | (2 Br gain 2 electrons) |

Equalize the loss and gain of electrons:

| | | |
|---|---|---|
| Oxidation | $2\,Fe^{2+} \longrightarrow 2\,Fe^{3+} + 2\,e^-$ | (2 Fe lose 2 electrons) |
| Reduction | $Br_2 + 2\,e^- \longrightarrow 2\,Br^-$ | (2 Br gain 2 electrons) |

Finally,

$$2\,Fe^{2+} + Br_2 \longrightarrow 2\,Fe^{3+} + 2\,Br^- \quad \text{(Balanced)}$$

Net charge: $(+4) + (0) = +4 \quad (+6) + (-2) = +4$

The balanced equation contains the same number of each kind of atom and the same electrical charge on each side of the equation. The charge on each side is +4.

---

*Example 6* Try the more complex equation

$$MnO_4^- + S^{2-} + H^+ \longrightarrow Mn^{2+} + S^0 + H_2O \quad \text{(Unbalanced)}$$

First, assign oxidation numbers:

$$MnO_4^- + S^{2-} + H^+ \longrightarrow Mn^{2+} + S^0 + H_2O$$

$+7 \quad -2 \quad\quad +2 \quad 0$

| | | |
|---|---|---|
| Oxidation | $S^{2-} \longrightarrow S^0 + 2\,e^-$ | ($S^{2-}$ loses 2 electrons) |
| Reduction | $Mn^{7+} + 5\,e^- \longrightarrow Mn^{2+}$ | ($Mn^{7+}$ gains 5 electrons) |

Multiply the oxidation step by 5 and the reduction step by 2 to balance the loss and gain of electrons:

| | | |
|---|---|---|
| Oxidation | $5\,S^{2-} \longrightarrow 5\,S^0 + 10\,e^-$ | ($5\,S^{2-}$ lose 10 electrons) |
| Reduction | $2\,Mn^{7+} + 10\,e^- \longrightarrow 2\,Mn^{2+}$ | ($2\,Mn^{7+}$ gain 10 electrons) |

Transfer balanced redox coefficients to the original equation:

$$2\,MnO_4^- + 5\,S^{2-} + H^+ \longrightarrow 2\,Mn^{2+} + 5\,S^0 + H_2O \quad \text{(Unbalanced)}$$

At this point there remain to be balanced the electrical charge, the H atoms, and the O atoms. First the electrical charges are balanced by the use of additional $H^+$ ions. The H and O atoms are then balanced by the use of $H_2O$ molecules as needed. We find that 16 $H^+$ ions are needed and 8 $H_2O$ molecules are needed to bring the equation into balance.

$$2\,MnO_4^- + 5\,S^{2-} + 16\,H^+ \longrightarrow 2\,Mn^{2+} + 5\,S^0 + 8\,H_2O \quad \text{(Balanced)}$$

$(-2) + (-10) + (+16) = +4 \quad (+4) + (0) + (0) = +4$

*Check:* Both sides have a net charge of +4 and contain the same number of atoms of each element. The equation is balanced.

---

The $H^+$ ions in the equation show that the reaction of Example 6 is in an acid solution. Therefore, $H^+$ ions and water molecules are used in balancing the ionic equation. For a reaction in an alkaline solution, $OH^-$ ions and $H_2O$ molecules are used as needed to balance the ionic equation.

## 17.5 Activity Series of Metals

Knowledge of the relative chemical reactivity of the elements is useful for predicting the course of many chemical reactions.

Calcium reacts with cold water and magnesium reacts with steam to produce hydrogen in each case. Calcium, therefore, is considered to be a more reactive metal than magnesium.

$$Ca(s) + 2\,H_2O(l) \longrightarrow Ca(OH)_2(aq) + H_2\uparrow$$

$$Mg(s) + \underset{\text{Steam}}{H_2O(g)} \longrightarrow MgO(s) + H_2\uparrow$$

The difference in their activity is attributed to the relative ease with which each loses its two valence electrons. It is apparent that calcium loses these electrons more easily than magnesium and is therefore more reactive and/or more readily oxidized than magnesium.

When a strip of copper is placed in a solution of silver nitrate ($AgNO_3$), free silver begins to plate out on the copper. After the reaction has continued for some time, we can observe a blue color in the solution, indicating the presence of copper(II) ions. If a strip of silver is placed in a solution of copper(II) nitrate [$Cu(NO_3)_2$], no reaction is visible. The equations are

$$Cu^0 + 2\,AgNO_3(aq) \longrightarrow 2\,Ag^0 + Cu(NO_3)_2(aq)$$

$$Cu^0 + 2\,Ag^+ \longrightarrow 2\,Ag^0 + Cu^{2+} \qquad \text{Net ionic equation}$$

$$Cu^0 \longrightarrow Cu^{2+} + 2\,e^- \qquad \text{Oxidation of } Cu^0$$

$$Ag^+ + e^- \longrightarrow Ag^0 \qquad \text{Reduction of } Ag^+$$

$$Ag^0 + Cu(NO_3)_2(aq) \longrightarrow \text{No reaction}$$

In the reaction between Cu and $AgNO_3$, electrons are transferred from $Cu^0$ atoms to $Ag^+$ ions in solution. Since copper has a greater tendency than silver to lose electrons, an electrochemical force is exerted upon silver ions to accept electrons from copper atoms. When an $Ag^+$ ion adds an electron, it is reduced to a $Ag^0$ atom and is no longer soluble in solution. At the same time, $Cu^0$ is oxidized and goes into solution as $Cu^{2+}$ ions. From this reaction we can conclude that copper is more reactive than silver.

Metals such as sodium, magnesium, zinc, and iron, which react with solutions of acids to liberate hydrogen, are more reactive than hydrogen. Metals such as copper, silver, and mercury, which do not react with solutions of acids to liberate hydrogen, are less reactive than hydrogen. By studying a series of reactions such as those given above, we may list metals according to their chemical activity, placing the most active at the top and the least active at the bottom. This list is called the **Activity Series of Metals**. Table 17.3 shows some of the common metals in the series. The arrangement corresponds to the ease with which the elements listed are oxidized or lose electrons. The most easily oxidizable element is listed first. More extensive tables are available in chemistry reference books.

activity series of metals

*Table 17.3.* Activity Series of Metals.

Ease of oxidation ↑ (increases from bottom to top)

|  | |  | |  |
|---|---|---|---|---|
| K | → | $K^+$ | + | $e^-$ |
| Ba | → | $Ba^{2+}$ | + | $2\,e^-$ |
| Ca | → | $Ca^{2+}$ | + | $2\,e^-$ |
| Na | → | $Na^+$ | + | $e^-$ |
| Mg | → | $Mg^{2+}$ | + | $2\,e^-$ |
| Al | → | $Al^{3+}$ | + | $3\,e^-$ |
| Zn | → | $Zn^{2+}$ | + | $2\,e^-$ |
| Cr | → | $Cr^{3+}$ | + | $3\,e^-$ |
| Fe | → | $Fe^{2+}$ | + | $2\,e^-$ |
| Ni | → | $Ni^{2+}$ | + | $2\,e^-$ |
| Sn | → | $Sn^{2+}$ | + | $2\,e^-$ |
| Pb | → | $Pb^{2+}$ | + | $2\,e^-$ |
| $H_2$ | → | $2\,H^+$ | + | $2\,e^-$ |
| Cu | → | $Cu^{2+}$ | + | $2\,e^-$ |
| As | → | $As^{3+}$ | + | $3\,e^+$ |
| Ag | → | $Ag^+$ | + | $e^-$ |
| Hg | → | $Hg^{2+}$ | + | $2\,e^-$ |
| Au | → | $Au^{3+}$ | + | $3\,e^-$ |

The general principles governing the arrangement and use of the activity series are as follows:

1. The reactivity of the metals listed decreases from top to bottom.
2. A free metal can displace the ion of a second metal from solution provided that the free metal is above the second metal in the activity series.
3. Free metals above hydrogen react with nonoxidizing acids in solution to liberate hydrogen gas.
4. Free metals below hydrogen do not liberate hydrogen from acids.
5. Conditions such as temperature and concentration may affect the relative position of some of these elements.

Two examples of the application of the Activity Series follow.

*Example 1* Will zinc metal react with dilute sulfuric acid?

From Table 17.3, we see that zinc is above hydrogen; therefore, zinc atoms will lose electrons more readily than hydrogen atoms. Hence, zinc atoms will reduce hydrogen ions from the acid to form hydrogen gas and zinc ions. In fact, these reagents are commonly used for the laboratory preparation of hydrogen. The equation is

$$Zn(s) + H_2SO_4(aq) \longrightarrow ZnSO_4(aq) + H_2\uparrow$$

$$Zn + 2\,H^+ \longrightarrow Zn^{2+} + H_2\uparrow \qquad \text{Net ionic equation}$$

*Example 2* Will a reaction occur when copper metal is placed in an iron(II) sulfate solution?

No, copper lies below iron in the series, loses electrons less easily than iron, and therefore will not replace iron(II) ions from solution. In fact, the reverse is true. When an iron nail is dipped into a copper(II) sulfate solution, it becomes coated with free copper. The equations are

$$Cu(s) + FeSO_4(aq) \longrightarrow \text{No reaction}$$

$$Fe(s) + CuSO_4(aq) \longrightarrow FeSO_4(aq) + Cu\downarrow$$

From Table 17.3 we may abstract the following pair in their relative position to each other.

$$Fe \longrightarrow Fe^{2+} + 2\,e^-$$
$$Cu \longrightarrow Cu^{2+} + 2\,e^-$$

According to the second principle listed above on the use of the activity series, we can predict that free iron will react with copper(II) ions in solution to form free copper metal and iron(II) ions in solution.

$$Fe + Cu^{2+} \longrightarrow Fe^{2+} + Cu$$

---

## 17.6 Electrolytic and Voltaic Cells

electrolysis
electrolytic cell

The process in which electrical energy is used to bring about a chemical change is known as **electrolysis**. An **electrolytic cell** uses electrical energy to produce a nonspontaneous chemical reaction. There are many applications of electrical energy in the chemical industry—for example, in the production of sodium, sodium hydroxide, chlorine, fluorine, magnesium, aluminum, and pure hydrogen and oxygen, and in the purification and electroplating of metals.

What happens when an electric current is passed through a solution? Let us consider a hydrochloric acid solution in a simple electrolysis cell, as shown in Figure 17.2. The cell consists of a battery connected to two electrodes that are immersed in a solution of hydrochloric acid. The cathode is attached to the negative pole of the battery and becomes the negative electrode. The anode is attached to the positive pole and becomes the positive electrode. The battery supplies electrons to the cathode.

When the switch is closed, the electric circuit is completed; positive hydronium ions ($H_3O^+$) migrate toward the cathode, where they pick up electrons and evolve hydrogen gas. At the same time, the negative chloride ions ($Cl^-$) migrate toward the anode, where they lose electrons, completing the cycle, and evolve chlorine gas.

Reaction at the cathode
$$H_3O^+ + 1\,e^- \longrightarrow H^0 + H_2O$$
$$H^0 + H^0 \longrightarrow H_2\uparrow$$

Reaction at the anode
$$Cl^- \longrightarrow Cl^0 + 1\,e^-$$
$$Cl^0 + Cl^0 \longrightarrow Cl_2\uparrow$$

Net reaction
$$2\,HCl \xrightarrow{\text{Electrolysis}} H_2\uparrow + Cl_2\uparrow$$

Note that oxidation–reduction has taken place. Chloride ions lose electrons (are oxidized) at the anode, and hydronium ions gain electrons (are reduced) at the cathode. In electrolysis, oxidation always occurs at the anode and reduction at the cathode.

When sodium chloride brines are electrolyzed, the products are sodium hydroxide, hydrogen, and chlorine. The overall reaction is

$$2\,Na^+ + 2\,Cl^- + 2\,H_2O \xrightarrow{\text{Electrolysis}} 2\,Na^+ + 2\,OH^- + H_2\uparrow + Cl_2\uparrow$$

The net ionic equation is

$$2\,Cl^- + 2\,H_2O \longrightarrow 2\,OH^- + H_2\uparrow + Cl_2\uparrow$$

During the electrolysis, $Na^+$ ions move toward the cathode and $Cl^-$ ions move toward the anode. The anode reaction is similar to that of hydrochloric acid; chlorine is liberated.

$$2\,Cl^- \longrightarrow Cl_2\uparrow + 2\,e^-$$

Even though $Na^+$ ions are attracted by the cathode, the facts show that hydrogen is liberated there. No evidence of metallic sodium is found, but the area around the cathode tests alkaline from accumulated $OH^-$ ions. The reaction at the cathode is believed to be

$$2\,H_2O + 2e^- \longrightarrow H_2\uparrow + 2\,OH^-$$

If the electrolysis is allowed to continue until all the chloride is reacted, the solution remaining will contain only sodium hydroxide. Large tonnages of sodium hydroxide and chlorine are made by this process.

When molten sodium chloride (without water) is subjected to electrolysis, metallic sodium and chlorine gas are formed:

$$2\,Na^+ + 2\,Cl^- \xrightarrow{\text{Electrolysis}} 2\,Na + Cl_2\uparrow$$

An important electrochemical application is electroplating of metals. Electroplating is the art of covering a surface or an object with a thin adherent electrodeposited metal coating. Electroplating is done for protection of the surface of the base metal or for a purely decorative effect. The layer deposited is surprisingly thin, varying from as little as $5 \times 10^{-5}$ cm to $2 \times 10^{-3}$ cm, depending on the metal and the intended use. The object to be plated is set up as the cathode and is immersed in a solution containing ions of the metal to be plated. When an electric current passes through the solution, metal ions migrating to the cathode are reduced, depositing on the object as the free metal. In most cases the metal deposited on the object is replaced in the solution by using an anode of the same metal. The following equations show the chemical changes in the electroplating of nickel:

| | | |
|---|---|---|
| Reaction at the cathode | $Ni^{2+} + 2e^- \longrightarrow Ni(s)$ | Ni plated out on an object |
| Reaction at the anode | $Ni(s) \longrightarrow Ni^{2+} + 2\,e^-$ | Ni replenished to solution |

Metals most commonly used in commercial electroplating are copper, nickel, zinc, lead, cadmium, chromium, tin, gold, and silver.

In the electrolytic cell shown in Figure 17.2, an electric current flows through the circuit when the switch is closed. The driving force responsible for the current is supplied by the battery (or other source of direct current). Electrons are moving through the wires and electrodes, and ions ($H_3O^+$ and $Cl^-$) are moving in the solution. As a result of the transfer of electrons and ions, hydrochloric acid is converted to hydrogen and chlorine. In electrolytic

$$2\,HCl(aq) \longrightarrow H_2\uparrow + Cl_2\uparrow$$

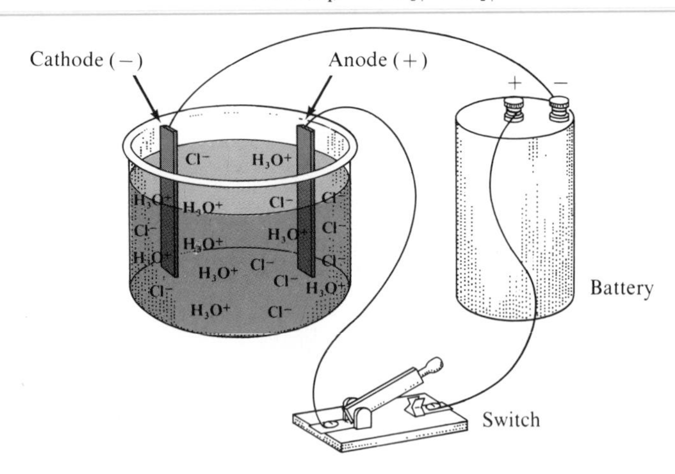


*Figure 17.2.* Electrolysis: During the electrolysis of a hydrochloric acid solution, positive hydronium ions are attracted to the cathode, where they gain electrons and form hydrogen gas. Chloride ions migrate to the anode, where they lose electrons and form chlorine gas. The equation for this process is

$$2\,HCl \longrightarrow H_2\uparrow + Cl_2\uparrow$$

processes of this kind, electrical energy is used to bring about nonspontaneous redox reactions. The hydrogen and chlorine produced have more potential energy than was present in the hydrochloric acid before electrolysis.

Conversely, some spontaneous redox reactions can be made to supply useful amounts of electrical energy. When a piece of zinc is put in a copper(II) sulfate solution, the zinc quickly becomes coated with metallic copper. We expect this to happen because zinc is above copper in the activity series; copper(II) ions are therefore reduced by zinc atoms:

$$Zn^0 + Cu^{2+} \longrightarrow Zn^{2+} + Cu^0$$

This reaction is clearly a spontaneous redox reaction. But simply dipping a zinc rod into a copper(II) sulfate solution will not produce useful electric current! However, when we carry out this reaction in the cell shown in Figure 17.3, an electric current is produced. The cell consists of a piece of zinc immersed in a zinc sulfate solution and connected by a wire through a voltmeter to a piece of copper immersed in copper(II) sulfate solution. The two solutions are connected by a salt bridge. Such a cell produces an electric current and a potential of about 1.1 volts when both solutions are 1.0 *M* in concentration. A cell that produces electric current from a spontaneous chemical reaction is called a **voltaic cell**.

voltaic cell

Although the zinc–copper voltaic cell is no longer used commercially, it was used to energize the first transcontinental telegraph lines. Such cells are the direct ancestors of the many different kinds of "dry" cells that operate portable radio and television sets, automatic cameras, tape recorders, and so on.

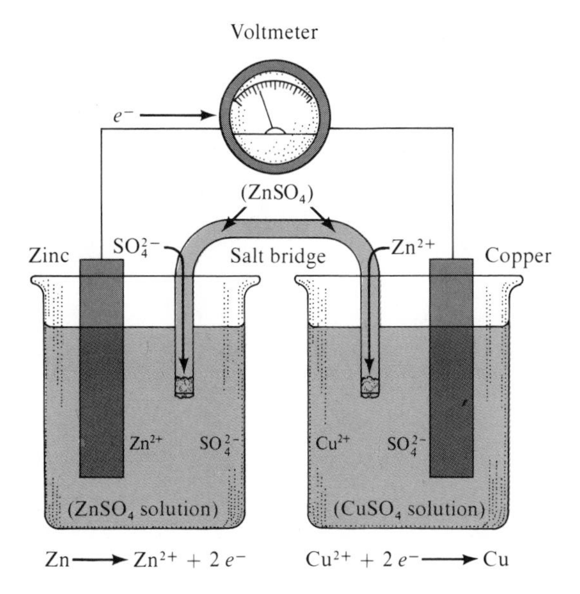


*Figure 17.3.* Zinc–copper voltaic cell. The cell has a potential of 1.1 volts when $ZnSO_4$ and $CuSO_4$ solutions are 1.0 *M*. The salt bridge provides electrical contact between the two half-cells.

The driving force responsible for the electric current in the zinc–copper cell originates in the great tendency of zinc atoms to lose electrons relative to the tendency of copper(II) ions to gain electrons. In the cell shown in Figure 17.3, zinc atoms lose electrons and are converted to zinc ions at the zinc electrode surface; the electrons flow through the wire to the copper electrode. Here, copper(II) ions pick up electrons and are reduced to copper atoms, which plate out on the copper electrode. Sulfate ions flow from the $CuSO_4$ solution via the salt bridge into the $ZnSO_4$ solution to complete the circuit. The equations for the reactions of this cell are

| | | |
|---|---|---|
| Anode | $Zn^0 \longrightarrow Zn^{2+} + 2e^-$ | (Oxidation) |
| Cathode | $Cu^{2+} + 2e^- \longrightarrow Cu^0$ | (Reduction) |
| Net ionic | $Zn^0 + Cu^{2+} \longrightarrow Zn^{2+} + Cu^0$ | |
| Overall | $Zn + CuSO_4 \longrightarrow Cu + ZnSO_4$ | |

The redox reaction, the movement of electrons in the metallic or external part of the circuit, and the movement of ions in the solution or internal part of the circuit of the copper–zinc cell are very similar to the actions that occur in the electrolytic cell of Figure 17.2. The only important difference is that the reactions of the zinc–copper cell are spontaneous. This is the crucial difference between all voltaic and electrolytic cells. Voltaic cells use chemical reactions to produce electrical energy, and electrolytic cells use electrical energy to produce chemical reactions.

An ordinary automobile storage battery is an energy reservoir. The charged battery acts as a voltaic cell and through chemical reactions furnishes electrical energy to operate the starter, lights, radio, and so on. When the engine

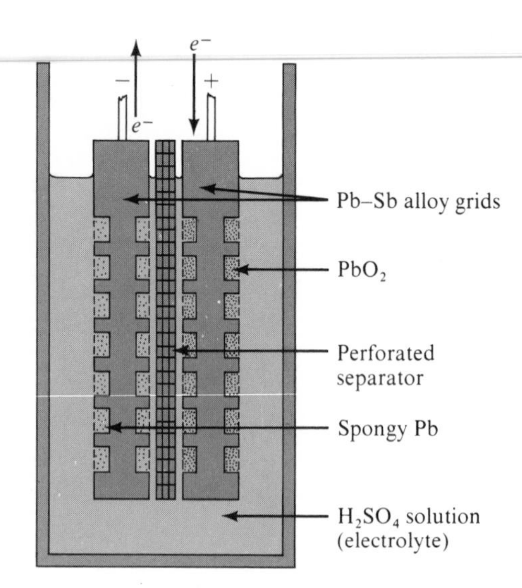


*Figure 17.4.* Cross-sectional diagram of a lead storage battery cell.

is running, a generator, or alternator, produces and forces an electric current through the battery and, by electrolytic chemical action, restores the battery to the charged condition.

The cell unit consists of a lead plate filled with spongy lead and a lead dioxide plate, both immersed in dilute sulfuric acid solution, which serves as the electrolyte (see Figure 17.4). When the cell is discharging, or acting as a voltaic cell, these reactions occur.

| | | |
|---|---|---|
| Pb plate (anode) | $Pb^0 \longrightarrow Pb^{2+} + 2e^-$ | (Oxidation) |
| $PbO_2$ plate (cathode) | $PbO_2 + 4H^+ + 2e^- \longrightarrow Pb^{2+} + 2H_2O$ | (Reduction) |
| Net ionic redox reaction | $Pb^0 + PbO_2 + 4H^+ \longrightarrow 2Pb^{2+} + 2H_2O$ | |
| Precipitation reaction on plates | $Pb^{2+} + SO_4^{2-} \longrightarrow PbSO_4$ | |

Since lead(II) sulfate is insoluble, the $Pb^{2+}$ ions combine with $SO_4^{2-}$ ions to form a coating of $PbSO_4$ on each plate. The overall chemical reaction of the cell is

$$Pb + PbO_2 + 2H_2SO_4 \xrightarrow[\text{cycle}]{\text{Discharge}} 2PbSO_4\downarrow + 2H_2O$$

If the active material on both plates is converted to $PbSO_4$, the cell is discharged and no more electrical energy is to be had. The cell can be recharged by reversing the chemical reaction. This is accomplished by forcing an electric current through the cell in the opposite direction. Lead sulfate and water are reconverted to lead, lead dioxide, and sulfuric acid:

$$2\,PbSO_4 + 2\,H_2O \xrightarrow[\text{cycle}]{\text{Charge}} Pb + PbO_2 + 2\,H_2SO_4$$

The electrolyte in a lead storage battery is a 38% by weight sulfuric acid solution having a density of 1.29 g/ml. As the battery is discharged, sulfuric acid is removed, thereby decreasing the density of the electrolyte solution. The state of charge or discharge of the battery can be estimated by measuring the density of the electrolyte solution with a hydrometer. When the density has dropped to about 1.05 g/ml, the battery needs recharging.

In an actual battery, each cell consists of a series of cell units of alternating lead–lead dioxide plates separated and supported by wood, asbestos, or glass wool. The energy storage capacity of a single cell is limited and its electrical potential is only about 2 volts. Therefore, a bank of six cells is connected in series to provide the 12 volt output of the usual automobile battery.

## Questions

A. *Review the meanings of the new terms introduced in this chapter.*

1. Oxidation–reduction
2. Redox
3. Oxidation
4. Reduction
5. Oxidizing agent
6. Reducing agent
7. Activity Series of Metals
8. Electrolysis
9. Electrolytic cell
10. Voltaic cell

B. *Answers to the following questions will be found in tables and figures.*

1. Would you expect nickel to replace hydrogen in either cold water or steam? Explain. (See Table 17.3.)
2. The reaction between powdered aluminum and iron(III) oxide (in the Thermite process), producing molten iron, is very exothermic.
   (a) Write the equation for the chemical reaction that occurs.
   (b) Explain in terms of Table 17.3 why a reaction occurs.
   (c) Would you expect a reaction between aluminum and chromium(III) oxide ($Cr_2O_3$)?
3. Write equations for the chemical reaction of each of the following metals with dilute solutions of hydrochloric and sulfuric acids: aluminum, chromium, gold, iron, magnesium, mercury, and zinc. If a reaction will not occur, write "no reaction" as the product. (See Table 17.3.)
4. A $NiCl_2$ solution (instead of HCl) is placed in the apparatus shown in Figure 17.2. Write equations for:
   (a) The anode reaction
   (b) The cathode reaction
   (c) The net electrochemical reaction

C. *Review questions.*

1. Which of the following half-reactions are oxidation and which are reduction? Supply the proper number of electrons to balance each equation.
   (a) $SO_3^{2-} + H_2O \rightarrow SO_4^{2-} + 2\,H^+$
   (b) $Al^{3+} \rightarrow Al$
   (c) $2\,H_2O \rightarrow O_2 + 4\,H^+$
   (d) $NO_3^- + 10\,H^+ \rightarrow NH_4^+ + 3\,H_2O$
   (e) $Mn^{2+} + 4\,H_2O \rightarrow MnO_4^- + 8\,H^+$
   (f) $Cr_2O_7^{2-} + 14\,H^+ \rightarrow 2\,Cr^{3+} + 7\,H_2O$

2. For the following equations:
   (a) Identify the oxidizing agent, the reducing agent, and the substances oxidized and reduced.
   (b) Assign oxidation numbers to each element in each oxidation and reduction step.
   (c) Write equations for the half-reactions of each oxidation and reduction.
      (1) $Cu + S \rightarrow CuS$
      (2) $Zn + HgO \rightarrow ZnO + Hg$
      (3) $Fe + Br_2 \rightarrow FeBr_3$
      (4) $H_2O + Cl_2 \rightarrow HCl + HOCl$
      (5) $(NH_4)_2Cr_2O_7 \xrightarrow{\Delta} Cr_2O_3 + H_2O + N_2$
      (6) $KBr + H_2SO_4 \rightarrow Br_2 + K_2SO_4 + SO_2 + H_2O$
      (7) $H^+ + NO_3^- + CuS \rightarrow Cu^{2+} + H_2O + NO + S$
      (8) $H^+ + CH_3OH + MnO_4^- \rightarrow MnO_2 + CO_2 + H_2O$
      (9) $H_2O + CrI_3 + Cl_2 \rightarrow CrO_4^{2-} + IO_3^- + Cl^-$
      (10) $H_2O + Si + OH^- \rightarrow SiO_3^{2-} + H_2$
3. Balance the following redox equations:
   (a) $Fe_2O_3 + CO \rightarrow Fe + CO_2$
   (b) $Ag + HNO_3 \rightarrow AgNO_3 + NO + H_2O$
   (c) $MnO_2 + HBr \rightarrow MnBr_2 + Br_2 + H_2O$
   (d) $CuO + NH_3 \rightarrow N_2 + Cu + H_2O$
   (e) $CuO + NH_3 \rightarrow Cu + N_2 + H_2O$
   (f) $H_2C_2O_4 + KBrO_3 \rightarrow CO_2 + KBr + H_2O$
   (g) $Cl_2 + KOH \rightarrow KClO_3 + KCl + H_2O$
   (h) $H_2O_2 + KMnO_4 + H_2SO_4 \rightarrow O_2 + MnSO_4 + K_2SO_4 + H_2O$
   (i) $HNO_3 + H_2S \rightarrow S + NO + H_2O$
   (j) $Cu_2S + HNO_3 \rightarrow Cu(NO_3)_2 + NO_2 + H_2O + S$
   [*Note:* Three elements are changing oxidation numbers.]
4. Balance the following ionic redox equations: [*Note:* Supply $H^+$ or $OH^-$ ions and $H_2O$ molecules if needed to balance.]
   (a) $H^+ + Cu + SO_4^{2-} \rightarrow Cu^{2+} + SO_2 + H_2O$
   (b) $Zn + Cr_2O_7^{2-} \rightarrow Zn^{2+} + Cr^{3+} + H_2O$
   (c) $H^+ + Cu + NO_3^- \rightarrow Cu^{2+} + NO$
   (d) $H^+ + Zn + AsO_4^{3-} \rightarrow AsH_3 + Zn^{2+}$
   (e) $H_2O + Al + OH^- \rightarrow AlO_2^- + H_2$
   (f) $H_2O + CrO_4^{2-} + Fe(OH)_2 \rightarrow Cr(OH)_3 + Fe(OH)_3 + OH^-$
   (g) $H^+ + ClO_3^- + Cl^- \rightarrow Cl_2$
   (h) $NO_3^- + I^- \rightarrow NO + I_2$
5. Why are oxidation and reduction said to be complementary processes?
6. When molten $CaCl_2$ is electrolyzed, calcium metal and chlorine gas are produced. Write equations for the two half-reactions that occur at the electrodes. Label the anode half-reaction and the cathode half-reaction.
7. Why is direct current instead of alternating current used in the electroplating of metals?
8. What property of lead dioxide and lead(II) sulfate makes it unnecessary to have salt bridges in the cells of a lead storage battery?

9. In one type of alkaline cell used to power devices such as portable radios, $Hg^{2+}$ ions are reduced to metallic mercury when the cell is being discharged. Does this occur at the anode or at the cathode? Explain.
10. Explain why the density of the electrolyte in a lead storage battery decreases during the discharge cycle.
11. Which of these statements are correct?
    (a) An atom of an element in the uncombined state has an oxidation number of zero.
    (b) The oxidation number of molybdenum in $Na_2MoO_4$ is $+4$.
    (c) The oxidation number of an ion is the same as the electrical charge on the ion.
    (d) The process in which an atom or an ion loses electrons is called reduction.
    (e) The reaction $Fe^{3+} + e^- \rightarrow Fe^{2+}$ is a reduction reaction.
    (f) In the reaction $2\,Al + 3\,CuCl_2 \rightarrow 2\,AlCl_3 + 3\,Cu$, aluminum is the oxidizing agent.
    (g) In a redox reaction the oxidizing agent is reduced and the reducing agent is oxidized.
    (h) $Cu^0 \rightarrow Cu^{2+}$ is a balanced oxidation equation.
    (i) In the electrolysis of sodium chloride brine (solution), $Cl_2$ gas is formed at the cathode and hydroxide ions at the anode.
    (j) In any cell, electrolytic or voltaic, reduction takes place at the cathode and oxidation occurs at the anode.
    (k) In the Zn–Cu voltaic cell, the reaction at the anode is An $\rightarrow Zn^{2+} + 2e^-$.

    *The statements in (l) to (o) pertain to the Activity Series:*

    Ba Mg Zn Fe H Cu Ag

    (l) The reaction $Zn + MgCl_2 \rightarrow Mg + ZnCl_2$ is a spontaneous reaction.
    (m) Barium is a more reactive metal than copper.
    (n) Silver metal will react with acids to liberate hydrogen gas.
    (o) Iron is a better reducing agent than zinc.

D. *Review problems.*

1. What weight and volume of $Cl_2$ will react with 40.0 g of KOH in the equation given in Question C.3, part (g)?
2. How many moles of $SO_2$ are formed by reacting 3.0 moles of Cu with sulfuric acid? [See the equation given in Question C.4, part (a).]
3. How many millilitres of 0.100 *M* $KMnO_4$ solution are needed to oxidize 8.50 g of $H_2O_2$? What volume of $O_2$ gas will be formed? [See the equation given in Question C.3, part (h).] Assume STP conditions.
4. What weight of copper is formed when 16.0 litres of ammonia gas are reacted with copper(II) oxide? [See the equation given in Question C.3, part (d).] Assume STP.
5. What weight of bromine (at STP) can be obtained from 50.0 g of HBr according to the equation in Question C.3, part (c)?
6. A sample of crude potassium iodide was analyzed and found to contain 72.0% KI. What maximum weight of iodine can be obtained by oxidizing 500 g of this iodide?

   $$KI + H_2SO_4 \rightarrow I_2 + H_2S + K_2SO_4 + H_2O$$

7. How many moles of $Fe^{2+}$ can be oxidized to $Fe^{3+}$ by 0.75 mole of $Cl_2$ according to the following equation?

   $$Fe^{2+} + Cl_2 \rightarrow Fe^{3+} + Cl^-$$

# Appendix I
# Mathematical Review

1. Multiplication. Multiplication is a process of adding any given number or quantity a certain number of times. Thus, 4 times 2 means 4 added two times, or 2 added together four times, to give the product 8. Various ways of expressing multiplication are

$$ab \qquad a \times b \qquad a \cdot b \qquad a(b) \qquad (a)(b)$$

All mean $a$ times $b$, or $a$ multiplied by $b$, or $b$ times $a$.

When $a = 16$ and $b = 24$, we have $16 \times 24 = 384$.

The expression $°F = (1.8 \times °C) + 32$ means that we are to multiply 1.8 times °C and add 32 to the product. When °C equal 50,

$$°F = (1.8 \times 50) + 32 = 90 + 32 = 122°F$$

The result of multiplying two or more numbers together is known as the *product*.

2. Division. The word *division* has several meanings. As a mathematical expression, it is the process of finding how many times one number or quantity is contained in another. Various ways of expressing division are

$$a \div b \qquad \frac{a}{b} \qquad a/b$$

All mean $a$ divided by $b$.

When $a = 15$ and $b = 3$, $\frac{15}{3} = 5$.

The number above the line is called the *numerator*; the number below the line is the *denominator*. Both the horizontal and the slanted (/) division signs also mean "per." For example, in the expression for density, the mass per unit volume:

$$\text{Density} = \text{Mass/Volume} = \frac{\text{Mass}}{\text{Volume}} = \text{g/ml}$$

The diagonal line still refers to a division of grams by the number of millilitres occupied by that weight.

The result of dividing one number into another is called the *quotient*.

3. Fractions and decimals. A fraction is an expression of division, showing that the numerator is divided by the denominator. A *proper fraction* is one in which the numerator is smaller than the denominator. In an *improper fraction*, the numerator is the larger number. A decimal or a decimal fraction is a proper fraction in which the denominator is some power of 10. The decimal fraction is determined by carrying out the division of the proper fraction. Examples of proper fractions and their decimal fraction equivalents are shown in the table.

| Proper fraction | | Decimal fraction | | Proper fraction |
|---|---|---|---|---|
| $\frac{1}{8}$ | = | 0.125 | = | $\frac{125}{1000}$ |
| $\frac{1}{10}$ | = | 0.1 | = | $\frac{1}{10}$ |
| $\frac{3}{4}$ | = | 0.75 | = | $\frac{75}{100}$ |
| $\frac{1}{100}$ | = | 0.01 | = | $\frac{1}{100}$ |
| $\frac{1}{4}$ | = | 0.25 | = | $\frac{25}{100}$ |

4. Adding or subtracting fractions. We cannot directly add $\frac{1}{2} + \frac{1}{4}$, but we can add these two fractions if they both have the same denominator. A *common denominator* is a number which, when divided by each denominator in a series of fractions, gives whole-number quotients. We usually use the smallest possible common denominator. Thus, $\frac{1}{2} = \frac{2}{4}$, and $\frac{2}{4} + \frac{1}{4} = \frac{3}{4}$. Therefore, to add or subtract fractions, first change the denominator of each fraction to be the same number; that is, make each fraction have a common denominator. Then add the numerators of each fraction and place this sum over the common denominator. It may then be possible to reduce the final fraction to a simpler fraction or to a decimal number.

$$\frac{3}{8} + \frac{4}{8} = \frac{3+4}{8} = \frac{7}{8}$$

$$\frac{13}{18} - \frac{3}{18} = \frac{13-3}{18} = \frac{10}{18} = \frac{5}{9}$$

$$\frac{1}{4} + \frac{2}{3} = \frac{3}{12} + \frac{8}{12} = \frac{3+8}{12} = \frac{11}{12}$$

$$\frac{2}{3} + \frac{5}{6} - \frac{1}{5} = \frac{20}{30} + \frac{25}{30} - \frac{6}{30} = \frac{20+25-6}{30} = \frac{39}{30} = 1\frac{9}{30} = 1.3$$

$$\frac{3}{2} + \frac{3}{4} + \frac{3}{5} = \frac{30}{20} + \frac{15}{20} + \frac{12}{20} = \frac{57}{20} = 2\frac{17}{20} = 2.85$$

5. Multiplication of fractions. A fraction may be multiplied by another fraction by first multiplying the numerators together, and then multiplying the denominators together and placing the product of the numerators over the product of the denominators. This fraction may then be reduced to its lowest terms.

$$\frac{4}{5} \times \frac{1}{3} = \frac{4 \times 1}{5 \times 3} = \frac{4}{15}$$

$$\frac{3}{8} \times 4 = \frac{12}{8} = 1\frac{4}{8} = 1\frac{1}{2} = 1.5$$

$$\frac{1}{2} \times \frac{5}{3} \times \frac{4}{7} = \frac{1 \times 5 \times 4}{2 \times 3 \times 7} = \frac{20}{42} = \frac{10}{21}$$

6. Division by a fraction. Dividing a number or a fraction by a fraction can be accomplished in this manner: Invert the denominator (which is a fraction) and multiply this inverted expression by the numerator, using the usual multiplication methods. For example, divide 6 by $\frac{3}{4}$.

$$\frac{6}{\frac{3}{4}}$$

Invert the denominator $\frac{3}{4}$ to give $\frac{4}{3}$. Then multiply:

$$6 \times \frac{4}{3} = \frac{24}{3} = 8$$

Divide $\frac{2}{3}$ by $\frac{3}{4}$.

$$\frac{\frac{2}{3}}{\frac{3}{4}}$$

Invert $\frac{3}{4}$ to give $\frac{4}{3}$. Then multiply:

$$\frac{2}{3} \times \frac{4}{3} = \frac{8}{9}$$

7. Addition of numbers with decimals. To add numbers with decimals, we use the same procedure as that used when adding whole numbers, but always line up the decimal points in the same column. For example, add 8.21 + 143.1 + 0.325.

```
      8.21-
  + 143.1--
  +   0.325
  ---------
    151.635
```

When adding numbers expressing units of measurement, always be certain that the numbers added together represent the same units. For example, what is the total length of these three pieces of glass tubing: 10.0 cm, 125 mm, 8.4 cm? If we add these directly, we obtain a value of 143.4, but we are not certain what the unit of measurement is. To add these lengths correctly, first change 125 mm to 12.5 cm. Now all the lengths are expressed in the same units and can be added.

$$\begin{array}{r} 10.0 \text{ cm} \\ 12.5 \text{ cm} \\ 8.4 \text{ cm} \\ \hline 30.9 \text{ cm} \end{array}$$

8. Subtraction of numbers with decimals. To subtract numbers containing decimals, we use the same procedure as for subtracting whole numbers, but always line up the decimal points in the same column. For example, subtract 20.60 from 182.49.

$$\begin{array}{r} 182.49 \\ -\quad 20.60 \\ \hline 161.89 \end{array}$$

9. Multiplication of numbers with decimals. To multiply two or more numbers together that contain decimals, we first multiply as if they were whole numbers. To locate the decimal point in the product, we add together the number of digits to the right of the decimal in all the numbers multiplied together. The product should contain this total number of digits to the right of the decimal point.

Multiply $2.05 \times 2.05$ (total of four digits to the right of the decimal):

$$\begin{array}{r} 2.05 \\ \times\ 2.05 \\ \hline 1025 \\ 4100\phantom{0} \\ \hline 4.2025 \end{array} \quad \text{(Four digits to the right of the decimal)}$$

$14.25 \times 6.01 \times 0.75 = 64.231875$ (Six digits to the right of the decimal)

$39.26 \times 60 = 2355.60$ (Two digits to the right of the decimal)

[*Note*: When at least one of the numbers that is multiplied is a measurement, the answer must be adjusted to contain the correct number of significant figures. (See Section 12 on significant figures.)]

10. Division of numbers with decimals. To divide numbers containing decimals, we first relocate the decimal points of the numerator and denominator by moving them to the right as many places as needed to make the denominator a whole number. (Move the decimal of both the numerator and the denominator the same amount and in the same direction.) For example,

$$\frac{136.94}{4.1} = \frac{1369.4}{41}$$

The decimal point adjustment in this example is equivalent to multiplying both numerator and denominator by 10. Now we carry out the division normally, locating the decimal point immediately above its position in the dividend.

$$\begin{array}{r} 33.4 \\ 41\overline{)1369.4} \\ 123\phantom{.4} \\ \hline 139\phantom{4} \\ 123\phantom{4} \\ \hline 164 \\ 164 \\ \hline \end{array} \qquad \frac{0.441}{26.25} = \frac{44.1}{2625} = \begin{array}{r} 0.0168 \\ 2625\overline{)44.1000} \\ 2625\phantom{00} \\ \hline 17850\phantom{0} \\ 15750\phantom{0} \\ \hline 21000 \\ 21000 \\ \hline \end{array}$$

[*Note*: When at least one of the numbers in the division is a measurement, the answer must be adjusted to contain the correct number of significant figures. (See Section 12 on significant figures.)]

The examples above are merely guides to the principles used in performing the various mathematical operations illustrated. There are, no doubt, shortcuts and other methods, and the student will discover these with experience. Every student of chemistry should use either an electronic calculator or a slide rule for solving problems. The use of these devices will save many hours of time that would otherwise be spent in doing tedious longhand calculations. After solving a problem, the student should check for errors and evaluate the answer to see if it is logical and consistent with the data given.

11. Algebraic equations. Many mathematical problems that are first encountered in chemistry fall into the following algebraic forms. Solutions to these problems are simplified by first isolating the desired term on one side of the equation. This is accomplished by treating both sides of the equation in an identical manner (so as not to destroy the equality) until the desired term is isolated.

(a) $$a = \frac{b}{c}$$

To solve for $a$, simply divide $b$ by $c$.
To solve for $b$, multiply both sides of the equation by $c$.

$$a \times c = \frac{b}{\cancel{c}} \times \cancel{c}$$

$$b = a \times c$$

To solve for $c$, multiply both sides of the equation by $\frac{c}{a}$.

$$\cancel{a} \times \frac{c}{\cancel{a}} = \frac{b}{\cancel{c}} \times \frac{\cancel{c}}{a}$$

$$c = \frac{b}{a}$$

(b) $\frac{a}{b} = \frac{c}{d}$

To solve for $a$, multiply both sides of the equation by $b$.

$$\frac{a}{\cancel{b}} \times \cancel{b} = \frac{c}{d} \times b$$

$$a = \frac{c \times b}{d}$$

To solve for $b$, multiply both sides of the equation by $\frac{b \times d}{c}$.

$$\frac{a}{\cancel{b}} \times \frac{\cancel{b} \times d}{c} = \frac{\cancel{c}}{\cancel{d}} \times \frac{b \times \cancel{d}}{\cancel{c}}$$

$$b = \frac{a \times d}{c}$$

(c) $a \times b = c \times d$

To solve for $a$, divide both sides of the equation by $b$.

$$\frac{a \times \cancel{b}}{\cancel{b}} = \frac{c \times d}{b}$$

$$a = \frac{c \times d}{b}$$

(d) $\frac{(b - c)}{a} = d$

To solve for $b$, first multiply both sides of the equation by $a$.

$$\frac{\cancel{a}(b - c)}{\cancel{a}} = d \times a$$

$$b - c = d \times a$$

Then add $c$ to both sides of the equation.

$$b - \cancel{c} + \cancel{c} = d \times a + c$$

$$b = (d \times a) + c$$

When $a = 1.8$, $c = 32$, and $d = 35$,

$$b = (35 \times 1.8) + 32 = 63 + 32 = 95$$

12. Significant figures. Every measurement that we make has some inherent error due to the limitations of the measuring instrument and the experimenter. The numerical value recorded for a measurement should give some indication of the reliability (precision) of that measurement. In measuring a temperature using a thermometer calibrated at one-degree intervals, we can easily read the thermometer to the nearest one degree, but we normally estimate and record the temperature to the nearest tenth of a degree (0.1°C). For example,

a temperature falling between 23°C and 24°C might be estimated at 23.4°C. There is some uncertainty about the last digit, 4, but an estimate of it is better information than simply reporting 23°C or 24°C. If we read the thermometer as "exactly" twenty-three degrees, the temperature should be reported as 23.0°C, not 23°C, because 23.0°C indicates our estimate to the nearest 0.1°C. Thus, in recording any measurement, we retain one uncertain digit. The digits retained in a physical measurement are said to be significant, and are called **significant figures** or **digits**.

Some numbers are exact and therefore have an infinite number of significant figures. Exact numbers occur in simple counting operations, such as 5 bricks, and in defined relationships, as 100 cm = 1 m, 24 hr = 1 day, and so on. Because of their infinite number of significant figures, exact numbers do not limit the number of significant figures in a calculation.

*Counting significant figures.* Digits other than zero are always significant. Depending on their position in the number, zeros may or may not be significant. There are several possible situations.

1. All zeros between other digits in a number are significant. For example: 3.076, 4002, 790.2. Each of these numbers has four significant figures.
2. Zeros to the left of the first nonzero digit are used to locate the decimal point and are not significant. Thus, 0.013 has only two significant figures (1 and 3).
3. Zeros to the right of the last nonzero digit and to the right of the decimal point are significant, for they would not have been included except to express precision. For example, 3.070 has four significant figures; 0.070 has two significant figures.
4. Zeros to the right of the last nonzero digit, but to the left of the decimal, as in the numbers 100, 580, 37,000, may or may not be significant. For example, in 37,000 the measurement might be good to the nearest 1000, 100, 10, or 1. There are two conventions that may be used to show the intended precision. If all the zeros are significant, then an expressed decimal may be added, as 580., or 37,000. But a better system, and one which is applicable to the case when some but not all of the zeros are significant, is to express the number in exponential notation, including only the significant zeros. Thus, for 300, if the zero following 3 is significant, we would write $3.0 \times 10^2$. For 17,000, if two zeros are significant, we would write $1.700 \times 10^4$. The number we correctly expressed as 580. can also be correctly expressed as $5.80 \times 10^2$. With exponential notation there is no doubt as to the number of significant figures.

The mass of an object given as 28.2 grams (g) is expressed to the nearest 0.1 g. This indicates that the weighing was done on a balance having a precision of 0.1 g. It also means that the true value actually lies between 28.15 and 28.25 g. The value 28.2 has three significant figures (2, 8, 2). This same mass, weighed on the same balance, but expressed in milligrams (mg) is 28,200 mg. This value, 28,200 mg, also has three significant figures (2, 8, 2); the zeros are needed to express the magnitude of the number in milligrams. This same mass expressed as 0.0282 kilograms (kg) still contains three significant figures; the zeros are used to locate the decimal point. Better expressions showing three significant figures for these masses are $2.82 \times 10^4$ mg and $2.82 \times 10^{-2}$ kg.

Additional examples illustrating the significant figures in a number are given in this table.

| Number | Number of significant figures |
|---|---|
| 2.45 | 3 |
| 2.450 | 4 |
| 2.045 | 4 |
| 0.245 | 3 |
| 0.0245 | 3 |
| 245.0 | 4 |
| 245 | 3 |

13. Exponents; powers of 10; expression of large and small numbers. In scientific measurements and calculations, we often encounter very large and very small numbers; for example, 0.00000384 and 602,000,000,000,000,000,000,000. These numbers are troublesome to write and awkward to work with, especially in calculations. A convenient method of expressing these large and small numbers in a simplified form is by means of exponents or powers of 10. This method of expressing numbers is known as **scientific** or **exponential notation.**

An *exponent* is a number written as a superscript following another number; it is also called a *power* of that number, and it indicates how many times the number is used as a factor. In the number $10^2$, 2 is the exponent and the number means 10 squared, or 10 to the second power, or $10 \times 10 = 100$. Three other examples are

$$3^2 = 3 \times 3 = 9$$

$$3^4 = 3 \times 3 \times 3 \times 3 = 81$$

$$10^3 = 10 \times 10 \times 10 = 1000$$

For ease of handling, large and small numbers are expressed in powers of 10. Powers of 10 are used because multiplying or dividing by 10 coincides with moving the decimal point in a number by one place. Thus, a number multiplied by $10^1$ would move the decimal point one place to the right; $10^2$, two places to the right; $10^{-2}$, two places to the left. To express a number in powers of 10, we move the decimal point in the original number to a new position, placing it so that the number is a value between 1 and 10. This new decimal number is multiplied by 10 raised to the proper power. For example, to write the number 42,389 in exponential form (powers of 10), the decimal point is placed between the 4 and the 2 (4.2389) and the number is multiplied by $10^4$; thus, the number is $4.2389 \times 10^4$. The power of 10 (4) tells us the number of places that the decimal point must be moved to restore it to its original position. The exponent of 10 is determined by counting the number of places that the decimal point is moved from its original position. If the decimal point is moved to the left, the exponent is a positive number; if it is moved to the right, the exponent is a negative number. To express the number 0.00248 in exponential notation (as a power of 10), the decimal point is moved three places to the right; the exponent of 10 is $-3$, and the number is $2.48 \times 10^{-3}$. Study the examples below.

$$1237 = 1.237 \times 10^3$$
$$988 = 9.88 \times 10^2$$
$$147.2 = 1.472 \times 10^2$$
$$2{,}200{,}000 = 2.2 \times 10^6$$
$$0.0123 = 1.23 \times 10^{-2}$$
$$0.00005 = 5 \times 10^{-5}$$
$$0.000368 = 3.68 \times 10^{-4}$$

The use of powers of 10 in multiplication and division greatly simplifies locating the decimal point in the answer. In multiplication, first change all numbers to powers of 10, then multiply the numerical portion in the usual manner, and finally add the exponents of 10 algebraically, expressing them as a power of 10 in the product. In multiplication, the exponents (powers of 10) are added algebraically.

$$10^2 \times 10^3 = 10^{(2+3)} = 10^5$$
$$10^2 \times 10^2 \times 10^{-1} = 10^{(2+2-1)} = 10^3$$

Multiply: $40{,}000 \times 4200$
Change to powers of ten: $4 \times 10^4 \times 4.2 \times 10^3$
Rearrange: $4 \times 4.2 \times 10^4 \times 10^3$
$16.8 \times 10^{(4+3)}$
$16.8 \times 10^7$ or $1.68 \times 10^8$ (Answer)

Multiply: $380 \times 0.00020$
$3.80 \times 10^2 \times 2.0 \times 10^{-4}$
$3.80 \times 2.0 \times 10^2 \times 10^{-4}$
$7.6 \times 10^{(2-4)}$
$7.6 \times 10^{-2}$ or $0.076$ (Answer)

Multiply: $125 \times 284 \times 0.150$
$1.25 \times 10^2 \times 2.84 \times 10^2 \times 1.50 \times 10^{-1}$
$1.25 \times 2.84 \times 1.50 \times 10^2 \times 10^2 \times 10^{-1}$
$5.325 \times 10^{(2+2-1)}$
$5.32 \times 10^3$ (Answer)

In division, after changing the numbers to powers of 10, move the 10 and its exponent from the denominator to the numerator, changing the sign of the exponent. Carry out the division in the usual manner and evaluate the power of 10. The following is a proof of the equality of moving the power of 10 from the denominator to the numerator.

$$1 \times 10^{-2} = 0.01 = \frac{1}{100} = \frac{1}{10^2} = 1 \times 10^{-2}$$

In division, the exponents in the denominator are subtracted algebraically from the exponents in the numerator.

$$\frac{10^5}{10^3} = 10^5 \times 10^{-3} = 10^{(5-3)} = 10^2$$

$$\frac{10^3 \times 10^4}{10^{-2}} = 10^3 \times 10^4 \times 10^2 = 10^{(3+4+2)} = 10^9$$

Divide: $\frac{2871}{0.0165}$

Change to powers of 10: $\frac{2.871 \times 10^3}{1.65 \times 10^{-2}}$

Move $10^{-2}$ to the numerator, changing the sign of the exponent. This is mathematically equivalent to multiplying both numerator and denominator by $10^2$.

$$\frac{2.87 \times 10^3 \times 10^2}{1.65}$$

$$\frac{2.87 \times 10^{(3+2)}}{1.65} = 1.74 \times 10^5 \quad \text{(Answer)}$$

Divide: $\frac{0.000585}{0.00300}$

$$\frac{5.85 \times 10^{-4}}{3.00 \times 10^{-3}}$$

$$\frac{5.85 \times 10^{-4} \times 10^3}{3.00} = \frac{5.85 \times 10^{(-4+3)}}{3.00}$$

$1.95 \times 10^{-1}$ or $0.195$ (Answer)

Calculate: $\frac{760 \times 300 \times 40.0}{700 \times 273}$

$$\frac{7.60 \times 10^2 \times 3.00 \times 10^2 \times 4.00 \times 10^1}{7.00 \times 10^2 \times 2.73 \times 10^2}$$

$$\frac{7.60 \times 3.00 \times 4.00 \times 10^2 \times 10^2 \times 10^1}{7.00 \times 2.73 \times 10^2 \times 10^2}$$

$4.77 \times 10^1$ or $47.7$ (Answer)

14. Rounding off numbers. When numbers are added, subtracted, multiplied, or divided, we often obtain answers with more figures than we are justified in using. Numbers are rounded off so that we can retain a specific number of digits consistent with the accuracy that they represent.

We round off a number by dropping digits from the end of the number, adjusting the last digit retained either to remain the same or be increased by one number.

Rules for rounding off numbers:

1. When the first digit after those being retained is less than 5, all digits retained remain the same.
2. When the first digit after those being retained is larger than 5, the last digit retained is rounded off by increasing it one number.
3. When the first digit after those being retained is 5 and all others beyond it are zeros, the last digit retained remains the same if it is an even number or is increased by one number if it is an odd number.

In the examples illustrating these rules, all numbers are rounded off to four digits.

| | | |
|---|---|---|
| *Rule 1.* | 1.0263 | Round off to 1.026 |
| | 23.04193 | Round off to 23.04 |
| *Rule 2.* | 1.0268 | Round off to 1.027 |
| | 23.04728 | Round off to 23.05 |
| | 18.998 | Round off to 19.00 |
| *Rule 3.* | 140.25 | Round off to 140.2 |
| | 63.3750 | Round off to 63.38 |

15. Significant figures in calculations. The result of a calculation based on experimental measurements cannot be more precise than the measurement that has the greatest uncertainty.

*Addition and subtraction.* The result of an addition or subtraction should contain no more digits to the right of the decimal point than are contained in that quantity which has the least number of digits to the right of the decimal point.

Perform the operation indicated and then round off the number to the proper significant figures.

$$\begin{array}{r} 142.8 \\ 18.843 \\ 36.42 \\ \hline 198.063 \end{array} \qquad \begin{array}{r} 93.45 \\ -18.0 \\ \hline 75.45 \end{array}$$

198.1 (Answer) 75.4 (Answer)

*Multiplication and division.* In calculations involving multiplication or division, the answer should contain the same number of significant figures as the measurement that has the least number of significant figures. In multiplication or division the position of the decimal point has nothing to do with the number of significant figures in the answer. Study the following examples:

| | Round off to |
|---|---|
| $2.05 \times 2.05 = 4.2025$ | 4.20 |
| $18.48 \times 5.2 = 96.096$ | 96 |
| $0.0126 \times 0.020 = 0.000252$ or | |
| $1.26 \times 10^{-2} \times 2.0 \times 10^{-2} = 2.52 \times 10^{-4}$ | $2.5 \times 10^{-4}$ |
| $\dfrac{1369.4}{41} = 33.4$ | 33 |
| $\dfrac{2268}{4.20} = 540$ | 540 |

16. Dimensional analysis. Many problems of chemistry can be solved readily by dimensional analysis using the factor-label or conversion factor method. Dimensional analysis involves the use of proper units of dimension on all factors that are multiplied, divided, added, or subtracted in setting up and solving a problem. Dimensions are physical quantities such as length, mass, and time, which are expressed in such units as centimetres, grams, and seconds, respectively. In solving a problem, these units are treated mathematically just as though they were numbers, giving us an answer that contains the correct dimensional units.

A measurement or quantity given in one kind of unit can be converted to any other kind of unit having the same dimension. To convert from one kind of unit to another, the original quantity or measurement is multiplied or divided by a conversion factor. The key to success lies in choosing the correct conversion factor. This general method of calculation is illustrated in the following examples.

Suppose we want to change 24 ft to inches. We need to multiply 24 ft by a conversion factor containing feet and inches. Two such conversion factors can be written relating inches and feet.

$$\frac{12 \text{ in.}}{1 \text{ ft}} \quad \text{or} \quad \frac{1 \text{ ft}}{12 \text{ in.}}$$

We choose the factor that will mathematically cancel feet and leave the answer in inches. Note that the units are treated in the same way we treat numbers, multiplying or dividing as required. Two possibilities then arise to change 24 ft to inches:

$$24 \cancel{\text{ft}} \times \frac{12 \text{ in.}}{1 \cancel{\text{ft}}} \quad \text{or} \quad 24 \text{ ft} \times \frac{1 \text{ ft}}{12 \text{ in.}}$$

In the first case (the correct method), feet in the numerator and the denominator cancel, giving us an answer of 288 in. In the second case, the units of the answer are $\text{ft}^2/\text{in.}$, the answer being $2.0 \text{ ft}^2/\text{in}$. In the first case, the answer is reasonable since it is expressed in units having the proper dimensions. That is, the dimension of length expressed in feet has been converted to length in inches according to the mathematical expression

$$\cancel{\text{ft}} \times \frac{\text{in.}}{\cancel{\text{ft}}} = \text{in.}$$

In the second case, the answer is not reasonable since the units ($ft^2$/in.) do not correspond to units of length. The answer is therefore incorrect. The units are the guiding factor for the proper conversion.

The reason we can multiply 24 ft times 12 in./ft and not change the value of the measurement is because the conversion factor is derived from two equivalent quantities. Therefore, the conversion factor 12 in./ft is equal to unity. And when you multiply any factor by 1, it does not change the value.

$$12 \text{ in.} = 1 \text{ ft} \qquad \text{and} \qquad \frac{12 \text{ in.}}{1 \text{ ft}} = 1$$

Convert 16 kg to milligrams. In this problem it is best to proceed in this fashion:

$$\text{kg} \rightarrow \text{g} \rightarrow \text{mg}$$

The possible conversion factors are

$$\frac{1000 \text{ g}}{1 \text{ kg}} \quad \text{or} \quad \frac{1 \text{ kg}}{1000 \text{ g}} \qquad \frac{1000 \text{ mg}}{1 \text{ g}} \quad \text{or} \quad \frac{1 \text{ g}}{1000 \text{ mg}}$$

We use the conversion factor that leaves the proper unit at each step for the next conversion. The calculation is

$$16 \cancel{\text{kg}} \times \frac{1000 \cancel{\text{g}}}{1 \cancel{\text{kg}}} \times \frac{1000 \text{ mg}}{1 \cancel{\text{g}}} = 1.6 \times 10^7 \text{ mg}$$

Many problems may be solved by a sequence of steps involving unit conversion factors. This sound, basic approach to problem solving, together with neat and orderly setting up of data, will lead to correct answers having the right units, fewer errors, and considerable saving of time.

17. Graphical representation of data. A graph is often the most convenient way to present or display a set of data. Various kinds of graphs have been devised, but the most common type uses a set of horizontal and vertical coordinates to show the relationship of two variables. It is called an $x$–$y$ graph because the data of one variable are represented on the horizontal or $x$ axis (abscissa) and the data of the other variable are represented on the vertical or $y$ axis (ordinate). (See Figure I.1.)

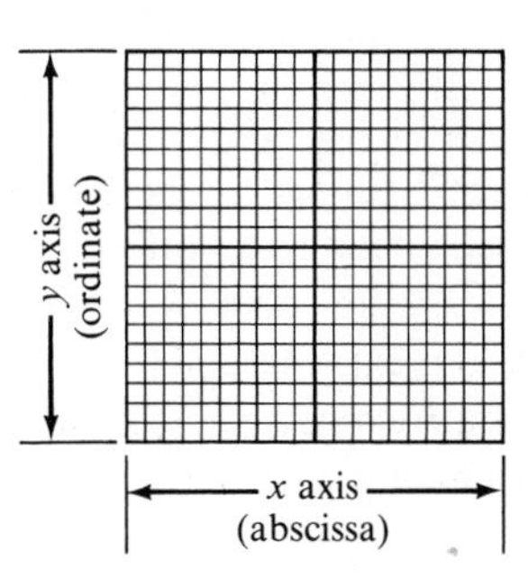


*Figure I.1*

As a specific example of a simple graph, let us graph the relationship between Celsius and Fahrenheit temperature scales. Assume that initially we have only the information in the table.

| °C | °F |
|---|---|
| 0 | 32 |
| 50 | 122 |
| 100 | 212 |

On a set of horizontal and vertical coordinates (graph paper), scale off at least 100 Celsius degrees on the $x$ axis and at least 212 Fahrenheit degrees on the $y$ axis. Locate and mark the three points corresponding to the three temperatures given and draw a line connecting these points (see Figure I.2). Here is how a point is located on the graph: Using the 50°C–122°F data, trace a vertical line up from 50°C on the $x$ axis and a horizontal line across from 122°F on the $y$ axis and mark the point where the two lines intersect. This process is called *plotting*. The other two points are plotted on the graph in the same way. [*Note*: The number of degrees per scale division was chosen to give a graph of convenient size. In this case there are 5 Fahrenheit degrees per scale division and 2 Celsius degrees per scale division.]

The graph in Figure I.2 shows that the relationship between Celsius and Fahrenheit temperature is that of a straight line. The Fahrenheit temperature

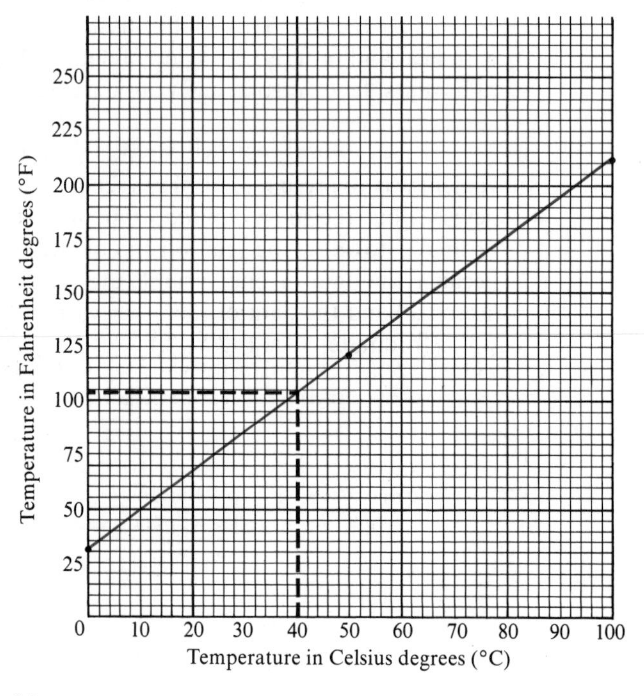


*Figure I.2*

corresponding to any given Celsius temperature between 0° and 100° can be determined from the graph. For example, to find the Fahrenheit temperature corresponding to 40°C, trace a perpendicular line from 40°C on the $x$ axis to the line plotted on the graph. Now trace a horizontal line from this point on the plotted line to the $y$ axis and read the corresponding Fahrenheit temperature (104°F). See the dotted lines on Figure I.2. In turn, the Celsius temperature corresponding to any Fahrenheit temperature between 32° and 212° can be determined from the graph. This is accomplished by tracing a horizontal line from the Fahrenheit temperature to the plotted line and reading the corresponding temperature on the Celsius scale directly below the point of intersection.

The mathematical relationship of Fahrenheit and Celsius temperatures is expressed by the equation $°F = 1.8°C + 32$. Figure I.2 is a graph of this equation. Since the graph is a straight line, it can be extended indefinitely at either end. Any desired Celsius temperature can be plotted against the corresponding Fahrenheit temperature by extending the scales along both axes as necessary. Negative, as well as positive, values can be plotted on the graph (see Figure I.3).

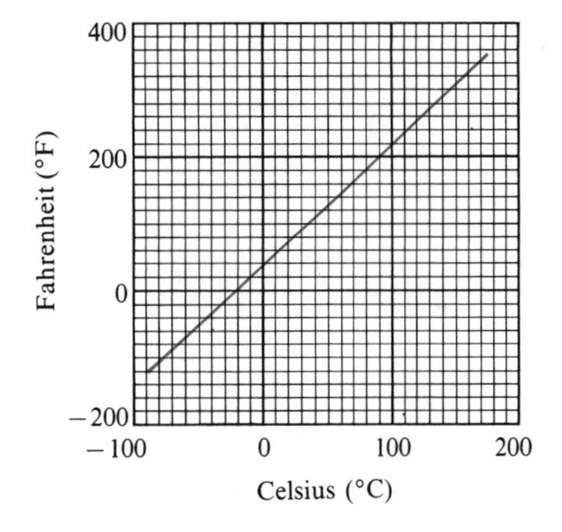


*Figure I.3*

Figure I.4 is a graph showing the solubility of potassium chlorate in water at various temperatures. The solubility curve on this graph was plotted from the data in the following table.

| Temperature (°C) | Solubility (g $KClO_3$/100 g water) |
|---|---|
| 10 | 5.0 |
| 20 | 7.4 |
| 30 | 10.5 |
| 50 | 19.3 |
| 60 | 24.5 |
| 80 | 38.5 |

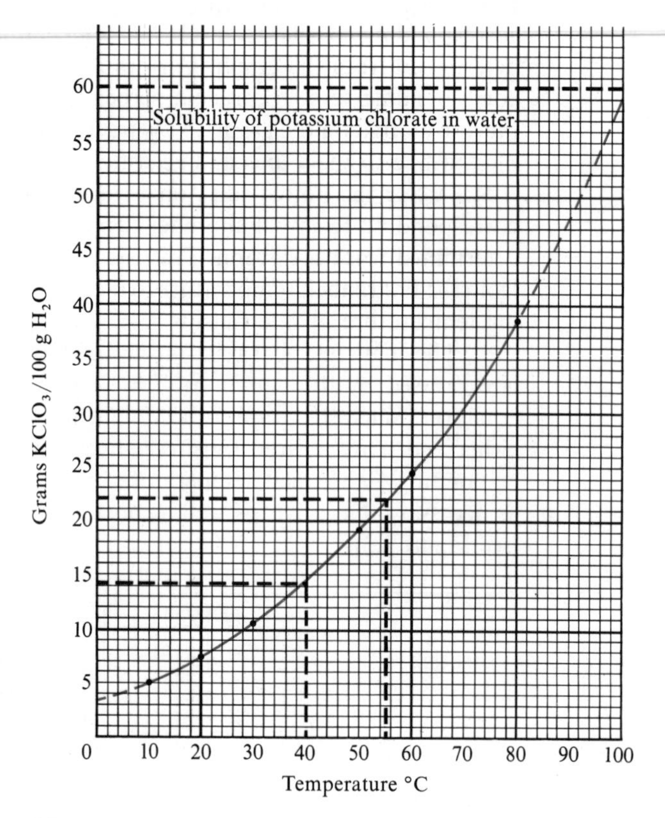


*Figure I.4*

In contrast to the Celsius–Fahrenheit temperature relationship, there is no known mathematical equation that describes the exact relationship between temperature and the solubility of potassium chlorate. The graph in Figure I.4 was constructed from experimentally determined solubilities at the six temperatures shown. These experimentally determined solubilities are all located on the smooth curve traced by the unbroken line portion of the graph. We are therefore confident that the unbroken line represents a very good approximation of the solubility data for potassium chlorate covering the temperature range from 10 to 80°C. All points on the plotted curve represent the composition of saturated solutions. Any point below the curve represents an unsaturated solution.

The dotted line portions of the curve are *extrapolations*; that is, they extend the curve above and below the temperature range actually covered by the plotted solubility data. Curves such as this are often extrapolated a short distance beyond the range of the known data, although the extrapolated portions may not be highly accurate. Extrapolation is justified only in the absence of more reliable information.

The graph in Figure I.4 can be used with confidence to obtain the solubility of $KClO_3$ at any temperature between 10°C and 80°C, but the solubilities between 0°C and 10°C and between 80°C and 100°C are less reliable. For example, what is the solubility of $KClO_3$ at 55°C, at 40°C, and at 100°C?

First draw a perpendicular line from each temperature to the plotted solubility curve. Now trace a horizontal line to the solubility axis from each point on the curve and read the corresponding solubilities. The values that we read from the graph are

55° 22.0 g $KClO_3$/100 g water
40° 14.2 g $KClO_3$/100 g water
100° 60 g $KClO_3$/100 g water

Of these solubilities, the one at 55°C is probably the most reliable because experimental points are plotted at 50°C and at 60°C. The 40°C solubility value is a bit less reliable because the nearest plotted points are at 30°C and 50°C. The 100°C solubility is the least reliable of the three values because it was taken from the extrapolated part of the curve, and the nearest plotted point is at 80°C. Actual handbook solubility values are 14.0 and 57.0 g of $KClO_3$/100 g of water at 40°C and 100°C, respectively.

The graph in Figure I.4 can also be used to determine whether a solution is saturated or unsaturated. For example, a solution contains 15 g of $KClO_3$/100 g of water and is at a temperature of 55°C. Is the solution saturated or unsaturated? *Answer*: The solution is unsaturated because the point corresponding to 15 g and 55°C on the graph is below the solubility curve—and all points below the curve represent unsaturated solutions.

# Appendix II
# Vapor Pressure of Water at Various Temperatures

| *Temperature* (°C) | *Vapor Pressure* (mm Hg) | *Temperature* (°C) | *Vapor Pressure* (mm Hg) |
|---|---|---|---|
| 0 | 4.6 | 26 | 25.2 |
| 5 | 6.5 | 27 | 26.7 |
| 10 | 9.2 | 28 | 28.3 |
| 15 | 12.8 | 29 | 30.0 |
| 16 | 13.6 | 30 | 31.8 |
| 17 | 14.5 | 40 | 55.3 |
| 18 | 15.5 | 50 | 92.5 |
| 19 | 16.5 | 60 | 149.4 |
| 20 | 17.5 | 70 | 233.7 |
| 21 | 18.6 | 80 | 355.1 |
| 22 | 19.8 | 90 | 525.8 |
| 23 | 21.2 | 100 | 760.0 |
| 24 | 22.4 | 110 | 1074.6 |
| 25 | 23.8 | | |

# Appendix III
# Units of Measurements

*Numerical Value of Prefixes Used with Units*

| | | |
|---|---|---|
| tera | 1,000,000,000,000 | $10^{12}$ |
| giga | 1,000,000,000 | $10^{9}$ |
| mega | 1,000,000 | $10^{6}$ |
| kilo | 1,000 | $10^{3}$ |
| hecto | 100 | $10^{2}$ |
| deca | 10 | $10^{1}$ |
| deci | 0.1 | $10^{-1}$ |
| centi | 0.01 | $10^{-2}$ |
| milli | 0.001 | $10^{-3}$ |
| micro | 0.000001 | $10^{-6}$ |
| nano | 0.000000001 | $10^{-9}$ |
| pico | 0.000000000001 | $10^{-12}$ |
| femto | 0.000000000000001 | $10^{-15}$ |
| atto | 0.000000000000000001 | $10^{-18}$ |

*Length*

1 in. = 2.54 cm
10 mm = 1 cm
100 cm = 1 m
1000 mm = 1 m
1000 m = 1 km
1 mile = 1.61 km
1 Å = $10^{-8}$ cm

*Mass*

1 lb = 453.6 g
1000 mg = 1 g
1000 g = 1 kg
1 ounce = 28.3 g
2.20 lb = 1 kg

*Volume*

1 ml = 1 $cm^3$
1000 ml = 1 litre
1 fluid ounce = 29.6 ml
1 qt = 0.946 litre
1 gal = 3.785 litres

*Temperature*

$°F = 1.8°C + 32$

$$°C = \frac{(°F - 32)}{1.8}$$

$K = °C + 273$

$°F = 1.8(°C + 40) - 40$

Absolute zero = −273.18°C or −459.72°F

# Appendix IV
# Solubility Table

| | $F^-$ | $Cl^-$ | $Br^-$ | $I^-$ | $O^{2-}$ | $S^{2-}$ | $OH^-$ | $NO_3^-$ | $CO_3^{2-}$ | $SO_4^{2-}$ | $C_2H_3O_2^-$ |
|---|---|---|---|---|---|---|---|---|---|---|---|
| $Na^+$ | *S* | *S* | *S* | *S* | *S* | *S* | *S* | *S* | *S* | *S* | *S* |
| $K^+$ | *S* | *S* | *S* | *S* | *S* | *S* | *S* | *S* | *S* | *S* | *S* |
| $NH_4^+$ | *S* | *S* | *S* | *S* | – | *S* | *S* | *S* | *S* | *S* | *S* |
| $Ag^+$ | *S* | *I* | *I* | *I* | *I* | *I* | – | *S* | *I* | *I* | *I* |
| $Mg^{2+}$ | *I* | *S* | *S* | *S* | *I* | *d* | *I* | *S* | *I* | *S* | *S* |
| $Ca^{2+}$ | *I* | *S* | *S* | *S* | *I* | *d* | *I* | *S* | *I* | *I* | *S* |
| $Ba^{2+}$ | *I* | *S* | *S* | *S* | *s* | *d* | *s* | *S* | *I* | *I* | *S* |
| $Fe^{2+}$ | *s* | *S* | *S* | *S* | *I* | *I* | *I* | *S* | *s* | *S* | *S* |
| $Fe^{3+}$ | *I* | *S* | *S* | – | *I* | *I* | *I* | *S* | *I* | *S* | *I* |
| $Co^{2+}$ | *S* | *S* | *S* | *S* | *I* | *I* | *I* | *S* | *I* | *S* | *S* |
| $Ni^{2+}$ | *s* | *S* | *S* | *S* | *I* | *I* | *I* | *S* | *I* | *S* | *S* |
| $Cu^{2+}$ | *s* | *S* | *S* | – | *I* | *I* | *I* | *S* | *I* | *S* | *S* |
| $Zn^{2+}$ | *s* | *S* | *S* | *S* | *I* | *I* | *I* | *S* | *I* | *S* | *S* |
| $Hg^{2+}$ | *d* | *S* | *I* | *I* | *I* | *I* | *I* | *S* | *I* | *d* | *S* |
| $Cd^{2+}$ | *s* | *S* | *S* | *S* | *I* | *I* | *I* | *S* | *I* | *S* | *S* |
| $Sn^{2+}$ | *S* | *S* | *S* | *s* | *I* | *I* | *I* | *S* | *I* | *S* | *S* |
| $Pb^{2+}$ | *I* | *I* | *I* | *I* | *I* | *I* | *I* | *S* | *I* | *I* | *S* |
| $Mn^{2+}$ | *s* | *S* | *S* | *S* | *I* | *I* | *I* | *S* | *I* | *S* | *S* |
| $Al^{3+}$ | *I* | *S* | *S* | *S* | *I* | *d* | *I* | *S* | – | *S* | *S* |

*Key:* *S* = soluble in water
*s* = slightly soluble in water
*I* = insoluble in water (less than 1 g/100 g $H_2O$)
*d* = decomposes in water

# Appendix V
# Answers to Problems

*Chapter 2*

C.9 The following statements are correct: b, c, e, f, g, i, j, l, p, r.

D.1 (a) 1 (b) 3 (c) 4 (d) 4 (e) 6 (f) 5 (g) 3 (h) 3

D.2 (a) 3.001 (b) 9.378 (c) 41.13 (d) 25.56 (e) 2144 (f) 82.36 (g) 20.00

D.3 (a) $8.47 \times 10^2$ (b) $5.86 \times 10^{-4}$ (c) $2.24 \times 10^4$ (d) $8.8 \times 10^{-2}$ (e) $6.11 \times 10^{-5}$ (f) $4.286 \times 10^3$ (g) $6.50 \times 10^{-2}$

D.4 (a) 38.2 (b) 148.8 (c) 104 (d) $2.44 \times 10^4$ (e) $\frac{5}{9}$ (0.556) (f) 0.429, 0.733, 0.667, 0.784 (g) $\frac{8}{9}$ (h) $7.5X + 30$ (i) 93 (j) 19.4 (k) 0.9357

D.5 (a) 0.12 m (b) 0.142 km (c) $2.5 \times 10^8$ Å (d) 424 mm (e) 30.5 cm (f) 8.05 km (g) $2.10 \times 10^2$ cm (h) $3.0 \times 10^3$ m (i) $1 \times 10^{-7}$ cm (j) 40 cm (k) 8.66 in. (l) 43.5 miles

D.6 89 km/hr

D.7 $1.86 \times 10^5$ miles/s

D.8 $4.9 \times 10^2$ s

D.9 $2 \times 10^5$ m$^2$

D.10 (a) $1.200 \times 10^3$ mg (b) 0.454 kg (c) $1 \times 10^{-3}$ kg (d) $1.16 \times 10^3$ g (e) $5.0 \times 10^{-2}$ g (f) $2.2 \times 10^3$ g (g) $3.50 \times 10^5$ mg (h) $5.64 \times 10^{-2}$ lb

D.11 77.3 kg

D.12 0.32 g

D.13 $2863

D.14 $4.83 \times 10^3$ cm$^3$; 4.83 litres

D.15 $19.87

D.16 39 miles/gal

D.17 $3.0 \times 10^3$ times as heavy

D.18 $3 \times 10^4$ mg

D.19 $8.8 \times 10^4$ tons/day; $8.0 \times 10^7$ kg/day

D.20 $1 \times 10^8$ g/day

D.21 (a) 0.145 litre (b) 81.9 cm$^3$ (c) 568 litres (d) $2.50 \times 10^3$ ml (e) $6.00 \times 10^3$ ml (f) 3.54 in.$^3$ (g) 0.661 gal (h) $2.24 \times 10^4$ ml

D.22 16 litres

D.23 $15.21

D.24 74 litres; 19 gal

D.25 (a) 60°C (b) −18°C (c) 255 K (d) −24°C (e) 77°F (f) 10°F (g) 546 K (h) 273 K

D.26 98.6°F

D.27 −135°F is colder than −90°C

D.28 (a) −40°F = −40°C (b) 11.4°F and −11.4°C

D.29 $3.0 \times 10^3$ cal

D.30 339 cal

D.31 0.0920 cal/g °C

D.32 16.7°C

D.33 1.039 g/ml

D.34 3.12 g/ml

D.35 7.1 g/ml

D.36 $1.19 \times 10^3$ g

D.37 72 g

D.38 680 g Hg

D.39 A is Mg; B is Al; C is Ag

D.40 $1.94 \times 10^3$ g Ag

D.41 0.965 g/ml at 90°C

D.42 (a) 3.20 g/ml (b) 3.20

D.43 100 g ethyl alcohol

D.44 (a) 1.16 g/ml (b) 77.7 ml $H_2SO_4$

Chapter 3

C.17 The following statements are correct: b, c, d, e, g, j.
D.1 −30.3°F
D.2 69.4 g mercury
D.3 (a) 6.25 g oxygen (b) 60.3% magnesium
D.4 (a) $1.1 \times 10^{14}$ cal (b) $3.6 \times 10^{8}$ gal

Chapter 4

D.1 9.82 g
D.2 78% Cu; 22% Zn
D.3 16.8 g CaO
D.4 2.67 g/ml
D.5 (a) 44.4% S (b) An atom of Ca has a greater mass. (c) 16.0 g S
D.6 18 carat gold
D.7 25% H in methane

Chapter 5

C.16 The following statements are correct: b, c, e.
C.23 The following statements are correct: a, d.
C.29 (a) $6.02 \times 10^{23}$ (b) $6.02 \times 10^{23}$ (c) $12.04 \times 10^{23}$ (d) 16.0 (e) 32.0
C.31 The following statements are correct: a, b, e, h, i, l, m, o, p, q, r, s.
D.1 (a) $6.4 \times 10^{5}$ (b) $5.68 \times 10^{-4}$ (c) $1 \times 10^{-4}$ (d) $1.25 \times 10^{2}$
D.2 (a) 4,200,000 (b) 0.00009 (c) 0.0001 (d) 35,000
D.3 $K = 2$; $L = 8$; $M = 18$; $N = 32$; $O = 50$; $P = 72$
D.4 Ca, 20; Ni, 31; Sn, 69; Pb, 125; U, 143; No, 152
D.5 (a) $3.14 \times 10^{23}$ atoms Na (b) $1.5 \times 10^{22}$ atoms P (c) $8.03 \times 10^{24}$ atoms C
(d) $2.37 \times 10^{23}$ atoms C (e) $3.01 \times 10^{23}$ atoms Cd (f) $6 \times 10^{19}$ atoms H
D.6 Atomic number 9: F; 9; 19.0; 19.0 g; $3.16 \times 10^{-23}$ g/atom
Atomic number 33: As; 33; 74.9; 74.9 g; $1.24 \times 10^{-22}$ g/atom
Atomic number 82: Pb; 82; 207.2; 207.2 g; $3.44 \times 10^{-22}$ g/atom
D.7 (a) $5.33 \times 10^{-23}$ g/atom S (b) $1.97 \times 10^{-22}$ g/atom Sn
(c) $3.33 \times 10^{-22}$ g/atom Hg (d) $6.64 \times 10^{-24}$ g/atom He
D.8 (a) $7 \times 10^{-17}$ g Ar (b) 122 g Al (c) 106 g Cl (d) 0.108 g Ag
D.9 (a) 31.4 g C (b) 9.52 g Cu (c) 2.43 kg Ag (d) $2.1 \times 10^{2}$ g Ag
(e) 2.24 moles Fe (f) $1.68 \times 10^{-2}$ mole Sn (g) 1.00 mole $N_2$
(h) 1.69 moles Hg
D.10 (a) $6.02 \times 10^{23}$ molecules $H_2O$ (b) $6.02 \times 10^{23}$ atoms O
(c) $1.20 \times 10^{24}$ atoms H
D.11 $1.2 \times 10^{24}$ atoms P
D.12 3.64 g Mg
D.13 137 g/mole
D.14 $1.5 \times 10^{14}$ dollars/person
D.15 $1.5 \times 10^{19}$ miles
E.6 2.87 g Na
E.7 58.7 g

Chapter 6

C.28 The following statements are correct: a, d, e, f, g, k, l.

Chapter 7

C.25 The following statements are correct: a, d, e, f, h, j, k, l, m, n, p, q, t, u.

Chapter 9

B.6 The following statements are correct: a, b, e, g, h.
C.1 (a) 149.9 (b) 106.8 (c) 174.3 (d) 228.7 (e) 342.3 (f) 64.5
(g) 180.0 (h) 159.8 (i) 136.1
C.2 (a) 60.0 g/mole (b) 331.2 g/mole (c) 123.0 g/mole (d) 63.0 g/mole
(e) 162.1 g/mole (f) 136.4 g/mole (g) 352.0 g/mole (h) 329.1 g/mole
(i) 237.9 g/mole
C.3 (a) 0.800 mole NaOH (b) 0.643 mole $N_2$ (c) 1.56 moles $CH_3OH$
(d) 0.244 mole $Ca(NO_3)_2$ (e) 0.0100 mole $MgCl_2$ (f) 6.1 moles KCl
C.4 (a) 0.30 mole Mg atoms (b) 1.00 mole Ar atoms (c) 0.338 mole Cl atoms
(d) 0.500 mole F atoms

C.5 (a) 900 g $H_2O$ (b) 19.0 g $SnCl_2$ (c) 119 g $H_3PO_4$ (d) 16.0 g $O_2$
(e) 0.0416 g $NH_4Br$ (f) 72.0 g $CH_4$
C.6 (a) $6.0 \times 10^{23}$ molecules $F_2$ (b) $2.1 \times 10^{23}$ molecules $N_2$
(c) $6.0 \times 10^{22}$ molecules $C_2H_6$ (d) $3.76 \times 10^{23}$ molecules $SO_3$
C.7 (a) $3.33 \times 10^{-22}$ g/Hg atom (b) $2.99 \times 10^{-23}$ g/$H_2O$ molecule
(c) $6.64 \times 10^{-24}$ g/He atom (d) $7.31 \times 10^{-23}$ g/$CO_2$ molecule
C.8 (a) $1.66 \times 10^{-21}$ mole $C_6H_6$ (b) $2 \times 10^{-12}$ mole Zn
(c) $1.66 \times 10^{-21}$ mole $CH_4$ (d) $9.97 \times 10^{-21}$ mole $NO_2$
(e) $2 \times 10^{-24}$ mole Mg (f) $3 \times 10^{-18}$ mole $H_2O$
C.9 (a) $2.41 \times 10^{24}$ atoms C (b) $6.92 \times 10^{22}$ atoms C (c) $1.80 \times 10^{21}$ atoms C
(d) $2.41 \times 10^{24}$ atoms C (e) $7.5 \times 10^{22}$ atoms C (f) $3.00 \times 10^{10}$ atoms C
C.10 (a) 30.1 g Ag (b) 7.99 g Br (c) 144 g S (d) 6.69 g Cr
C.11 0.0618 mole $K_2CrO_4$; 11.1 moles $H_2O$
C.12 10.3 moles $H_2SO_4$
C.13 1.77 moles $HNO_3$
C.14 (a) 60.3% Mg; 39.7% O (b) 7.79% C; 92.2% Cl
(c) 29.4% Ca; 23.6% S; 47.0% O (d) 38.7% K; 13.8% N; 47.5% O
(e) 65.9% Al; 34.1% N (f) 2.74% H; 97.3% Cl
(g) 63.5% Ag; 8.24% N; 28.3% O (h) 52.2% Fe; 44.9% O; 2.81% H
C.15 $Mg^{2+}$, $C^{4+}$, $Ca^{2+}$, $K^+$, $Al^{3+}$, $H^+$, $Ag^+$, $Fe^{3+}$
C.16 (a) 77.7% Fe (b) 69.9% Fe (c) 72.3% Fe (d) 15.2% Fe
C.17 (a) $Li_2O$, 53.7% O (b) MgO, 39.7% O (c) $Bi_2O_3$, 10.3% O
(d) $TiO_2$, 40.1% O
C.18 93.1% Ag, 6.9% O
C.19 Empirical formulas: (a) CuS (b) $Cu_2S$ (c) $CaC_2$ (d) $N_2O_3$
(e) $Cl_2O_7$ (f) $K_2MnO_4$ (g) $Na_2SO_4$ (h) $ZnCO_3$ (i)HClO
(j) $C_3H_8O$
C.20 (a) 65.2% Cd (b) 56.3% C (c) 34.7% Mn (d) 35.0% N
C.21 (a) $H_2O$ (b) $Na_2MnO_4$ (c) $K_2Cr_2O_7$ (d) Both the same
(e) $Na_2SO_4$
C.22 $Ga_2O_3$
C.23 No, not enough Mg to react with all the $N_2$
C.24 $C_6H_6O_2$
C.25 $C_6H_{12}O_6$
C.26 (a) $CCl_4$ (b) $C_2Cl_6$ (c) $C_6Cl_6$ (d) $C_3Cl_8$

Chapter 10

C.12 The following statements are correct: a, d, e, f, h, i, j.

Chapter 11

B.1 (a) 0.288 mole $MnO_2$ (b) 3.57 moles $H_2SO_4$ (c) 0.282 mole $Br_2$
(d) $6.49 \times 10^{-3}$ mole $CCl_4$ (e) 17.1 moles NaCl (f) 0.250 mole $C_2H_6O$
(g) 9.32 moles $CO_2$ (h) 1.88 moles $O_2$ (i) 0.100 mole $HNO_3$
(j) 3.0 moles HCl
B.2 (a) 17.6 g $C_3H_8$ (b) 40.5 g Al (c) 15.0 g $H_2$ (d) 17.0 g $AgNO_3$
(e) 152 g $FeSO_4$ (f) 33.4 g $AlCl_3$ (g) 118 g Au (h) 6.6 g $Ni(NO_3)_2$
B.3 (a) $\frac{2\text{ Mg}}{O_2}, \frac{2\text{ Mg}}{2\text{ MgO}}, \frac{O_2}{2\text{ Mg}}, \frac{O_2}{2\text{ MgO}}, \frac{2\text{ MgO}}{2\text{ Mg}}, \frac{2\text{ MgO}}{O_2}$

(b) $\frac{2\text{ Al}}{6\text{ HCl}}, \frac{2\text{ Al}}{2\text{ AlCl}_3}, \frac{2\text{ Al}}{3\text{ H}_2}, \frac{6\text{ HCl}}{2\text{ Al}}, \frac{6\text{ HCl}}{2\text{ AlCl}_3}, \frac{6\text{ HCl}}{3\text{ H}_2}, \frac{2\text{ AlCl}_3}{2\text{ Al}}, \frac{2\text{ AlCl}_3}{6\text{ HCl}}$

$\frac{2\text{ AlCl}_3}{3\text{ H}_2}, \frac{3\text{ H}_2}{2\text{ Al}}, \frac{3\text{ H}_2}{6\text{ HCl}}, \frac{3\text{ H}_2}{2\text{ AlCl}_3}$

(c) $\frac{3\,Zn}{N_2}, \frac{3\,Zn}{Zn_3N_2}, \frac{N_2}{3\,Zn}, \frac{N_2}{Zn_3N_2}, \frac{Zn_3N_2}{3\,Zn}, \frac{Zn_3N_2}{N_2}$

(d) $\frac{2\,C_2H_6}{7\,O_2}, \frac{2\,C_2H_6}{4\,CO_2}, \frac{2\,C_2H_6}{6\,H_2O}, \frac{7\,O_2}{2\,C_2H_6}, \frac{7\,O_2}{4\,CO_2}, \frac{7\,O_2}{6\,H_2O}, \frac{4\,CO_2}{2\,C_2H_6}, \frac{4\,CO_2}{7\,O_2},$

$\frac{4\,CO_2}{6\,H_2O}, \frac{6\,H_2O}{2\,C_2H_6}, \frac{6\,H_2O}{7\,O_2}, \frac{6\,H_2O}{4\,CO_2}$

B.4 (a) 8.0 g $CH_4$ (b) 4.0 g CO B.5 3.4 moles HCl
B.6 4.0 moles $CO_2$ + 8 moles $SO_2$ B.7 16 moles $O_2$
B.8 0.80 mole $Na_2S_2O_3$, 2.1 moles $KMnO_4$, 0.27 mole $H_2O$
B.9 21.0 g $HNO_3$ B.10 55.6 moles $O_2$
B.11 (a) 1.0 mole $FeCl_3$ (b) 5.8 moles HCl (c) 25 g $K_2Cr_2O_7$
(d) 1.08 g $CrCl_3$ (e) 2.3 moles $H_2O$ (f) 7.68 g $FeCl_3$
B.12 560 g CaO B.13 323 kg C
B.14 77.8 g $H_2O$, 181 g Fe B.15 3.54 g weight loss (as $O_2$)
B.16 (a) 8.96 g $H_3PO_4$ (b) 9.98 g $Fe_2O_3$ (c) 15.2 g $SiF_4$ (d) 2.13 g $B_2O_3$
B.17 $MgCl_2$
B.18 (a) KOH is the limiting reagent; HCl is in excess
(b) $Bi(NO_3)_3$ is the limiting reagent; $H_2S$ is in excess
(c) $H_2O$ is the limiting reagent; Fe is in excess
(d) $C_2H_6$ is the limiting reagent; $O_2$ is in excess
B.19 571 g $CH_3OH$; CO is the limiting reagent
B.20 94.0% yield B.21 890 g $CaC_2$
B.22 372 kg $Li_2O$
B.23 The following statements are correct: a, c, e.

Chapter 12

B.4 $6.6 \times 10^3$ g B.8 56.0 litres
C.12 The following statements are correct: a, d, g, i, j, l, m, n, o.
D.1 (a) 0.829 atm (b) 24.8 in. Hg (c) 12.2 lb/in.$^2$ (d) 630 torr
(e) 840 mbar
D.2 0.691 atm D.3 (a) 150 ml (b) 600 ml
D.4 (a) 514 mm Hg (b) $1.03 \times 10^3$ mm Hg
D.5 80 atm
D.6 (a) 3.24 litres (b) 2.29 litres (c) 2.37 litres (d) 3.56 litres
D.7 317°C D.8 27 lb/in.$^2$
D.9 $1.54 \times 10^3$ ml D.10 $2.11 \times 10^3$ mm Hg
D.11 $1.75 \times 10^5$ litres D.12 36.9 litres
D.13 112 litres D.14 2.37 moles
D.15 0.112 mole $H_2$ D.16 46.5 g/mole
D.17 87.8 g/mole D.18 0.255 litre
D.19 (a) 1.34 g/litre (b) 3.58 g/litre (c) 3.17 g/litre (d) 0.179 g/litre
D.20 (a) 2.32 g/litre (b) −78°C D.21 79.5 g/mole
D.22 (a) 22.4 litres (b) 0.267 litre (c) 11.2 litres
(d) 3.68 litres
D.23 6.99 litres D.24 1350 mm Hg
D.25 $1.07 \times 10^{24}$ molecules D.26 719 mm Hg
D.27 383 ml D.28 0.365 g $CH_4$
D.29 He = 380 mm Hg; Ne = 228 mm Hg; Ar = 152 mm Hg
D.30 121 litres $H_2$
D.31 (a) 5 moles $NH_3$ (b) 6.2 moles $O_2$ (c) 100 litres $NH_3$ and 125 litres $O_2$
(d) 165 litres $O_2$ (e) 179 litres NO

D.32 (a) 513 litres $O_2$ (b) 373 litres $SO_2$
D.33 No CO; 2.5 moles $O_2$; 15 moles $CO_2$
D.34 3 $ft^3$ $H_2$; 1 $ft^3$ CO
D.35 47.6 litres air
D.36 65.6% $KClO_3$
D.37 (a) 21.3 litres $O_2$ (b) 798 g $C_2H_6$ (c) 9.94 g $N_2$ (d) 53.5 litres $CO_2$

*Chapter 13*

C.42 The following statements are correct: a, b, c, f, h, l, m, o.
D.1 (a) 2.00 moles $H_2O$ (b) 0.819 mole $H_2O$
D.2 62.9% $H_2O$
D.3 $9.00 \times 10^4$ cal
D.4 $7.88 \times 10^3$ cal
D.5 (a) $9.72 \times 10^3$ cal/mole (b) $4.43 \times 10^4$ cal
D.6 Yes, there is sufficient ice
D.7 80.4 g $H_2O$
D.8 (a) 2.8 moles $O_2$ (b) 79.2 litres $O_2$
D.9 9.88 litres $O_2$
D.10 $CdBr_2 \cdot 4H_2O$
D.11 61.0 g $CuSO_4 \cdot 5H_2O$
D.12 $1.24 \times 10^3$ litres $H_2$
D.13 (a) 18.0 g $H_2O$ (b) 18.0 g $H_2O$ (c) 0.783 g $H_2O$ (d) 0.460 g $H_2O$
D.14 0.851 atm, or 646 mm Hg
D.15 $7 \times 10^{18}$ molecules/sec (7 billion billion molecules/sec)
D.16 92 g $H_2SO_4$

*Chapter 14*

C.25 The following statements are correct: a, c, d, e, h, j, k, m
D.1 16.7% $CuSO_4$
D.2 (a) 9.1% KCl (b) 17% KCl (c) 17% $MgCl_2$ (d) 6.2% $KMnO_4$
D.3 (a) 3.0 g NaCl (b) 37.5 g KCl
D.4 3.6 g NaCl
D.5 88.6 g $H_2O$ must evaporate
D.6 Yes, there is sufficient KOH solution
D.7 240 g sugar
D.8 208 g $H_2SO_4$ solution; 178 ml $H_2SO_4$ solution
D.9 (a) 0.700 *M* $CaBr_2$ (b) 2.24 *M* $NH_4Cl$ (c) 1.50 *M* NaOH
(d) 0.0682 *M* $BaCl_2 \cdot 2H_2O$
D.10 (a) 6.00 moles $CaCl_2$ (b) 0.338 mole $KC_2H_3O_2$ (c) 0.500 mole $AgNO_3$
(d) 0.180 mole $HNO_3$ (e) 0.0800 mole NaOH (f) 3.75 moles KF
D.11 (a) 18.6 g KCl (b) 26.6 g $Na_2SO_4$ (c) 98 g $H_3PO_4$
(d) 0.455 g $Zn(NO_3)_2$
D.12 8.90 *M* HBr
D.13 (a) 33 ml 12 *M* HCl (b) 75 ml 16 *M* $HNO_3$ (c) 167 ml 18 *M* $H_2SO_4$
(d) 4.4 ml 17 *M* $HC_2H_3O_2$
D.14 91.5 ml
D.15 (a) 0.25 *M* (b) 0.65 *M* (c) 0.68 *M*
D.16 (a) 1.0 mole $FeCl_3$ (b) 0.33 mole $CrCl_3$ (c) $6.7 \times 10^{-3}$ mole $K_2Cr_2O_7$
(d) 83 ml 0.080 *M* $K_2Cr_2O_7$ (e) 16 ml 6 *M* HCl
D.17 (a) 1.0 mole $MnCl_2$ (b) 2.5 moles $Cl_2$ (c) 8.0 moles HCl
(d) 1000 ml 0.100 *M* HCl (e) 2.10 litres $Cl_2$
D.18 (a) 3.17 g $BaCrO_4$ (b) 12 ml 1.0 *M* $K_2CrO_4$
D.19 (a) 9.52 g Cu (b) 2.24 litres NO
D.20 0.150 mole $H_2$
D.21 2.14 *M* HCl
D.22 $Al(OH)_3$ will neutralize more acid
D.23 (a) 37.0 g/eq wt (b) 26.0 g/eq wt (c) 29.2 g/eq wt (d) 23.9 g/eq wt
(e) 56.1 g/eq wt

D.24 (a) 85.5 g/eq wt (b) 12.2 g/eq wt (c) 32.7 g/eq wt
(d) 103.6 g Pb/eq wt; 31.8 g Cu/eq wt (e) 18.6 g Fe/eq wt;
(f) 23.2 g/eq wt
D.25 (a) $-3.72$°C (b) 2.00 *m*
D.26 (a) $-12.0$°C (b) 103.4°C
D.27 (a) $-0.6$°C (b) 83.3°C
D.28 258 g/mole
D.29 $C_6H_{12}O_6$
D.30 (a) $6.67 \times 10^3$ g ethylene glycol (b) $6.01 \times 10^3$ ml

## Chapter 15

C.37 The following statements are correct: a, b, c, d, e, f, i, k, l, m, o, p
D.1 (a) 0.01 $M$ $Na^+$, 0.01 $M$ $Cl^-$ (b) 0.32 $M$ $K^+$, 0.32 $M$ $NO_3^-$
(c) 2.50 $M$ $Na^+$, 1.25 $M$ $SO_4^{2-}$ (d) 0.68 $M$ $Ca^{2+}$, 1.36 $M$ $Cl^-$
(e) 0.22 $M$ $Fe^{3+}$, 0.66 $M$ $Cl^-$ (f) 0.75 $M$ $Mg^{2+}$, 0.75 $M$ $SO_4^{2-}$
(g) 0.15 $M$ $NH_4^+$, 0.050 $M$ $PO_4^{3-}$ (h) 0.100 $M$ $Al^{3+}$, 0.150 $M$ $SO_4^{2-}$
D.2 0.263 $M$ $MgBr_2$
D.3 $6 \times 10^{18}$ $Al^{3+}$ ions; $2 \times 10^{19}$ $Cl^-$ ions
D.4 (a) 0.210 $M$ HCl (b) 0.243 $M$ HCl (c) 1.19 $M$ HCl (d) 0.257 $M$ NaOH
(e) 0.0637 $M$ NaOH
D.5 1.68 $M$ $Mg^{2+}$; 3.36 $M$ Cl; 0.840 mole AgCl
D.6 0.131 $M$ $Ba(OH)_2$
D.7 2.69 moles $C_6H_6$; 6.49 moles $CCl_4$
D.8 239 ml $C_6H_6$; 627 ml $CCl_4$
D.9 0.118 $M$ HCl
D.10 0.468 litre $H_2$
D.11 (a) 3 (b) 1.0 (c) 7 (d) 9.07 (e) 0.30
D.12 (a) 3.7 (b) 2.80 (c) 5.20 (d) 10.49
D.13 126 ml
D.14 20.0 weight percent $BaCl_2$; 1.15 $M$ $BaCl_2$
D.15 3.33 litres 18.0 $M$ $H_2SO_4$
D.16 2.23 g Agl

## Chapter 16

C.23 The following statements are correct: b, c, d, e, f, g, i, l, m, n
D.1 2.40 moles HI
D.2 (a) 2.4 moles HI (b) 0.22 mole $H_2$; 0.42 mole $I_2$; 2.6 moles HI
D.3 1.38 moles $H_2$; 0.269 mole $I_2$; 0.250 mole HI
D.4 0.52 mole $H_2$; 0.52 mole $I_2$; 4.0 (3.95) moles HI
D.5 256 times faster at 100 than at 20°C.
D.6 (a) 0.422 mole $Cl_2$ (b) 16.4 litres $Cl_2$
D.7 $K_{eq} = 57$
D.8 Propanoic $1.4 \times 10^{-5}$; hydrofluoric, $7.9 \times 10^{-4}$; hydrocyanic, $4.0 \times 10^{-10}$
D.9 $[H^+] = 1.9 \times 10^{-3}$; pH = 2.72
D.10 $1.1 \times 10^{-2}$ ($1.06 \times 10^{-2}$) M $NO_2^-$
D.11 $1.2 \times 10^{-3}$ $MH^+$
D.12 4.2≅ ionization
D.13 3.38
D.14 (a) $2.50 \times 10^{-10}$ (b) $1.2 \times 10^{-23}$ (c) $7.9 \times 10^{-26}$ (d) $8.19 \times 10^{-12}$
(e) $1.0 \times 10^{-15}$ (f) $2.4 \times 10^{-5}$ (g) $5.13 \times 10^{-17}$ (h) $1.81 \times 10^{-18}$
D.15 $\sim 4 \times 10^{-2}$ ($4.5 \times 10^{-2}$) mole/litre; $\sim$0.7 (0.75) g/100 ml
D.16 (a) $3.9 \times 10^{-8}$ mole/litre (b) $8.7 \times 10^{-4}$ mole/litre
(c) $1.6 \times 10^{-2}$ mole/litre (d) $1.3 \times 10^{-4}$ mole/litre
D.17 (a) $CaCO_3$ (b) $Ag_2CO_3$
D.18 (a) $[H^+] = 3.6 \times 10^{-5}$; pH = 4.4 (4.44) (b) $[H^+] = 1.8 \times 10^{-5}$;
pH = 4.7 (4.74)

Chapter 17

C.11 The following statements are correct: a, c, e, g, j, k, m
D.1 25.3 g $Cl_2$; 7.99 litres $Cl_2$
D.2 3.0 moles $SO_2$
D.3 1,000 ml $KMnO_4$ solution; 5.60 litres $O_2$
D.4 68.0 g Cu
D.5 24.7 g $Br_2$
D.6 275 g $I_2$
D.7 1.5 moles $Fe^{2+}$

Chapter 18

C.22 The following statements are correct: c, e, f, g, i, l, m, n
D.1 6844 A.D.
D.2 $190,000
D.3 0.391 g Sn-119
D.4 ~22,700 years old
D.5 24 minutes
D.6 (a) mass defect = 0.0305 g (b) binding energy = $6.7 \times 10^{11}$ cal

Chapter 19

C.1 (a) 12.7 kg NaCl (b) 109 moles $Cl_2$
C.2 (a) 664 g NaCl (b) 127 litres $Cl_2$
C.3 $2.4 \times 10^6$ gals.
C.4 The mineral was magnetite
C.5 328 tons bauxite ore

Chapter 20

C.1 105 litres
C.2 29.7 litres
C.3 60 g $Cl_2$ per hour
C.4 28.1 kg $HNO_3$
C.5 0.536 mole $Br_2$
C.6 (a) 18 M $H_2SO_4$ (b) 16 M $HNO_3$ (c) 12 M HCl

Chapter 21

C.23 The following statements are correct: a, b, c, e, f, i, k, m, n, r, s

Chapter 22

B.23 The following statements are correct: b, c, d, g, i, j, l, m, n

Chapter 23

B.28 The following statements are correct: a, e, f, g, h, i, k, l, m, o, q, r, u, v

Chapter 24

B.34 The following statements are correct: a, b, e, g, h, i, j, m, o

Chapter 25

B.21 The following statements are correct: b, c, f, h, i, k, m, o

Chapter 26

B.38 The following statements are correct: a, b, f, g, h, i, j, m, n, o, p, r, s, t, w

Chapter 27

B.21 The following statements are correct: a, b, d, f, g, i, k

Chapter 28

B.20 62.5°
B.23 The following statements are correct: a, f, g, h, i, k, l

Chapter 29

B.31 About 2000 glucose anhydride units
B.34 The following statements are correct: b, d, f, g, h, j, l, n, o, p

Chapter 30

B.41 37.5% protein
B.42 $6 \times 10^4$
B.43 The following statements are correct: b, c, d, f, g, i, l, m, n, r

Chapter 31

B.34 The following statements are correct: a, b, d, g, h, j, m, p

Chapter 32

B.38 The following statements are correct: a, c, d, f, h, i, j, l, m, o, p

# Index